T0251779

Nonlinear Dynamical Systems Analysis for the Behavioral Sciences Using Real Data

Nonlinear Dynamical Systems Analysis for the Behavioral Sciences Using Real Data

Edited by

Stephen J. Guastello and Robert A.M. Gregson

CRC Press
Taylor & Francis Group
Boca Raton London New York

CRC Press is an imprint of the
Taylor & Francis Group, an **informa** business

MATLAB® is a trademark of The MathWorks, Inc. and is used with permission. The MathWorks does not warrant the accuracy of the text or exercises in this book. This book's use or discussion of MATLAB® software or related products does not constitute endorsement or sponsorship by The MathWorks of a particular pedagogical approach or particular use of the MATLAB® software.

CRC Press
Taylor & Francis Group
6000 Broken Sound Parkway NW, Suite 300
Boca Raton, FL 33487-2742

© 2011 by Taylor and Francis Group, LLC
CRC Press is an imprint of Taylor & Francis Group, an Informa business

No claim to original U.S. Government works

ISBN 13: 978-1-4398-1997-5 (hbk)

This book contains information obtained from authentic and highly regarded sources. Reasonable efforts have been made to publish reliable data and information, but the author and publisher cannot assume responsibility for the validity of all materials or the consequences of their use. The authors and publishers have attempted to trace the copyright holders of all material reproduced in this publication and apologize to copyright holders if permission to publish in this form has not been obtained. If any copyright material has not been acknowledged please write and let us know so we may rectify in any future reprint.

Except as permitted under U.S. Copyright Law, no part of this book may be reprinted, reproduced, transmitted, or utilized in any form by any electronic, mechanical, or other means, now known or hereafter invented, including photocopying, microfilming, and recording, or in any information storage or retrieval system, without written permission from the publishers.

For permission to photocopy or use material electronically from this work, please access www.copyright.com (http://www.copyright.com/) or contact the Copyright Clearance Center, Inc. (CCC), 222 Rosewood Drive, Danvers, MA 01923, 978-750-8400. CCC is a not-for-profit organization that provides licenses and registration for a variety of users. For organizations that have been granted a photocopy license by the CCC, a separate system of payment has been arranged.

Trademark Notice: Product or corporate names may be trademarks or registered trademarks, and are used only for identification and explanation without intent to infringe.

Visit the Taylor & Francis Web site at
http://www.taylorandfrancis.com

and the CRC Press Web site at
http://www.crcpress.com

Contents

Preface

The study of nonlinear dynamical systems (NDS) dates back to the late nineteenth century, although the majority of the progress has occurred in the last 30 years, and the majority of the applications to theory and research in the behavioral sciences solidified in the last 10–15 years. Two of its central concepts are (1) that most forms of meaningful change are not proportional to inputs, and are thus nonlinear; and (2) that there are many types of continuous and discontinuous change processes, not simply the one linear form, all of which contribute to the understanding of the phenomenon at hand. The change constructs include attractors, bifurcations, chaos, entropy, self-organization, Markov processes, catastrophes, and fractals, with "complex systems" dynamics being closely related. The substantive progress in psychology was captured recently by prominent researchers in psychology (Guastello et al., 2009), economics (Dore and Rosser, 2007; Rosser, 2004), and medicine (West, 2006), and regularly appear in the peer-reviewed journal, *Nonlinear Dynamics, Psychology, and Life Sciences* (*NDPLS*).

There are several challenges to doing research in this area. One is the lack of a suitable text that explains and compares analytic techniques that are actually pertinent to research problems in the social and life sciences. Another is that the right material actually exists, but it is strewn about the literature in mathematics, physics, biology, economics, and psychology at so many levels of accessibility that the spirited graduate student and committed researcher would be hard-pressed to find what they really need.

A third challenge is that the analytic requirements for problems faced by social scientists are often very different from those of mathematicians and natural scientists. The former require a much stronger layer of statistical reasoning and procedures to cope with the level of noise and instability in their NDS applications. A fourth challenge is to develop skill and expertise in framing hypotheses dynamically and building viable analytic models to test them. Although the latter is also inherent in the conventional linear framework, it is a more complex task here. Fortunately, *NDPLS* has channelled a substantial amount of information about new and relevant techniques, which is a good start from the standpoint of growth at the cutting edge of the field.

Thus, we have brought together constructive work on new practical examples of methods and applications that are built on nonlinear dynamics. It has become possible with the help of a range of valued researchers whose names are listed in this book and who are warmly thanked for their contributions. Psychology, for example, necessarily studies processes that generate errors and instabilities as we explore our capacities to evolve, learn, and forget. Those obfuscations could partly be avoided by adopting the best practices of the simpler physical sciences that are at the core of our approach to the analysis of real data. Doing so creates a need for care against the results of spurious solutions, and we discover such cautions here in our book. In discussing the use of real psychological and life sciences data, a review of contexts and applications makes things more readable and encourage further speculations.

The prominent biologists, social scientists, mathematicians, and statisticians from the late nineteenth century, writing on many examples of human behavior and its diversities, established a tradition that we still respect. We have since strengthened our resources and methods of computing, which were not available until the late twentieth century. We have valued the opportunity to collaborate in this endeavor; it will inevitably evolve to embody the growing resources that widen our perspectives of dynamical behavioral science.

Before moving onward, we would like to thank Kevin J. Dooley, one of our colleagues in nonlinear dynamics, for contributing the cover image from his provocative series of time lapse photographs.

Stephen J. Guastello

Robert A.M. Gregson

MATLAB® is a registered trademark of The MathWorks, Inc. For product information, Please contact:

The MathWorks, Inc.
3 Apple Hill Drive
Natick, MA, 01760-2098 USA
Tel: 508-647-7000
Fax: 508-647-7001
E-mail: info@mathworks.com
Web: www.mathworks.com

References

Dore, M. H. I. and Rosser, J. B. Jr. (2007). Do nonlinear dynamics in economics amount to a Kuhnian paradigm? *Nonlinear Dynamics, Psychology, and Life Sciences, 11*, 119–148.

Guastello, S. J., Koopmans, M., and Pincus, D. (Eds.). (2009). *Chaos and Complexity in Psychology: The Theory of Nonlinear Dynamical Systems*. New York: Cambridge University Press.

Rosser, J. B. Jr. (Ed.). (2004). *Complexity in Economics*, Vols. I, II, and III. Northampton, MA: Edward Elgar.

West, B. J. (2006). *Where Medicine Went Wrong*. Singapore: World Scientific.

Editors

Stephen J. Guastello is a professor of psychology at Marquette University, Milwaukee, Wisconsin, where he specializes in organizational behavior and human factors engineering. His published applications in the field of nonlinear dynamics extends across a wide range of topics, including economics. His more extensive consultancies involved psychological test development and related issues in personnel selection, human–computer interaction, and expert systems for power plants and other facilities, as well as expert testimony in accident litigation. He has previously authored three books: *Chaos, Catastrophe, and Human Affairs* (1995, Erlbaum/Taylor & Francis), *Managing Emergent Phenomena* (2002, Erlbaum/Taylor & Francis), and *Human Factors Engineering and Ergonomics: A Systems Approach* (2006, Erlbaum/Taylor & Francis). He has also coedited *Chaos and Complexity in Psychology: The Theory of Nonlinear Dynamical Systems* (with M. Koopmans and D. Pincus, 2009, Cambridge University Press). He is the founding editor in chief of the quarterly research journal, *Nonlinear Dynamics, Psychology, and Life Sciences*, published by the Society for Chaos Theory in Psychology & Life Sciences.

Robert A.M. Gregson is an emeritus professor at the Australian National University, Canberra, where he specializes in cognitive psychology, brain functioning, time series analysis, and related problems in engineering. He holds degrees in psychology, statistics, and engineering, and has held teaching and research positions in Antarctica, Australia, Germany, England, New Zealand, Sweden, and the United States. His more extensive consultancies included the psychophysics of chemical senses, problems in management, and expert testimony on criminal evidence. He has previously authored six books: *Psychometrics of Similarity* (1975, Academic Press), *Time Series in Psychology* (1983, Erlbaum/Taylor & Francis), *Nonlinear Psychophysical Dynamics* (1988, Erlbaum/Taylor & Francis), *n-Dimensional Nonlinear Psychophysics* (1992, Erlbaum/Taylor & Francis), *Cascades and Fields in Perceptual Psychophysics* (1995, World Scientific), and *Informative Psychometric Filters* (2006, Australian National University E-Press). He is a member of several journal editorial boards, including *Nonlinear Dynamics, Psychology*, and *Life Sciences*.

Contributors

Deborah J. Aks
Rutgers University
Piscataway, New Jersey

Jonathan Butner
University of Utah
Salt Lake City, Utah

Didier Delignières
University of Montpellier I
Montpellier, France

Oscar Diez-Martínez
Universidad de las Américas Puebla
Puebla, Mexico

David Elliott
The University of Newcastle
Newcastle, New South Wales,
   Australia

Matthew S. Fairbanks
University of Oregon
Eugene, Oregon

Rainer Feistel
Baltic Sea Research Institute
Rostock, Germany

Barbara Bruhns Frey
Vanderbilt University
Nashville, Tennessee

Robert A.M. Gregson
Australian National University
Canberra, Australian Capital
   Territory, Australia

Stephen J. Guastello
Marquette University
Milwaukee, Wisconsin

Andrew Heathcote
The University of Newcastle
Newcastle, New South Wales,
   Australia

Joseph P. Huston
University of Düsseldorf
Düsseldorf, Germany

Miguel A. Jiménez-Montaño
University of Veracruz
Veracruz, Mexico

Joachim Krauth
University of Düsseldorf
Düsseldorf, Germany

David M. Kreindler
University of Düsseldorf
Düsseldorf, Germany

Loïc Lemoine
University of Montpellier I
Montpellier, France

Jay-Shake Li
National Chung Cheng University
Minxiong, Taiwan

Charles J. Lumsden
University of Toronto
Toronto, Ontario, Canada

Sean R. McDade
PeopleMetrics
Philadelphia, Pennsylvania

Patrick E. McSharry
University of Oxford
Oxford, United Kingdom

Stephen J. Merrill
Marquette University
Milwaukee, Wisconsin

Annette M. Metten
Chapman University
Orange, California

Terence A. Oliva
Temple University
Philadelphia, Pennsylvania

David L. Ortega
Chapman University
Orange, California

David Pincus
Chapman University
Orange, California

Mark Shelhamer
The Johns Hopkins Medical
  School
Baltimore, Maryland

T. Nathan Story
University of Utah
Salt Lake City, Utah

Richard P. Taylor
University of Oregon
Eugene, Oregon

Kjerstin Torre
University of Montpellier I
Montpellier, France

1

Introduction to Nonlinear Dynamical Systems Analysis

Robert A.M. Gregson and Stephen J. Guastello

CONTENTS

Since the 1970s, the literature on nonlinear dynamical systems (NDS) and related applications has exploded. Specialist journals on chaos and associated topics such as fractals and bifurcations provide us with methods and applications to many disciplines—engineering, economics, biology, neural networks, or ecology, to list a few—but psychology lags behind and does not match in statistical mathematics the developments that have been partly inspired previously from physics. We think that enough can now be brought together to compensate for this lag in the behavioral sciences, done in no short measure to the contributions in the journal *Nonlinear Dynamics, Psychology, and Life Sciences*, up to 2008, which we cite partly in this book. Viable methods of measures to help us model, analyze, and predict are scattered and need bringing together from freely available but little-known programs. We are no longer tied to a tradition of linear static stimulus–response packages, which have been dominant in statistics texts for social sciences, not even giving the simple use of the Bayesian methods that are necessary in mathematical statistics, let alone being extended to Markov Chain Monte Carlo and Metropolis–Hastings procedures (Gilks, Richardson, & Spiegelhalter, 1966).

We have become aware of the need to model the data analysis regimens that are essential to describe the intricate and constantly evolving dynamics that we encounter in real samples. As May (1976) seminally pointed out, simple models can have surprisingly subtle and complex outputs.

The problem we face is to create a structure of symbols and relationships among symbols that are written in one fixed form, but represents a collection of structures that change over time, and that also change their structural interdependence. Each structure is in a limited set that could be written independently but could be coupled to a variable number of other structures.

This remark at first seems both vague and pompous, but it may help briefly to compare it with the familiar linear statistical model that has been used almost exclusively in experimental psychology for the last half century. Most models started from linear regression and a residual of unknown or locally undefined causality that is treated as Gaussian noise. The regression analyses, uni- or multidimensional, had one fixed model with scalar weights, which does not by definition use variable coupled dynamics.

The dynamics that are employed in neural networks to represent brain sensory processes, or the ecology in competing animal species, or the economics of the computing production and consumption under partial constraints imposed by political or power resources, are partially unstable. All of such models represent a world with a limited life in space and time; they may be evolving or getting extinguished by their competitors. In neuropsychophysiology, the life of a system can be extinguished by tumors or disease. In ecology, the predators can extinguish prey and hence extinguish themselves by eliminating their own food supply, or by accidents of climate change. In conflicts between socioeconomic subsystems such as socialist and liberal market forces, each is partially sustained by ideological power distributions and constraints; they are ubiquitous but not modeled by simple linear regressions.

The use of contrasts between deterministic, stochastic, cyclical, chaotic, fractal, and transient systems has been offered often, particularly in the last 20 years. There are partly and optimistically modeled in scales and bounded time series that describe real systems. Questions about how much information is available and what resolution of detailed encoding is needed to create testable models are often unsettled.

The very ideas about what constitutes proof are fluid, and the creating of dynamical models can leave proof itself ambiguous. A few competing statistical frameworks have emerged in company with the profusion of dynamical systems. In a particular case, statistical inference has evolved both as a framework to select models but also as a model in its own right as a depiction of human cognition, and as a model of variation in the rationality and shortcomings in cognitive inference itself (Chater & Oaksford, 2008).

When data are regarded as a sample from a time series, then local jumps, gaps, or abrupt reversals are often treated as arising from corrupted records, probably due to defects in the recording equipment. Gaps in series are replaced by estimates or substitutes with local average values, or weighting the gap substitutes by estimates based on the variances of local autocorrelations of the whole series; fractal polynomials have been proposed for fitting nonlinear trends over time. This assumes that the process that generates data is stationary in a wide sense and sufficiently summarized by anything between linearity with random noise, up to complicated wavelet-based mixed models.

NDS analyses, such as those used in cardiac pathologies, can have local gaps or stationarities; if the heart sequence lands in a gap (a myocardial infarction) then the beat stalls and the process has to be kicked across to restart the fractal cyclic patterns (Winfree, 1987). Other purely mathematical models such as the Shilnikov (1966) attractor have jumps between slow and rapid transitions within a pattern, and jumps between patterns that can be induced by the input of contextual stimuli but are themselves continuous with some irregular autocorrelations. Such gaps can be thought of as bifurcations as in the fold or cusp catastrophes. When we have a dynamic gap, we do not seek to plug or mask it with the local autocorrelation but rather to identify it when a bifurcation arises.

Essentials of Nonlinear Dynamics

NDS is a general systems theory for describing, modeling, and predicting change processes. Whereas linear models of change assume proportionality between inputs and outputs and assume that there is only one type of change, NDS allows the possibilities that small inputs at the right time can produce a dramatic impact, large inputs at the wrong time can produce nothing at all, and that there are many possible patterns of change. The identification of those types of change contributes substantially to the understanding of the phenomenon at hand. They do require proper forms of analysis that permit the literal identification of the dynamics. In studies where it has been possible to compare proportions of variance in an outcome measure that were explained by an accepted nonlinear model and its alternative theoretical counterpart, which was usually linear, the NDS model outperformed the alternative by a ratio of 2:1 (Guastello, 1995, 2002). Thus, at least half of our understanding of a phenomenon is associated with identifying the correct dynamics.

The central concepts of NDS are attractors, bifurcations, structural stabilities and instabilities, chaos, fractals, catastrophes, self-organization, emergence, cellular automata, agent-based models, genetic algorithms, and related evolutionary processes. Many of the foregoing constructs are eminently useful in the analysis of real data, and are thus the concern of this book. Some of the foregoing concepts invoke simulation techniques, which fall outside the present boundaries except to say perhaps that we know the complete story does not end here. Many of the NDS constructs were introduced to the behavioral sciences in the late 1970s, with the data analysis strategies entering shortly afterward. They have since become influential on most of the major areas of psychological research today, ranging from neuroscience to organizational behavior (Barton, 1994; Guastello, 2009; Guastello, Koopmans, & Pincus, 2009). The basic concepts are described below with the understanding that substantially more expansion is planned for later chapters in the book.

Attractors and Stability

Attractors are spatial structures that characterize the motion of points when they enter the space. Three common varieties are the fixed-point, periodic, and chaotic attractors. Collectively they represent specific varieties of constructs that were known in the past as *equilibria*. In the fixed-point attractor, a point may enter the space, and once it does, it remains at a fixed point. Other points that are close enough to the epicenter of the attractor are pulled into the attractor space. Points traveling outside the boundaries or basin of the attractor are not pulled into the attractor region. Fixed-point functions are observed as time series showing changes of a system toward an asymptotic upper or lower boundary, as in learning curves. Limit cycle attractors are oscillations, which are, in many cases, amenable to spectral analysis or modeling with sine and cosine functions. Chaotic attractors are discussed below. In between the three prototypic groups are the quasi-periodic functions and dampened oscillators.

Repellors and saddles are attractor structures that work differently. Repellors are in essence the opposite of attractors: a space that points avoid. When points get too close to the basin of a repellor, they are kicked out of the space. At that stage, the reject points can go anywhere, so long as the destination is "out." Saddles are mixed dynamics consisting of attractor functions and repellors. Points are attracted to the saddle temporarily, but they are ultimately deflected in two or more, sometimes many, possible directions.

Structural stability describes the situation where all the points are following the same rules, which are usually the rules or mathematical description of an attractor's behavior. Thus fixed points, oscillators, chaos, and the things in between are usually structurally stable. Repellors, saddles, and bifurcations are not.

Bifurcations

A bifurcation is a change in a dynamic field from one type of dynamics to another. For instance, a field can contain two or more attractors, if not also repellors and such. Alternatively, an attractor can change from a fixed point to something more complicated. Changes in dynamics are often observed in the time series data, which constitute one reason why a time series could be regarded as nonstationary.

When bifurcations occur, the analyst is looking for critical points where the dynamics change. The critical point can be as simple as a single point, or it could be a more complex pattern such as a trajectory in and of itself.

Chaos and Turbulence

A chaotic time series shows highly complex patterns that appear random, but are actually the result of a relatively simple and deterministic, but non-repeating, function. The property of non-repetition is actually a matter of degree. One can specify the minimum small difference in a metric that is meaningful; observed values that fall within that small radius can be interpreted as the same for all intents and purposes.

The hallmark of chaos is the principle of sensitivity to initial conditions (Lorenz, 1963): If a small random shock were introduced into a time series (from outside the system), the subsequent segment of the time series would be propelled into a new pattern that would not have otherwise occurred. In other words, if two points that are arbitrarily close together are iterated through the same function, they eventually become exponentially separated from each other. In spite of all the opportunities for instability and nonrepeatability, chaotic functions are bounded; they stay within numerical limits.

Not all instances of chaos indicate a chaotic *attractor*, however. A chaotic attractor exhibits two additional characteristics in a time series. The first is a structurally stable attractor basin. Although the trajectories appear to be near random when viewed over time, all trajectories within the attractor are performing according to the same rules. There is also a firm but permeable boundary to the basin, not unlike the basin of a fixed-point attractor; points may enter if they veer close enough. When they do enter, they follow the same chaotic regimen as the other points already inside the attractor. A pattern of folding and expanding takes place within the chaotic attractor itself. Trajectories that travel too close to the epicenter are pushed outward. Trajectories that travel from within the attractor to the edge of the basin are pulled inside. The folding and expanding movement forms the basis of the *Lyapunov exponent* that is used to assess the level of turbulence in the attractor.

Fractals

A fractal is a geometric structure that has a non-integer dimension associated with it. One of its prominent features is sometimes known as the *scaling principle*: the pattern that is observable at a broad scale is also observable on a smaller scale if one were to zoom in on a detail of the larger image. For the most part, visual images that contributed to making fractals famous are not widely used in the analysis of real data in the behavioral sciences. Far more common are analyses of time series data that culminate, in part, in the calculation of a fractal dimension for the series. The fractal dimension in turn is used as a measure of complexity for a dynamical process. A good many of the analytic techniques described in this book have revolved around problems associated with calculating a fractal dimension on real data.

An important turning point in the development of NDS theory was the discovery that the basins of chaotic attractors are fractal. That finding led to many attempts to calculate the fractal dimension as a proof of chaos: If the dimension is fractional, then the function must be chaotic. It turned out that one does not guarantee the other. The presence of a fractal dimension in a time series is good evidence that chaos could be present, but the determination of chaos requires a different approach, which usually takes the form of calculating a Lyapunov exponent. If the Lyapunov exponent is positive, not only does one have a better case for chaos, but exponent converts to a fractional dimension under limiting conditions.

Phenomena known in shorthand as "$1/f$" are also closely related to fractal principles and gaining a great deal of attention in recent years. A $1/f$ phenomenon is one that conforms to a frequency distribution $Freq(X) = aX^B$, $B < 0$. The distribution is also known as a *Pareto distribution* or an *inverse power law distribution*. Sometimes, credit goes to Zipf (1949), who found that the distribution is flagrant in linguistics; or to Bak (1996) for discovering its role in self-organized phenomena (see below). The concept is developed in Chapter 3 in the context of statistical distributions that are associated with NDS phenomena and clearly defy the assumptions of the normal or Gaussian distribution. The punch line for present purposes is that B is a fractal dimension.

Self-Organization and Emergence

Systems in a state of chaos, or far-from-equilibrium conditions, self-organize by building feedback loops among the subsystems thereby shaping their own structures. The feedback loops control and stabilize the system in a state of lower entropy. There are a few different processes by which self-organization occurs, but the common feature, nonetheless, is that the feedback loops are essentially patterns of information flow (Haken, 1988). For instance in organizations or long-standing human groups, explicit and implicit cultural norms evolve to permit or require certain actions, or facilitate, catalyze, or inhibit other actions. Positive feedback loops facilitate growth, development, or radical change in the extreme. Negative feedback loops have the net impact of inhibiting change.

Other phenomena can be observed in data when self-organization occurs. One is phase shift that results from the formation of new internal structures in the system. The phase shift is similar in principle to the change of ice to water or water to vapor; this type of discontinuity is captured well in the catastrophe models (for discontinuous change). Another possible outcome is the formation of an inverse power law distribution of objects or values of a variable X. A third possibility, which is not exclusive of the other two outcomes, is a sudden drop in the variability of X. Changes in variability are analytically interesting in NDS, and are considered here in the context of entropy statistics.

The earliest concept of emergence predates NDS, but it required NDS to fully appreciate, operationalize, and study (Sawyer, 2005). In the early twentieth century, Durkheim wanted to identify sociological phenomena that could not be reduced to the psychology of individuals, e.g., groups, organizational and social institutions. Bilateral interactions among individuals eventually give rise to norms, deeper and consistent patterns of behavior and relationships, and other forms of superordinate structure. Emergence became an NDS principle when Holland's (1995) computer simulations illustrated how the local interaction among agents gave rise to self-organized structures. In stronger cases of emergent order, the *supervenience* principle engages whereby the superordinate structure maintains a downward influence on the behaviors and interactions among the lower-level agents.

Complex Adaptive Systems

A *complex adaptive system* (CAS) is a living system that maintains a readiness to adapt to new situations. Although the system might have self-organized its resources to engage in a strategy for survival, it is ready to adapt to environmental or internal stimuli at a moment's notice. When it adapts, it reorganizes its communication, feedback, or workflow patterns to respond to the new situation and engage in a pertinent response. Many actions of a living system, no matter how well rehearsed or stereotypic, harbor a modicum of variability in their execution. The variability is not error, measurement or otherwise; rather it serves the purpose of permitting adaptation when necessary. In this regard, the well-rehearsed behavior is *metastable* (Hollis, Kloos, & Van Orden, 2009).

One would observe relatively greater levels of entropy in the behavior of a healthy CAS, and less entropy, or more rigidity and stereotypic behavior in a less functional system. Time series of events from healthy versions of systems thus gravitate toward fractal dimensions between 1.0 and 2.0 (Goldberger et al., 2002). This is also the range of B associated with self-organized criticality (Bak, 1996), and is also the range known as *low-dimensional chaos*.

Catastrophes

Catastrophes are models for describing and predicting discontinuous changes in events (Thom, 1975). The models involve attractors, bifurcations, saddles, repellors, and control variables that play different roles in the behaviors that are captured by a catastrophe model. There is actually a set of catastrophe models that vary in complexity. As one can imagine, the simpler models have seen a wider range of uses than the complex models.

Three Traditions of Analysis

The types of analyses that are developed in this text arise from a confluence, if not a collision from three sources—mathematics, biology and physics, and social science—each having its own concerns for what should be analyzed and how it should be done (Guastello & Liebovitch, 2009). The mathematicians would generate numbers from the chaotic attractor of their choice, and analyze the number series for the properties of their choice. They had no shortage of data in the time series because they essentially made up the numbers. The underlying model was known for the same reasons. They had no issues with noise because there was no noise in the data unless they put it in deliberately to test a new procedure. For the most part, the analytic concerns of mathematicians fall outside the realm of real data such as one might collect in a laboratory experiment, thus we only pause here briefly.

The biologists and physicists stuck close to the mathematicians' approach although they were working with data that came from models that were unknown, and which they had to identify. There is an immediate problem: Sprott (2003) catalogued 62 different chaotic systems in 6 categories. Where does one start, and what heuristics does one use to identify an optimal description of the phenomenon? A strong theory about the phenomenon would greatly reduce the number of models that should be tested. By the same token, there is no reason to believe that the theoretical models originating in mathematics would be observed in their idyllic forms in real situations (Ruelle, 1990).

Noise is problematical, furthermore; there were some early disappointments when what was thought to be chaos turned out to be noise generated by the laboratory equipment. The remedy for noise was to filter the data first and then proceed with direct calculations on the results.

The social scientists' strategy arose from its earlier history of developing linear models and separating them from "error." A nonlinear model, chaotic or otherwise, needed to be tested statistically, and separated from residual variance. Filtering is done only on rare occasions, as when laboratory equipment with known properties is involved. Most types of data encountered in the social sciences, however, are not generated by machines that have separable error functions. Perhaps for that reason alone a vibrant branch of mathematical statistics grew up during the twentieth century known as *psychometrics*. There are other complications in the dynamics itself. There could be transient effects operating, whereby the dynamics change for a period of time and, maybe, revert to the original dynamic pattern. Transients could be difficult to identify if the data sets are relatively short, which they often are (Gregson, 2000).

Given the foregoing trepidations and the need to find useful information in one's lifetime, numerical techniques were developed or deployed to capture NDS properties of the data that were not specifically constrained to 62 possible chaotic models plus other non-chaotic models. Some widely used choices were types of dimension analysis, information and entropy functions, the Grassberger–Procaccia algorithm for computing a fractal dimension from a time series, and graphic displays of data with phase space diagrams and recurrence plots. Noise still had a distracting impact on these analyses, however, and solutions were devised as one might anticipate. One solution was to use surrogate data and random shuffling (Theiler, Eubank, Longtin, Galdrikian, & Farmer, 1992) as a crude test of the null hypothesis; no inferential statistic was used, just a visual comparison of results from real and surrogate series. A second class of solutions involved filtering techniques, some of which could become uncomfortably obtuse.

A third solution was to go beyond the single time series design, which is essentially an elaborate case study approach, and use the familiar experimental designs with manipulated levels of variables, treatment and control groups, and so forth. The nonlinear indicators would be dependent measures in the analysis (Delignières, Torre, & Lemoine, 2005). Comparisons between values in the experimental cells would be the most important source of inference. Any aberrations in the data that could affect the dimension analysis would be consistent across cells and thus controlled in that fashion.

A fourth class of solutions involved developing techniques from a more basic theoretical origin that captured the dynamics of interest and either navigated the issues of error and sampling or were immune to at least some of them. Thus we arrive at our introduction to the chapters in this book. The methodologies selected for exposition here focus on problems and constraints encountered in psychology, physiology and medicine, economics, and other social sciences. Some attention is also paid to the available computer programs that run the analyses and the commands that operate them.

Elements of Nonlinear Time Series Analysis

Chapter 2 covers some basic issues that we inherit from the more conventional linear time series analysis such as interpreting causal relationships, building models, and comparing them. The search for models with fractals components introduces some new complications to the building and assessment of model structures.

Chapter 3 addresses frequency distributions and error functions. It begins with the problem of the relentless assumption of Gaussian distributions and incomplete specification of sufficient statistics in the social sciences. Power law and exponential distributions, which are much more evident in NDS processes, are then described. The next section discusses the parsing of variance between linear and nonlinear components and IID (independent and identically distributed) and non-IID error. The Brock-Dechert-Scheinkman (BDS) statistic is presented as a basic means of determining nonlinearity in a time series. Also important is the contrast between global versus local dynamics, and the importance of a proper time lapse between measurements.

The chapter concludes by describing computational filtering techniques such as z-filtering, moving averages and medians, polynomial prediction models, and singular value decomposition.

The subsequent chapters have been grouped into four main sections: traditional NDS indicators and graphic techniques, statistically driven techniques, Markov processes and symbolic dynamics, and broad issues. In each section, the chapters combine theoretical reviews and worked examples drawn from papers that have appeared in *Nonlinear Dynamics, Psychology, and Life Sciences*. The methods described are not mutually exclusive, and for any real data problem one has to proceed by drawing on relevant prior knowledge about how the data might be generated, and the intrinsic accuracy or inaccuracy of the data encoding. For exploratory data analysis, one may use a mix of methods, taking care to consider the order in which they are applied.

If a method reveals local irregularities in time or space, or indications of nonstationarity in the dynamics, then those features may in themselves necessitate examination of what is happening around the apparent discontinuities. In contrast to using preset methods that rest on the general linear model, caution should be applied before pooling data across subjects and tests of homogeneity may be necessary. The whole area of methodology for nonlinear dynamics is still evolving, so new methods that become available should be surveyed and added to the repertoire presented here, when comparative reviews of their efficiency are published and appear relevant.

Traditional NDS Indicators and Graphic Techniques

Phase diagrams, described in Chapter 4, are graphic plots of velocity as a function of position. The pictorial results are often indicative of particular dynamics occurring in the data. There was some initial attempt in the social sciences to use the plots without inferential statistics, but it quickly became apparent that noise and misestimating the number of dimensions produced distorted representations of the dynamics. Analytic techniques were thus developed to respond to these issues. The principles of dimension analysis and their calculation are concomitant to this exposition. Phase space diagrams have been useful in physiological and psychological research.

Chapter 5 investigates how some of the traditional NDS computations withstand noise or psychometric error. The analysis and results are organized around some of the subtleties encountered in response time data, which are widely used in cognitive studies. The explanation includes a description of the algorithms contained in the TISEAN (nonlinear time series analysis) program and operation commands. Similarly, Chapter 6 investigates how some of the traditional NDS computations withstand missing data. The peculiarities of physiological time series are used as a reference point.

It is probably inevitable that experimenters would find a need to make some adaptations either in their laboratory procedures or in harnessing an analytic technique to make the two parts of the job fit together. Chapter 7 describes a variation of the phase space diagram that was particularly useful in studying learning processes. Chapter 8 describes some adjustments in laboratory procedure that were designed to get the most out of some traditional NDS indicators. It also provides a working example of the BDS and other statistics.

Entropy measures (Chapter 9) are particularly interesting because they capture the nature and type of variability in data, and have been found to relate closely to each other and to some of the traditional nonstatistical NDS metrics. The concept of entropy evolved in four epochs beginning with thermodynamics. The next major installments arose from statistical mechanics, the concept of information, and, most recently, self-organization. This chapter describes the Shannon, Kolmogorov–Sinai, Boltzmann, mutual, topological, diffusion, Kullbeck–Liebler, approximate and sample entropy metrics, and detrended fluctuation analysis; the foregoing are essentially descriptive statistics or procedures to produce them. The chapter concludes with the state-space grid, which conjoins ideas from phase space analysis and entropy constructs.

Recurrence plots (Chapter 10) are another form of graphic analysis. They are created by plotting the value of a point (continuous variable) along two dimensions—the point in time when it first appeared against the point in time when it appeared next. A completely random series would produce a homogeneous square patch of black. A periodic series would produce stripes, and a deterministic, but chaotic, series would produce an interesting pattern loosely resembling that of an oriental rug. Historically, recurrence plots were interpreted only by comparing their visual patterns. Statistical indicators have been developed for recurrence plots in recent years. Examples show where they have been enlightening in sensation research.

Statistically Driven Techniques

The next suite of analyses expands NDS analysis in two directions. In one direction they invoke strongly defined mathematical models. In another direction they use inferential statistics to determine the extent to which the model fits the data and to assess the viability of the component parts of the models. Importantly, they allow point estimation based on the models that have been ascertained by the analysis.

Chapter 11 features techniques for analyzing discontinuities that can be accomplished with the use of the general linear model. SETAR parses a time series into sections where each section contains a different linear difference model. SETAR is an example of spline regression in which the critical points where the changes in a function occur are of central interest and whose location has to be estimated. Catastrophe models describe and predict discontinuous changes of events. They invoke bifurcation structures and support critical points analysis. Examples are excerpted from published studies. Chapter 11 is one of three chapters describing different techniques for analyzing catastrophes.

Nonlinear regression is a technique for testing the goodness of fit for *any* differentiable nonlinear function. Although it is nearly a half-century old and familiar in biology, it remains unfamiliar to the vast majority of social scientists. Chapter 12 begins by describing the nature of nonlinear regression and how to use it. The second section of the chapter describes a hierarchical series of structural nonlinear equations introduced by Howell Tong (1990) that are exponential in nature and capable of determining a point estimation model for chaotic functions. The third section of the chapter describes a hierarchical set of structural nonlinear models developed by the author that captures and distinguishes a range of dynamical phenomena that includes point attractors, chaos, bifurcation functions, transfer, and coupling functions. The chapter includes examples that have been excerpted from published studies along with a comparison of the two modeling systems.

Chapter 13 combines principles from previous chapters to define nonlinear regression models that can test hypotheses concerning catastrophes. Although perhaps not as appealing intuitively as the polynomial difference method in Chapter 11, these static models, which are based on the unique probability density functions associated with catastrophe models, are useful when variables are measured at only one point in time. Once again, examples are excerpted from published studies.

The distinguishing feature of the GEMCAT technique for catastrophe analysis is that it can search for an optimal combination of variables in a large data set that corresponds to a dependent measure. The dependent measure, in turn, is a catastrophe function of another optimized set of variables that function as control variables. The technique is described in Chapter 14 in the context of an application where the authors required a suitable nonlinear analysis to decide between two theoretical explanations for a decision phenomenon.

The next two chapters are concerned with models that involve oscillators as their central dynamic. Spectral analysis is a time-honored technique in engineering for separating multiple oscillators in a complex wave form or time series. The technique produces optimal results for stable periodic functions. Chapter 15 explains the technique and how it has been relevant to the analysis of NDS models, particularly where the data are organized in two-dimensional spatial arrays.

Chapter 16 describes an approach that uses differential equations to identify single oscillations, dampened oscillators, coupled oscillators, and tri-oscillators. The structures are built by predicting the acceleration of a variable from its velocity, position, Rayleigh, van der Pol, and Duffing functions; the latter three are polynomial functions. Because the components are additive, they can be assessed through a variant of the general linear model.

Markov Processes and Symbolic Dynamics

The next suite of analyses address NDS processes that involve nominal or categorical data that define system states that change over time. Markov processes (Chapter 17) begin with an array of categories that represent system states and then predict the odds of a point moving from any category to any other category. The transition matrix that describes the pattern of motion can remain constant throughout the series or itself undergo transition. The transitions can illustrate a wide range of nonlinear dynamics. Hidden Markov models can be a stochastic extension of the basic matrix concepts. Chapter 18 presents an example of Markov chain analysis in a cognitive experiment that involves ambiguous figures.

Chapter 19 underscores the conceptual connections between Markov chain analyses and symbolic dynamics, which occupy the next three chapters. Several types of research problems are presented where the data are more amenable to symbolic dynamics than to metric analyses. The chapter also foreshadows a connection between symbol sequences and entropy measures that are also developed in later chapters.

Chapter 20 describes the WinGramm suite of programs that isolate symbol sequences, and calculate entropies and related features of a categorical NDS time series. Examples from problems in computational neuroscience and the organization of macromolecules are presented.

Orbital decomposition (Chapter 21) is a symbolic dynamics technique that is based on the principle that chaos is composed of unstable periodic orbits that are all intertwined. It requires categorical data that is coded in some fashion of interest to the research project. The temporal pattern can be as complex as chaos, but chaos is not required. The analysis identifies patterns of events and pattern lengths, and produces inferential statistics, measures of topological and Shannon entropy, Lyapunov exponents, and fractal (Lyapunov) dimensions. The chapter explains the analysis of single time series with both simple categorical coding and complex coding for multiple attributes, which can be accomplished with a simple spreadsheet, frequency distributions, hand calculations, and a little patience. The chapter concludes with some frontier applications where metrics from complex sources, electromyographs and functional magnetic resonance imaging, can actually be broken down into categories and analyzed in symbolic dynamic form.

Chapter 22 explains the use of orbital decomposition where the properties of multiple time series, e.g., from different experimental conditions, are compared. It considers the role of power law distributions in the array of patterns that is produced by an analysis, and how those distributions further characterize the sets of patterns that are extracted from different experimental conditions.

Broad Issues

So you think you found chaos in your data? Now what? Chapter 23 recounts some of the difficulties researchers have had answering that question, often because the problems that are associated with the most popular traditional methods can lead to incorrect conclusions. The chapter makes a case for exploring the nontraditional techniques covered in this book that have stronger grounding in statistical theory. Chapter 24, on the other hand, suggests that the merits of the traditional techniques can be recovered if researchers would only get beyond analyzing single time series and expand their thinking in the direction of broader experimental designs.

There is always more, as Chapter 25 explains future developments that are based on multistability phenomena. Problems have begun to emerge in the mathematical literature where a collection of attractors interact in diverse and transient couplings to generate complex and variable patterns with considerable qualitative diversity. As processes at various levels of psychological interest, from brain psychophysiology to networks in social organizations, appear to be multistable, they can encourage models that incorporate switching in time and space. Problems in simulation and prediction for multistable dynamics can be expected to become active areas of research in the twenty-first century.

References

Bak, P. (1996). *How nature works: The science of self-organized criticality*. New York: Springer-Verlag/Copernicus.

Barton, S. (1994). Chaos, self-organization, and psychology. *American Psychologist, 49*, 5–14.

Chater, N., & Oaksford, M. (2008). *The probabilistic mind*. Oxford: Oxford University Press.

Delignières, D., Torre, K., & Lemoine, L. (2005). Methodological issues in the application of monofractal analysis in psychological and behavioral research. *Nonlinear Dynamics, Psychology, and Life Sciences, 9*, 435–461.

Gilks, W. R., Richardson, S., & Spiegelhalter, D. J. (Eds.). (1966). *Markov chain Monte Carlo in practice*. London: Chapman & Hall.

Goldberger, A. L., Amaral, L. A. N., Hausdorff, J. M., Ivanov, P. C., Peng, C. K., & Stanley, H. E. (2002). Fractal dynamics in physiology: Alterations with disease and aging. *Proceedings of the National Academy of Sciences, 99*, 2466–2472.

Gregson, R. A. M. (2000). Elementary identification of nonlinear trajectory entropies. *Australian Journal of Psychology, 52*, 94–99.

Guastello, S. J. (1995). *Chaos, catastrophe, and human affairs: Nonlinear dynamics in work, organizations, and social evolution*. Mahwah, NJ: Lawrence Erlbaum Associates.

Guastello, S. J. (2002). *Managing emergent phenomena: Nonlinear dynamics in work organizations*. Mahwah, NJ: Lawrence Erlbaum Associates.

Guastello, S. J. (2009). Chaos as a psychological construct: Historical roots, principal findings, and current growth directions. *Nonlinear Dynamics, Psychology, and Life Sciences, 13*, 289–310.

Guastello, S. J., Koopmans, M., & Pincus, D. (Eds.). (2009). *Chaos and complexity in psychology: The theory of nonlinear dynamical systems*. New York: Cambridge University Press.

Guastello, S. J., & Liebovitch, L. S. (2009). Introduction to nonlinear dynamics and complexity. In S. J. Guastello, M. Koopmans, & D. Pincus (Eds.), *Chaos and complexity in psychology: Theory of nonlinear dynamical systems* (pp. 1–40). New York: Cambridge University Press.

Haken, H. (1988). *Information and self-organization: A macroscopic approach to self-organization*. New York: Springer-Verlag.

Holland, J. H. (1995). *Hidden order: How adaptation builds complexity*. Cambridge, U.K.: Perseus.

Hollis, G., Kloos, H., & Van Orden, G. C. (2009). Origins of order in cognitive activity. In S. J. Guastello, M. Koopmans, & D. Pincus (Eds.), *Chaos and complexity in psychology: The theory of nonlinear dynamical systems* (pp. 206–241). New York: Cambridge University Press.

Lorenz, E. N. (1963). Deterministic periodic flow. *Journal of the Atmospheric Sciences, 20*, 130–141.

May, R. M. (1976). Simple mathematical models with very complicated dynamics. *Nature, 261*, 459–467.

Ruelle, D. (1990). Deterministic chaos: The science and the fiction. *Proceedings of the Royal Society of London, A, 427*, 241–248.

Sawyer, R. K. (2005). *Social emergence: Societies as complex systems*. New York: Cambridge University Press.

Shilnikov, L. P. (1966). Birth of a periodic movement departing from the trajectory that goes to an equilibrium state of a saddle-point-saddle-point and then returns [In Russian]. *Proceedings of the Academy of Sciences of the USSR, 170*, 49–52.

Sprott, J. C. (2003). *Chaos and time-series analysis*. New York: Oxford University Press.

Theiler, J., Eubank, S., Longtin, A., Galdrikian, B., & Farmer, J. D. (1992). Testing for nonlinearity in time series: The method of surrogate data. *Physica D, 58*, 77–94.

Thom, R. (1975). *Structural stability and morphogenesis*. New York: Benjamin-Addison-Wesley.

Tong, H. (1990). *Non-linear time series: A dynamical system approach*. New York: Oxford University Press.

Winfree, A. T. (1987). *When time breaks down*. Princeton: Princeton University Press.

Zipf, G. K. (1949). *Human behavior and the principle of least effort*. Reading, MA: Addison-Wesley.

2

Principles of Time Series Analysis

Robert A.M. Gregson

CONTENTS

In this study of nonlinear dynamical systems (NDS), as with any other approach to building statistical models, we have to be mindful of three sorts of questions. What are the events in the physical world and in our mental covariates of that world we seek to model? What can we choose as representational structures in symbols, algebra, or logical relations that are intrinsically tractable? And what rules can we adopt for evaluating the relative worth of alternative models of the real world? It is possible to build general theories about finding a hierarchical order for alternative models, but these are incomplete (Speekenbrink, 2003). These three areas of discourse, data, models, and mapping between data and models, all involve decisions about what is thought to be scientifically relevant and potentially fruitful to predict in a wider range of apparently related scenarios.

If we are very restrictive in space and time when choosing events to model, then we may achieve simplicity at the expense of being falsely static. Outcomes then take precedence over processes. Relationships between things we can approximately measure, where the direction and persistence of causality seems credible, can be mapped into linear models, with a residual component of failing to fit data that can be treated as effectively random. This has been done in experimental psychology, and its offshoots into social studies, since Gauss first created and adopted such statistical structures when surveying the streets of Hannover being rebuilt after the end of the Napoleonic wars. A long tradition that works in a limited area is not lightly given up until we aspire to understand not only static but changing relations; the focus shifts to asking how do things come into transient stability and what will happen over longer times and wider contexts? Two recurrent problems when choosing both data and the representations in models of data will recur in our examples: they are the problems of choosing resolutions of time and of space. How often should we sample the evolution of processes, in milliseconds, hours, months, to seek stabilities, instabilities, trends, discontinuities, recursions, and asymptotic death states? How precisely do we measure the levels of stimulation, and seek evidence of intrinsic continuities? There is evidence of discontinuities in some perceptual or psychophysical processes, so that models of stimulus–response relations with a break in them at around some stimulus level fit data much better (Eisler & Eisler, 2009). How might we choose coarse scalings of variables to smooth out the second-order irregularities, superimposed variations of small response magnitude as compared with major trends that mask smoother relationships between variables of interest? Are underlying processes, both in space and in time, continuous or discrete?

These questions, which are at the focus of philosophy of science over a range of disciplines, move us to consider nonlinear dynamics as a family of models that attempt to address change and at the same time create relatively simple and tractable structures to act as models. They also mean that data in the form of time series, and not steady-state descriptions, are the essential core and starting point for investigations. Providing examples of how this approach has been used in the behavioral sciences, particularly over the last two decades, and the availability and use of statistical methods that are necessary to implement the study of change in its widest sense, are the purpose of this book.

Any coherent nonlinear dynamic model that does correspond to some psychological processes will generate, as local limiting conditions, static linear representations if we truncate (and if you like, freeze) the processes in space and time. Engineers have known this for years, and reduce a variable process to a series of short segments by what is called piecewise linearization, each short segment is then used to predict and control the process, and segments are linked by what are called *splines*. This functions well if the system is under efficient control intermittently or continuously. One can effect reductions from nonlinear to linear structures, but one cannot do the opposite. There are no clues within the structure of a static linear model as to what its unique generating nonlinear structure would be, it has been filtered out. What however we can sometimes do, given a nonlinear model, is see if its structure and hence its predictive properties are common to a class of models with diverse psychological application. Some writers have argued that in fact this is what higher level cognition is about. An example is that of Kepler's model of the planetary system, which according to Gentner et al. (1997) was derived in part by analogical reasoning from other models of simpler processes.

For a long time, problems of measurement and prediction in the behavioral sciences have been centered uncomfortably on the desirability of being able to scale in continuous numerical measures with metric properties. If one cannot do this, then there are profound difficulties in using the familiar differential equations of classical analysis, and defining and measuring rates of change becomes problematic. One may instead turn to using transition probabilities between states, in what is called symbolic dynamics. Failure to do this, except by fiat, restricted modeling to what are called stationary processes or steady-state input–output psychometrics. For example, catalogued descriptions of what a learning process looks like when it has reached asymptote, and what are the errors made during learning, are not sufficient in themselves to model the dynamics that generate the errors and the learning, but approaching the problem the other way round, a model of the dynamics should be able to incorporate both as intrinsic to the same process.

The arguments about the possibility of climbing from ordinal to interval scales dominated psychometrics (Michell, 2009), and left little room for considering if this was crucial for predicting the dynamics of behavior. As measurement in itself is purely a descriptive exercise, and prediction and control are the desiderata of science, prediction was limited to those aspects that were not relegated to random noise, though that randomness itself was almost always treated as Gaussian and hopefully second order. Interestingly, more statisticians than psychologists were at times uncomfortable with this assumption, made because of considerations of mathematical tractability.

Two principles have been repeatedly invoked in creating rules to decide between alternative models of processes in scientific explanation, sufficiency, and simplicity; one can be traced back to William of Occam (circa 1282–1349) and called Occam's Razor, though usually misquoted, and the other attributed to Bertrand Russell.[1] Sufficiency can be expressed in an aphorism credited to Russell ("if an account of something is sufficiently detailed, then it is more probably true"), which he might have written, as he subscribed to the correspondence theory of truth (Russell, 1940, p. 289). We shall see later that a number of statistical indices used to compare the goodness-of-fit of competing alternative models attempt to balance between sufficiency and simplicity. Trade-offs between structural complexity on the one hand, and the number of free parameters or degrees of freedom on the other, became a challenge to the theoretician. There is not one single definition of complexity that is universally applicable, and a fundamental distinction has to be made between structural and dynamical complexity (Wackerbauer, Witt, Atmanspacher, Kurths, & Scheingraber, 1994).

In the late twentieth century, an increasing interest in the dynamics of physical, biological, and social processes meant that a requirement of model construction became that it should show how change evolved through time, and not leave as undescribed the complex nonlinear components of dynamics by relegating them to a basket category of random noise. The magnitude of secondary effects, as expressed in variance components, is not a sure guide to their relative importance in nonlinear systems that evolve over time. Systems can function simultaneously at different levels of resolution, and the levels can be related to one another in a hierarchical manner (Haken, 1987). It turned out that much of what is critical in describing dynamics can be achieved without satisfying the requirements of metric scaling, but instead other problems of setting the proper resolution of data at varying space-times becomes central.

Abandonment of the principle that output changes should be proportional to input changes within the range that biological systems usually function, and hence that model structure should not automatically be built on that linearity assumption, was highlighted in an influential paper by May (1976). Communicating these concepts to a wider audience, at least in terms of their aesthetics, became widespread after Mandelbrot in the 1970s created a computer-generated map of the Fatou and Julia sets of the logistic function, and showed pictorially mathematical objects that had never been seen in their fullness before. This expansion of pictorial representations, and the accompanying mathematics, with potential applications to represent processes in the natural world, was powerfully extended by Barnsley (1988, 2006). The contrast between the diversity of pictorial representations of nonlinear dynamics, and the simple if not simplistic graphical depictions of linear relationships in much of psychological experimental work, is indeed dramatic. However, it is not the case that a pictorial representation of an attractor can always be generated; some are not computable. Mandelbrot fortunately picked a tractable example (Braverman & Yampolsky, 2008).

Given that two or more models of comparable plausibility have been, by elimination, identified, using only the criteria of sufficiency and simplicity, is there any further basis to choose between them? At this point, we can use the prior probabilities of classes of models as additional information (Hoijtink, Klugkist, & Boelen, 2008). It is conceptually difficult to construct indices of maximal relative simplicity, but turning the problem around, modern methods use instead indices of complexity, which can be minimized, though complexity admits of a range of different mathematical definitions. Kolmogorov complexity is the most used, with modifications (Li & Vitanyi, 2008). There are conflicting views concerning whether the regularity of a string of symbols is in any way related to its complexity. Adami and Cerf (2000, p. 63) take the position that randomness "is a meaningless concept without reference to an environment.... A sequence can be random with respect to one environment while perfectly 'meaningful' with respect to another." That position implies that complexity has to be assessed with respect to an environment with which it is correlated.

Time Series

A time series is a collection of outcomes labeled by at least two symbols x, x', one of which may be a blank, drawn from a finite set of symbols $\{X\}$. The series has been or is generated by a sequential rule. We have to know what the full set {X} consists of, even if not all its symbols are used in a given sample from the time series, if we are to compute statistics of importance. It is important to know if missing terms are gaps due to imperfect recording or part of the process having inbuilt pauses (blanks) of some sort, The rule is what we seek to discover, if it is not pre-defined. The rule may be itself a random walk.

For example, a dealer of a shuffled deck of cards produces a time series of ♣♦♥♠. The four symbols are neither letters nor numbers but it is still possible to determine if the deck has been shuffled properly or is stacked by a cheating croupier.

Many statistical properties of time series that do not depend on numerical symbols can be used as exploratory diagnostics. Each symbol may be a single undecomposable entity or may be an ordered set, expressed as a vector. If the latter is the case, then it is usual to call the process a multivariate time series (mts). Any psychological experiment or sampled life history that runs through time with repeated trials is an mts composed of stimuli and responses. Both the stimuli and the responses may be unitary or a vector. We may only be able to observe the responses, as happens in some behavioral series with a hidden physiological substrate.

Associated with the time series, and may be known or assumed, is a series of points in time, t_x, and this associated temporal series can have in turn a series of differences (the time intervals) $\delta t = |t_x - t_{x-k}|$ where k ranges over $1, 2, \ldots,$ and for $k = 1$ we have first differences or lagone differences. In the simplest situation δt = constant, so we assume that the symbols are equally spaced in real time. Many econometric time series, for example, are data points spaced at once per month or per year. We may, as in engineering or high-speed physiology, choose to have all time intervals very small and constant as this may not cause us to incur a loss of information, and it makes the subsequent analyses easier. Time series with unequal intervals can be difficult to interpret (Brillinger, Fienberg, Gani, Hartigan, & Krickenberg, 1984).

In a full process description, then, we have two series running in parallel, the symbol string and the associated time points. If all the δt are the same constant, then the two series are causally independent. If we have only one non-blank symbol and variable δt then the series is called by some statisticians a point process. If there are two symbols, then we can have switching and a situation like the perpetual processes called figure–ground reversal. The switching does not have to be regular and periodic.

Sequential Properties

In the simplest of processes, with symbols taking numerical values, a series is called IID, meaning *independent and identically distributed*. This is the inbuilt assumption in most Anova models, and it is generally false unless an experiment is so spaced out over time that successive observations are supposedly not influenced by what has gone before. That is, there is no learning or forgetting in the process. If we have reason to believe that successive observations are not independent, then a variety of tests are available. The simplest is to look for the frequency distribution of lengths of runs of equivalent symbols. There are tests of persistence or anti-persistence (the Hurst index), and of autoregression over a range of lags, called the autocorrelation spectrum. Spectral analysis may be used to identify recurrent periodic components, but can be fatally misleading if there are local spikes.

Collateral Variables

In psychology, we are often interested in whether a response is said to be correct or false, which means that either the response set of symbols is partitioned or that a mapping from stimuli to responses exists. This use of the collateral variable, correctness, does not often have a parallel in the physical sciences unless as in engineering we are exploring component defects. Other collateral variables, such as cost or risk, or subjective confidence, may be used. A collateral variable may reveal changes over time, in the underlying process, and be a valuable index of nonstationarity, which is not detected in the original symbol string itself.

Analysis in States

We always have a metric on t, but not always on psychological variables. If numerical values are not known, or suspected not to be well behaved in a metric sense (as in arbitrary ordered encodings such as Likert scales), we can replace them with an unordered set of labels called states. The interest then is in the transition probabilities between states, and the dwell times within any state. If it is possible to get, through a number of steps, from any one state to any other, the system is said to be *ergodic*. When the system has a limited memory, only one step back or one step forward, it is said to be *Markovian*. In the long run, a Markovian system runs to a probabilistic stability, the stationary state vector, and this may be computed and tested against long-term probabilistic observations.

Time Dependencies

The causal relationships that exist may involve the δt_x driving the x or being driven by x. For example, response latencies may be a function of the previous stimulus, or of task difficulty involving the choice of response due to the number of potential alternatives in $\{X\}$. If the underlying process that determines sequential dependencies is known, it is commonly represented by simultaneous difference or differential equations, with implicit parameters. Commonly quoted examples have two or three implicit parameters, with cross-products of the parameters in one or more generating equation; if there are three or more parameters, chaos may ensue.

Model Identification

Theoretical time series generated by mathematical models are called *trajectories*, so the problem of model identification and fitting is the problem of matching trajectories to empirical realizations. Each realization is a relatively short sample time series, and is matched to segments of theoretical trajectories than can be infinite in length. Time series that make jumps between different stationary processes, or have interpolated subseries, require careful analysis, and may be evidence of multistable dynamics (Tong, 1983). Recursive loops, if identified, are evidence of feedback and make linear models very improbable.

Fractals

Many short samples from time series may be difficult to identify in terms of their internal generating dynamics, and if irregular and complicated may be wrongly labeled as random. In the past, there was no other way for statisticians to characterize them in order to operate on data pooling or prediction exercises. In recent years, there has been a shift to considering model identification and comparison, rather that statistical significance testing against some null hypothesis, as the methodological objective, and fractal analysis has thus come to center stage, because the time evolution of processes gives us more insight into dynamics, by definition. Fractal properties of time series samples can be defined by their characteristics (Delignieres et al., 2006), which include: a complex pattern of correlations created by interpenetrated time scales, long memory and current dependence of fluctuations in the remote past, self similarity or scale independence, that is made pictorially convincing by the figures of Julia sets first popularized by Mandelbrot, and an independence from the need to establish metric space representations as opposed to symbolic dynamics (Adami & Cerf, 2000).

Giving a formal definition of what a fractal is, as opposed to describing its properties, is a task that Mandelbrot eschewed in his pioneering work, and Feder (1988) follows the same cautious approach. Already there exists a number of diverse areas of psychology, motor performance, reaction times, mood fluctuations, self esteem, and psychophysics in which fractal properties of sequential data have been proved to be present. The internal dynamic structure generating a fractal is impossible unequivocally to identify if the sample is too short, or seriously corrupted by exogenous effects. However, using a method called *detrended fluctuation analysis* (DFA), originated by Peng et al. (1992), and further developed by Taqqu, Teverovsky, and Willinger (1995), some important progress can be made (Arianos & Carbone, 2007).

These methods relate to the use of the Hurst index, which we consider later. DFA has been used in cardiac dynamics, economics, and ethology, to name a few areas. As Hu, Ivanov, Chen, Carpena, and Stanley (2001, p. 01114-1) note, "Traditional approaches such as the power-spectrum and correlation analysis are not suited to accurately quantify long-range correlations in nonstationary signals. DFA is a scaling analysis method providing a simple quantitative parameter, the scaling exponent α—to represent the correlation properties of a signal."

It is necessary and possible under some conditions to correct for both trends and nonstationarities (Chen, Ivanov, Hu, & Stanley, 2002). The existence of local spikes, or singularities, may interact with the antipersistence of a series to create what are called cross-over effects; which occur when there are transitions between anticorrelated behavior and random behavior. These depend in part on how a series is broken into subsections of a chosen length, each locally fitted by a linear trend component, and how the examination of data using a number of different partitions, or scalings, is performed.

> Any approach to scientific inference which seeks to legitimize *an* answer in response to complex uncertainty is, for me, a totalitarian parody of a would-be rational learning process Smith (1984, p. 247).

Some Nonlinear Indices

The construction of indices to identify the evolution of dynamics has been an active area of nonlinear mathematics and, in turn, such indices are used as approximate descriptions of real data samples through time. Statistical indices perform two tasks, deciding that a process running through time is deterministic to some degree, or creating a basis to compare two or more models. There are three aspects of an index that are important in practical applications: (a) that it is intended to measure a property of the dynamics, and the limitations of the conditions under which it is a valid measure; (b) the mathematical foundations of the index, as discussed for example by Abarbanel (1996) or Davies (2003); and (c) the availability of off-the-shelf approximate software for computer implementation.

In linear time series analysis (Gregson, 1983), with Gaussian assumptions, the first two moments of the process and the autocorrelation spectra are almost always sufficient statistics, provided the process is stable in the mean. In nonlinear dynamics that is not the case, and there has come into use a diversity of indices. Some of these indices are based on entropy measures, others on dimensional or correlational analyses. It may be necessary to use a number of different indices, in a hierarchical sequence, particularly in fractal analysis (Delignieres et al., 2006), and for more accurate estimations of the Hurst exponent (see Carbone, 2007). We list in the following collection indices that have been drawn on in psychological research, but do not pretend to be exhaustive:

1. The fractal dimension, D_F, is an expression of the space-filling property of a trajectory in its embedding space, called the topological dimension. The dimension of the embedding space is called the box-counting dimension. The Hausdorff–Besicovitch dimension is (1) for lines, (2) for planes and surfaces, and (3) for finite volumes. D_F is not restricted to integer values, unlike the more familiar ideas of degrees of freedom. A fractal set is one for which the Hausdorff–Besicovitch dimension strictly exceeds the topological dimension.

2. The Lyapunov exponents model whether there is expansion or contraction on an underlying dimension of a process over time (Chapter 9). In order to compute them for a multivariable process, we need to identify the dimensions of an attractor, then compute the Jacobian matrix, and then find its eigenvalues. A rule that is almost universally applicable, given a sufficient data base, is that the sum of the Lyapunov exponents is less than unity, and the largest exponent is positive if the attractor dynamics are chaotic. This is often advocated as the sufficient rule to distinguish chaos from stochastic evolution. For real data time series, the largest Lyapunov exponent (LLE) is a variable, this can arise either for exogenous reasons or because the attractor trajectory varies in its local properties. Detecting such fluctuations requires a long sample and sensitive computation. Some examples are given by Gregson (2006, chap. 10).

3. The Hurst exponent was invented (Hurst, 1951) to study the flow of the Nile River, and called the *rescaled range analysis*, a function of the dimensionless ratio R/S. The empirical relation is $R/S = (\tau/2)^H$, where τ is the time span considered.

   It is expressed by an exponent of the time span considered, more or less symmetrically distributed about a mean of 0.73, with a standard deviation of about 0.09 (Feder, 1988, p. 153). A high value indicates that the variational statistics are strongly non-Gaussian (Feder, 1988, p. 194), and are interpreted as evidence of anti-persistence. The Hurst index can be used for distinguishing between fractional Gaussian noise and white noise in first-order autoregression (Davies & Harte, 1987).

4. Minimum description length (MDL) (Rissanen, 1987) is a measure of one sort of complexity, related to Kolmogorov complexity. Its primary purpose is the compare alternative models. MDL rests on what is called stochastic complexity (Grünwald, 2000) and the principle that underlying regularities can be used to construct a coding basis that enables us to compress the representation of data strings. It performs at least as well as the significance-testing approach and

comparable to a Bayesian procedure in a study of the rate problem (Lee & Pope, 2006). Computationally, it can be prohibitive, and has been simplified by Lee (2004) for practical use.

5. Exit times. This method (Abel et al., 2000) arises from the need to estimate the complexity of a time series as part of the exercise of distinguishing pure stochastics from chaotic determinism. The basic idea is to look at a sequence of data not at fixed sampling time but only when the fluctuation in the signal is larger than some fixed threshold ε.

6. Approximate entropy was devised by Pincus (1991) and is a compromise that only uses some sequential information (see Gregson, 2006, p. 145 et seq) but is more readily computed than some of the other possible indices of entropy (see Chapter 9). It has been applied in medical and psychophysiological research. Unsurprisingly, it is less sensitive to second-order fluctuations in dynamics than some other measures that draw on higher-order sequential properties.

7. The Entropic Analogue of the Schwarzian Derivative (ESf), and a bivariate extension using higher-order dependencies (BESf) are examples of coarser and finer entropy filtering introduced by Gregson (2006, chap. 5). Examples on psychometric data of diverse types have been given in that source book.

Causality and Model Identification

Theoretical time series generated by nonlinear mathematical models based on attractors are called trajectories, so the problem of model identification and fitting is the problem of matching trajectories to empirical realizations. Each realization is a relatively short sample time series, and is matched to segments of theoretical trajectories than can be infinite in length. Time series that make jumps between different stationary processes, or have interpolated subseries, require careful analysis, and may be evidence of multistable dynamics. Recursive loops if identified are evidence of feedback and make linear models very improbable. If there are two time series running alongside one another, as in experimental psychology, we commonly have a stimulus and a response series, then it is necessary to address the questions of the direction of causality, and the lags in time from cause to effect.

Mathematically, there is more than one definition of causality, and Granger causality (Granger, 2001) is a special case that is mathematically tractable, and of interest to economists. More recently, it has found application in neurophysiology and brain-behavior relationships (Gourévitch, Le Bouquin-Jeannès, & Faucon, 2006; Kaminski, Ding, Truccolo, & Bressler, 2001). Granger causality is based on stochastic predictions, that is to say time runs only forward, as is usual in most causal models. Time going in a reverse direction is a curious phenomenon found in some theories in physics, and not of course in economics or neurophysiology. However, we have to be careful here, as human behavioral or social processes can involve stochastic expectations about the future being used to create present choice decisions, as can happen in politics, economics, or religion. The algebra to represent future expectations modifying present behavior resembles time running in reverse, but does not have that existential meaning. Problems in causal inference and their rectification have been reviewed by Imai, King, and Stuart (2009). Granger causality can give meaningless results in nonlinear or multivariate contexts, and should be interpreted with extreme caution (Gourévitch et al., 2006). If we model in the frequency domain instead of the time domain, by using Fourier transforms, and frequency spectra instead of autocorrelation spectra, then the analogue of Granger causality is Gersh causality (Gersh, 1970; Gersh & Goddard, 1970) and similar warnings apply; a time series that is maximally noise free is not necessarily one that drives the dynamics of a group of interlinked channels (Albo et al., 2004). Nonstationarity and multivariate dynamics continue to offer serious analysis problems, and short-window methods have been developed in visual psychophysiology, which offer promise (Ding, Bressler, Yang, & Liang, 2000). An index that can be shown to be a special case of Granger causality, but is nonlinear, is the directed transfer function, which has been applied in multivariate neurobiological contexts, such as human sleep data.

Matching model predictions to data, or between models, using only symbolic dynamics, can be represented by Kullback–Leibler statistics, for data with n elements x and a model with corresponding elements θ, the disparity (but not a metric distance) between data and model, and thus a measure of mismatch of fit for two or more models in turn, is given by

$$\mathrm{KL}(x, \theta) \equiv \sum_{i=1}^{n} p(x_i) \log \left\{ \frac{p(x_i)}{p(\theta_i)} \right\}$$

and necessary corrections for model complexity are introduced in Akaike Information Criteria (AIC; Akaike, 1973), or Bayes Information Criteria (BIC; Schwarz, 1978), which in turn lead to MDL indices (Lee & Pope, 2006; Rissanen, 1987). AIC and BIC exist in a number of algebraic forms, and there are complicated risks in using AIC in nonlinear contexts (Gourévitch et al., 2006). The AIC is a truncated version of a Bayes posterior odds criterion (Zellner, 1978). A comprehensive review of the interrelation of these measures, and their refinement and development over time, is given within Grünwald, Myung, and Pitt (2005).

The point about these indices, which is seen by some authors as them hence being modern versions of Occam's razor, is that they penalize overfitting. AIC penalizes the deviance between model and data by subtracting a term twice the number of free parameters from the likelihood. BIC also involves the sample size n, and penalizes the deviance by $\log n$ times the number of free parameters. There is evidence that a BIC can be helpful in identifying some nonlinear models (Huang & Yang, 2004). The BIC has also been extended to resolve some interactive effects in genetic and multiple regression applications (Bogdan, Ghosh, & Doerge, 2004).

This search for simplicity is in some psychological applications not enough in itself, Chater (1996) considers that to offer adequate accounts of perception, we also need to invoke other principles, involving, for example, constraints on representation and process, the perceptual system must be able to separate genuine structure from noise, and to do this, the system has to be able to learn and to use information from past examples input over time. If we have, for example, two postulated structures that are about equivalent in AIC, BIC, or MDL terms, then the one that can be better computed by a neurological network is going to be more plausible as a real model.

It is now known that AIC is inadequate and can even be undefined, whereas minimum message length (MML) is always defined, is statistically invariant, and performs better on small sample sizes (Dowe, Gardner, & Oppy, 2007). MML was developed by Wallace (Wallace, 2005; Wallace & Dowe, 1999) and can be extended to multiple-factor analysis and classification tasks, it rests on a Bayesian definition of simplicity.

Note

1. The correction to the misquotations is to emphasize that Ockham (a) is not recorded as having, and (b) wouldn't have, said "Entia non sunt multiplicanda praeter necessitatem" (Don't multiply entities except by necessity). He did say "Pluralitas non est ponenda sine necessitate" (plurality shouldn't be posited without necessity). Similar forms are found in the writings of his teacher Duns Scotus (c 1266–1308). Dr. David Dowe of Monash University (Victoria, Australia) has compiled an extensive Web-page bibliography of work on Occam's razor and its modern extensions.

References

Abarbanel, H. D. I. (1996). *Analysis of observed chaotic data.* New York: Springer.

Abel, M., Biferale, L., Cencini, M., Falcioni, M., Vergni, D., & Vulpiani, A. (2000). Exit-times and ε–entropy for dynamical systems, stochastic processes, and turbulence. *Physics Review Letters, 84*(26), 6002–6005.

Adami, C., & Cerf, N. J. (2000). Physical complexity of symbolic sequences. *Physica D, 137*, 62–69.

Akaike, H. (1973). Information theory as an extension of the maximum likelihood principle. In B. N. Petrov & F. Csaki (Eds.), *Second International Symposium on Information theory* (pp. 267–281). Budapest: Akademiai Kiado.

Albo, Z., Di Prisco, G. V., Chen, Y., Rangarajan, G., Truccolo, W., Feng, J., Vertes, R. P., & Ding, M. (2004). Is partial coherence a viable technique for identifying generators of neural oscillations? *Biological Cybernetics, 90*, 318–326.

Arianos, S., & Carbone, A. (2007). Detrending moving average (DMA) algorithm: A closed form approximation of the scaling law. *Physica A, 382*, 9–11.

Barnsley, M. F. (1988). *Fractals everywhere*. Boston, MA: Academic Press.

Barnsley, M. F. (2006). *Superfractals: Patterns of nature*. Cambridge: Cambridge University Press.

Bogdan, M., Ghosh, J. K., & Doerge, R. W. (2004). Modifying the Schwarz Bayesian information criterion to locate multiple interacting quantitative trait loci. *Genetics, 167*, 989–999.

Braverman, M., & Yampolsky, M. (2008). Computability of Julia sets. *Moscow Mathematical Journal, 8*, 185–231.

Brillinger, D., Fienberg, S., Gani, J., Hartigan, J., & Krickenberg, K. (Eds.). (1984). *Time series analysis of irregularly observed data* (Lecture Notes in Statistics, Vol. 25). New York: Springer-Verlag.

Carbone, A. (2007). Algorithm to estimate the Hurst exponent of high-dimensional fractals. *Physical Review E, 76*, 056703-1–056703-7.

Chater, N. (1996). Reconciling simplicity and likelihood principles in perceptual organization. *Psychological Review, 103*, 566–581.

Chen, Z., Ivanov, P. Ch., Hu, K., & Stanley, H. E. (2002). Effects of nonstationarities on detrended fluctuation analysis. *Physical Review E, 65*, 041107.

Davies, B. (2003). Nonlinearity and complexity: An introduction. In R. Ball & N. Akhmediev (Eds.), *Nonlinear dynamics: From lasers to butterflies* (pp. 1–75). Singapore: World Scientific.

Davies, R. B., & Harte, D. S. (1987). Tests for Hurst effects. *Biometrikia, 74*, 95–101.

Delignieres, D., Ramdani, S., Lemoine, L., Torre, K., Fortes, M., & Ninot, G. (2006). Fractal analyses for 'short' time series: A re-assessment of classical methods. *Journal of Mathematical Psychology, 50*, 525–544.

Ding, M., Bressler, S. L., Yang, W., & Liang, H. (2000). Short-window spectral analysis of cortical event-related potentials by adaptive multivariate autoregressive modelling: Data preprocessing, model validation, and variability assessment. *Biological Cybernetics, 83*, 35–45.

Dowe, D. L., Gardner, S., & Oppy, G. (2007). Bayes not bust! Why simplicity is no problem for Bayesians. *British Journal of Philosophy of Science, 58*, 709–754.

Eisler, A. D., & Eisler, H. (2009). Experienced speed of time in durations of known and unknown length. *NeuroQuantology, 7*, 66–76.

Feder, J. (1988). *Fractals*. New York: Plenum Press.

Gersh, W. (1970). Spectral analysis of EEG's by autoregressive decomposition of time series. *Mathematical Biosciences, 7*, 205–222.

Gersh, W., & Goddard, G. V. (1970). Epileptic focus location: Spectral analysis method. *Science, 169*, 701–702.

Gentner, D., Brem, S., Ferguson, R., Markman, A. B., Levidow, B. B., Wolff, P., & Forbus, K. (1997). Conceptual change via analogical reasoning. A case study of Johannes Kepler. *Journal of the Learning Sciences, 6*, 3–40.

Gourévitch, B., Le Bouquin-Jeannès, R., & Faucon, G. (2006). Linear and nonlinear causality between signals: Methods, examples and neurophysiological applications. *Biological Cybernetics, 95,* 349–369.

Granger, C. W. J. (2001). *Essays in econometrics: The collected papers of Clive W. J. Granger.* Cambridge: Cambridge University Press.

Gregson, R. A. M. (1983). *Time series in psychology.* Hillsdale, NJ: Lawrence Erlbaum Associates.

Gregson, R. A. M. (2006). *Informative psychometric filters.* Canberra: ANU E-Press.

Grünwald, P. (2000). Model selection based on minimum description length. *Journal of Mathematical Psychology, 44,* 133–152.

Grünwald, P., Myung, I. J., & Pitt, M. A. (2005). *Advances in minimum description length: Theory and applications.* Cambridge, MA: MIT Press.

Haken, H. (1987). *Advanced synergetics: Instability hierarchies of self-organized systems and devices.* Berlin: Springer-Verlag.

Hoijtink, H., Klugkist, I., & Boelen, P. A. (2008). *Bayesian evaluation of informative hypotheses.* New York: Springer.

Hu, K., Ivanov, P. Ch., Chen, Z., Carpena, P., & Stanley, H. E. (2001). Effect of trends on detrended fluctuation analysis. *Physical Review E, 64,* 011114.

Huang, J. Z., & Yang, L. (2004). Identification of non-linear additive autoregressive models. *Journal of the Royal Statistical Society: Series B (Methodology), 66,* 463–477.

Hurst, H. E. (1951). The long-term storage capacity of reservoirs. *Transactions of the American Society of Civil Engineers, 116,* 770–808.

Imai, K., King, G., & Stuart, E. A. (2009). Misunderstandings between experimentalists and observationalists about causal inference. *Journal of the Royal Statistical Society: Series A (Statistics in Society), 171,* 481–502.

Kaminski, M., Ding, M., Truccolo, W. A., & Bressler, S. L. (2001). Evaluating causal relations in neural systems: Granger causality, directed transfer function and statistical assessment of significance. *Biological Cybernetics, 85,* 145–157.

Lee, M. D. (2004). An efficient method for the minimum description length evaluation of deterministic cognitive models. In K. Forbus, D. Gentner, & T. Regier (Eds.), *Proceedings of the 26th Annual Conference of the Cognitive Science Society* (pp. 807–812). Mahwah, NJ: Erlbaum.

Lee, M. D., & Pope, K. J. (2006). Model selection for the rate problem: A comparison of significance testing, Bayesian, and minimum description length statistical inference. *Journal of Mathematical Psychology, 50,* 193–202.

Li, M., & Vitanyi, P. (2008). *An introduction to Kolmogorov complexity and its applications* (3rd ed.). New York: Springer.

May, R. M. (1976). Simple mathematical models with very complicated dynamics. *Nature, 261,* 459–467.

Michell, J. (2009). The psychometricians' fallacy: Too clever by half? *British Journal of Mathematical and Statistical Psychology, 62,* 41–56.

Peng, C.-K., Buldyrev, S. V., Goldberger, A. L., Havlin, S., Sciortino, F., Simons, M., & Stanley, H. E. (1992). Long-range correlations in nucleotide sequences. *Nature, 356,* 168–170.

Pincus, S. M. (1991). Approximate entropy as a measure of system complexity. *Proceedings of the National Academy of Sciences of the USA, 88,* 2297–2301.

Rissanen, J. (1987). Stochastic complexity. *Journal of the Royal Statistical Society, Series B, 49,* 223–239.

Russell, B. (1940). *An inquiry into meaning and truth.* London: George Allen & Unwin.

Schwarz, G. (1978). Estimating the dimension of a model. *Annals of Statistics, 6,* 461–464.

Smith, A. F. M. (1984). Present position and potential developments: Some personal views of Bayesian statistics. *Journal of the Royal Statistical Society, Series A, 147,* 245–259.

Speekenbrink, M. (2003). The hierarchical theory of justification and statistical model selection. In Y. Kano & J. J. Muelman (Eds.), *New developments in psychometrics* (pp. 331–338). Tokyo: Springer.

Taqqu, M. S., Teverovsky, V., & Willinger, W. (1995). Estimators for long-range dependence: An empirical study, *Fractals, 3,* 785–798.

Tong, H. (1983). *Threshold models in non-linear time series analysis* (Lecture Notes in Statistics, Vol. 21). New York: Springer-Verlag.

Wackerbauer, R., Witt, A., Atmanspacher, H., Kurths, J., & Scheingraber, H. (1994). A comparative classification of complexity measures. *Chaos, Solitons & Fractals, 4,* 133–173.

Wallace, C. S. (2005). *Statistical and inductive inference by minimum message length.* Berlin: Springer.

Wallace, C. S., & Dowe, D. L. (1999). Minimum message length and Kolmogorov complexity. *Computer Journal, 42,* 270–283.

Zellner, A. (1978). Jeffreys-Bayes posterior odds ratio and the Akaike information criterion for discriminating between models. *Economics Letters, 1,* 337–342.

3

Frequency Distributions and Error Functions

Stephen J. Guastello

CONTENTS

Biologists in the nineteenth century discovered many examples of normally (Gaussian) distributed variables. They asked mathematicians whether there was some law of mathematics that explained the prevalence of normal distributions. The mathematicians replied that they knew of no such law, and that any explanation would have to be a law of biology (Zeeman, 1977). One might question whether normal distributions actually exist in real data, or whether they are simply approximations that allow the statistical mathematics to become more tractable.

Psychologists and others in the twentieth century also found many examples of normally distributed variables. There are so many of them that modern-day psychology students often walk away from their statistics courses thinking that there are only four types of distributions: normal distributions, distributions that will become normal when they grow up, skewed distributions that want to be normal, and nameless grotesque aberrations that one must analyze using statistics based on chi-square.

The objective of this chapter is to open a few horizons regarding probability distributions, which are prominent in nonlinear dynamical systems (NDS). The first section recounts some well-known points about the Gaussian distribution that we contrast later on. The second and third sections describe two important distributions in NDS, the power law and the exponential distributions along with a generalization of the exponential distribution that can reframe almost any differential equation as a probability density function. A generalizable point is that instead of assuming a normal distribution versus no distribution at all, it is advantageous to consider assuming particular alternative distributions to the normal distribution.

The fourth section revisits the standard psychometric assumptions regarding mental measurements containing true score variance and error score variance. When a measurement is taken from a time series produced by a dynamical system, both elementary parts of the measurement expand into more complex components. One of the expansions is the distinction between *independently and identically distributed* (IID) error and non-IID error. Two statistics are then introduced that were devised to address non-IID error.

The other expansion of the standard psychometric assumption pertains to the substance of the model itself and its linear and nonlinear components, which is elaborated in the fifth section. Here we encounter issues related to time lags, sample size, and statistical power. Neyman–Pearson, Fisherian, and Bayesian types of hypotheses are potentially involved in assessing the efficacy of a nonlinear model. The final section describes filtering processes that are often used to remove error variance before making nonstatistical calculations on nonlinear time series.

Normal Distribution

The normal or Gaussian distribution is the familiar bell-shaped curve. Its peak density is located directly in its center. The shape is symmetrical with low-density areas appearing in each of its tails. For a population represented by this distribution, the mean, mode, and median are all located at the same value of the variate X.

One normal distribution differs from another on the basis of two parameters: the mean and the standard deviation. According to the *central limit theorem*, as a sample becomes large, the values of the mean and standard deviation approach the population values. The mean of a sample is an unbiased estimate of the population mean. The standard deviation of a sample is a systemically biased estimate of its population value, however, such that the standard deviation becomes larger as the sample size increases. Thus the calculation for the standard deviation of a sample contains $N - 1$ in its denominator, whereas the population value contains N in the denominator.

Role in the General Linear Model

Normal distributions play an intimate role with the general linear model (GLM) in its univariate form

$$Y = B_0 + B_1 X, \tag{3.1}$$

its multivariate form

$$Y = B_0 + B_1 X_1 + B_2 X_2 + \cdots + B_n X_n, \tag{3.2}$$

and the analysis of variance (ANOVA) models. In Equation 3.1, X and Y are assumed to be normally distributed as are all Xs and Y in Equation 3.2. In ANOVA designs, the Xs are multichotomous variables and the conditional distributions of Y are assumed to be normally distributed.

In the linear regression analyses, the assumption of normality leads to a further assumption that all conditional distributions of Y are normally distributed around the regression line and that the variances of the conditional distributions are equal (homoscedasticity assumption). The cost of not meeting this assumption is that point estimation and the confidence intervals around the predicted points would not be consistent for all values of X; predicted conditional means would depend on conditional variances. The homoscedasticity assumption becomes more severe in the multivariate case where, conceivably, each of the variables could have different deviations from the main assumption. The assumption is convenient because it permits the standard errors for all the regression weights to be calculated the same way.

A similar assumption appears in ANOVA analyses. Here, the inferences about differences in means are based on separating sources of variance between and within experimental conditions. If it can be assumed that all conditional variances are equal, then statistically significant differences in means can be attributed to differences in means. If variances are unequal, however, one cannot tell if the significant effect is the result of differences in means or differences in variance. Thus, several tests for homogeneity of variance were developed over the years. The assumption of homogeneity of variance becomes both untenable and unnecessary in many instances of NDS analysis, as the story unfolds later in this chapter.

Distributions Growing Up

The binomial and t distributions are two distributions that become normal distributions when sample sizes become large enough. The binomial distribution is used for dichotomous events, such that its mean is p, the probability of event X, and its standard deviation is $(pq)^{1/2}$, where q is the probability of event not-X.

The t-distribution is used to evaluate an inferential statistic that compares a sample's parameters with those of its population. The most common conditions involve a comparison between a sample and a population where both the population mean and standard deviation are known, a comparison between a sample and a population where the population mean is known but the population standard deviation is not, a comparison between two independent sample means, a distribution of differences between paired observations in one sample, and the significance test on a regression weight in a simple linear model (with Gaussian marginal distributions of each of the two variates), or a multiple linear regression model.

Neither distribution is particularly worrisome in NDS analysis. The point of mentioning them is that they represent types of familiar distributions that are more similar to the normal distribution than they are to distributions that are more germane to NDS.

Random numbers and normal distributions deserve some mention as well. The actual distribution of random numbers, such as those produced by random number generators, is flat or rectangular rather than centrally peaked. It is only when repeated samples are taken that the distribution of sample means becomes normal. Thus, numbers that are randomly selected from a normal distribution have become the common representation of random processes. The implicit thinking is that a normal distribution could be the result of all sorts of unknown random variables that are not (yet) taken into account in a piece of research.

Random numbers that are produced by random number generators are not really random. They are produced by an algorithm that respects the boundary conditions of selecting an X in a range between two endpoints and a self-avoidance feature that ensures each successive number selected by the algorithm is as different as possible from the number selected at the previous step. As a result, differences in sequential random numbers take the form of a simple quadratic process (Szpiro, 1994). This contrast between the appearance of randomness and the underlying deterministic process is an important feature of chaos (Casti, 1991).

Skewed Distributions

The idea of a skewed distribution is a matter of viewpoint. If the analytic viewpoint insists on normal distributions, then the two distributions shown in Figure 3.1 would be called skewed. The distribution on the left is ambiguously a power law distribution or a lognormal distribution. Transformations such as $X \to \log_{10}X$ or $X \to X^{1/2}$ are often recommended as a means of forcing the distribution to appear normal. The distribution on the right is ambiguously an exponential distribution or a Poisson distribution, which often appears in conjunction with rate data. They often meet with a transformation of $X \to 1/X$.

It is also a standard advisement to warn the user that if one were to transform X in one of these ways and then use the transformed variable in a linear correlation analysis with a dependent variable Y, the analysis would actually be testing for a particular *nonlinear* relationship between X and Y. Thus, it is recommended that the user have good theoretical reason for choosing a transformation.

Grotesque Aberrations

The nonparametric statistics are so named because they do not rely on the assumption of the normal distribution, but may rest on order or permutation statistics (Walsh, 1962–1968). It is also possible to perform analyses such as ANOVA by using information measures instead of raw data, and avoid the Gaussian assumptions that way (Walsh, 1968, Vol. III). They rely on matching

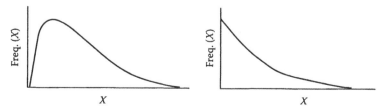

FIGURE 3.1
Two distributions that are often regarded as skewed.

frequency distributions of observed variables with expected distributions, and use the chi-square statistic to assess whether a deviation has occurred. Frequency-based statistics, such as the Kruskal–Wallis test can make the same types of comparisons as offered by ANOVA, but without the assumptions. The price is sometimes paid in statistical power.

The odds ratio is a widely used univariate test that is based on frequencies. It produces results such as "the odds of X occurring are three times as great when condition A exits compared to when A does not exist." Those types of statements sound good at first blush, but the meaning could be very different if the base rate of X is 0.10 or 0.0001. The odds ratio is indifferent to those distinctions. Odds ratios lend themselves to simplistic forms of risk analysis in accident analysis or medicine where the analysis typically ignores the role of any other variable in the medical or sociotechnical process that could in fact be nonlinear and dynamic (Gregson, 2009; Guastello, 2006).

Multivariate versions of frequency-based statistics have also existed for some time. Logistic regression and linear structural relations (LISREL; Joreskog & Sorbom, 1988; Schumacker & Marcoulides, 1998) are perhaps the most widely used examples, followed by maximum likelihood regression. Logistic regression is essentially the odds ratio for multiple independent variables.

LISREL ("linear" was eventually dropped from its name) is based on the older idea of path analysis. An elementary example is shown in Figure 3.2, where A, B, C, and D are all variables that have different levels of linear correlation with each other. The links between C and B, D and B, and B and A are strong, whereas the links between C and A and D and A are weak. Results like these would lead to the conclusion that B acts as a mediating variable between C or D and A. This form of analysis was an attempt to infer patterns of causality where variables were measured at one point in time, rather than at different points in time. One would not say that one variable caused another, but rather that the pattern of correlation is consistent with a hypothetical causal pattern.

Once can see the rudiments of a complex systems analysis, and in recent years variables have been transformed to permit logarithmic relationships for instance. One limitation of the technique is that it does not permit reciprocal causality or closed loops such as those that are typically implicated in complex systems analysis.

Maximum likelihood theory, as used in LISREL and elsewhere, contrasts with least squares, as used in ordinary linear regression and ANOVA, in

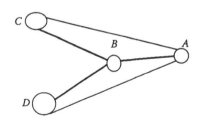

FIGURE 3.2
Basic path diagram for four variables.

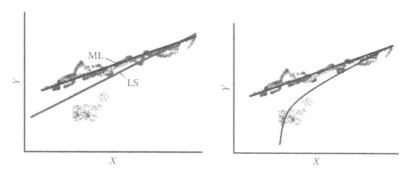

FIGURE 3.3
Ambiguous point clouds that could be fit by different lines or a curve.

the manner by which the error function is calculated. Consider the cloud of data points in left panel of Figure 3.3 where the goal is to fit a linear relationship between X and Y. There are actually two clouds in the figure. Least squares analysis would attempt to place the line where it would minimize the squared distance between all the points and the line. The position of the line would reflect some gravity of the smaller cluster of points. The maximum likelihood method would respect the frequency of points and place the line in a different position that would ignore the smaller cloud.

If the line marked ML were less distinct and less obvious to the naked eye due to a greater dispersion of points overall, one might prefer the maximum likelihood analysis because it would have greater power for finding the line of choice. On the other hand, others have argued that in the long run, maximum likelihood methods tend to be biased in favor of the preferred alternative hypothesis, which could be cause for concern (Brock, Hseih, & LeBaron, 1991; Tauchen, 1985). If the system portrayed in Figure 3.3 is regarded as possibly multistable, then two regression lines would be fitted and a switching rule between them sought for in a time series analysis (see Chapter 25). The distinction between the two error functions also arises in nonlinear analysis. Consider the right panel of Figure 3.3. In this example, if one took the same clouds of points and gave a nonlinear regression program a model equation that could result in parameters that denoted either the line or the curve, the algorithm would be hard-pressed to conclude a set of parameters using either maximum likelihood or least squares. One might then argue that if one method were just a little stronger for making a determination that would be a better result and the preferred conclusion. The alternative argument is that if the determination is so ambiguous, or the results are not strong enough, neither answer is really correct. One would do well to investigate why there ever were two separate clouds in the data and fashion a model that accounted for both principles simultaneously. A bifurcation could be involved (see Chapter 11).

There is a growing myth that least squares analysis should be used for linear models and maximum likelihood should be used for nonlinear models. The proposition is simply untrue. The historical link between linear models and least squares is simply the result of the two ideas being stranded in time together for so many years, dating back to the nineteenth century. Polynomial regression and other nonlinear transformations were in use for quite some time with least squares. Nonlinear regression (see Chapters 12 and 13) did not exist until the early 1960s. Chi-square is about as old as linear regression, but the preponderance of maximum likelihood statistics did not congeal until the 1970s. There was never a formal theoretical link between either error function and linear versus nonlinear models.

Power Law Distributions

A power law distribution is asymmetric with a fixed endpoint where the variate $X = 0$ and can become infinitely long in its tail. The equation of the power law distribution is

$$\Pr(X) = a\left(\frac{X}{s}\right)^{B},\tag{3.3}$$

where the frequency (or probability) of a value of the variate X is a function of the value of X raised to an exponent (Evans, Hastings, & Peacock, 1993). Parameter s is the standard deviation of X. Parameter a is simply a constant to improve the fit between the left side and right side of the equation by ensuring that the values of X match up with a total probability density of 1.0.

Note that the value of B for an observed variable X is no different from B associated with the standardized variate $Z = X/s$. Power law distributions are thus regarded as "scale free." If one were to change the scale of measurement, e.g., from inches to centimeters, both X and s would be multiplied by the same constant, and the constants would cancel each other out. Thus, the power law distribution has only one important parameter, the shape parameter B.

An *inverse* power law is simply the set of cases where $B < 0$. Figure 3.4, left panel, shows two different shapes, which were obtained for a positive B and a small negative B. The J-shape persists for all $B < 0$, and changes shape when $B > 0$. The J-shapes in the functions become more severe as B becomes more strongly negative, as shown in Figure 3.4, right panel.

Fractal Dimensions

The scale-free property of power law distributions makes it an attractive characterization for fractal phenomena. There are two methods for determining the fractal dimension from a power law distribution. One is to assess Equation 3.3 using nonlinear regression. In practice, s is dropped from Equation 3.3 (or treated as 1.0).

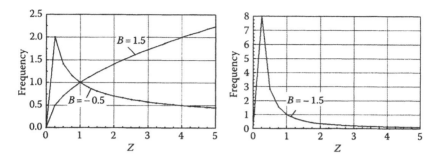

FIGURE 3.4
Comparison of power law distributions showing positive and negative shape parameters (left) and an example from the pink noise range (right).

The more common method for assessing the fractal dimension associated with a power law distribution is to take the log of both sides of Equation 3.3:

$$\log\left(\Pr[X]\right) = \log\left(a\right) - B\log\left(X\right). \tag{3.4}$$

and calculate the parameters using the ordinary product-moment correlation, where $\log(a)$ is the regression constant, and B is the raw score regression weight. For problems involving fractal dimensions, $B < 0$, and $D_F = B$ (The fractal dimension is a positive number, and the regression weight is negative.) The two means of calculating B are mathematically equivalent. Statistical equivalence only exists in the asymptotic case where there is no error in the measurement, however. In a correlation analysis, the regression weight is systemically lowered by error in the data. This hazard can be corrected, however, if the reliability of the measurement, r_{xx}, is known by using the standard correction for attenuation on the standardized correlation coefficient:

$$r' = \frac{r}{r_{xx}^{1/2}}. \tag{3.5}$$

Then take the corrected correlation, r', and multiply it by the ratio of standard deviations, $S[\log(X)]/S[\log(\Pr[X])]$, to produce the corrected value of B. The underlying r is usually greater than .90 for a good fit.

Reports using either form of calculation for the fractal dimension often include log–log plots such as the one shown in Figure 3.5, left panel. Liebovitch (1998) noted that the scale of measurement that is used on a fractal object will affect the size of the measurements X. In one of Mandelbrot's (1983) classic problems, two adjacent countries have been known to measure their common boundary, be in full agreement as to where the boundary is located, but come up with different measurements for the length of that boundary. The differences arose from the two agents using different scales of measurement, perhaps 10-m increments instead of 1-m increments;

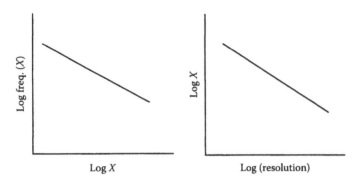

FIGURE 3.5
Log–log plots of power law relationships.

the one that used the smaller scale of measurement was able to measure more precisely the twists and turns along the nonlinear boundary and thus cover more distance when all the twists and turns were stretched out. The relationship between the scale of measurement and the resulting measurement is also logarithmic.

Other Similar Distributions

The examples of power law distributions in Figure 3.4 used $X = 0.25$ as the increment of measurement. If the increment were larger, then the mode would be shifted to the right. As the scale of measurement becomes infinitesimally small, however, the peak is shifted to the left and could be visually indistinguishable from an exponential distribution. The assumption of one distribution or another makes the case for calculating a fractal dimension, however, versus calculating something else. By the same token, any assumptions that permit a calculation of dimensionality should be made with regard to the possible real-world meaningfulness of the dimensionality; specious uses of dimensionality calculations do not create fractals or dynamics where none really exist.

If the peaks on the examples of the power law distributions were a bit more rounded, they might be visually indistinguishable from a lognormal distribution. One might also find cogent arguments favoring the lognormal over the power law in some situations (Philippe, 2000), and it is often associated with response time data in cognitive studies. Mandelbrot (1997) would characterize many of the uses of lognormal distributions as sloppy approximations of what should be treated as a power law. The power law can render a D_F whereas the lognormal cannot.

Interpretation of Dimensions

Here we encounter questions regarding the relationship between the fractal dimension (D_F), that is associated with a variable and the interpretation of the underlying dynamics if any. When it was first discovered that the basin of a chaotic attractor was fractal, the D_F was often calculated for a time series variable and interpreted as the presence of chaos (Theiler & Eubank, 1993). It was eventually shown, however, that the presence of a fractal dimension does not guarantee chaos. Rather the (largest) Lyapunov exponent λ must be calculated instead. If λ is positive, then it is possible to conclude the presence of chaos (Ruelle, 1991; Wolf, Swift, Swinney, & Vastano, 1985). On the other hand, the positive Lyapunov is a necessary, but not sufficient, condition to determine the presence of chaos (Schuster, 1984). It could be very difficult to detect statistically (Lardjane, 2004), although a lot depends on how one approaches the problem. Nonetheless, it is also known by way of the Kaplan–Yorke conjecture that $D_F = e^{\lambda}$ (Ott, Sauer, & Yorke, 1994), and it is analytically advantageous to assume an exponential distribution for a continuous variable for which the D_F is being calculated from a time series (Guastello, 1995, 2002, 2005).

Nicolis and Prigogine (1989) recommended that conclusions about chaos be reserved for cases where $D_F > 3$. This was probably a pretty safe recommendation at a time when the early explorers were finding chaos and not knowing exactly what to do with that finding. Although it still serves as a guideline today, some well-known chaotic attractors show $D_F < 3$ (Sprott, 2003).

Shortly afterward, it appeared useful to distinguish low-dimensional chaos from high-dimensional chaos. There was at least an intuitive understanding that the two levels of dimensionality were phenomenologically different (Dooley & Van de Ven, 1999). In current understanding, the presence of a fractal dimension could suggest chaos, but does not guarantee it. The presence of a fractal dimension between 1 and 2 could also suggest a self-organizing process (Bak, 1996), but does not guarantee that either. Self-organizing processes often take the form of discontinuities and phase shifts and produce distributions of events that do not resemble the power law or chaos (Guastello, 2005). Chaos and self-organization are usually regarded as different events that are both present at one time or another in a complex adaptive system (Dooley, 1997). The zone where $0 < D_F < 1$ is characterized as white noise; in signal processing, white noise consists of equal representations of random frequencies from all the audible octave bands. The zone where $1 < D_F < 2$ is now known as *pink noise*, which encompasses all the known properties of this region of dimensional complexity.

The zone where $2 < D_F < 3$ is known as *brown noise*, or sometimes *black noise*. Its power law distribution is more peaked than the one associated with pink noise, and its tail is thinner. The distribution is dominated by a small range of values as if it were a relatively monotone sound with occasional variations. In a time series, it would be dominated again by a small range of values with occasional bursts into high-range values and back again. These bursts are departures from a self-organized state, and are known as *Levy flights* (Sprott, 2003).

Numerical Example

Figure 3.6 shows a distribution of frequencies that were encountered in a real problem where X varied from 1 to 8 with the relative frequencies as shown, totaling 133 observations. For a *weighted* analysis, all 133 observations would be used such that there would be 80 observations of [1, 80] 21 observations of [2, 21] and so forth. The power law can be calculated through nonlinear regression using Equation 3.3, where a and B are treated as nonlinear regression weights (see Chapter 14 for the details on how nonlinear regression works and Statistical Package for the Social Sciences (SPSS; Chicago, IL) commands). S can be ignored by treating it as a constant = 1. The weight for B is -1.653; $R^2 = .991$. The fractal dimension is thus 1.653.

The same set of 133 cases can be processed as a linear regression problem by applying a log transformation to X and Frequency of X and computing Equation 3.4. The results are: $B = -1.639$, $r^2 = .983$, and thus $D_F = 1.639$. The difference between the results of the two procedures is trivial, amounting to a difference of only 0.014 dimensions, and readily explained by the influence of error on linear regression weights.

Next consider the unweighted analysis where each value of X and its corresponding frequency are represented in the data only once. The N is very small, of course, but the important result for present purposes is the regression weight B. Using nonlinear regression, $B = 1.688$, $R^2 = .983$. This result is only trivially different from the result obtained from the weighted analysis, amounting to 0.035 dimensions. If the analysis were conducted using Equation 3.4, however, $B = 2.036$, which is larger than the other three values, and could be regarded as substantial because it would change the interpretation of the system from pink noise and self-organizing to brown noise. r^2 for this analysis was .881, which would be considered large relative to the conventional analyses used in the social sciences, but questionable when

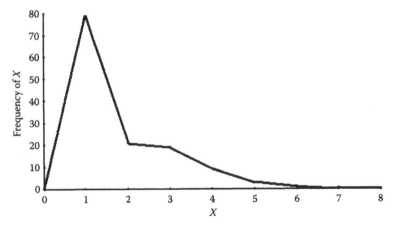

FIGURE 3.6
Example of a possible power law observed in real data.

assessing the fit of a frequency distribution. Thus the option with the highest R^2 looks like the best choice of the four possibilities.

Many data sets that could induce power law analysis involve more specific values of X than the eight values used in this example. The standard procedure of aggregating the values of X into 15–20 bins would be recommended; the distribution often loses its shape when the frequency of every possible value is observed separately. If the data come from a time series, the additional question of stationarity complicates the analysis: Is there one distribution throughout the data set, or is the distribution shifting between two or more forms? The means of isolating fractal dimensions from time series are considered in several chapters in this book. At the risk of getting ahead of ourselves, however, it should be mentioned parenthetically that Clauset, Rohila, Shalizi, and Newman (2009) recently introduced a maximum likelihood method for calculating power laws in time series where the optimal bin size is unknown and stationarity cannot be assumed.

Central Limit Theorem

The nature of the fractal generated some interesting insights regarding the role of the central limit theorem. A true fractal illustrates the same patterns of events at progressively larger or smaller scales. Thus some writers would remark that more data does not produce population values of the distribution parameters. For instance:

> The proof of the central limit theorem requires that the variance exist(s) and that it has a finite value. For a fractal, the variance is not defined, and thus the central limit theorem does not apply. As more data are included, the fractal distributions do not become smoother or more Gaussian. Rather the distributions become rougher because the correlations that link the deviations together in a self-similar way become more noticeable over longer times (Liebovitch, 1998, p. 86).

West (2006, p. 131) made a similar observation. Both occurred in the context of explaining how physiological data sets with a fractal structure involving thousands of observations did not produce a normal distribution. It would be useful, however, to decompose the issues involved. Some distributions become Gaussian in the long run, as mentioned earlier. Samples taken from a Gaussian population will become Gaussian if enough samples are taken, even if particular samples are "skewed." Many distributions do not become Gaussian in the long run, the power law and the exponential distributions are two examples, and there is no reason to suppose that any particular non-Gaussian distribution would do so unless there is a proof to support it. Here again it is important to distinguish between a skewed-normal distribution and distributions of distinctly different types.

The other point is the finite variance. A fractal's inherent structure is repeated and observable at smaller and larger scales. "More data" in this context is not the same as collecting more samples. If "more data" means taking measurements from a time series at progressively smaller time increments, the variance of the new time series of observations becomes biased toward the smaller values. A good illustration is the distribution of bubbles in Figure 3.7 (from Taylor, 2005). If we allow smaller-size bubbles to be counted, we have more bubbles, and their mean size becomes smaller, and the variance changes along with it.

A central idea in statistical inference is that computations made on properties of the sample permit conclusions about the population. Given the unusual properties of fractal data, is it still possible to apply statistical analysis with them? The short answer is "yes." First, if one wants "more" data, it is important to maintain a stratified sampling technique where the same proportions of large and small bubbles are represented in the new samples. Second, if different data sources produce different distributions of bubble sizes when all data are measured on the same scale, one has a basis of comparison between experimental conditions.

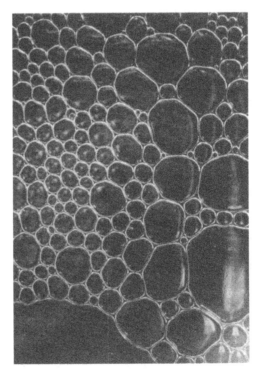

FIGURE 3.7
Bubbles. (Reprinted from Taylor, R.P., *Nonlinear Dynamics Psychol. Life Sci.*, 9, 115, 2005. With permission.)

Exponential Distributions

The basic exponential distribution is characterized by a high density of values at the zero point with a decaying density toward the higher values. Like the power law distribution, it is one-tailed. The equation of the exponential distribution itself is

$$\Pr(X) = \left(\frac{1}{s}\right) \exp\left(-\frac{X}{s}\right), \tag{3.6}$$

where s is the standard deviation of the variate X. Distributions with different s will have different shapes, but if we confine ourselves to standardized variates, Equation 3.6, simplifies to

$$\Pr(Z) = \exp(-z), \tag{3.7}$$

and all distributions look like the one shown in Figure 3.8. The exponential distribution also has a convenient property in that its mean is asymptotically equal to its standard deviation.

The exponential distribution and the Poisson distribution are one and the same for distributions where $s = 1$. Thus for standardized variates, the two distributions are always the same (Evans et al., 1993). The exponential distribution has an interesting relationship to the power law distribution also. The two are one and the same when s in Equation 3.6 = $1/B$ in Equation 2.3 (Evans et al., 1993).

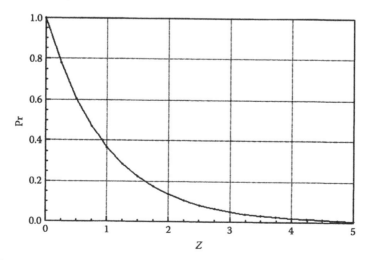

FIGURE 3.8
The exponential distribution.

Numerical Example

If we take the distribution shown in Figure 3.6 and fit it to the exponential distribution (unweighted method) ignoring the point [0, 0], the resulting function will be frequency of $X = 0.269 \exp(-0.168X)$; $r^2 = .880$. The fit is not as good was what was obtained from some of the power law analysis.

If we were to change the frequency of $X = 0$–138, which is what it was in the original data (Guastello, 1995, p. 246), the results are frequency of $X = 153.0 \exp(-0.794)$; $r^2 = .974$. Note that it would be impossible to calculate a power law from this scenario using the log-linear approach unless one were to add a small constant, such as 0.1, to X to get rid of $X = 0$. The degree of fit would be very poor, however; $r^2 = .077$.

Implications for ANOVA Designs

Exponentially distributed variables are often epidemic in nature. The majority of people in a given sample or population would not have experienced the event or condition at all, a smaller number would have experienced it once or experience it at a relatively low level of severity, a still-smaller number would have experienced it twice or at the second level of severity, and so on. There is a great interest in predicting these relatively rare events, and it is not surprising that ANOVA-type research designs are attractive. The researcher usually relies on two options: One is to use a log transformation on the dependent measure; usually a small constant needs to be added to the value before making the log transformation in order to recapture undefined values. The other is to use a logistic or other analysis that does not rely on the normal distribution.

There is a third option, however, which is to assume an exponential distribution. Whereas it was problematic to determine whether differences in experimental conditions were the result of differences between means if conditional variances differed, the problem goes away by assuming an exponential distribution. Because the mean and the standard deviation are the same value, differences in means and standard deviations are one and the same and it is no longer necessary to rely on that separation.

Complex Exponential Distributions

Exponential functions and distributions have been very useful for determining whether a hypothesized nonlinear dynamical function actually fits a set of observations. This facility was made possible by the Ito–Wright formula by which any differential function of a standardized variate $f(z)$ can be stated as a probability density function (pdf):

$$\text{pdf}(z) = \xi \exp\left[-\int f(z)\right], \tag{3.8}$$

where

    z is the order parameter

    pdf(z) is the cumulative probability associated with a particular value of z

    ξ is a constant that is introduced to ensure unit density (Cobb, 1978, 1981; Guastello, 1995, 2002)

In this general model, the Gaussian distribution becomes a special case or subset within the more complex function.

The cusp catastrophe model serves as an example. The equation of its response surface for a process that changes over time is

$$\frac{dy}{dt} = y^3 - by - a, \tag{3.9}$$

where

    y is the dependent measure

    b is the bifurcation variable

    a is the asymmetry variable

The catastrophe probability density functions (pdfs) are multimodal distributions of the exponential family. Note that none of the catastrophe models or their pdfs contains a fractional exponent. After applying the Ito–Wright formula, the cusp probability density function is

$$\text{pdf}(z) = \xi \exp\left[\frac{-z^4}{4} + \frac{bz^2}{2} + az\right], \tag{3.10}$$

where

    pdf(z) signifies the cumulative probability associated with the standardized variate

    ξ is a constant that ensures unit density

The Ito–Wright formula can accommodate any integratable function, define a probability density function associated with it, and allow the researcher to determine whether the function fits a static distribution of the data or not, usually in conjunction with nonlinear regression. The discussion on testing catastrophe models using this method is resumed in Chapter 13.

Location and Scale

Standardizing the variate to produce z in Equations 3.7, 3.8, and 3.10 is almost as simple as the common procedure of subtracting the mean and dividing by the standard deviation. More generally, however, the order parameter y is corrected for *location* and *scale* before analysis through nonlinear regression:

$$z = \frac{(y - y_0)}{\sigma_s}. \tag{3.11}$$

In Equation 3.11, location parameter, y_0, refers to the lowest value of the data set. Fixing this parameter defines a zero point for the nonlinear function that accounts for the entire data set. The mean makes a convenient zero-point in a two-tailed normal distribution, but in an exponential distribution it does not. The scale parameter, σ_s, is usually the standard deviation of the data set; other options are considered in later chapters. The scale correction eliminates bias in the nonlinear regression weights between two or more variables with different standard deviations (Evans, 1991). The resulting z-scores represent data in terms of moments.

Structure of Behavioral Measurements

In the classic definition, a mental measurement, X, consists of a true score, T, plus an error score, ε. Thus the variance structure for a population of scores is

$$\sigma^2(X) = \sigma^2(T) + \sigma^2(\varepsilon). \tag{3.12}$$

(Lord & Novick, 1968). The classic assumptions are that errors are independent of true scores and other error scores and that the mean of error scores is zero. Reliability, furthermore, is the ratio of true score variance to total score variance, $\sigma^2(T)/\sigma^2(X)$. There is a variety of methods for calculating reliability for mental measurements, which depend in part on characteristics of the measurements themselves; a full exposition of those methods is well beyond the scope of the present discussion, and better saved for standard textbooks on psychological testing.

In a time series of measurements, error variance that meets the classic assumption is known as *independently and identically distributed* error. For an NDS process, however, the two components of a measurement are split into two further components each:

$$\sigma^2(X) = \sigma^2(L) + \sigma^2(NL\text{-}L) + \sigma^2(\text{non-IIDe}) + \sigma^2(\text{IIDe}), \tag{3.13}$$

where the function that is intended to predict X is composed of a linear component, a nonlinear component, an IID error, and a non-IID or dependent error. The third variance component could be non-IID because the errors are dependent on the variate X; this condition occurs either through autoregression or because X is exponentially distributed. If they are exponentially distributed as most dynamic variables tend to be (Wei, 1998), the same principle that the conditional variance in a regression models is dependent on the conditional means applies as it does in ANOVA. The variance component could be non-IID by simply being non-Gaussian. Alternatively, the third variance component could be non-IID because the prediction model is nonstationary.

Equation 3.13 was first introduced in Guastello (2002, p. 61) as a preamble to a discussion of structural equations for studying NDS phenomena such as those covered in Chapters 11 through 13. Gregson (1994) introduced the distinction and use of two variance components in time series somewhat earlier, however, in conjunction with studies of nonlinear psychophysics. In nonlinear psychophysics, a signal can be unidimensional or multidimensional, and the source of the signal can move through space and over time. Similarly, the human perceiving the signal could be moving through space or time and not simply sitting in a chair pushing buttons. The structural equation that describes the human response contains one or more control parameters that reflect environmental conditions of some sort. The variance associated with the control parameter is expected to be normally distributed. The variance associated with the perception process, response time, and all the stages therein is exponentially distributed. Gregson noted that the perception processes are not entirely separable, but not entirely integrated into a single process either.

The concept of two variance components was anticipated to a limited extent in ANOVA for experimental designs with repeated measures, in which variance is separated into two sources: variance between subjects and variance within subjects manifest over time. The limitation, however, is that ANOVA assumes that both variance components are normally distributed, and time is used as an exogenous variable rather than an endogenous function, as it is in NDS processes. The issues connected to the error terms are discussed next, followed by issues connected to the model definition.

Error Components

Errors in NDS models could be additive as they are in the standard static measurements and their associated prediction models, $Y = f(X) + \varepsilon$, where ε is IID. Alternatively, they could be proportional

$$Y = w_1 f(X) + w_2(X)\varepsilon, \qquad (3.14)$$

or multiplicative

$$Y = f(X)(1 + \varepsilon). \qquad (3.15)$$

The nonadditive forms of error take the form of autocorrelation of residuals. For instance if a linear model were applied to a time series and the predicted values of Y were subtracted from the actual values of Y, one might find any one of the three possibilities shown in Figure 3.9. The third figure may have more than one interpretation.

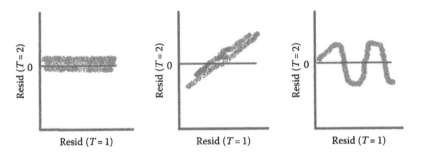

FIGURE 3.9
Plots of residuals that are uncorrelated (left), positively correlated (center), and negatively correlated (right).

Multiplicative errors should be regarded as likely suspects in NDS processes because of the way most of them are structured. Rather than predicting Y from an exogenous variable X or time, they involve affine transformations and iterative functions such that $Y_2 = f(Y_1)$. Here Y_1 is plugged into the equation to produce Y_2, Y_2 is plugged into the equation to produce Y_3, and so on. Next, suppose the process starts without error variance, but at time-3, an exogenous shock of some sort enters the system. That bit of shock would register as IID error at time-3, but it would be added to the true score of Y_3 and carried through the iteration into time-4 and subsequently. Thus what was once an IID error becomes a dependent error as its value is governed by the function that is applied to true scores. For this reason, it was conjectured (Guastello, 1995) that the R^2 coefficients for nonlinear models that are unusually high, sometimes could be so because they account for dependent error in ways that linear models cannot.

Durbin–Watson Statistic

The Durbin–Watson statistic was introduced to time series analysis at a time when models were heavily dominated by linear relationships. The typical model would look like Equation 3.1, where Y and X are time series variables lagged by a fixed amount. If all errors are IID, then the resulting statistical model can be accepted as is. If errors are non-IID, however, the R^2 for the model becomes inflated; the two variables are connected by a common third variable, which is time. If both variables change over time independently at the same time, it will appear that one is related to the other when in fact they are not. The Durbin–Watson statistic, d, was thus devised as a test for autocorrelation of residuals:

$$d = \frac{\left[\sum^{t} (\varepsilon_{t+1} - \varepsilon_t)^2 \right]}{\sum^{t} \varepsilon_t^2},$$

(3.16)

where ε is a residual for each point in the time series. The range of d is 0–4, with the null hypothesis supported with values close to 2. If autocorrelation is positive, d is small; when the autocorrelation is negative., d is large (Ostrom, 1978).

The critical values for d depend on T the length of the time series, and k the number regression parameters including the constant. The recommended lower and upper critical values at $p < .01$ of d are Q and $4\text{-}Q$, respectively, where

$$Q = \frac{(T-1)}{(T-K)} - \frac{2.3264}{(T+2)^{1/2}}. \tag{3.17}$$

For $p < .05$, the value 1.6449 replaces 2.3264. If the test is significant, one is then advised to rebuild the prediction model, perhaps in the form of Equation 3.18

$$Y_2 = B_0 + B_1Y_1 + X_1. \tag{3.18}$$

Equation 3.18 contains an autoregressive component as well as the exogenous variable X. If X has an effect over and above the autoregressive component, one might then conclude that X has some relevance for predicting Y.

Of course there is no guarantee that adopting a model such as Equation 3.18 will make the non-IID error go away, which brings us back to the conjecture introduced earlier that the high levels of accuracy that are often (but not always) reported for nonlinear models compared to linear alternative models are such because they actually account for the non-IID variance, and thus turn it from variance unaccounted for to variance accounted for. There were two illustrations of this point that relied on the Durbin–Watson statistic (Guastello, 1995). In one example, the time series was analyzed with a cusp catastrophe model, based on Equation 3.9; there was a theory and hypothesis leading up to the analysis, as one might expect. The linear alternative looked just like Equation 3.18 except that there were two exogenous variables. The Durbin–Watson statistic was significant for the linear alternative model, but not for the cusp.

The second example used the Durbin–Watson statistic as a diagnostic for a nonlinear model. The model was an exponential structure from population dynamics and applied to a problem involving the management of a small work force, where the availability of workers was relatively scarce. For one data set, which had some built-in rules for longer-range planning in a volatile environment, the Durbin–Watson statistic was significant, which led to an inspection of the data points that were furthest away from their predicted values. It turned out that the points were aberrant for reasons that were connected to random shocks and seasonal adjustments that were deliberately placed in the data series. For the other data set, which did not have the long-range planning rule in place, the Durbin–Watson statistic was not significant.

The conclusion from this particular test was that long-range planning has practical value in the long run, but an adaptive response to an external shock is not one of them.

The Durbin–Watson statistic is not part of any regular arsenal of NDS analytic techniques and was only used in that context briefly for exploratory purposes. If anything, it faded from prominence in linear time series analysis with the advent of autoregressive integrated moving averages (ARIMA) methods (Box & Jenkins, 1970), which made the calculation of autoregression more or less routine. It is, nonetheless, available as an option on linear regression subprogram on SPSS. It is mentioned here, nonetheless, because of its conceptual contribution to the understanding of error variances in time.

BDS Statistic

The Brock–Dechert–Scheinkman (BDS) statistic is a test for nonlinearity in a time series (Brock, 1991; Brock, Dechert, & Scheinkman, 1986; Brock, Dechert, & Scheinkman & LeBaron, 1996; Brock et al., 1991). It is a blunt tool in the sense that it does not distinguish any particular type of nonlinearity; it only shows that one exists. Its charm and relevance, however, is that it makes its determination on the basis of whether a non-IID error is present. Its supporting proofs showed that if there is a non-IID error present, then nonlinearity is present. Thus the proposition concerning the relationship between dependent error and nonlinearity is no longer a conjecture.

The first step to calculating the BDS statistic is to calculate a linear autoregression of the time series variable X_T with itself. Second step: Once the linear model is obtained, subtract the predicted values of X_2 from actual values of X_T to produce a time series of residuals. The residuals are used in the third and fourth steps.

The third step is to calculate the correlation dimension (e.g., using the Grassberger and Procaccia (1983)) algorithm for the time series of residuals using an embedding dimension of 1 ($C_{1,T}$) and a larger embedding dimensions, $m(C_{m,T})$, where T is the length of the time series.

The value 4 is a recommended value of m that should be large enough for most time series (Brock, 1991). The BDS statistic is then

$$\text{BDS} = T^{1/2} \frac{\left[C_{1,T} - C_{m,T}{}^m \right]}{\sigma_{m,T}}, \tag{3.19}$$

where $\sigma_{m,T}$ is the standard deviation of the time series of residuals. BDS is normally distributed under the null hypothesis. Its value can be interpreted as a z-score that can be translated into a p-value on a standard normal distribution table.

Monte Carlo studies indicate that the power of the BDS statistics is sketchy for time series of 200 observations or less. The results are considered acceptable for single-mode distributions of residuals for times series of 500 observations or more, but close to 1000 observations are needed for bimodal distributions of residuals (Brock et al., 1991). In this regard, the BDS statistic inherited some of the unfortunate properties of the correlation dimension, which some of the techniques considered in later sections of this book purport to correct.

The correlation dimension and the BDS statistic are available in the *Chaos Data Analyzer* (Sprott & Rowlands, 1995) and in other programs. Correlation dimensions are also available in nonlinear time series analysis (TISEAN) (Hegger & Kanz, 1999, chap. 5); once the correlation dimensions are calculated, the BDS is easy to calculate by hand. Frey presents an example in Chapter 8 where the BDS is used in a study of response times collected in a cognitive psychology experiment.

Linear and Nonlinear Determinism

A prevailing thought thus far is that, if the variance of a time series is not IID, then the time series stands a good chance of being a nonlinear dynamic process and thus has a deterministic explanation. The concept of chaos is iconic in this regard: seemingly random events are actually predictable from simple deterministic functions. The challenge, of course, is to identify which functions apply to which phenomena. We would go so far as to suggest that the hesitation of social scientists to adopt nonlinear modeling as part of their theory and research can be traced to the challenge of figuring out which models to use when. After all "nonlinear" can be anything that is not a straight line, and psychology, in contrast, has a very limited history of using nonlinear mathematical models to describe and predict events— applications of NDS in recent years notwithstanding.

Distinguishing Linear and Nonlinear Model Components

Equations 3.1 and 3.2 are the classic linear functions. Equation 3.3 is a nonlinear function. Not only are the changes in the left-hand side not proportional to changes in the right-hand side, but it is not possible to estimate the parameter in the exponent using ordinary (multiple) linear regression. Some nonlinear functions, however, can be *linearized* into a form that is friendly to the general linear model, hence the popularity of transforming Equation 3.3 into Equation 3.4 to find the fractal dimension associated with a power law relationship. The point is that we are adopting a distinction between linear and nonlinear models based on the proportionality of the left-hand side of the equation to the right-hand side or lack of it, as opposed to defining "linear" as a set of additive components analyzable by linear regression.

Equation 3.13 strongly suggests that an NDS function has a linear and a nonlinear component to it. Are the two parts separable? Perhaps the polynomial case is the one where the components are the most separable:

$$Y = B_0 + B_1 X + B_2 X^2 + B_3 X^3 + \cdots + B_n X^n. \tag{3.20}$$

Equation 3.20 is not dynamical in any sense because X is an exogenous variable and the model has no expression of time. Nonetheless, it is a classic procedure in multiple linear regression to transform X a number of times into powers of X, and using those as independent variables. The same objective can be obtained by using trend analysis in ANOVA. A multiple regression analysis of Equation 3.20 would tell us whether there is a nonlinear component that is independent of a linear component.

Unfortunately, the most convenient case bears little resemblance to actual NDS models. We could get closer, however, if Y were a time series such that a change in Y is a polynomial function of itself:

$$\Delta Y = B_0 + B_1 Y_1 + B_2 Y_1^2 + B_3 Y_1^3 + \cdots + B_n Y_1^n. \tag{3.21}$$

Equation 3.21 is a generic nonlinear process, but it bears no resemblance to any actual NDS process. Furthermore, if there were any exogenous variables in the process, whatever it is, the model offers no place to put them.

On the other hand, we could easily work with an important NDS function such as

$$\Delta Z = B_0 + B_1 Z^3 + B_2 Z^2 + B_3 bZ + B_4 a, \tag{3.22}$$

where Y was transformed into Z using Equation 3.11, and a and b are exogenous variables. Here, we have a situation where the NDS model can be tested through the polynomial variety of multiple regression. Exogenous variables appear in the model exactly where they are meant to appear. An R^2 coefficient can be calculated to determine the extent to which the model fits the data overall, and significance tests can be calculated on the regression weights to determine if each component is present as hypothesized.

Comparing Linear and Nonlinear Models

When the first uses of structural equations for testing NDS hypotheses were introduced (Guastello, 1982a, 1982b), the objective was to compare the NDS concept, its means of framing a problem, and its ensuing structural model with an alternative means of framing the same problem that was forthcoming from conventional (or some other previously existing) theory. Sometimes the problem might be framed as a simple multiple regression such as Equation 3.2. Sometimes there was no preexisting theory, so simple linear models such as

$$\Delta Y = B_0 + B_1 a + B_2 b \tag{3.23}$$

or

$$Y_2 = B_0 + Y_1 + B_2 a + B_3 b \tag{3.24}$$

would be common choices by default. Sometimes the conventional theory would use a bilinear interaction term ab, which is nominally nonlinear; so it might be reasonable to include $B_i ab$ to the end of Equation 3.23 or 3.24. Sometimes the alternative model is not linear at all, but a complex nonlinear function of a different type; for some examples see Guastello, Nathan, and Johnson (2009), Guastello and Philippe (1997) and Philippe (2000).

Whatever the alternative explanation happens to be, it is possible to compare R^2 coefficients for the competing models and inspect the regression weights associated with the components. There are actually two types of hypothesis testing taking place. One is the Neyman–Pearson variety where null hypotheses are defined and rejected based on a pre-decided level of rarity associated with the event. The other is a Bayesian approach in which the odds of an accurate prediction based on one theory outweigh the odds of an accurate prediction based on another. Tests of statistical significance are typically not used to compare Model R^2 for two reasons: (a) In many cases, the differences between the two values is so obvious that a significance test is pointless. (b) It is really only necessary to show that the hypothesized nonlinear model is at least as good as a linear alternative. In so doing, the qualitative properties of the nonlinear model can be accepted, at least provisionally, and examined again in future research designs.

Berliner (1991) determined that it was possible to build statistical predictions of nonlinear time series containing chaotic data; the Duffing two-well oscillator was used as an example. The strategy was to specify a probability function associated with the Lyapunov exponent, and then apply Bayesian calculations to predict the next Lyapunov exponent in the series, and thereby estimate the point. By all appearances, he was able to estimate points accurately, but it would have been helpful if the technique had concluded with a measure of effect size, such as R^2. Effect sizes have become very necessary in social science applications because of the notorious level of error in the system and relatively small effect sizes on record for most types of problems. Procedures such as meta-analysis and validity generalization gained prominence in psychology while NDS was developing, resulting in a demand for the reporting of effect sizes. The focus on the Lyapunov exponent and the use of an underlying exponential distribution was promising nonetheless; this topic is resumed in Chapter 12.

Note throughout this discussion that the emphasis is on theoretically driven modeling, not curve fitting. Strong point estimation procedures are desirable when they are available. Some statisticians have put their priorities in a different place, nonetheless. For instance, Judd and Mees (1995) developed a strategy for identifying a model in a given set of data that was based on a universe of possible nonlinear models that could be grouped into classes. The pseudo-linear models contain additive components such as the types one might find in polynomial regression or other linearized models. The fully nonlinear models contain parameters within exponents and other sundry places and require nonlinear regression to estimate. In principle, the strategy would encompass testing a sampling of structural forms of models within each class, and the structural forms that show the most promise would be refined in a later stage of analysis. The Judd–Mees strategy produces a plethora of possible models but with little theoretical basis for narrowing the field. Metrics such as minimum description length (Rissanen, 1989), which also assumes an exponential distribution of parameters, would determine which models were best. Again, there was no mention of a summary statistic characterizing effect size. For researchers who do prefer a shotgun approach to analysis, Spiegelhalter, Best, Carlin, and van der Linde (2002) have been investigating statistical means for comparing models.

Lehman (1990) distinguished interpolatory formulae from explanatory models, a distinction that actually has a long history. Interpolatory formulae require a convenient flexible family of distributions or models given a priori. Explanatory models reflect the mechanisms underlying the phenomenon. There is continuity between the two types of models, and one might go as far as to suggest that a strong theoretical explanation would greatly assist in the identification of an optimal means of point prediction. Smith (2007, p. 120), writing as a meteorologist, suggested that it might even be easier to develop a strong theoretical model than to predict points.

There are going to be occasions where the NDS model will look viable when considered in isolation, but will not do as well as expected in comparison with a linear alternative. Two classes of problems are considered next, local linearity and sample size, where conditions exist that give the linear alternative explanations an unfair advantage.

Global versus Local Dynamics

Equations 3.23 and 3.24 are only different for having moved Y_1 from the left to the right-hand side of the equation. There could be a large impact on R^2, however, if the relative amount of stability exceeds the relative amount of change. An unduly high proportion of stability could result from not allowing the system enough time to change. Alternatively, if the interesting changes only occur within narrow regions of system parameters (Y, a, and b in this example), the amount of change could be underrepresented if the sampled values of the system parameters are too wide, including more non-changing regions

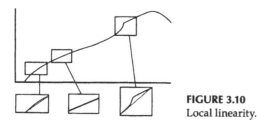

FIGURE 3.10
Local linearity.

than necessary, or too narrow and not sampling enough of the regions where the interesting events occur.

Consider first the case where not enough time has been allowed. Y_1 and Y_2 are too close together in time. Figure 3.10 shows what could happen. Any two points on a curve can be locally linear if they are close enough together; this often happens when the points are observed close to the center of an attractor (Wiggins, 1988). The figure shows a curve that assumes no particular NDS properties. Observations taken over the small ranges marked by the boxes left and center are approximated well by a linear function. More change occurs in the box on the right, so the linear approximation is not as good. It is not until the whole set of segments is available for view that one sees a globally nonlinear path.

If a curve were represented by a statistical model that concluded with an R^2 coefficient, and a nonlinear model were compared with a linear model, one might see the pattern of results shown in Figure 3.11. As the time lapse between measurements widens, the linear autocorrelation loses predictive power, while the nonlinear model gains some. There will be a point, however, where the forecasting limit of the nonlinear model is reached, and the accuracy of the nonlinear model declines also.

Theiler and Eubank (1993) made a similar observation in conjunction with sampling rates for physiological data that might be subjected to an analysis for a correlation dimension. There is a temptation for a researcher to increase the sample size by taking the observations more closely together in time;

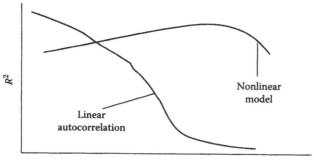

Time lapse between measurements

FIGURE 3.11
Change in R^2 for linear and nonlinear models as the time lapse increases.

oversampling occurs where the observations are too close together in time, producing an artifactual locally linear function. They found that calculations of a correlation dimension will change if under- or oversampling occurs.

Undersampling occurs where the data points are separated too far in time. Undersampling issues often arise in conjunction with other sampling issues, and are considered below under the heading of sample size.

An artifactual linear function that is produced by oversampling or perhaps by nonstationarity could lead a researcher to remove the linear trend before analysis. Theiler and Eubank (1993) recommended against this *bleaching* process, which we saw is routine in the calculation of the BDS statistic. Bleaching the linear trend is done under the assumption that the correlation dimension itself does not change if the linear trend is removed. They showed that the assumption was false; correlation dimensions change after bleaching.

There are two solutions to oversampling. One involves the calculation of a Theiler window (Theiler, 1986), which is an optimal lag length. The calculation is based on a topological theorem by Whitney (1936), which holds that, for any non-Euclidean space, it is possible to define Euclidean subspaces that are interconnected. The calculation of the window searches for distances between points that are connected but not proximal (Shelhamer, 2007). The calculation is part of some of the algorithms that are used to produce phase space diagrams, which are discussed in Chapter 4.

The other solution, which was offered by Theiler and Eubank (1993) is to use lag lengths that have an intrinsic meaning in the context of the problem that is being studied. For instance, if economic policies are adjusted every quarter of the year, then a 3-month lag would be appropriate for studying dynamics related to those policies. In another type of example involving the performance of human groups in a card game task in which observations of performance were made at discrete intervals where the group generated a new number (e.g., Guastello & Bond, 2007), the interval between observations is the interval to use.

There have been other situations, however, where the natural interval was not immediately apparent, and the problem had not been studied enough for the researchers to know what it was. In those cases, the relationships shown in Figure 3.11 were an effective guide. In one example, human participants in an experiment recorded life events at 15 min intervals in diary logs, and there was a critical measurement that was expected to display dynamical properties (Guastello, Johnson, & Rieke, 1999). The linear explanation overwhelmed the nonlinear linear explanation at intervals of 15 min, but the nonlinear explanation overtook the linear one at intervals of 1 h. Further thought about the contents of the logs suggested that events did not change much in 15 min, but most of the changes in the respondents' activities occurred at 1 h intervals. Note that it was still possible to use measurements taken every 15 min; the lag between any two points was 1 h, however.

A similar problem and solution was encountered in the analysis of galvanic skin responses for participants in a different experiment (Guastello,

Pincus, & Gunderson, 2006). In that case time intervals of 10 s were involved, which were the fastest speed at which the equipment could generate numbers. Preliminary tests indicated that the linear explanation overwhelmed the nonlinear explanation at intervals of 10 s, but the nonlinear explanation overtook the linear one at intervals of 20 s. The analyses for the experiment were thus done at 20 s intervals.

Sample Size

In conventional statistical analyses, the concern for sample size is based on questions of statistical power and the generalization from the sample to the population. Statistical power is the odds of rejecting the null hypothesis when it is indeed false. Statistical power is better when the sample size, the alpha level for rejecting the null hypothesis, and the expected effect size are all larger. One can compensate for a more stringent test or a small effect size by using a larger sample. In the extreme, tiny effects that could be practically useless look significant if the sample size is large enough. For multiple regression and other multivariate analyses, the demands are stronger for larger sample sizes because of the variance of Y that could be shared by two or more predictors. The issues for univariate tests are handled well in standard statistics books, and the case of multiple regression has been analyzed in detail by Green (1991) and Maxwell (2000).

Generalization to the population is limited to the extent that multiple regression weights capitalize on chance, particularly when many predictors are involved. Thus, Wainer (1978) recommended using equally weighted predictors for doing point estimations based on results from small samples. Several cross-validation techniques have also been produced over the years, including split-sample or out-of-sample tests, bootstrapping or leave-out-one techniques, or formula calculations of the overall R (Darlington, 1991).

The issues are not entirely the same in nonlinear analysis, however. Nonstatistical calculations such as the correlation dimension, other indicators of dimension, the Hurst exponent, and Lyapunov exponent are not tied to any notions of probability, thus the usual reasoning does not apply for those indicators. Statistical versions of these techniques do inherit the issues associated with regression techniques. The concept of restriction of topological range, which is common to both nonstatistical and statistical NDS metrics, is described next and is followed by yet another curiosity in the determination of sample size.

Restriction of Topological Range

Restriction of topological range (Guastello, 1995) is similar to restriction of range in ordinary linear regression: The R^2 for the model is larger (the model fits better) to the extent that the full range of Xs and Y are sampled. In NDS, the concern is that the full ranges of the hypothesized *dynamics* are sampled. For instance, it was shown already that a restricted range of the curve shown in Figure 3.9 might support a linear function without any support for a nonlinear component.

This restriction of topological range could be more dramatic, however. Consider the time series shown in Figure 3.12, which contains two epochs with different dynamics in each one. The dependent variable is stable, if not unmoving in Epoch 1, but is either oscillating or doing something else in Epoch 2. What type of model would explain the full result? It *might* be a result of a logistic map function where the control parameter moved the system from Period 1 to Period 2; the oscillations are not regular but could be explained by the noise in the system.

On the other hand, it might be a cusp catastrophe model where the system moved from one of its stable states into hysteresis. A second control variable would be needed to explain the large and small zigzags in Epoch 2, in which case, Epoch 2 is not showing oscillations in the true sense. At the same time, the data do not show evidence of the second stable state of the model, so the support for the cusp model would be compromised.

A logistic map should show period doubling, but we do not see that either, let alone chaos. Or do we have a model that bifurcates directly into chaos without stopping for oscillations or period doubling? There could be other dynamics occurring in later epochs if we waited long enough. The point is we do not know what we are looking at unless we have some assurance that the full range of the dynamics has been captured. The best way to do so is to start with a strong *theory* concerning the dynamics for a situation. A strong theory would tell us what dynamics should be present and what model to use to evaluate them.

The fit between the theoretical model and the data is not expected to be perfect, but it might come very close. The attenuation of the model could be the result of all the usual forms of error or nonstationarity. Nonstationarity

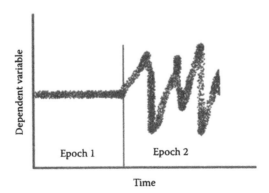

FIGURE 3.12
A time series showing two apparently different dynamics.

could be simple such as the drift in the location parameter y_0 along the scale of y. It may be possible to identify a variable that predicts the drift, put that variable into the model, and thus the error is no longer error.

Nonstationarity could be more dramatic, however, where a completely different dynamical regime intrudes on the time series, which could leave the time series in a very different place from where the intrusion stated. Sensitivity to initial conditions could possibly rule here when the first function resumes its activity. There are problems in determining if a shift in dynamics is reversible or irreversible; in some cases the system can become trapped in an attractor after a period in transients (Gregson, 2006). One might then consider searching for an explanation for the intrusive regime.

Myth of the Million Data Points

A nasty rumor sprang up in the early years of applied NDS that NDS analyses required millions of data points. Because such data sets rarely exist in social science data, it was impossible to test NDS models. Thus, social science research should continue on its own merry way building and testing linear models with all other business running as usual.

The allegation about sample size originated with concerns about the calculation of dimensions and the portrayal of data in phase space plots. Liebovitch (1998, p. 211) compiled a non-exhaustive list of seven very different calculations, such that a six-dimensional attractor would require between as few as 10 and as many as 5 billion data points! Fortunately, any real system reported in the behavioral sciences seldom surpasses four dimensions. Thus the range of required observations using the same set of formulae would fall between 3.24 points and 3.1 million. Of course the estimate of 3.24 points required to display a four-dimensional object looks impossibly wrong. The next smaller estimate (Ding, Grebogi, Ott, Sauer, & Yorke, 1993) would be $10^{(D/2)}$, or 1000 observations for a six-dimensional system and 100 observations for a four-dimensional system. The latter estimates are a lot friendlier.

The disparity in estimates, nonetheless, gives one pause to consider whether the sample size calculations for the nonstatistical renditions of dimension were off track in a fundamental way. Thus the hunt began for methods that were robust with respect to non-stationarities, varieties of noise, singularities, catastrophes, similarity across scales, edge-of-chaos phenomena, and violations of metric axioms (Gregson & Guastello, 2005).

Computational Filtering Techniques

Some data require filtering out noise generated by the equipment that is producing the time series that are going to be subjected to nonlinear analysis. If it is possible to identify the type of noise generated by the equipment, it is a relatively straightforward procedure to define a filter that will remove it. For other sources of data where the nature of the noise is not easily characterized, such as the psychometric error that we have been using as the prototype throughout the chapter, there is a temptation to use a filter there as well. One reason for filtering is that the nonstatistical metrics such as the correlation dimension, Lyapunov exponent, and Hurst exponents can give inaccurate results for noisy data. Another reason is that phase space diagrams, which are intended to give visual renditions of the dynamics can become very cloudy when data are noisy. The objective of this section is not to recount every filtering technique ever offered, but to explain the reasoning behind some of the more preeminent varieties and the limitations of the same.

The literature on filtering (e.g., Abarbanel, 1996; Aguire & Billings, 1995; Broomhead, Huke, & Muldroon, 1992; Kantz & Schreiber, 1997) does not appear to follow from a math–physics counterpart of psychometric theory such as Lord and Novick (1968). The reasoning, nonetheless, is consistent with the distinction between additive Gaussian noise and non-IID error. It would be helpful to compare a basic point about psychometric reliability with a counterpart in signal processing. Recall that the reliability of a measurement is the ratio of true score variance to total score variance. Whereas psychometric reliability is classically defined in static form, signal processing makes use of, or grapples with, the possibility that the signal strength $f(S)$ can fluctuate over time while the noise $f(N)$ could remain constant, or move independently of the signal (Sprott, 2003, p. 252). It is thus possible to say, nonetheless, that the two relationships are equivalent:

$$\frac{\sigma^2(T)}{\sigma^2(X)} \approx \frac{f(S)}{f(S)+f(N)}. \tag{3.25}$$

The additive nature of the noise is considered realistic in signal processing.

z-Filtering

Earlier in the chapter, we made note of the z transformation for location and scale (Equation 3.11) for reasons other than filtering. Broomhead et al. (1992) did propose a very similar transformation as a filtering method if the additive nature of the noise could be assumed. It actually works as a filter to some extent if one considers that the mean of the error components is zero, and that the contribution of the noise is all variance. The variance contribution would disappear in the z-transformation so long as it could be assumed that the error is homogenous throughout the time series. Its benefits are greater when two or more variables are involved for reasons given earlier. Broomhead et al. did recognize the need for a version of this filter that work with nonadditive errors. The utility of revisiting this complex method is low at this point in history, however, given the relationships that are now known between non-IID error and the deterministic function itself. Thus we move on to other filtering methods.

Moving Averages and Medians

Moving averages are a traditional technique in linear forecasting. The technique involves taking a small number of successive points, calculating the average for those points, and replacing the observed values with averaged values. Thus for a series of observations at $t = 1, 2, 3, 4, 5$, and 6, the point at $t = 2$ would be replaced by the average values of points at $t = 1, 2$, and 3; the average point at $t = 3$ would be replaced by the average values of points at $t = 2, 3$, and 4; and so on. Points at the beginning and end of the series are lost using moving averages.

The assumption behind the technique is that deviations from local linearity are the result of noise, error, or exogenous shock. The technique favors a linear model, but could dampen or obscure a true observation resulting from an implicit nonlinear function. The use of a three-point moving average in the example is arbitrary but common. ARIMA time series analysis optimizes the width of the moving average for a linear model. Here again one might conceivably be trading a chance effect that works against a linear hypothesis for a chance effect that unduly favors one.

Averages are susceptible to extreme values, which might occur in functions with wide or irregular local conditional distributions or have long tails as might be the case if the conditional distributions are exponentially distributed. Thus Press, Flannery, Teukolsky, and Vetterling (1992) recommended using the moving median instead of the moving average. The biases toward linear versus nonlinear model results still apply, however.

Use of Prediction Models

Another class of filtering models relies on the relationship expressed in Equation 3.13. Prediction techniques involve the following steps: (a) Predict the order parameter y to the maximum extent possible. (b) Use the prediction model to calculate predicted values of each y_2 from its y_1. (c) Use the set of predicted values to calculate the correlation dimension and so forth or to construct a phase space diagram. (d) Throw the residual error away.

So how do the prediction models work? Proponents of the filtering techniques only went so far as defining the prediction model as a generic polynomial function akin to Equation 3.20, except that the right-hand side contains polynomials of y_1 instead of X_i, and multiple lags of y could be involved. The reasoning behind the polynomial approach relies on two principles. First, any curve can be modeled by a polynomial of a high-enough order. The unfortunate extreme case is where there are as many polynomial terms as there are data points and thus all regression weights capitalize maximally on chance. Second, all the information needed to construct the dynamics of a time series is contained in the time series itself (Packard, Crutchfield, Farmer, & Shaw, 1980), in much the same way as the groove of a phonograph record contains all the reproducible sound. Thus, the full compliment of dynamic information can be expressed by functions containing a sufficient array of lag functions (Takens, 1981).

Aguirre and Billings (1995) developed a polynomial function for filtering with the general structure:

$$y(t) = \Psi_{yu}^{T}(t-1)\Theta_{yu} + \Psi_{yu\varepsilon}^{T}(t)\Theta_{yu\varepsilon} + \Psi_{\varepsilon}^{T}(t)\Theta_{\varepsilon}, \qquad (3.26)$$

which they named the *nonlinear autoregressive moving averages model with exogenous inputs* (NARMAX). In Equation 2.26, "where the superscript T denotes transposition and $\Psi_{yu}^{T}(t-1)$ includes a constant term and all nonlinear combinations of the output and input terms up to degree l and up to time $t-1$" (p. 240). The exogenous variable is designated as u. Θ_{yu}, $\Theta_{yu\varepsilon}$, and Θ_{ε} are weighted lag coefficients corresponding to components of the function, non-IID error, and IID error, respectively.

Their numerical example for a time series generated from the Chua attractor produced the following set of coefficients:

$$z(k) = 2.017z(k-1) - 0.598z(k-2) - 0.020z(k-1)z(k-3)z(k-4) + 0.514z(k-5)$$

$$- 0.434z(k-3) - 0.314z(k-4) + 0.063z(k-1)z(k-2)z(k-3)$$

$$- 0.078z(k-1)^{3} + 0.093z(k-1)z(k-2)z(k-3)$$

$$- 0.082z(k-1)z(k-5)^{2} - 0.145z(k-1)^{2}z(k-5) - 0.052z(k-5)^{2}$$

$$+ 0.097z(k-3)z(k-5) - 0.050z(k-3)^{3} + 0.066z(k-1)^{2}z(k-3)$$

$$+ 0.029z(k-1)z(k-3)z(k-5) + \Psi_{\varepsilon}^{T}(k-1)\Theta_{\varepsilon} + \varepsilon(k), \qquad (3.27)$$

where y was converted to z. For clarity, the coefficients in Equation 3.27 were rounded, and the lag designations are shown as normal-sized type characters in parentheses instead of subscripts. The algorithm explored all possible combinations of polynomials and cross products of lag terms. Decision rules based on the Akaike information function, model entropy and Schwarz information function were employed to determine when the model should be truncated. The final two terms of the model represent the variance of z that is not recoverable by the stochastic model.

The first observation that one might make is that Equation 3.27 contains 16 parameters and the modeling system was positioned to obtain very many more. The luxury of so many parameters is affordable for data sets where it is possible to generate thousands of observations from a generating equation, in which case it would be possible to obtain parameters that were not inflated over 0.0 by chance. Nonstationarity is not a potential issue when analyzing numbers generated from known formal functions. Relatively short data sets with unknown functions, such as those that exist in the real world are another matter entirely.

A second observation that one might make is that Equation 3.27, should it have been obtained from the analysis of a real-world data set bears no resemblance to the equations for a Chua attractor or to any other known NDS function. It is regrettably an uninformative gibberish of lag functions and polynomial terms. NARMAX could possibly work as a filter, but it has little utility as a descriptive, theoretically useful modeling strategy.

Kanz and Schreiber (1997) proposed a different polynomial filtering that had the same objectives as NARMAX, but was much simpler. The stochastic function involved moving windows of five points:

$$y_6 = B_1 y_5 + B_2 y_4^2 + B_3 y_3^3 + B_4 y_2^4 + B_5 y_1^5. \tag{3.28}$$

As a prediction filter, it would be used in the same way are NARMAX. The system is greatly simplified, however, because it always tests for the same lags and same polynomial orders in the same way each time it is used. Because of the way the polynomials are organized, it is possible to gain some sense for the topological complexity of the function, although the same question is destined to be answered by computing a correlation dimension on the filtered time series. Otherwise, the models produced by Equation 3.28 are not any more informative dynamically than those produced by NARMAX. The compatibility with real-world data series is much greater, however. This polynomial filtering system is part of the suite of programs in TISEAN, which is described in further detail in Chapter 5.

Singular Value Decomposition

The use of principal components analysis as the basis of a filtering system arose in conjunction with the production of phase space diagrams, which are covered in Chapters 7 through 9. Suppose, by some analysis such as a correlation dimension, we suspected that a time series variable should be portrayed in three dimensions. If we have a single time series variable, what exactly are the coordinates that we plot in three-dimensional space? This question was raised on top of the usual filtering question of how to get rid of the noise in a time series in order to produce a visually interpretable plot.

A general understanding of principal component analysis is assumed here; otherwise standard textbooks (e.g., Tabachnick & Fidell, 2007) have excellent explanations of the procedure. Principal components analysis can be done using the popular statistical packages such as SPSS. The analysis for singular value decomposition (Abarbanel, 1996; Broomhead & King, 1986) can be started with either a covariation matrix or a matrix of autocorrelations among the time series at lags 1, 2, 3, ..., 10. For the present discussion, the latter procedure is both easier to explain and simpler to use, particularly when long time series are involved. The solution to a principal component analysis takes the form of a matrix of correlations between the original variables (which is the same variable at different lags) and the underlying components. The dimensions of this *factor matrix* are N variables (10 in this example) by P components; P should be much less than N if only the most meaningful components are retained.

In an ordinary principal components analysis, the matrix of variables and components is further subjected to a geometric rotation, which can either be a linear (orthogonal) or nonlinear (oblique) transformation. Rotations are used to enhance the interpretability of the component analysis and the content of each component. According to the principle of *simple structure*, a variable should load (correlate larger than some set criterion) on only one component, and be only trivially correlated with the other components. Given that different rotation strategies are available, the choice rests on the interpretability of the components (Fabrigar, Wegener, McCallum, & Strahan, 1999).

Once the final solution has been adopted, one can then create component scores for each of the resulting components. SPSS and other packages can do this task automatically with a simple command. The component scores take the form of a new time series for each component, which indicates a separate geometric axis. The set of components would denote common variance that was retained; "unique variance" would be discarded.

Principal component analysis usually involves several heuristics about procedures, assumptions, and caveats that go with them. Heuristics are used

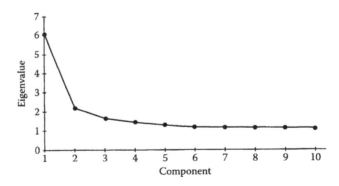

FIGURE 3.13
Scree test for single-value decomposition problem. (Reprinted from Guastello and Bock 2001, p. 187, With permission from the Society for Chaos Theory in Psychology & Life Sciences.)

to determine the correct number of components, and the choice of rotation strategies. Two important assumptions concern Gaussian residuals and the sampling space of the time series. To determine the number of components, the common practice is to retain components with eigenvalues greater than 1.00. That practice is controversial because it has a tendency to retain too few factors (Fabrigar et al., 1999). One might, therefore (a) raise or lower the eigenvalue criterion, (b) arbitrarily set the number of components, (c) base the choice on the percentages of variance accounted for by the components, and (d) use a scree test. The percentages of total variance accounted for by a component are a function of the eigenvalue, which simplifies things a bit.

A scree test (Catell, 1966) is a plot of the serial order of a component with its eigenvalue. The scree plot should show a precipitous decline in variance associated with the first few components. The decline is followed by a sharp bend in the curve, beyond which only very little variance is associated with the component. The breakpoint in the curve serves as a demarcation between components that are retained or discarded. An example of a scree test that was produced during a single-value decomposition (Guastello & Bock, 2001) appears in Figure 3.13. Here, there were only two strong dimensions; the third was weak.

Abarbanel (1996) did not state whether orthogonal or oblique rotation of the component solution should be used. It would appear, however, that orthogonal rotation was intended, since that is the closest in meaning to an embedding dimension. He did mention, however, that if synchronization occurs between order parameters, points across dimensions are correlated and the embedding space is non-integer. Synchronization implies oblique rotation.

References

Abarbanel, H. D. I. (1996). *Analysis of chaotic data*. New York: Springer-Verlag.

Aguire, L. A., & Billings, S. A. (1995). Identification of models for chaotic systems from noisy data: Implications for performance and nonlinear filtering. *Physica D, 85,* 239–258.

Bak, P. (1996). *How nature works*. New York: Springer-Verlag.

Berliner, L. M. (1991). Likelihood and Bayesian prediction of chaotic systems. *Journal of the American Statistical Association, 86,* 938–952.

Box, G. E. P., & Jenkins, G. M. (1970). *Time series analysis: Forecasting and control*. San Francisco, CA: Holden-Day.

Brock, W. A. (1991). Explanation and prediction in economics and finance. In J. L. Casti & A. Karlqvist (Eds.), *Beyond belief: Randomness, precision, and explanation in science* (pp. 230–279). Boca Raton, FL: CRC Press.

Brock, W. A., Dechert, W., & Scheinkman, J. (1986). *A test for independence based on the correlation dimension* (Working paper). Madison, WI: University of Wisconsin, Department of Economics.

Brock, W. A., Dechert, W., Scheinkman, J., & LeBaron, B. (1996). A test for independence based on the correlation dimension. *Economic Reviews, 15,* 197–235.

Brock, W. A., Hseih, D. A., & LeBaron, B. (1991). *Nonlinear dynamics, chaos, and instability: Statistical theory and economic evidence.* Cambridge, MA: MIT Press.

Broomhead, D. S., & King, G. P. (1986). Extracting qualitative dynamics from experimental data. *Physica D, 20,* 217–236.

Broomhead, D. S., Huke, J. P., & Muldoon, M. R. (1992). Linear filters and non-linear systems. *Journal of the Royal Statistical Society, B, 54,* 373–382.

Cattell, R. B. C. (1966). The scree test for the number of factors. *Multivariate Behavioral Research, 1,* 245–276.

Casti, J. L. (1991). Chaos, G'del, and truth. In J. L. Casti & A. Karlqvist (Eds.), *Beyond belief: Randomness, prediction and explanation in science* (pp. 280–328). Boca Raton, FL: CRC Press.

Clauset, A., Shalizi, C. R., & Newman, M. E. J. (2009). Power-law distributions in empirical data. Retrieved October 7, 2009, from http://www.arXiv.org/abs/0706-1062

Cobb, L. (1978). Stochastic catastrophe models and multimodal distributions. *Behavioral Science, 23,* 360–378.

Cobb, L. (1981). Parameter estimation for the cusp catastrophe model. *Behavioral Science, 26,* 75–78.

Darlington, R. B. (1991). *Regression and linear models.* New York: McGraw-Hill.

Ding, M., Grebogi, C., Ott, E., Sauer, T., & Yorke, J. A. (1993). Plateau onset for correlation dimension: When does it occur? *Physics Review Letters, 70,* 3872–3875.

Dooley, K. J. (1997). A complex adaptive systems model of organization change. *Nonlinear Dynamics, Psychology, and Life Sciences, 1,* 69–97.

Dooley, K. J., & Van de Ven, A. H. (1999). Explaining complex organizational dynamics. *Organization Science, 10,* 358–372.

Evans, M., Hastings, N., & Peacock, B. (1993). *Statistical distributions* (2nd ed.). New York: Wiley.

Evans, M. G. (1991). The problem of analyzing multiplicative composites. *American Psychologist, 46,* 6–15.

Fabrigar, L. R., Wegener, D. T., MacCallum, R. C., & Stahan, E. G. (1999). Evaluating the use of exploratory factor analysis in psychological research. *Psychological Methods, 4,* 272–290.

Grassberger, P., & Procaccia, I. (1983). Characterization of strange attractors. *Physics Review Letters, 50,* 346–349.

Green, S. B. (1991). How many subjects does it take to do a regression analysis? *Multivariate Behavioral Research, 26,* 499–510.

Gregson, R. A. M. (1994). Similarities derived from 3-D nonlinear psychophysics: Variance distributions. *Psychometrika, 59,* 97–110.

Gregson, R. A. M. (2006). *Informative psychometric filters.* Canberra, Australia: ANU E-Press.

Gregson, R. A. M. (2009). Conceptual problems in cardiological prediction. *Nonlinear Dynamics, Psychology, and Life Sciences, 13,* 207–221.

Gregson, R. A. M., & Guastello, S. J. (2005). Introduction to nonlinear methodology, part 1: Challenges we face and those that we offer. *Nonlinear Dynamics, Psychology, and Life Sciences, 9,* 371–374.

Guastello, S. J. (1982a). Color matching and shift work: An industrial application of the cusp-difference equation. *Behavioral Science, 27,* 131–137.

Guastello, S. J. (1982b). Moderator regression and the cusp catastrophe: Application of two-stage personnel selection, training, therapy and program evaluation. *Behavioral Science, 27,* 259–272.

Guastello, S. J. (1995). *Chaos, catastrophe, and human affairs: Applications of nonlinear dynamics to work, organizations, and social evolution*. Mahwah, NJ: Lawrence Erlbaum Associates.

Guastello, S. J. (2002). *Managing emergent phenomena: Nonlinear dynamics in work organizations*. Mahwah, NJ: Lawrence Erlbaum Associates.

Guastello, S. J. (2005). Statistical distributions and self-organizing phenomena: What conclusions should be drawn? *Nonlinear Dynamics, Psychology, and Life Sciences, 9*, 463–478.

Guastello, S. J. (2006). *Human factors engineering and ergonomics: A systems approach*. Mahwah, NJ: Lawrence Erlbaum Associates.

Guastello, S. J., & Bock, B. R. (2001). Attractor reconstruction with principal components analysis: Application to work flows in hierarchical organizations. *Nonlinear Dynamics, Psychology, and Life Sciences, 5*, 175–191.

Guastello, S. J., & Bond, R. W., Jr. (2007). The emergence of leadership in coordination-intensive groups. *Nonlinear Dynamics, Psychology, and Life Sciences, 11*, 91–117.

Guastello, S. J., Johnson, E. A., & Rieke, M. L. (1999). Nonlinear dynamics of motivational flow. *Nonlinear Dynamics, Psychology, and Life Sciences, 3*, 259–274.

Guastello, S. J., Nathan, D. E., & Johnson, M. J. (2009). Attractor and Lyapunov models for reach and grasp movements with application to robot-assisted therapy. *Nonlinear Dynamics, Psychology, and Life Sciences, 13*, 99–121.

Guastello, S. J., & Philippe, P. (1997). Dynamics in the development of large information exchange groups and virtual communities. *Nonlinear Dynamics, Psychology, and Life Sciences, 1*, 123–149.

Guastello, S. J., Pincus, D., & Gunderson, P. R. (2006). Electrodermal arousal between participants in a conversation: Nonlinear dynamics for linkage effects. *Nonlinear Dynamics, Psychology, and Life Sciences, 10*, 365–399.

Hegger, R., & Kantz, H. (1999). Practical implementation of nonlinear time series methods: The TISEAN package. *Chaos, 9*, 413–435.

Joreskog, K. G., & Sorbom, D. (1988). *LISREL 7: A guide to program and applications*. Chicago, IL: SPSS, Inc.

Judd, K., & Mees, A. (1995). On selecting models for nonlinear time series. *Physica D, 82*, 426–444.

Kantz, H., & Schreiber, T. (1997). *Nonlinear time series analysis*. New York: Cambridge University Press.

Liebovitch, L. S. (1998). *Fractals and chaos, simplified for the life sciences*. New York: Oxford.

Lardjane, S. (2004). Consistent Lyapunov exponent estimation for one-dimensional dynamical systems. *Computational Statistics, 19*, 159–168.

Lehman, E. L. (1990). Model specification: The views of Fisher and Neyman, and later developments. *Statistical Science, 5*, 160–168.

Lord, F. M., & Novick, M. R. (1968). *Statistical theories of mental test scores*. Reading, MA: Addison-Wesley.

Mandelbrot, B. B. (1983). *The fractal geometry of nature*. New York: Freeman.

Mandelbrot, B. B. (1997). *Fractals and scaling in finance: Discontinuity, concentration, risk*. New York: Springer.

Maxwell, S. E. (2000). Sample size and multiple regression analysis. *Psychological Methods, 5*, 434–358.

Nicolis, G., & Prigogine, I. (1989). *Exploring Complexity*. New York: Freeman.

Ott, E., Sauer, T., & Yorke, J. A. (Eds.). (1994). *Coping with chaos*. New York: Wiley.

Ostrom, C. W. Jr. (1978). *Time series analysis: Regression techniques*. Beverly Hills, CA: Sage.

Packard, N. H., Crutchfield, J. P., Farmer, J. D., & Shaw, R. S. (1980). Geometry from a time series. *Physics Review Letters, 45,* 712–716.

Philippe, P. (2000). Epidemiology and self-organized critical systems: An analysis in waiting times and disease heterogeneity. *Nonlinear Dynamics, Psychology, and Life Sciences, 4,* 275–295.

Press, W. H., Flannery, B. P., Teukolsky, S. A., & Vetterling, W. T. (1992). *Numerical recipes: The art of scientific computing* (2nd ed.). New York: Cambridge University Press.

Rissanen, J. (1989). *Stochastic complexity in statistical inquiry.* Singapore: World Scientific.

Ruelle, D. (1991). *Chance and chaos.* Princeton, NJ: Princeton University Press.

Schumacker, R. E., & Marcoulides, G. A. (Eds.). (1998). *Interaction and nonlinear effects in structural equation modeling.* Mahwah, NJ: Lawrence Erlbaum Associates.

Schuster, H.-G. (1984). *Deterministic chaos.* Weinheim, Germany: Physik-Verlag.

Shelhamer, M. (2007). *Nonlinear dynamics in physiology: A state-space approach.* Singapore: World Scientific.

Smith, L. A. (2007). *Chaos: A very short introduction.* New York: Oxford University Press.

Spiegelhalter, D. J., Best, N. G., Carlin, B. P., & van der Linde, A. (2002). Bayesian measures of model complexity and fit. *Journal of the Royal Statistical Society B, 64,* 583–639.

Sprott, J. C. (2003). *Chaos and time series analysis.* New York: Oxford.

Sprott, J. C., & Rowlands, G. (1995). *Chaos data analyzer: The Professional version.* New York: American Institute of Physics.

Szpiro, G. G. (1994). *Like sparrows on a clothes line: The self-organization of random number series* (Technical Report). Jerusalem: Hebrew University.

Tabachnick, B. G., & Fidell, L. S. (2007). *Using multivariate statistics* (5th ed.). Boston, MA: Allyn and Bacon.

Takens, F. (1981). Detecting strange attractors in turbulence. In D. A. Rand & L. S. Young (Eds.), *Dynamical systems and turbulence* (pp. 366–381). New York: Springer-Verlag.

Tauchen, G. (1985). Diagnostic testing and evaluation of maximum likelihood models. *Journal of Econometrics, 30,* 415–443.

Taylor, R. P. (2005). Fractal aesthetics: Covert art for 2005. *Nonlinear Dynamics, Psychology, and Life Sciences, 9,* 115–116.

Theiler, J. (1986). Spurious dimension from correlation algorithms applied to limited time series data. *Physical Review A, 34,* 2427–2432.

Theiler, J., & Eubank, S. (1993). Don't bleach chaotic data. *Chaos, 3,* 771–782.

Wainer, H. (1978). On the sensitivity of regression and regressors. *Psychological Bulletin, 85,* 267–273.

Walsh, J. E. (1962–1968). *Handbook of nonparametric statistics* (Vols. I–III). Princeton, NJ: Van Nostrand.

Wei, B.-C. (1998). *Exponential family nonlinear models.* New York: Springer.

West, B. J. (2006). *Where medicine went wrong.* Singapore: World Scientific.

Whitney, H. (1936). Differentiable manifolds. *Annals of Mathematics, 27,* 645–680.

Wiggins, S. (1988). *Global bifurcations and chaos.* New York: Springer-Verlag.

Wolf, A., Swift, J. B., Swinney, H. L., & Vastano, J. A. (1985). Determining Lyapunov exponents from a time series. *Physica D, 16,* 285–317.

Zeeman, E. C. (1977). *Catastrophe theory: Selected papers 1972–1977.* Reading, MA: Addison-Wesley.

4

Phase Space Analysis and Unfolding

Mark Shelhamer

CONTENTS

Introduction to Phase Space

Nonlinear systems are inherently high-dimensional. The meaning and implications of this will become clear in this chapter. For now, we note that the dimension of a system can be thought of as the number of independent mathematical variables that are needed to fully describe its behavior. For a continuous-time system, at least three such variables are needed for anything other than the simplest behaviors (steady-state, periodic). In other words, chaotic behavior (and randomness) requires at least a three-dimensional system.

With this in mind, the problem becomes one of analyzing, and attempting to visualize, the behavior of nonlinear systems in high-dimensional spaces. This chapter presents some techniques for doing this, and for developing intuition for how nonlinear systems behave in these spaces.

Definition, Example

The *phase space* (or *state space*) is an abstract space in which a trajectory is formed by the *state* of the system as it propagates through time. The state is defined by the instantaneous values of a set of key variables, known as *state variables*. A complete set of state variables will uniquely and completely describe the current state of a system. The phase space is usually created simply by plotting the state variables along orthogonal axes, showing the progression of the state variables *with respect to each other* (not with respect to time as with the typical time-series plot). We distinguish state variables from system parameters, which are (generally) fixed properties of a system. Nonstationary and time-varying systems confound this clear distinction, as system parameters may then change over time, and it is often a matter of judgment and insight as to whether a given quantity should be treated as a parameter or a state variable. This can best be attained through a thorough knowledge of the system under study, with an appreciation of the constraints under which the system is allowed to operate.

As a simple example, consider a damped oscillator (e.g., pendulum with friction). The angular position at time t is $x(t)$. The pendulum traces an arc that is a sinusoidal function with respect to time, but one that decays exponentially:

$$x(t) = e^{-at} \sin(\omega t).$$

To look at this system in the phase space, we first identify a set of state variables. While there are many possible choices, the simplest for this system turn out to be position and velocity, with velocity given by

$$\dot{x}(t) = \omega e^{-at} \cos(\omega t) - ae^{-at} \sin(\omega t).$$

The math for this example can be made simpler by setting $a = 1$ and $\omega = 20$. As $\omega \gg a$ we ignore the second term in the derivative:

$$\dot{x}(t) \approx \omega e^{-at} \cos(\omega t).$$

Figure 4.1 shows this system's trajectory in the phase space: the values of $\dot{x}(t)$ versus $x(t)$ are plotted to the right, with the time series $x(t)$ plotted at left.

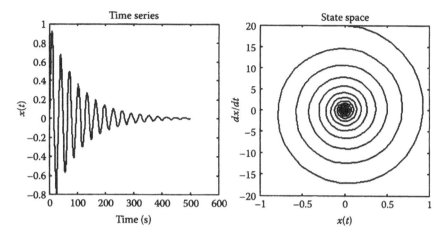

FIGURE 4.1
Damped oscillation. Time series (left) and trajectory in the phase space (right).

One can gain insight into the dynamics from this graphical representation. Starting at a maximum velocity of 20 and position zero, the system traces out a behavior that is almost periodic, but slowly decaying.

Attractors and Dimension

Several important points can be seen even in this simple example. Whatever the beginning state of the system (the initial condition), it always comes to rest at (0,0). This point is an *attractor*: all trajectories are attracted to it. Attractors are in general much more complex than this. It is important to keep in mind that "attractor" is a term with a specific technical meaning—trajectories will be attracted to it, no matter what the starting point (there is some mathematical subtlety to the formal definition, but that is the essence). It is not always easy to verify that an attractor exists in any given system. One experimental approach to confirming the existence of an attractor is to perturb the system and see if the trajectories return to a well-defined subset of the phase space (Roux, Simoyi, & Swinney, 1983).

Another general concept demonstrated here is the *dimension* of a system. The concept of dimension is multi-faceted, and we will revisit it below in different guises. At this point, we note that this system requires two state variables in order to fully describe its current state. One state variable is not sufficient. For any given position, the system can have more than one value of velocity, and vice versa. Thus, both position and velocity must be specified in order to uniquely define the system's state at a given time. This can only be depicted graphically with a phase space that is, likewise, at least two-dimensional. If the dimension of the phase space is too small, then the trajectory will cross itself. If this occurs, then once the system state comes to the point of intersection, it cannot be determined which path the trajectory will follow. Since this is a *deterministic* system, such uncertainty is not legitimate. Thus, trajectory crossings indicate that the system is random, or that the dimension of the phase space is not high enough. This relationship between the number of state variables and the dimension of the phase space can be used to advantage in the analysis of systems for which the number of state variables is not known *a priori*.

Time-Delay Reconstruction: Background

Implicit in our discussion so far has been the assumption that we *know* the state variables of the system that we are interested in, and have *access* to all of them (i.e., they can be measured). This is not always the case. Especially with high-dimensional systems, being able to identify and measure all relevant state variables can be difficult or impossible. Exacerbating this problem is the fact that even the number of relevant variables is not always known. Fortunately, there is a very powerful and clever way to deal with this when analyzing systems based on the phase space: *time-delay reconstruction*.

Packard and colleagues (Packard, Crutchfield, Farmer, & Shaw, 1980) were apparently the first to suggest the essence of this method in print: time-delayed values can be used to reconstruct a phase space. Although that paper is widely credited with being the first to make this observation, both it and one by Ruelle (1990) mention that it was actually Ruelle who made this suggestion initially.

What Is Time-Delay Reconstruction?

The basis of time-delay reconstruction is that, instead of the actual state variables, successively delayed values of a single measured quantity $x(t)$ are used in creating the phase space. Let the original time series be $x(i)$, which is a signal that we measure from the system under study. Within the limits discussed below, it does not matter what signal is used. The reconstructed phase space consists of a set of M-dimensional points $y(i)$, generated from the time series $x(t)$

$$y(1) = \left[x(1), x(1+L), \ldots, x(1+(M-1)L) \right],$$

$$y(2) = \left[x(1+J), x(1+J+L), \ldots, x(1+J+(M-1)L) \right],$$

$$y(N) = \left[x(1+(N-1)J), x(1+(N-1)J+L), \ldots, x(1+(N-1)J+(M-1)L) \right].$$

N is the number of points on the trajectory in the reconstructed phase space. M is the *embedding dimension*, the dimension of the reconstructed phase space (and the number of putative "state variables"). Each point $y(i)$ on the reconstructed trajectory can be thought of as a vector whose M elements are values from the time series $x(i)$, separated by time-delay L. The interval between the first elements of successive points is J, usually set to 1.

An example will demonstrate the simplicity of this method. With time signal $x(i) = \{1, 2, 3, 4, 5, 6, 7, 8, 9\}$, we let $M = 3$, $L = 2$, and $J = 1$. The points on the reconstructed trajectory are then

$$y(1) = \left[x(1)x(3)x(5)\right] = [1\,3\,5],$$

$$y(2) = \left[x(2)x(4)x(6)\right] = [2\,4\,6],$$

$$y(3) = \left[x(3)x(5)x(7)\right] = [3\,5\,7].$$

A note on terminology is in order. It is perhaps a matter of semantics as to whether our definition above is "time-delay" reconstruction or "time-ahead" reconstruction. The successive elements of each reconstructed point $y(i)$ are further ahead in time; whether this represents a delay or an advance can be debated. In fact it makes no difference for the techniques discussed here, but the reader should be aware of the two ways to define this reconstruction.

An example of this reconstruction technique is shown in Figure 4.2. The model system is the damped oscillator shown in Figure 4.1. Each graph shows the reconstructed phase space, using the time-delay method. A different value of the time delay L is used in each case. The values of $x(t)$ are plotted on the abscissa, and the values $x(t + L)$ on the ordinate. Note that these reconstructed trajectories closely resemble those in Figure 4.1, where

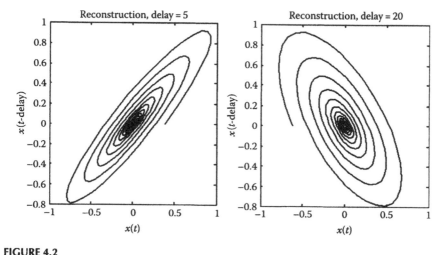

FIGURE 4.2
Damped oscillation from Figure 4.1, shown in the phase space using time-delay reconstruction with a delay L of 5 (left) and a delay of 20 (right).

the actual state variables were used. (We will discuss in a subsequent section what we mean by "resembles.") In each case the trajectories converge to the fixed point $(0,0)$, without any intersections.

Theoretical Basis in Brief, and Some Topology

The true importance of time-delay reconstruction lies not in its ability to create graphs for qualitative analysis, but in a much deeper mathematical reality. A properly performed reconstruction preserves the *topological properties* of the object being reconstructed—in this case, the object being the set of trajectories in phase space. One of the important topological properties preserved in this manner is *dimension*, which, as noted, is related to the complexity of the system under study. Here, we discuss briefly some relevant aspects of topology, and in a later section some definitions and interpretations of dimension.

We begin with some concepts in differential topology: calculus on multi-dimensional curved surfaces and spaces (as opposed to "flat" Euclidean space). A *manifold* is a space with specific mathematical characteristics. It is sometimes easiest to think of a manifold as a curved surface (such as a saddle) within space, but in fact defining certain properties and operations on the surface make it into a *space* in the technical sense. This space (surface) can be of any finite dimension. A space is a manifold if, near every point in that space, the local region resembles Euclidean space (i.e., a local coordinate system can be defined). This means that we can specify locations and directions on the manifold. The concepts that follow from this enable us to perform calculus on manifolds, by making rigorous such concepts as smoothness, continuity, and inverse.

Now assume that we have two such spaces or manifolds, X and Y, and a function f that *maps* points in X to points in Y. This function is called a *homeomorphism* if it has these properties: (a) each point in X is associated with one and only one point in Y (f is single-valued and has a single-valued inverse f^{-1}), (b) f is continuous, and (c) f^{-1} is continuous. By *continuous* we mean, intuitively, that the function is smooth (there are no abrupt breaks or jumps in the function).

To help solidify the concept of homeomorphism, some examples are provided in Figure 4.3. Each graph shows a time-delay reconstruction of the damped oscillation from Figure 4.1 after it has gone through a mapping function f. The reconstruction is carried out in each case with a delay of 20, as on the right side of Figure 4.2. Four different mapping functions are examined. Although these examples are described in terms of a simple function of two variables (the function acts on the x and y values of points on the original trajectory), a more inclusive way to think of this operation is that the function maps the first space (where the reconstructed trajectory exists) to a new space, by stretching or otherwise modifying the original space in one or more directions. The nature of these modifications determines whether or not the function or map is a homeomorphism.

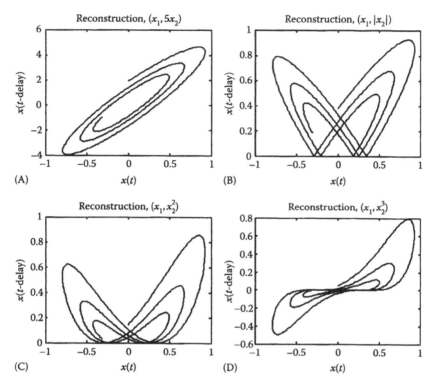

FIGURE 4.3

Examples of the effects of four different mapping functions. In each case, the indicated function is applied to the damped oscillation of Figure 4.1 and a time-delay reconstruction of the transformed data is graphed. The transformations or maps in panels A and D are homeomorphisms.

In panel A is the reconstruction after the map or transformation $(x_1,x_2) \rightarrow (x_1,5x_2)$ has been applied; in other words, the values along the ordinate have been scaled up by a factor of 5. This simply stretches the trajectories vertically. The two-valued function f in this case is a homeomorphism, since all three properties above apply.

Panel B shows the reconstruction for the function $(x_1,x_2) \rightarrow (x_1,|x_2|)$. This function is *not* a homeomorphism since, by virtue of the absolute value, two different points in the original trajectory space can be mapped to a single point in the new space. This leads to trajectory crossings in the new space. The function shown in panel C is $(x_1,x_2) \rightarrow (x_1,x_2{}^2)$ is also not a homeomorphism, for the same reason. The function $(x_1,x_2) \rightarrow (x_1,x_2{}^3)$ in panel D is a homeomorphism; although the re-mapped trajectories are compressed near the origin, there are no crossings, and since raising a number to the third power is continuous and invertible, the mapping is homeomorphic.

The purpose of these examples is to develop some intuition as to what a homeomorphism looks like in a two-dimensional space. Although a homeomorphic map or function might distort the object (the trajectory) or the space in some way, it retains the overall shape or essential nature of the original. Abrupt alterations, tearing, rending, or folding are not permitted. There is a sense in which graphs A and D in Figure 4.3 are more like Figure 4.2 than are graphs B and C; the former are homeomorphisms.

A *diffeomorphism* is a homeomorphism that preserves differentials (or increments, or derivatives): the mapping function *f* is differentiable, and so is its inverse. In other words, there is no loss of differential structure in the mapping function. Loosely speaking, the *flow* of a trajectory—its motion in different directions at different points—is preserved in such a mapping. An *embedding* is a function or mapping from the space X to the space Y that is a diffeomorphism, and is also smooth (derivatives exist at all points). An embedding is also, of course, a homeomorphism, which means that the topological properties (such as connectedness and neighborhoods of points) of the two spaces are identical.

As an aside, we note that the spaces we are dealing with (topological spaces) do not require a measurement specification—distances per se between points are not important. Rather, the crucial element is the notion of the *neighborhood* of a point—is that point connected to other points, or are there separate and disjoint subsets of points? Thus, it is *connectivity* and not *distance* that is a primary topological characteristic. As an example, there is a homeomorphism between any closed curve and a circle. The surface of a cube is homeomorphic to that of a sphere. In each case, a "smooth" distortion can change one into the other. A homeomorphism will not, on the other hand, make the surface of a sphere into a donut (torus), because to do so would require either separating points on the sphere that are close together, or placing together points that are initially separated.

Homeomorphic means *topologically equivalent*. A topological property of one space or object is unchanged (invariant) in a space (object) that is homeomorphic to the first. So, if we can find something that is homeomorphic to the underlying unobserved true phase space, and if it is easier to deal with, then we can make use of the new homeomorphic phase space for further analysis. This is precisely the point of time-delay reconstruction.

One useful topological property is dimension (see below). A topological property is one that is retained under transformations that can stretch, translate, or rotate an object, but that cannot tear it or connect points or regions together that are not already adjacent. The critical part of this characterization is that important properties such as dimension are *invariant* under a topological embedding and, since time-delay reconstruction (performed properly) is such an embedding, we can use the reconstructed phase space to measure properties that are of interest in the true (but often unobservable) phase space.

The basis of the formal proof that time-delay reconstruction is a proper embedding is the Whitney embedding theorem (Whitney, 1936), which states (roughly) that any smooth *m*-dimensional manifold can be embedded in a conventional Euclidean space of dimension 2*m*. In other words, any *m*-dimensional surface, no matter how complex, can be transformed into an object in familiar Euclidean space, and this transformed object will be topologically equivalent to the original. The new space (the embedding space) must be at least twice the dimension of the original space in order to guarantee a proper embedding (i.e., to guarantee that the transformation is a homeomorphism). This takes us to the specific theorem that makes time-delay reconstruction mathematically valid: the *Takens embedding theorem* (Takens, 1981). It says, basically, that almost any function that meets certain very broad criteria can serve as an embedding function or map *f*. These criteria are met by the time-delay reconstruction technique. Therefore, time-delay reconstruction creates a proper embedding from the original trajectory (actually, from the manifold on which the attractor trajectories lie) to the conventional Euclidean space. For this reason, the method is also referred to as *time-delay embedding*. The reconstructed trajectory (or attractor) is topologically equivalent to the original if the embedding space has a dimension of at least $2m + 1$, where m is the dimension of the original space in which the attractor trajectories exist. It is worth noting that although an embedding dimension of at least twice the dimension of the set of trajectories (attractor) under study is required to guarantee a proper embedding (reconstruction), it has been shown (Ding, Grebogi, Ott, Sauer, & Yorke, 1993) that an embedding dimension that is *at least equal to* the attractor dimension is sufficient for reliable computation of the fractal dimension (described below).

Intuitive Justification

There is a more intuitive way to look at time-delay reconstruction. Recall from calculus that the definition of the derivative of a function $x(t)$ with respect to time t is

$$f' = \frac{df(x)}{dt} = \lim_{h \to 0} \frac{f(x+h) - f(x)}{h}.$$

As a measure of change with respect to time, the derivative is based on how much $x(t)$ changes over a small time interval h as that interval becomes infinitesimally small. Time-delay reconstruction mimics this operation of taking the difference between values of $x(t)$ separated in time, but uses a fixed and finite value of time difference. One might conclude from this analogy that very small values of delay L are preferred in the reconstruction process, but this is typically not true, as seen below.

We noted above that it is common to use successive derivatives of some measured quantity as state variables: position, velocity, acceleration, and so on. These derivatives often form a legitimate and convenient set of state variables: they provide complete information about the system state at any time. Since these successive derivatives are often good state variables, and the use of consecutively time-delayed values approximates the differencing operation in the definition of a derivative, it follows that time-delay reconstruction is a reasonable way to recreate a set of trajectories in phase space.

Ability to Work in High Dimensions

One advantage of time-delay reconstruction is that it enables us to work in high-dimensional spaces ($M = 10$ is not uncommon) without having to identify all of the state variables and without having to take high-order derivatives. The issue of high-order derivatives with respect to time is especially important, since differentiation tends to increase noise in real data. (For those familiar with Laplace transforms and operators for linear systems, note that the transfer function of a derivative with respect to time is $H(s) = s$. The magnitude of this transfer function increases as frequency increases: the power spectrum is $S_{xx}(f) = f^2$. Noise typically has high frequencies relative to the desired signal, and therefore differentiation amplifies noise.) Time-delay reconstruction enables us to work in high dimensions, with many state variables, while avoiding noise from consecutive differentiations.

Example Using the Lorenz System

This reconstruction method provides—seemingly magically—complete information on the system state through the measurement of just one variable. In order for this to work, all of the actual state variables of the system under study must be coupled to each other. When the system is defined mathematically,

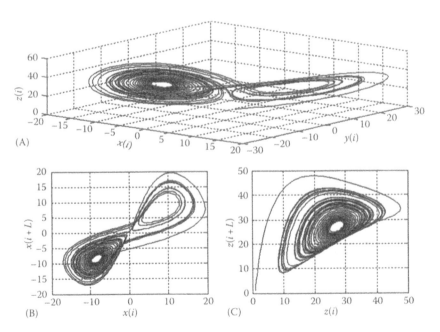

FIGURE 4.4
Lorenz attractor, formed from the three state variables in the differential equations (A), and from time-delay reconstruction using the x variable (B) or the z variable (C). Use of z in the time-delay reconstruction does not capture the dynamics as readily as does the use of x or y.

this is easy to verify. In experimental systems, however, it must be verified that there are no isolated states that evolve with dynamics that are independent from the dynamics of the quantity that has been measured and used in the reconstruction.

The well-known Lorenz system, a simplified model of atmospheric dynamics, was one of the first demonstrations of chaotic dynamics, and its "butterfly attractor" has become a standard test case for demonstrating principles and methods in the fields of chaos and fractals (Lorenz 1963). This system consists mathematically of three coupled nonlinear differential equations, and hence three state variables (and an embedding dimension of three) are needed to fully illustrate the dynamics. The trajectories in the phase space (which actually do form an attractor) are shown in Figure 4.4A, using the actual state variables (x, y, z). Time-delay reconstruction (in two dimensions) using the x variable is shown in panel B, and using the z variable in panel C. Reconstruction with the x variable closely resembles the actual dynamics in panel A, while reconstruction with the more loosely coupled z variable leaves something to be desired.

Aids to Visualization

One of the great values of the phase-space approach is that it lends itself to visualization of the data in a unique manner—in terms of the dynamical interactions of the key quantities. There is no substitute for actually *looking* at the data. It has been said that no amount of sophisticated statistical manipulation will convince a skeptical reader of one's conclusions if those conclusions are not apparent by looking at the data. While this may be debatable in the context of dynamical systems—especially in cases of high-dimensional systems where visualization of the complete dynamical space is impossible—it is nevertheless true that proper visualization can inform the choice of more rigorous computational procedures, and can aid in the interpretation of their results.

Two- and Three-Dimensional Views

The simplest approach to visualizing the phase space is to plot one of the state variables as a function of one of the others. The values of the two variables are plotted with respect to each other (one along the abscissa, one along the ordinate), with each point representing the corresponding values at a given point in time. In so doing, explicit reference to time is lost. The plot traces a trajectory that shows how the values of the variables change with respect to time, but time itself is implied as there is no explicit time scale. Nevertheless, it is possible to impose a time scale of sorts by plotting individual dots with no lines connecting them, the dots being evenly spaced in time, thereby creating a time reference. This is useful in seeing when and where the trajectory

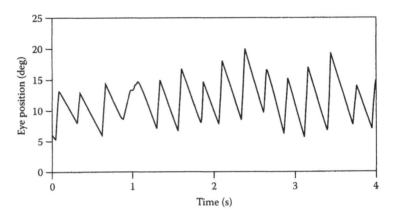

FIGURE 4.5
Example of optokinetic nystagmus eye movements (OKN). (From Shelhamer, M., Approaches to the nonlinear dynamics of reflexive eye movements, in *Nonlinear Dynamics in Human Behavior*, Sulis, W. and Combs, A. (eds.), World Scientific, Singapore, 1996. With permission.)

is traversing the space rapidly and where it is moving slowly. In dense areas of the plot, however, this method can create confusion, as there is no way to determine which dots are connected—*flow* information is lost.

To demonstrate this, we draw on an example from oculomotor research—a reflexive pattern of eye movements known as *optokinetic nystagmus* (Figure 4.5), in which the eyes track a visual scene by moving slowly in one direction, and occasionally make rapid resetting movements in the opposite direction. A time-delay reconstruction of 4s (2000 points) from this signal is shown in Figure 4.6, using a delay L of 150 points (0.3s). This signal was sampled at 500 Hz; the interval between each plotted point is thus 2 ms. The slow eye-tracking phases are easy to see, since the plotted points are close together; these phases are approximately parallel to each other and to the right-hand $(t–L)$ axis. The points are farther apart during the fast phases.

Three Dimensions, Stereo Viewing

There are some additional ways to assist in three-dimensional visualization on a two-dimensional image. Many software programs for mathematical and statistical analysis will plot a three-dimensional graph, so that three state variables can be plotted together. By changing the orientation of such a graph (the viewing angle), one can gain an appreciation of the geometry of the phase-space trajectories. An example is shown in Figure 4.7 for the OKN trajectory. If the software package allows the graph to be turned in real time, so much the better.

Another way to visualize a three-dimensional graph is through stereo viewing. As the eyes are separated horizontally, each one sees any object in space from a slightly different perspective. If this situation can be reproduced artificially, a vivid three-dimensional effect will be produced. One

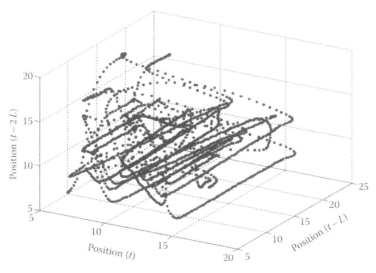

FIGURE 4.6
Time-delay reconstruction of OKN trajectory in the phase space. (Reprinted from Shelhamer, M. and Zalewski, S., *Phys. Lett. A*, 291, 349, 2001. With permission.)

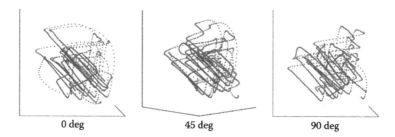

FIGURE 4.7
Time-delay reconstruction of OKN system, shown from three views to aid in visualization of the three-dimensional structure. (From Shelhamer, M., Approaches to the nonlinear dynamics of reflexive eye movements, in *Nonlinear Dynamics in Human Behavior*, Sulis, W. and Combs, A. (eds.), World Scientific, Singapore, 1996. With permission.)

way to do this is by creating two three-dimensional graphs, identical except that the viewing angle of one is shifted horizontally by a few degrees (it is rotated about its longitudinal or z axis). Place the plots side by side and insert an index card between them, perpendicular to the paper, so that each eye sees only one of the plots. If one is adept at fusing the two images (a skill that can be improved with practice), a striking three-dimensional effect will be obtained. This is demonstrated, again with the OKN system, in Figure 4.8.

While these techniques are sometimes entertaining, they can be used, as noted, to help understand the computational methods to follow. Anything that aids intuition in dealing with higher-dimensional spaces is valuable, and should be considered when dealing with an unfamiliar system or when

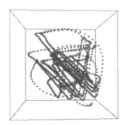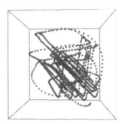

FIGURE 4.8
Stereogram of OKN trajectory. (From Shelhamer, M., *Biol. Cybern.*, 76, 237, 1997, Figure 2. With permission.)

faced with seemingly contradictory results from computational measures. While visualization techniques can aid in developing intuition about the underlying dynamics, they can serve another purpose as well—helping to select and verify parameters used in time-delay reconstruction. To this end, the approaches described below to select these parameters (time delay, embedding dimension) can be validated by seeing how the resulting trajectories appear in these visualizations.

Time-Delay Reconstruction and Unfolding: Practical Aspects

Parameter Selection: Delay Time, Embedding Dimension

It is not always a trivial matter to perform a proper time-delay reconstruction. The constraints of noise, limited resolution, and finite (often small) samples—which are the rule rather than the exception with real data—mean that the Takens theorem must be applied with care and more than a modicum of discretion. The important parameters in a delay-time reconstruction are the dimension of the embedding space (*embedding dimension M*), and the time delay L. The first addresses the question of how many state variables there are, and the second addresses how far apart in time each of the delayed elements should be when creating each point in the phase space. We discuss each of these points.

The key concept in the choice of time delay L is that the individual data values (vector elements) that make up a point $y(i)$ in the phase space should be close in time so that they approximate a derivative and are dynamically related, yet far enough apart in time that they are not redundant. Each of these points $y(i)$ provides some information on the system's dynamics, and if the individual elements $x(i)$ of $y(i)$ are too close in time then the information they provide will be redundant.

A simple but effective approach to choosing the delay L is that it be some small multiple (2 or 3) of the system's *correlation time*. The correlation time refers to the autocorrelation function $R_{xx}(\tau)$ of the measured time signal $x(i)$;

it is the shift or lag τ at which the autocorrelation has decayed to $1/e$ of its peak value. Values of $x(i)$ this far apart are partially correlated, and so this represents a way to select values that are "far apart but not too far apart." Although many other methods have been proposed, this remains a good starting point for the value of L, which can be explored more fully through visualization of the reconstruction, as noted above.

A more general approach to this same issue—identifying delay L—is based on mutual information. Unlike the autocorrelation, which assesses linear correlations, mutual information (Fraser & Swinney 1986) measures the "information" about one variable that is provided by knowledge of another variable. (By *information* is meant the *entropy* or statistical variation.) Mutual information quantifies how much the probability distribution of $x(i)$ tells about the distribution of $x(i + L)$. Mutual information between $x(i)$ and $x(i + L)$ is found at different values of L until a minimum is found. The minimum mutual information occurs when the consecutively delayed values of $x(i)$ are maximally unrelated (they give the least amount of information about each other), and this should reduce redundancy in the elements of each trajectory point $y(i)$. While mutual information is promising in this application, it is not simple to compute, and its value has been questioned (Martinerie, Albano, Mees, & Rapp, 1992).

In general, careful use of the autocorrelation criterion, verified by plotting the reconstructed trajectories in two and three dimensions, should suffice in most applications. The other major parameter to be determined is the embedding dimension M. It should be large enough so that the phase-space trajectories are properly reconstructed—if the system is deterministic, there should be no trajectory crossings. On the other hand, if M is too large, then not only is the computational burden greater than necessary for subsequent analyses, but noise can come to play an artificially large role. This is because once the embedding dimension is large enough to accommodate the actual trajectories, noise (being infinite-dimensional) will continue to occupy each additional dimension, implying a trajectory/attractor dimension larger than the true value.

False Nearest Neighbors

There is a more rigorous way to assess the size of the embedding dimension, and it demonstrates how simple geometric reasoning can be applied in a straightforward way to provide insights into the behavior of trajectories in high-dimensional spaces. This method is based on a simple realization: if the state-variable trajectories of a system are reconstructed in a space that is too small to properly contain it (i.e., if the embedding dimension M is too small), then points on the trajectories that are far apart may appear artificially close to each other. This is because the trajectories are squeezed into a space that is too small—they are not allowed to unfold fully and occupy the space they require.

One way to visualize this is to imagine that the trajectories form a helix, like a large spring with just a few coils. We might even allow the spring to be distorted so that it does not collapse on itself when compressed, but instead the consecutive coils are slightly misaligned. Such a spring is obviously a three-dimensional object, and to properly make out its geometry it must be allowed to expand and occupy three dimensions. If it is compressed so that it is almost flat, it might appear as a two-dimensional object. In this case, points on opposite ends of the spring, which are in reality far apart, will look like they are close together. If the distance between them is measured while the spring is flattened, and then when the spring is extended, there will be a large increase. This is the essence of the technique of *false nearest neighbors* (FNN: Kennel, Brown, & Abarbanel, 1992).

This concept can be quantified. First, we determine the standard Euclidean distance between two points $y(i)$ and $y(j)$ in M-dimensional space:

$$D_M(i,j) = \sqrt{\sum_{k=1}^{M} [y_k(i) - y_k(j)]^2}.$$

D_M denotes distance measured in an embedding space of dimension M. The subscript k is an index to the M time-delayed values that form each M-dimensional point on the trajectories. If point $y(j)$ is artificially close to point $y(i)$ because the embedding dimension is too small, then when the embedding dimension is increased the measured distance between these two points should increase substantially.

$$\frac{D_{M+1}(i,j)}{D_M(i,j)} > R_{thr}.$$

R_{thr} is a distance-ratio threshold that quantifies the term "substantially." If the inter-point distance increases by more than this factor, then the two points are *false neighbors*. A value of 10 for R_{thr} is recommended in most cases.

To implement this technique, first select a starting value for embedding dimension M; typically, 2 or 3. Then choose a point on the trajectory to serve as a reference point, find the distances to all other points, and find the minimum of these distances. This identifies the *nearest neighbor* to the reference point. Repeat this process, taking every point on the trajectory as a reference point and identifying its nearest neighbor and the corresponding distance. Next, find the distances between each of these reference points and the nearest neighbors just identified (in dimension M), but now using an embedding dimension of $M + 1$. Each reference point–nearest neighbor combination for which the distance increases by more than the factor R_{thr} identifies a *false nearest neighbor*, and the proportion of false nearest neighbors is plotted as a function of M. When M is large enough for the trajectories to completely unfold in the phase space, the proportion of false neighbors will reach a lower asymptote.

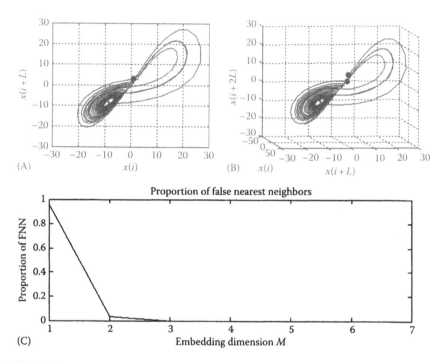

FIGURE 4.9

False nearest neighbor method applied to the Lorenz system. (A) Reconstruction with $M = 2$. Two points that are nearest neighbors are indicated with (overlapped) large dots. (B) Reconstruction with $M = 3$. The two neighbors are now farther apart. (C) Proportion of false nearest neighbors as a function of embedding dimension M.

The use of FNN is illustrated in Figure 4.9. The reconstructed trajectory of the Lorenz system (using time-delayed values of $x(t)$) is in panel A ($M = 2$). Filled circles indicated two points that are very close together in this two-dimensional reconstruction. When reconstructed in three dimensions (panel B), these same two points are seen to be actually somewhat far apart. With $M = 2$ these points are nearest neighbors, and since they are much farther apart in $M = 3$ they are false nearest neighbors. Panel C shows the proportion of false nearest neighbors for this system, as a function of embedding dimension. The fractal dimension of this system is slightly more than 2, and it requires three state variables, so we would expect the trajectories to be fully unfolded in an embedding space of dimension 3 or more, which is what we see.

Small data sets and noise can confound this test. If two nearest neighbors are not in fact close to each other (due to small N), then it is not possible for them to get as far apart as neighbors that are very close together, when adding another dimension to the embedding space. Thus, another criterion can be added for cases in which the initial distance between two nearest neighbors is on the order of the size of the entire reconstructed object trajectory (Kennel et al. 1992).

Dimension

Definitions, Topology

The dimension concept most familiar to us is *Euclidean dimension*. This is given by the number of independent coordinates required to specify a location on a given object (or, in the case of *Euclidean space*, the number of coordinates required to specify a location in space). Thus, an idealized point in space has a dimension of zero, a line has a dimension of one, and a plane has a dimension of two. Our common experience is that of Euclidean space with three dimensions, and time-delay reconstructions exist in M-dimensional Euclidean spaces.

A generalization of Euclidean dimension is *topological dimension*. Topological dimension is a *topological invariant*. As discussed above, this means that the dimension of an object is not altered if the object undergoes a homeomorphic transformation. One variant of topological dimension is the fractal dimension.

In a classic paper titled "How long is the coast of Britain?" Mandelbrot (1967) describes some of the basics of *fractals*. Imagine measuring a coastline with a measuring stick one mile long. This gross resolution means that small features such as inlets are passed over. If the measuring stick is reduced in length, more of these features will be resolved. As the measuring unit decreases in length, the apparent length of the coastline increases. Does the apparent length increase in a systematic manner as the size of the measuring unit decreases?

There is a mathematical law that describes this increase in apparent length. It is known as *power-law scaling*:

$$N \propto \varepsilon^{-D} = \frac{1}{\varepsilon^D}.$$

N is the total length, expressed as the number of units of basic length ε. As ε decreases, N increases, and the rate of increase is determined by the value of the exponent D.

If the data points for different pairs (of ε, N) are plotted on logarithmically scaled axes and connected by a best-fit line, the line is determined by the equation:

$$\log(N) = \log(K\varepsilon^{-D}) = \log(K) - D\log(\varepsilon),$$

where K is a constant of proportionality. The equation shows that a line formed by plotting $\log(N)$ as a function of $\log(\varepsilon)$ has a slope of $-D$. Values between 1 and 2 have been found for various coastlines. This power-law scaling is, of course, only true over a certain range of values of ε, as ε can become larger than the coastline itself or too small to make the measurements practical. Nevertheless, the *scaling region* can span several orders of magnitude (Mandelbrot 1983).

The power law reflects the fact that the coastlines in question are *self-similar*. Qualitatively, this means that, given an image of a coastline with no external length reference, one cannot determine the length scale of the image. This is a feature of a great many physical objects and natural processes (Bassingthwaighte, Liebovitch, & West, 1994), which are termed *fractals*.

Quantitatively, self-similarity means that if the measurement unit length ε decreases by a factor of a, then the total length in terms of the number of ε-units increases by a factor of a^D:

$$\varepsilon' = \frac{\varepsilon}{a} \quad \rightarrow \quad N' = K\varepsilon'^{-D} = K\left(\frac{\varepsilon}{a}\right)^{-D} = Ka^D\varepsilon^{-D} = Na^D.$$

What makes this "self-similar" is that this property holds true no matter where we start and no matter what we choose for N and ε. This gives *power-law scaling* its unique characteristics.

The value of the exponent D in our power-law formulation is the *fractal dimension*. It shows how the "bulk" (length, size, mass) of an object scales as a function of the size of the reference unit used to make the measurement.

We can use this definition to find the fractal dimension of a line that has length L. Clearly, if we measure the length of the line by comparing it to the length of a reference line of length ε, it takes $N = L/\varepsilon$ of these reference lines to cover the line being measured. However:

$$N = \frac{L}{\varepsilon} = L\varepsilon^{-D} \propto \varepsilon^{-D}, \quad \text{if } D = 1.$$

This plainly shows that the fractal dimension of a line is 1, as is its Euclidean dimension and its topological dimension. The same reasoning shows that any simple closed curve (whether it exists on a plane or in three-dimensional Euclidean space) also has a fractal dimension of 1. Similar reasoning shows that the number of ε-sided squares (each with area ε^2) that are needed to cover a square with side of length L (area L^2) grows as a power law with exponent 2. Thus, the fractal, topological, and Euclidean dimensions of a plane are identical. A "twisted" plane, on the other hand (like a floppy sheet of paper), has fractal and topological dimensions of 2 but a Euclidean dimension of 3.

Clearly, we do not need such an odd definition of dimension just to examine lines and planes. Let's examine an object with more interesting scaling properties. Construction of the *Koch curve* is demonstrated in Figure 4.10. Begin with an equilateral triangle with sides of unit length, in step zero. In step one, remove the middle third of each side of the triangle and tack on a triangle with a side length of one-third the original. Repeat this procedure without bounds. As the figure shows, the "curve" takes on finer and finer detail as the construction proceeds. It is truly self-similar, because if one

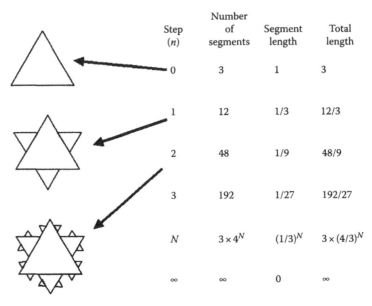

Step (n)	Number of segments	Segment length	Total length
0	3	1	3
1	12	1/3	12/3
2	48	1/9	48/9
3	192	1/27	192/27
N	3×4^N	$(1/3)^N$	$3 \times (4/3)^N$
∞	∞	0	∞

FIGURE 4.10
Construction of the Koch curve, or "Koch snowflake," which has infinite length but finite area.

small section is magnified (no matter by how much), it will look exactly like a larger section. The total length of the curve—the sum of the lengths of all of the tiny triangles—approaches infinity. Yet, the entire object can be circumscribed by a circle with a radius of less than one, and therefore the area is finite. It has infinite length, yet is bounded in space. This gives a sense of how strange a fractal can be.

The dimension can be found by setting N in the dimension definition equal to the number of segments in the construction:

$$N = 3 \cdot 4^n = \left[\left(\frac{1}{3} \right)^n \right]^{-D} = \left(\frac{1}{3} \right)^{-nD},$$

$$3 \cdot 4^n = \left(\frac{1}{3} \right)^{-nD},$$

$$\log(3) + n \log(4) = -nD \log\left(\frac{1}{3} \right) = nD \log(3),$$

$$D = \frac{\log(3) + n \log(4)}{n \log(3)} \quad \text{as } n \to \infty,$$

$$D = \lim_{n \to \infty} \frac{\log(3) + n \log(4)}{n \log(3)} = \frac{\log(4)}{\log(3)} \approx 1.2619.$$

Here, use is made of the fact that the scaling is evident for small segment sizes, which means large values of the step size n. The fractal dimension turns out to have a fractional value. The dimension is between the topological dimensions of a line and a plane, reflecting the intuitive sense that a line of infinite length is "more than a line" but "not quite a plane."

Phase-space attractors can arise from systems with chaotic dynamics; these are termed *strange attractors*. They occupy a well-defined and bounded region of the phase space, yet the behavior is aperiodic, so the trajectory never returns to the same location. How can we jam a potentially infinitely long trajectory, which never repeats or crosses itself, into a finite volume of space? One way to accomplish this is if the attractor forms a fractal, such that there is finer and finer detail as we look at it more and more closely; in this sense, no matter how "dense" the trajectory in any given area of the phase space, there is always room to squeeze in another trajectory passage. This illustrates the close connection between fractals and chaos, and is one reason why dimension in the phase space is so heavily studied in nonlinear dynamics.

Dimension Measures

How can this concept of dimension as a scaling process be applied to a set of trajectories or an attractor in phase space? We follow the reasoning above, where dimension is defined as the exponent in a power-law scaling process.

The *box-counting dimension* implements the idea of power-law scaling in a more general form. Given an object in an M-dimensional space, count the number N of M-dimensional boxes, each with side of length ε, that are needed to cover the object. (Another way to envision this is that the entire space is filled with a grid of ε-boxes, and a box is included in the count $N(\varepsilon)$ if the attractor trajectory visits that box.) If N increases as a power law function of ε, then we can define the dimension D_B:

$$N(\varepsilon) \propto (1/\varepsilon)^D \quad \text{as } \varepsilon \to 0,$$

$$\log(N) = \log(k) + D\log(1/\varepsilon) \quad \text{as } \varepsilon \to 0,$$

$$D = \frac{\log(N) - \log(k)}{\log(1/\varepsilon)} \quad \text{as } \varepsilon \to 0,$$

$$D = \lim_{\varepsilon \to 0} \frac{\log(N) - \log(k)}{\log(1/\varepsilon)} = \lim_{\varepsilon \to 0} \frac{\log(N)}{\log(1/\varepsilon)}.$$

Box-counting dimension has a clear relationship to power-law scaling, and hence an intuitive appeal. For conceptual and practical reasons (Greenside, Wolf, Swift, & Pignataro, 1982), *correlation dimension* is used in most applications rather than box-counting dimension—computational details are provided below.

Box-counting dimension is one of a series of dimensions based on a general form that places more or less weight on how often different locations in the phase space are visited by the trajectories. The most general form of these *Renyi dimensions* is

$$D_q = \frac{1}{q-1} \lim_{\varepsilon \to 0} \frac{\log I(q,\varepsilon)}{\log(\varepsilon)}.$$

Here, q indicates which in the series of dimensions is being considered ($q = 0$ is the box-counting dimension), ε is the size of a box as before, and

$$I(q,\varepsilon) = \sum_{i=1}^{M(\varepsilon)} [\mu(C_i)]^q,$$

$$\mu(C_i) = \lim_{T \to \infty} \frac{\eta(C_i, T)}{T}.$$

The quantity $\eta(C_i, T)$ is the amount of time that a trajectory spends in box C_i in the time span from 0 to T. Hence, $\mu(C_i)$ is the proportion of time that the trajectory spends in box C_i (in the long run, as T increases), and this is essentially the probability that the attractor trajectory passes through box C_i. If $q = 0$, $\mu(C_i)$ is raised to the power zero in the equation for $I(q, \varepsilon)$, so that $[\mu(C_i)]^q$ is zero if C_i is not visited at all, and one if it is visited by the trajectory, no matter how briefly. In other words, $I(q, \varepsilon)$ is a count of the number of boxes C_i visited by the trajectory. Noting the similarity of the equation for D_q to that above for the box-counting dimension (and noting that $\log(1/\varepsilon) = -\log(\varepsilon)$), it is clear that D_0 is the box-counting dimension.

The dimension when $q = 2$ is of most interest. It is the *correlation dimension*, and the equations above reduce to

$$D_2 = \lim_{\varepsilon \to 0} \frac{\log\left[\sum_i \mu^2(C_i)\right]}{\log(\varepsilon)},$$

Grassberger and Procaccia (1983) showed that the summation in this equation can be approximated by a *correlation integral*, which is much easier to compute from experimental data:

$$\sum_I \mu^2(C_i) \cong \frac{1}{N^2} \sum_i \sum_j U(\varepsilon, |y_i - y_j|) \quad (i \neq j),$$

$$U(\varepsilon, |y_i - y_j|) = \begin{cases} 1 & \text{if } |y_i - y_j| < \varepsilon, \\ 0 & \text{otherwise.} \end{cases}$$

Although expressed as a discrete summation, the quantity on the right in the upper equation is known as a *correlation integral*. The operator $U(\cdot)$ is a step function: it has a value of one if the points y_i and y_j are within distance ε of each other, and zero otherwise. Thus, the correlation integral counts the number of pairs of points that are within distance ε of each other, and divides this by N^2, the total number of pairs of points.

The demonstration of this equality can be found in Grassberger and Procaccia (1983), but an intuitive argument can be made to justify it. If, at a given box size ε, the (discretized) attractor visits box C_i for P points out of a total number N of points on the attractor, then $\mu^2(C_i) = (P/N)^2$. On the other hand, as there are P points in box C_i, there will be approximately P^2 pairs of points—that is, box C_i will contain P^2 pairs of points within distance ε of each other. By the definition of the function U, this means that C_i will contribute an amount P^2 to the correlation integral. Since this is divided by the total number of point pairs N^2, this contribution ($(P/N)^2$) is identical to that of the contribution of C_i to the summation of $\mu^2(C_i)$, and the two quantities are equal. Actually, the equality is an approximation, largely due to the fact that the correlation integral is expressed in terms of inter-point distances, and therefore implies a "ball" of radius ε to establish the criterion distance ε, while the original definition of the dimension is based on a cube with a side length of ε. For most practical applications the approximation is close, and improves as N increases.

Since it is so easy to compute, the correlation dimension has become a standard measure of the fractal dimension of attractors that have been reconstructed in the phase space. It approximates, and is a lower bound for, the box-counting dimension (it is less than or equal to the box-counting dimension, with equality in the case when all the boxes C_i are occupied equally).

Measuring Dimension, Correlation Dimension

We turn our attention to the practical problem of measuring the fractal dimension of a set of trajectories in the phase space. The correlation dimension can be used to approximate the fractal dimension:

$$D_2 = \lim_{r \to 0} \frac{\log[C(r)]}{\log(r)},$$

$$C(r) = \frac{1}{N(N-1)} \sum_i \sum_j U(r, |y_i - y_j|) \quad (i \neq j).$$

The notation has been changed to use r (radius) rather than ε to designate the criterion distance; when two points y_i and y_j are closer together than r, they are "spatially correlated" and contribute to the correlation integral $C(r)$. The divisor has also been changed to reflect the fact that $i = j$ is skipped in the summation (since distance is zero when $i = j$).

If $C(r)$ increases as a power-law function of r, then $C(r)$ versus r on a log-log plot should be a straight line, and the slope will be the correlation dimension D_2. The construction of $C(r)$ is straightforward. First, reconstruct the trajectories in an M-dimensional embedding space. Then, for all pairs of points, and a range of criterion distances r, compute the correlation integral. Finally, plot $\log[C(r)]$ versus $\log(r)$. The slope is the dimension D_2 (or D_{corr}). When the scaling region has been identified from the slope graph, go back to the graphs of $C(r)$ and find the slope over the scaling region with a linear regression. Several useful guidelines have been suggested to make these computations less subjective (Albano, Muench, Schwartz, Mees, & Rapp, 1988): less than 10% variation in the slope of $C(r)$ across the scaling region, less than 10% variation across consecutive values of M, and scaling region at least one log unit in length (one factor of 10). In addition, D_{corr} should be robust to variations in T_w, to changes in N, and to changes in the sample rate.

Another enhancement can be made to the dimension computations. It is our intent to measure the *spatial* properties of an attractor, as a reflection of the underlying temporal dynamics. The correlation dimension does this by finding points that are close together in space. However, the trajectory represents consecutive points in time, and these consecutive time points will, in many cases, also be close together in space along a given path of the trajectory. This can give rise to spurious correlations based on temporal rather than spatial or "dynamic" proximity. The way around this is to exclude from the computation of the correlation integral those pairs of points that are close together in time (Theiler, 1986).

Finally, for a reconstruction with N points and a minimum scaling region of one log unit (factor of 10), there is an upper bound on the value of D_{corr} (Eckmann & Ruelle, 1992): $2 \log(N)$.

In theory, the dimension definitions involve a limiting process, specifying power-law scaling as the criterion distance (r or ε) decreases. Yet the correlation dimension is determined over a *scaling region* that does not include the smallest available values of distance r. The correct way to think of the limiting process is that it specifies the existence of power-law scaling over a range of distances, and only in the idealized case of a mathematical construction will the scaling hold for infinitesimally small distances. This raises the question of what range of distances r to use. Obviously, one can increase r until all pairs of points are included and $C(r) = 1.0$, but the statistics become poor and the upper range is often not useful. If an estimate can be made of the minimum inter-point distance, then it can be used as the smallest value of r.

Example, Dimension of the Lorenz System

Correlation integrals from analysis of the Lorenz system are shown in Figure 4.11. The trajectories were reconstructed from 8000 values, using embedding dimensions of 5–10. There is a line for each of the six embedding dimensions, although they overlap almost entirely. The correlation integrals $C(r)$

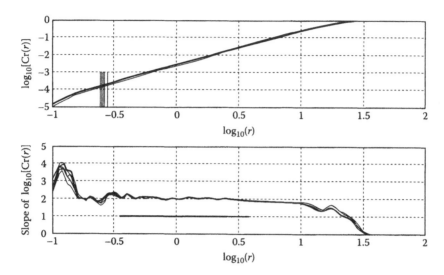

FIGURE 4.11
Correlation integrals (top) and their slopes (bottom), as steps in the determination of the correlation dimension of the Lorenz attractor.

have their maximum value when the criterion distance r is large enough to include all pairs of points; $C(r_{max}) = 1$ and so $\log[C(r_{max})] = 0$. Power-law scaling is evident as straight lines on this log-log plot, over a range of $\log_{10}(r)$ from approximately −0.5 to 1.0.

The smaller vertical lines near the left end of the graph give an indication of the smallest inter-point distances. For each correlation integral $C(r)$, the minimum inter-point distance ($|y(i) - y(j)|$) was found for each reference point $y(i)$, and the mean of these minima (across all reference points) was determined and plotted as a vertical marker. These markers indicate the distance r below which noise dominates the computation of $C(r)$. Computing and graphing this information provides a diagnostic criterion when looking for scaling: any power-law scaling at r values below these is suspect.

The bottom graph in the same figure shows the slopes of the correlation integrals. The slopes are approximately constant over the range given by the thick horizontal line; this is the *scaling region*. The slope over this scaling region, averaged over all six values of M, is 2.05, which matches well the established dimension of this system.

The embedding dimension in the case examined here was verified by finding the correlation dimension with different values of M until D_{corr} no longer increased with M, as shown in Figure 4.12. The false nearest neighbors method could be used as well.

The reconstruction time delay L was initially set based on the correlation time as discussed above. Correlation time for this system is 15 (in arbitrary units of sampled time), and so $L = 30$ is a reasonable starting point. However, in practice, these computations made use of a recommendation of Albano

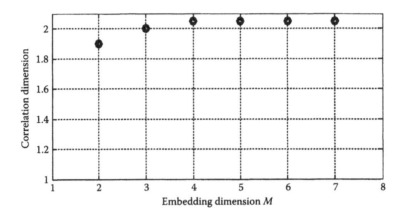

FIGURE 4.12
Saturation of the correlation dimension with increasing embedding dimension.

et al. (1988). Instead of determining a value for L and keeping it constant as M is varied, an *embedding window* T_w is established, and this is kept constant across changes in M. It is defined as $T_w = [M - 1]L$. Recall that each point in the phase space is given by: $y(i) = [x(i) \, x(i + L) \, x(i + 2L) \dots x(i + (M - 1)L)]$. The time spanned by the time-series values $x(i)$ that make up a single attractor point $y(i)$ is thus $[M - 1]L$, or T_w. The duration of the window T_w can be established based on the same reasoning as originally used for L: the window should span a time period that is long enough so that the elements of $y(i)$ are not too close in time and therefore redundant, and short enough that they approximate a time derivative in the sense of capturing the flow of the trajectory. As recommended for L, a suitable value is some small multiple of the correlation time. As M is increased in the course of the dimension estimation, L should be reduced in order to keep T_w approximately constant: $L = T_w/[M - 1]$.

Interpretation

Interpretation of dimension values depends on what one is trying to demonstrate about the system in question. Many early studies applied the correlation dimension in an attempt to "prove" that the underlying system exhibits chaotic behavior. If the attractor had a fractional dimension, it was a strange attractor, and must have come from a chaotic system. The problem is that, given the limitations of limited noisy data, it is very difficult to determine when the actual dimension is not an integer. That, plus the fact that dimension estimates depend on the specific computational parameters, make this a questionable endeavor. This is exacerbated by the fact that filtered noise can exhibit a finite correlation dimension, compatible with that from a chaotic system (Rapp, 1993; Rapp, Albano, Schmah, & Farwell, 1993). Purely random processes with power-law frequency spectra (spectra that decay as $1/f^\alpha$) can also yield finite dimensions

(Osborne & Provenzale, 1989). These are especially troubling because *noise is infinite-dimensional*, and ideally the correlation dimension should increase without limit with increasing embedding dimension.

Even when the "chaoticity" or dimension cannot be determined with certainty, there are cases in which relative dimension values—dimensions obtained under different stimulus conditions or health versus pathology—can be useful (Rapp, 1993). Also, by seeing how the dimension changes under certain data manipulations, one can use it as a tool to examine the degree of nonlinearity and randomness in a system, among other properties.

One of the more straightforward interpretations of the dimension is as the number of state variables needed to capture the dynamics of the system. This is a useful way to think of dimensions, but it has practical limitations because, even knowing the number of state variables, one cannot generally work back to the equations that describe the system. There are also cases in which signals generated by horrendously large and complicated systems produce almost absurdly small dimensions. Thus, it seems that using the dimension to indicate a more general sense of "dynamic complexity" may be more fruitful.

Software

A great variety of commercial and free educational software packages are available for analysis of dynamical systems, including graphical presentation. These are recommended for work that involves computation of dimension and other dynamical measures. For basic generation and visualization of trajectories in the phase space, almost any graphics or data-plotting program will suffice, even one as generic as Excel™. It is only necessary to plot one variable versus another to create a two-dimensional phase space, or three variables simultaneously to create a three-dimensional phase space. Time-delay reconstruction is easiest with a more general mathematical analysis program, which allows great flexibility in choosing the time delay and other parameters, but even here simply offsetting the time series by a number of values will allow plotting $x(i)$ and $x(i + L)$ together, where L is the delay or data offset.

A few specific software packages that have proven to be useful in this area are described below. This is not an exhaustive listing, and other packages are available that are more suitable for specific applications.

MATLAB®: The MathWorks, Inc. www.mathworks.com for mathematics and data analysis, is extremely extensive, standard for signal-processing applications, and has a large library of third-party commercial and user-contributed (free) packages for specific tasks. It is very flexible, but not specifically for nonlinear dynamics: user software or additional "toolboxes" are needed.

Mathcad: Mathsoft Engineering & Education, Inc. www.mathsoft.com is a general mathematics and analysis package in wide use.

Chaos Data Analyzer: Physics Academic Software http://webassign.net/pas/index.html is oriented toward educational use, but can be used profitably for research applications as well. It includes many computational tools that have become standard in nonlinear dynamical analysis.

Time Series Analysis (TISEAN): Institut für Physikalische und Theoretische Chemie, Universität Frankfurt (Main), and Max-Planck-Institut für Physik komplexer Systeme, Dresden.

  http://www.mpipks-dresden.mpg.de/~tisean/TISEAN_2.1/index.html is a free time-series analysis package, which is very extensive.

TSTOOL: Drittes Physikalisches Institut, Universität Göttingen. http://www.physik3.gwdg.de/tstool/ is a free, third-party add-on to MATLAB, for nonlinear analysis.

Acknowledgment

Portions of this chapter have been adapted with permission of the publisher from Chapters 4 and 5 of the author's book, *Introduction to Nonlinear Dynamics in Physiology: A State-Space Approach*, Singapore: World Scientific, 2007.

References

Albano, A. M., Muench, J., Schwartz, C., Mees, A. I., & Rapp, P. E. (1988). Singular-value decomposition and the Grassberger-Procaccia algorithm. *Physical Review A, 38*, 3017–3026.

Bassingthwaighte, J. B., Liebovitch, L. S., & West, B. J. (1994). *Fractal physiology*. Bethesda: American Physiological Society.

Ding, M., Grebogi, C., Ott, E., Sauer, T., & Yorke, J. A. (1993). Plateau onset for correlation dimension: When does it occur? *Physical Review Letters, 70*, 3872–3875.

Eckmann, J.-P., & Ruelle, D. (1992). Fundamental limitations for estimating dimensions and Lyapunov exponents in dynamical systems. *Physica D, 56*, 185–187.

Fraser, A. M., & Swinney, H. L. (1986). Independent coordinates for strange attractors from mutual information. *Physical Review A, 33*, 1134–1140.

Grassberger, P., & Procaccia, I. (1983). Measuring the strangeness of strange attractors. *Physica D, 9*, 189–208.

Greenside, H. S., Wolf, A., Swift, J., & Pignataro, T. (1982). Impracticality of a box-counting algorithm for calculating the dimensionality of strange attractors. *Physical Review A, 25*, 3453–3456.

Kennel, M. B., Brown, R., & Abarbanel, H. D. (1992). Determining embedding dimension for phase-space reconstruction using a geometrical construction. *Physical Review A, 45,* 3403–3411.

Lorenz, E. N. (1963). Deterministic non-periodic flow. *Journal of Atmospheric Science, 20,* 130–141.

Mandelbrot, B. (1967). How long is the coast of Britain? Statistical self-similarity and fractional dimension. *Science, 156,* 636–638.

Mandelbrot, B. B. (1983). *The fractal geometry of nature.* New York: WH Freeman.

Martinerie, J. M., Albano, A. M., Mees, A. I., & Rapp, P. E. (1992). Mutual information, strange attractors, and the optimal estimation of dimension. *Physical Review A, 45,* 7058–7064.

Osborne, A. R., & Provenzale, A. (1989). Finite correlation dimension for stochastic systems with power-law spectra. *Physica D, 35,* 357–381.

Packard, N., Crutchfield, J., Farmer, J., & Shaw, R. (1980). Geometry from a time series. *Physical Review Letters, 45,* 712.

Rapp, P. E. (1993). Chaos in the neurosciences: Cautionary tales from the frontier. *Biologist, 40,* 89–94.

Rapp, P. E., Albano, A. M., Schmah, T. I., & Farwell, L. A. (1993). Filtered noise can mimic low-dimensional chaotic attractors. *Physical Review E, 47,* 2289–2297.

Roux, J.-C., Simoyi, R. H., & Swinney, H. L. (1983). Observation of a strange attractor. *Physica D, 8,* 157–266.

Ruelle, D. (1990). Deterministic chaos: The science and the fiction. *Proceedings of the Royal Society A, 427,* 241–248.

Shelhamer, M. (1996). Approaches to the nonlinear dynamics of reflexive eye movements. In W. Sulis & A. Combs (Eds.), *Nonlinear dynamics in human behavior* (pp. 107–125). River Edge, NJ: World Scientific.

Shelhamer, M. (1997). On the correlation dimension of optokinetic nystagmus eye movements: Computational parameters, filtering, nonstationarity, and surrogate data. *Biological Cybernetics, 76,* 237–250.

Shelhamer, M., & Zalewski, S. (2001). A new application for time-delay reconstruction: Detection of fast-phase eye movements. *Physics Letters A, 291,* 349–354.

Takens F. (1981). Detecting strange attractors in turbulence. In *Lecture notes in mathematics, Vol. 898: Dynamical systems and turbulence* (pp. 366–381). Berlin: Springer.

Theiler, J. (1986). Spurious dimension from correlation algorithms applied to limited time-series data. *Physical Review A, 34,* 2427–2432.

Whitney, H. (1936). Differentiable manifolds. *Annals of Mathematics, 37,* 645–680.

5

Nonlinear Dynamical Analysis of Noisy Time Series[*]

Andrew Heathcote and David Elliott

CONTENTS

Empirical time series in the life sciences are often nonstationary and have small signal-to-noise ratios, making it difficult to accurately detect and characterize dynamical structure. The usual response to high noise is averaging, but time domain averaging is inappropriate, especially when the dynamics are nonlinear. We review alternative delay-space averaging methods based on the topology and short-term predictability of nonlinear dynamics and illustrate their application using the TISEAN software (Hegger, Kantz, & Schreiber, 1999). The methods were applied to a Lorenz series, which resembles the dynamics found by Kelly, Heathcote, Heath, and Longstaff (2001) in series of decision times. The Lorenz series was corrupted with up to 80% additive Gaussian noise, a lower signal-to-noise ratio than has been used in any previous test of these methods,

---

[*] The original article was edited for format and published as Heathcote, A. & Elliott, D. (2005). Nonlinear dynamical analysis of noisy time series. *Nonlinear Dynamics, Psychology, and Life Sciences, 9,* 399–433. Reprinted by permission of the Society for Chaos Theory in Psychology & Life Sciences.

but consistent with Kelly et al.'s data. Prediction methods performed the best for detecting nonstationarity and nonlinear dynamics, and optimal predictability provided an objective criterion for setting the parameters required by the analyses. Local linear filtering methods preformed best for characterization, producing informative plots that revealed the nature of the underlying dynamics. These results suggest that a methodology based on delay-space averaging and prediction could be useful with noisy empirical data series.

The study of mathematical chaos has provided useful new techniques for the analysis of time series (e.g., Hegger et al., 1999) and a framework for models of complex behavioral processes (e.g., Gregson, 1988, 1992, 1995; Guastello, 1995; Heath, 2000; Kelso, 1995; Newell, Liu, & Mayer-Kress, 2001). However, most of these techniques were developed for the physical sciences where the underlying determining equations are usually known and measurement noise can be minimized. In the behavioral sciences, in contrast, determining equations are rarely known, and it is difficult to eliminate measurement noise because measured systems cannot be sufficiently isolated to remove the effects of influential environmental variables.

This paper addresses the application of nonlinear dynamical analysis (NDA) techniques to behavioral data with particular reference to the problem of measurement noise. The usual response to noise in the behavioral sciences is averaging. However, when behavior is nonlinear, inappropriate averaging can introduce distortion so that the average is not representative of any individual's behavior (e.g., Brown & Heathcote, 2003; Heathcote & Brown, 2004; Heathcote, Brown, & Mewhort, 2000). Once the necessity of studying the individual is accepted, one is faced with a daunting level of variability in behavior. Gilden (1997), for example, found that changes in the mean accounted for only 10% of individual variance in a range of choice response time (RT) experiments. Discarding the remaining 90% of variation as "error" assumes that no systematic explanation is possible for the vast majority of individual behavior. Nonlinear dynamics, which can generate complex and apparently random behavior while obeying relatively simple deterministic equations, provides an alternative conceptualization of behavioral variability. As Luce (1995) states: "...the findings of the past 10 to 15 years about nonlinear dynamic systems call into question whether the actual source of the noise is randomness or ill-understood dynamics" (p. 24).

However, it is unlikely that chaotic dynamics can explain all of the variability of individual behavior. The research reported here was inspired by studies of sequential dependences in the times to make series of simple choices, which under the right conditions can show evidence for nonlinear dynamics (Kelly et al., 2001), but also clearly contained high levels of measurement noise. Where genuine noise is present, measurement of dynamics becomes problematic.

Nonlinear Dynamics

Figure 5.1a is an example of low-dimensional mathematical chaos, a time series from the Lorenz equations (see caption for details), which will be used to illustrate nonlinear dynamics throughout this paper. The Lorenz series was chosen because its dynamics resemble those observed by Kelly et al. (2001). The Lorenz series exhibits a pair of complex oscillatory states, with rapid transitions between these states occurring at apparently irregular intervals. Within each state, oscillations increase from small to large amplitudes before undergoing a state transition. The series in Figure 5.1a begins with a rapid state transition then spends a long time in the lower state, but later in the series the upper state has the majority of extended oscillations. The first major section of this paper reviews a range of NDA techniques and applies them to the Lorenz series. We then examine NDA for noisy series of the type that might be expected in behavioral data, by adding Gaussian noise to the series in Figure 5.1a.

It should be emphasized from the outset that is difficult to provide a definitive step-by-step guide that covers NDA for all types of nonlinear dynamics. The very complexity that makes nonlinear dynamics attractive for modelling also means that caution is required in generalizing findings about a particular type of chaotic series. Rather, the aim of this chapter is to illustrate an investigative methodology that readers can adapt in order to ascertain

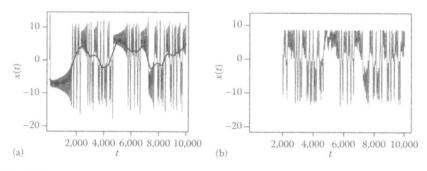

FIGURE 5.1

(a) 10,000 x values (thin line) for the Lorenz differential equations: $dx/dt = 10(y - x)$, $dy/dt = x(28 - z - y)$, and $dz/dt = xy - z8/3$, solved with step size = 0.01 and starting from $(x, y, z, t) = (0, 1, 0, 0)$, with a Lowess (Cleveland, 1979) smooth using a 1000 element moving window (thick line). The Lorenz series was generated using the default settings of the Smalldyn program (http://keck2.umd.edu/dynamics, see Nusse & Yorke, 1997). Analyses reported here examine the first 10,000 values (data file lorenz1.dat), values from $t = 2,001$ to 10,000 (lorenz2.dat) or values from $t = 2,001$ to 100,000 (lorenz3.dat). (b) Filtered noisy Lorenz series from $t = 2,001$ to 10,000. Noise was additive Gaussian with the same SD as the Lorenz series (i.e., 50% noise, n1.dat). The geometric filter was iterated six times, retaining two singular values and using a minimum neighborhood size of 400 in a 10 dimensions delay-one embedding (ghkss n1.dat -i6 -q2 -k400 -m10 -d1 -o n1q2k400.dat).

if NDA will be useful with their own data, perhaps aided by analysis of model time series. The analysis software used here (with commands set in Courier) is available free through the TISEAN project (Version 2.1, http://www.mpipks-dresden.mpg.de/~tisean, see Hegger et al., 1999).

Methods for Dynamical Analysis

Chaos can range from low dimensions to high dimensions. Although noise and high-dimensional chaos are conceptually distinct, they are almost impossible to distinguish in practice. Low-dimensional chaos can be distinguished from noise, but obtaining data of sufficient quality can be difficult. The time series must be long, sampled at regular time intervals, and stationary (stationarity is defined below). Initially, NDA techniques were developed for physical science applications that yielded such series. In the life sciences, however, long stationary series are difficult to obtain, due to processes such as learning and fatigue, and high levels of measurement noise are often present. Genuine noise, particularly linearly autocorrelated noise, is problematic as it can appear like chaos to some NDA techniques.

The Fourier power spectrum of the series in Figure 5.1a illustrates why noise is problematic for NDA. Fourier analysis is a traditional approach to such apparently structured oscillatory signals, allowing for dominant frequencies to be identified through peaks in the spectrum, and broadband noise to be removed through band-pass filtering. However chaotic signals can themselves be broadband, and so frequency domain filters are ineffective. Where noise is white (i.e., each sample is independent) the spectrum is flat. Colored noises, such as those produced by linear and nonlinear autocorrelation produce spectra, are characterized by a decrease in power with increasing frequency. The Lorenz series also displays a decrease in power as frequency increases. The similarity of spectra for chaos and colored noises makes them hard to distinguish.

Whereas Fourier decomposition transforms temporal information into frequency information, most NDA techniques rely on the transformation of temporal information into a geometrical representation: *delay embedding*. An m-dimensional delay embedding converts a one-dimensional time series, $x(t)$, to a set of m-dimensional points $(x(t + \delta), x(t + 2\delta), ..., x(t + m\delta))$, where δ is the time delay between samples. Takens' (1981) theorem states that a one-to-one image of the d-dimensional set of points visited by a stationary dynamical system (its attractor) can be reconstructed from an embedding in a delay space with dimension $m > 2d$. For the Lorenz system, for example, $d \approx 2.05$, with the attractor being a fractal subset of the space defined by the three Lorenz variables (i.e., (x, y, z) of which only x is shown in Figure 5.1a). The complex, fractal nature of this set for chaotic systems have led to them being described as "strange attractors." Since Takens' initial result, "embedology" has been an active area of theoretical development, with Sauer, Yorke, and Casdagli (1991) showing that more general schemes than simple delay coordinates can be used, including delay embeddings of series transformed by singular value decomposition and geometric filtering (Grassberger, Hegger, Kantz, Schaffrath, & Schreiber, 1993).

Stationarity is essential for establishing an embedding. For a stochastic process, $y(t)$, stationarity can be defined as invariance of all finite-dimensional joint distribution functions of $(y(t_1), y(t_2), ..., y(t_m))$ over shifts on the temporal dimension (e.g., $\{t_{1+\Delta}, ..., t_{m+\Delta}\}$; Rao & Gabr, 1984). In the context of NDA, Casdagli (1997, p. 12) stated that for practical purposes: "...a time series $x_1, x_2, ..., x_N$ is nonstationary if, for low m, there are variations in the estimated joint distribution of $x_i, x_{i+1}, ..., x_{i+m-1}$ that occur on time scales of order N." An embedded representation can only translate all temporal information to spatial information if the underlying process is stationary. Given stationarity, NDA can recover aspects of the dynamics using geometrical analyses of the embedded set. The analyses often involve estimating the properties of local neighborhoods or sets of points, with a neighborhood defined on a distance measure in the embedding space. Where such analyses aggregate local measures to estimate global properties of the attractor they strongly rely on the assumption of stationarity. Aggregation is particularly important to counter the effects of measurement noise.

Even where stationarity holds, it should be acknowledged that a particular data series may not be suitable for NDA if all regions of the dynamics are not sampled, and hence the full joint distribution cannot be estimated. As in conventional statistics, one needs a sufficient number of replicates to form a reliable model, so intermittent chaos or noisy periodic behaviors cannot be characterized when few intermittent events or periods are sampled. In some circumstances, even arbitrarily small amounts of noise can destroy the embedding property (Casdagli, Eubank, Farmer, & Gibson, 1991). However, delay coordinates are also useful when an embedding is not required, such as in the prediction of nonlinear time series with both deterministic and stochastic dynamical structure (Weigend & Gershenfeld, 1993). As will be discussed later, NDA based on prediction is especially attractive because it remains useful when noise levels are high.

Figure 5.2a illustrates the nonlinear dynamics of the Lorenz time series, by plotting each value against an estimate of its first derivative, obtained by successive ("lag one") differences. The derivative plot shows that the series changes quickly near the center of each oscillation state, and more slowly in the region between states, and at the positive and negative extremes. The spikes to the left and right of the main body of the plot are produced by very large changes, a large decrease for the positive state and three large increases for the negative state. The derivative plot also shows that the initial 2000 or so observations in the series have different dynamics to the remainder of the series. The series first exhibits an unusually large decrease then oscillates

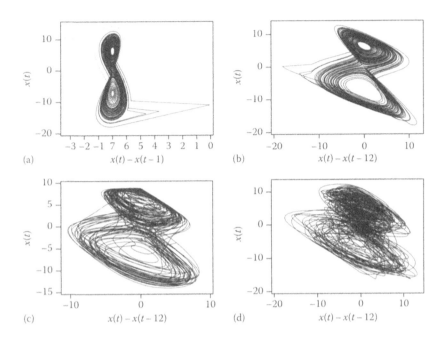

FIGURE 5.2
A plot of $x(t)$ from Figure 5.1a against its derivative estimated at (a) lag one, and (b) lag 12. Points in the derivative plot are joined to emphasize the trajectory of the dynamics. In (a) $t = 1 \ldots 10,000$, whereas for (b) and (c) $t = 2,001 \ldots 10,000$. In (a) points for $t = 1 \ldots 2000$ are joined by a dotted line, and the four "spikes," one pointing to the left from the top lobe and three pointing to the right from the bottom lobe, are caused by single points with unusually large differences. Lag 12 difference plots for (c) the filtered 50% noise series shown in Figure 5.1b, and (d) the 50% noise series filtered with the same parameters as (c), but using a three rather than two-dimensional projection.

for an extended period around the center of the lower state. The values of the series from $t = 2,001$ to $10,000$ will be used to illustrate the analysis of stationary dynamics, with the first 2,000 observations used to illustrate the detection of nonstationarity.

Most NDA techniques are based on time delay coordinates, so they require the user to select a delay, embedding dimension, and often other parameters. Parameter selection is often difficult because the best choice depends both on the data set and the technique. A range of methods have been devised to select parameters for a given data set; however, the methods can also require parameter values to be selected. Further, both the dynamical analysis and parameter selection methods can fail when noise is present. In practice, NDA for empirical data is still as much an art as a science.

The following sections survey a range of methods for performing linear and nonlinear dynamical analyses. The first section "Nonlinear Dynamics" examines linear methods, which are a necessary preliminary to nonlinear analysis. The following section "Methods for Dynamical Analysis" discusses measures of nonlinearity based on the ideas of recurrence and sensitive dependence. The final section "Nonlinear Dynamic Analysis of Noisy Data" discusses inferential testing for the presence of nonlinearity, and filtering methods that remove noise, enabling the graphical characterization of dynamical structure.

Linear Dynamical Analysis

Linear dynamical analyses can be accomplished using standard ARMA (AR = autoregressive, MA = moving average) modelling (Box & Jenkins, 1976; Cryer, 1986). Linear autoregressive models create future values from linear combinations of past values and noise, producing linear autocorrelations between past and future observations. Most chaotic models have the same type of recursive dynamics, except that the combination also contains non-linear components. As a result, chaos almost always produces strong linear autocorrelations. The best-fitting linear autoregressive model for the series in Figure 5.1a, as determined by the number of significant partial autocorrelation coefficients, is of order 12 and accounts for more than 99% of the variation in the series. However, the residuals for this model are structured, as it does not account for all of the increasing magnitude of oscillations within a state, with four sets of large deviations corresponding to the large differences in Figure 5.2a. Fits requiring large numbers of parameters (e.g., 12 compared to the 3 parameters of the Lorenz equations), and failure to account for fine-grained structure, are characteristic of ARMA models of chaotic series.

De-trending, through transformations such as taking successive differences or subtracting a regression estimate of the mean, is usually applied because ARMA analysis assumes second-order or "weak" stationarity: constancy of the mean, variance, and auto-covariance.[1] The Lorenz series also displays nonstationarity up to around 2000 observations, evident in an increase in the local mean of the series (the thick line in Figure 5.1a). However, despite a broad averaging window of 1000 observations, the local mean continues to fluctuate appreciably throughout the series. These fluctuations do not reflect nonstationarity, but rather the underlying nonlinear dynamics, which spends varying amounts of time in the upper and lower states. Similar observations can be made about the local variance of the series, so in this case, neither the local means nor variances are useful for detecting the initial nonstationarity.

For the Lorenz series, the initial nonstationarity did not make much difference to the fitted AR model, but the autocorrelation function was somewhat shorter when the first 2000 observations were omitted. The length of an autocorrelation function is often characterized by its first zero, or by the "correlation time," the lag at which it drops to $1/e$ (≈ 0.378) of its initial value. Chaotic series can have very long autocorrelation functions. For the full series in Figure 5.1a, autocorrelations up to a lag of 500 did not cross zero and had a correlation time of 41. With the first 2000 values omitted, the autocorrelation function crossed zero at lag 314 and had a correlation time of 31.

De-trending must be applied cautiously as a preliminary to NDA. For example, subtracting the local mean averaged at the scale illustrated in Figure 5.1a would destroy information about the underlying dynamics. In particular, NDA should not be carried out on residuals from ARMA analysis (Theiler & Eubank, 1993). Chaotic dynamics often produce linear components, and ARMA models with a large number of parameters can fit some variation due to nonlinear components, so that residuals do not contain the information necessary for NDA. However, NDA does rely on comparisons with linear models. Instead of examining residuals, an experimental series is compared to "surrogate" series, formed by randomizing the experimental series while maintaining structure dictated by a null hypothesis, such as linear structure (Theiler, Eubank, Longtin, Galdrikian, & Farmer, 1992), as described below.

Nonlinear Methods Based on Recurrence

Graphical methods are particularly useful for revealing recurrent, or close to recurrent, structure in time series. A delay plot graphs a time series, $x(t)$, against a delayed version of itself, $x(t + \delta)$, where δ is the delay. Periodic attractors display exact recurrence, returning to the same state or states at regular intervals. Hence, for appropriately chosen delay intervals, such systems can be represented by a single point or small set of points in a delay plot. Chaotic systems are not exactly recurrent because they contain many unstable periodic orbits, but delay plots can still reveal behavior that is close to recurrent, and so they are useful for understanding the qualitative dynamics of a chaotic system.

Note that delay plots and derivative plots (e.g., Figure 5.2) contain the same information. For example, the lag-one derivative estimate for a point in a lag-one delay plot equals the difference between vertical and horizontal distances from the right diagonal. The derivative plot has the advantage of a direct interpretation in terms of process dynamics, and can be extended using estimates of higher-order derivatives. However, estimates of higher-order derivatives can be very variable when noise is present, so attention is restricted here to the first derivative.

Figure 5.2 shows that an appropriately chosen delay provides a clear representation of temporal structure in derivative plots. When the delay is short (e.g., Figure 5.2a) orbits are too closely packed to be discriminable, whereas when the delay is too long, details may be lost in some regions. In general, no single delay necessarily provides the best detail of all regions of the dynamics (cf. Hegger et al., 1999, Figure 5.1). The embedding theorem should apply for any delay, but most NDA techniques require specification of an appropriate delay in order to avoid the redundancy evident in Figure 5.2a. A series of derivative or delay plots can be used to choose the delay producing the clearest structure. For the Lorenz data, a delay of 12 provided the best result, agreeing with the order of the AR model for this time series.

Methods of choosing delay based on autocorrelation, such as the correlation time or the first zero of the autocorrelation function, are sometimes recommended. However, these estimates, given earlier, were not useful for the Lorenz series. Time-delayed mutual information (i.e., the information shared by lagged versions of a time series) is also useful for estimating delay (Fraser & Swinney, 1986). Mutual information accounts for both linear and nonlinear structure, with the delay being set to its first minimum as a function of lag. For the stationary Lorenz series, the resulting estimate was 17 (mutual lorenz2.dat -D100 -o provides mutual information to 100 lags), somewhat longer than the linear estimate of 12, reflecting the longer time scale of the nonlinear interactions. For the following analyses of the deterministic Lorenz series, little difference was observed for delays ranging from 12 to 17.

Recurrence plots, which were originally proposed by Eckmann, Oliffson-Kamphorst, and Ruelle (1987), provide an alternative means of graphical analysis. They are constructed by plotting points (on a grid defined by two time axes) where recurrence nearly occurs in an m dimensional embedding. Recurrence was originally defined as being a kth nearest neighbor, but a definition based on a minimum distance is more often used (Koebbe & Meyer-Kress, 1991). The TISEAN command recurr produces an output file that can be used to create a recurrence plot of the latter type. Recurrence plots produce visual patterns that may be useful in detecting nonstationarity, but general guidelines for how to interpret the patterns are difficult to formulate (cf. Schreiber, 1999).[2]

Dimensionality estimates provide quantitative indices of nonlinear dynamics based on topological regularities such as near recurrence. According to Takens' (1981) theorem, the minimum dimension guaranteed to establish an embedding is given by the smallest integer $m > 2d$, where d is the possibly fractional dimension of the attractor that generated the time series. Simple periodic attractors have a dimension of 1, and quasi-periodic attractors, such as tori, have higher integer dimensions. Chaotic dynamics have attractors with non-integer dimensions, because they fill the phase space in a fractal manner with an invariant distribution of points on the attractor at different length scales.

The Lorenz equations operate in a three-dimensional space, but the Lorenz attractor has dimension, $d \approx 2.05$. A value of $m = 5$ will be adopted here for further analysis of the Lorenz data, in accord with Takens' theorem. Note, however, that Takens' theorem does not dictate that smaller values of m will not work. For the Lorenz system, for example, $m = 3$ can be sufficient to achieve an embedding. Heath (2000) favors the false nearest neighbor method (Kennel, Brown, & Abarbanel, 1992) to determine a minimum embedding dimension. Estimation of a minimum embedding dimension will be discussed once some background information has been established.

A number of different attractor dimensionality estimates are available, based on either the correlation sum or information theory measures.

For example, dimensionality may be estimated using D1, the information dimension, and the dynamics of two series may be compared using Kullback entropy (see Schreiber, 1999). We will examine the most commonly used estimate of dimensionality, D2. D2 is based on the correlation sum, $C(\varepsilon)$, which measures the proportion of embedded points that fall within a given distance (ε) of each other (Grassberger & Procaccia, 1983). The correlation sum measures recurrence at a given distance or scale, so it is equivalent to the proportion of points marked as recurrent at a given distance on a recurrence plot.

D2 summarizes the way the correlation sum changes as a function of distance. It can be estimated from the exponent of a power law relationship between the correlation sum and distance. The exponent is usually measured by the slope of a $\log(C(\varepsilon)) - \log(\varepsilon)$ plot. Because the power law relationship never applies for all distances in a finite series (technically the dimension of a finite series is zero), the slope is assessed locally, that is, it is assessed for a restricted range of distances. For chaotic systems, the slope increases with m, but then becomes constant once a proper embedding is achieved. The constant slope provides the D2 estimate. White noise, in contrast, consists of a series of independent random values, so it fills the delay space uniformly no matter what its dimension, and produces increasing slope estimates.

Early applications of NDA used finite D2 estimates as evidence for chaos. However, a finite D2 is a necessary but not sufficient condition for chaos. Colored noises, which result from stochastic dependencies between series of values, also produce finite D2 estimates. Osborne and Provenzale (1989) showed that an embedded random walk has a dimension of 2. Provenzale, Smith, Vio, and Murante (1992) discuss these problems and provide examples of finite D2 estimates for both linear and nonlinear autoregressive models. Theiler and Rapp (1996) concluded that much of the supposed evidence for chaos relying on finite estimated dimensionality suffered from these problems. Hence, D2 estimates may be best used as a method of characterizing rather than identifying nonlinear dynamics.

Even where nonlinear dynamics are present, measurement noise is particularly detrimental to D2 estimation, as it smears out the fine details of an attractor. As a result, accurate absolute dimensionality estimates are difficult to obtain. Relative measurements of dimensionality may, however, remain useful, as long as estimates are obtained with the same measurement function, $y = f(x)$, where y is the observed time series, x is the series produced by the dynamics, and f is a monotone function. Under ideal circumstances D2 is invariant with respect to the measurement function, but when only relative measurements are possible it may not be invariant. Hence, comparisons may be confounded if the measurement functions differ. Similarly, changes in the level and type of measurement noise may change D2 estimates, and so confound comparisons.

An important technical issue in D2 estimation, and other estimates based on the correlation sum, is that they assume that pairs of points are drawn randomly and independently according to the scale invariant measure of the attractor. Independence cannot apply for points occurring close in time, and if such points are included spuriously low estimates of D2 occur. To avoid the problem, points closer than some minimum time can be excluded from the correlation sum (Grassberger, 1987; Theiler, 1990). The number of points excluded, w, is called the Theiler window. An easy estimate of the Theiler window can be obtained by multiplying the correlation time by three (Heath, 2000). A more rigorous estimate is given by a space-time separation plot (Provenzale et al., 1992).

A space-time separation plot is related to the correlation sum and the recurrence plot. It shows equal probability contours for the distribution of distances between pairs of points as a function of time. For a chaotic series, the contours initially rise then oscillate around a constant value, whereas for colored noise they continue to rise. Figure 5.3 shows a space-time separation plot for the Lorenz data, which displays contours of constant probability for the spatial separation of pairs of points in an embedding as a function of their separation in time. The Theiler window, w, is chosen at a time beyond which the constant behavior is operating. For the following analyses, w was set at a value of 100, which approximately equals the estimate from three times the correlation time (3×31). Values in the range 50–300 were found to have a similar effect.

Figure 5.4 shows, for the stationary Lorenz series and a Theiler window of 100, local slope estimates of $\log(C(\varepsilon)) \sim \log(\varepsilon)$ as a function of $\log(\varepsilon)$ for 1–10 embedding dimensions. The slope in a "scaling region" (a range of distances with the same constant slope for all larger embedding dimensions) provides

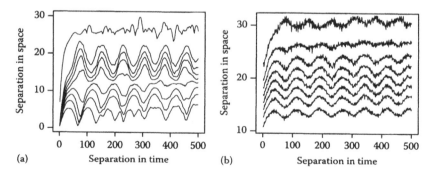

FIGURE 5.3

Space-time separation plots showing contours of constant probability for spatial separation as a function of time separation from 1 to 500 for pairs of embedded points from (a) the series in Figure 5.1a from $t = 2{,}001$ to 10,000 using a five-dimensional delay-17 embedding (stp lorenz2.dat -%0.1 -t500 -d17 -m5 -o) and (b) The Lorenz series from $t = 2{,}001$ to 10,000 with additive Gaussian noise of equal SD and a 10-dimensional delay-one embedding (stp n1.dat -d1 -m10 -%0.1 -t500 -o). For parameter –%f, the algorithm calculates k contours, where k is the nearest integer *less* than $1/f$, at probability values $1/k, 2/k...1$. A weakness of this algorithm is that it produces a noisy upper contour, as it is the sample maximum, which should not be given much weight in examining the pattern in the space-time plot. The contour for the maximum value has been omitted from (b) to provide clearer resolution of the lower contours.

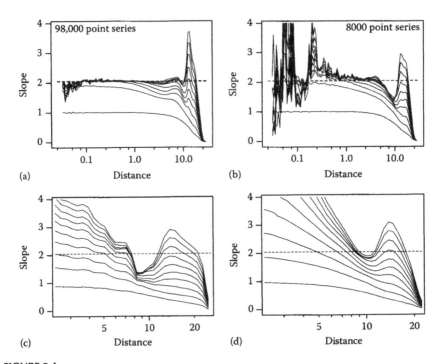

FIGURE 5.4
Local slopes of log($C(\varepsilon)$) vs. log(ε) plots for a range of distances, ε, and embedding dimensions from
1 to 10 in a delay-17 embedding with a Theiler window of 100 (a) for the Lorenz series for $t = 2{,}001$–
100,000 (`d2 lorenz3.dat -M1,10 -d17 -t100 -N0 -o`), (b) the Lorenz series for $t = 2{,}001$–10,000
(`d2 lorenz2.dat -M1,10 -d17 -t100 -N0 -o`), and the filtered noisy Lorenz series, and (c)
using two-dimensional projection, as in Figures 5.1b and 5.2c (`d2 n1q2k400i6.dat -M1,10 -d17
-t100 -N0 -o`) and (d) three-dimensional projections, as in Figure 5.2d (`d2 n1q3k400i6.
dat -M1,10 -d17 -t100 -N0 -o`). The -N0 option indicated that the computation should be
done on all possible pairs of points (the default is a maximum of 1000). Lower lines indicate lower
embedding dimensions. The dashed line in each panel indicates the true dimension, 2.05.

an estimate of dimensionality. The local slope plot is a useful alternative to
directly estimating D2 for the slope of log($C(\varepsilon)$) ~ log(ε) plots because it makes
evident the extent of, and deviations from, scaling behavior. Dimensionality
cannot be estimated unless scaling behavior occurs; the TISEAN program-
mers emphasize this point so strongly that they purposely do not provide any
automatic estimate of dimensionality, only outputs for constructing plots.

In Figure 5.4a, the scaling region is broad, because estimates were based
on 98,000 stationary points. Scaling is achieved by an embedding dimension
of 3, and the slope in the scaling region agrees closely with the true dimensionality. As the computation of D2 scales as the square of series length, it
can be quite time consuming (e.g., the estimate for the 98,000 length series
took 3 h on a 2.6 GHz AMD Athlon XP computer). In Figure 5.4b, the scaling
region is much narrower, as the estimates are based on only 8000 stationary
points, and was not achieved until an embedding dimension of 5. However,

it is still clearly identifiable around a distance of 2 and only slightly overestimates the true dimensionality.

Nonlinear Methods Based on Sensitive Dependence

An alternative approach to measures based on fractal topology comes from another defining characteristic of chaos: sensitive dependence on initial conditions. In chaotic systems, initially similar points soon move far apart as trajectories diverge exponentially. Divergence occurs at a rate described by the Lyapunov spectrum. Globally stable chaotic systems (such as the Lorenz) are bounded overall, as the sum of their Lyapunov exponents is negative, but trajectories diverge in one or more dimensions at rates determined by the set of positive exponents. Methods are available to estimate the full Lyapunov spectrum (e.g., the Netle software, http://www.sfu.ca/~rgencay/lyap.html). However, Heath (2000) notes that these methods are not noise tolerant, and favors a more robust approach based on estimating only the maximum Lyapunov exponent, with a positive maximum exponent indicating chaos. We found that even this method fails in noisy series, so Lyapunov-based methods are not further examined here.

The false nearest neighbor method (Kennel et al., 1992), which is used to determine a minimum embedding dimension, is also based on the idea of examining the divergence of neighboring points. For a range of embedding dimensions, the distance in the future (usually just one step) between each data point and its nearest neighbor is compared. If the ratio of the distance after one step to the original distance exceeds a criterion, the point is declared a false neighbor. The process is repeated for a range of embedding dimensions and the percentage of false neighbors plotted. For example, the TISEAN command `false_nearest lorenz2.dat -M10 -d17 -o -t100 -f5` calculates the number of nearest neighbors with a ratio greater than 5 for 1–10 dimensions. The result clearly indicates $m = 3$ is sufficient, a result that holds for a wide range of criterion ratios. A Theiler window of 100 was specified, as the false nearest neighbor technique makes similar assumptions to the dimensionality estimation techniques. A minimum embedding dimension can also be estimated from the D2 plots by the minimum number of dimensions necessary to attain a scaling region.

Sugihara and May (1990) suggested that predictability could be used to measure nonlinear dynamics in a time series. Although deterministic chaotic series are predictable in the short term, small initial differences due to measurement error are rapidly magnified by sensitive dependence, so that the series cannot be predicted in the long term. White noise, in contrast, is not predictable on any time scale. Sugihara and May suggested that decreasing predictability with time provides evidence for chaos. However, as with other NDA measures, stochastic temporal dependencies (e.g., colored spectra produced by linear and nonlinear AR models) can confound results as they also cause a gradual decrease in predictability.

To achieve prediction without knowledge of the underlying determining equations, Sugihara and May (1990) used a principle suggested by Lorenz (1969): similar present states in a deterministic system should evolve into similar future states. To compensate for the effects of noise, the evolution of a point in an embedding is predicted by the aggregate evolution of a set of neighboring points, rather than just the evolution of the single most similar point (Lorenz's "analogue" point). Aggregation can range from simple averaging to more sophisticated combinations, and the size of the neighborhood varied to match the level of noise. TISEAN provides the predict and zeroth programs, which use local averages, and onestep and nstep, which use local linear prediction. Barahona and Poon (1996) suggested a global polynomial prediction method for detecting chaos in short, noisy, time series. However, Schreiber and Schmitz (1997) report that when noise is high, local average techniques are better for this purpose.

Surrogate Testing, Time Asymmetry, and Geometric Filtering

Estimators for most quantitative indices assume nonlinear dynamics are present, and can produce misleading results when they are not. Hence, it is important to test for, rather than assume, nonlinear dynamics in a time series. However, most measures of nonlinear dynamics can vary with the static distribution of the data and are also sensitive to linear dynamics. A bootstrap (e.g., Davison & Hinkley, 1997) solution to these problems is provided by surrogate series testing (Theiler et al., 1992). Surrogate series are random variations on an experimental series that are constrained to preserve structure assumed under a null hypothesis. Permutations of the order of the experimental series, for example, produce surrogates that realize the null hypothesis of no temporal structure and the same static distribution of values as the experimental series.

A linear ARMA process provides a more interesting null hypothesis. For example, Theiler et al.'s (1992) amplitude adjusted Fourier transform (AAFT) surrogates have as a null hypothesis a Gaussian ARMA process observed through a static, possibly nonlinear, but invertible measurement function. Allowing for a monotonic nonlinear measurement function provides considerable flexibility to accommodate whatever static distribution characterizes the measured series. Schreiber and Schmitz (2000) developed an iteratively refined AAFT surrogate, computed by the TISEAN program surrogates, which is more accurate in matching both the spectrum and distribution of a finite data set.

A test is constructed by comparing the surrogate series to the experimental series on a measure sensitive to nonlinearity. Significance is determined by the percentile attained by the nonlinear measure for the experimental series in the distribution of estimates obtained from the surrogates. In a survey of measures of nonlinearity, Schreiber and Schmitz (1997) found that nonlinear prediction error, obtained using a local averaging technique such as the TISEAN predict algorithm, was the most consistent in discriminating a range of noisy chaotic times series (see also Tong, 1990, for a review of tests of linearity). For nonlinear prediction error, the test is one tailed, with prediction error for the experimental series at the pth percentile allowing rejection of the null hypothesis with confidence $100 - p$.

A two-tailed test can be made using a cubic time-reversal index, which measures the asymmetry of the distribution of differences between series values at a fixed delay using a statistic related to the third cumulant around zero (e.g., the TISEAN `timerev` command calculates the sum of cubes of the differences divided by their sum of squares). This statistic is based on the theory of polyspectra (Rao & Gabr, 1984), which generalizes Gaussian linear ARMA and Fourier models to include moments of order greater than two. Stationary linear Gaussian processes have a symmetric difference distribution and zero cumulants of order higher than two, so nonlinearity or non-stationarity is indicated by either a significantly positive or negative estimate relative to the surrogate distribution.

Schreiber and Schmitz (1997) found the cubic time-reversal index to have low power with noisy Lorenz x series data, although it performed well for other chaotic series. The Lorenz series shown in Figure 5.1a yielded cubic time-reversal indices of 0.396 and 0.381 for delays 1 and 12, respectively. At a delay of one, the difference distribution contains three very large positive values, which dominate the time-reversal index because of the cubing operation (the index is 0.007 with these values removed). The presence of such large outliers exacerbates the low efficiency of higher-order cumulants for estimating distribution properties (cf. Ratcliff, 1979), perhaps explaining the low power with the Lorenz x series found by Schreiber and Schmitz.

Diks, van Houwelingen, Takens, and DeGoede (1995) discuss indices based on the reversibility of linear time series and suggest a potentially more efficient kernel-based symmetry measure. Stam, Pijn, and Pritchard (1998) suggested an alternative approach that combines prediction and time asymmetry without the need for surrogates. Prediction is compared for the original ($x(t)$, $t = 1, \ldots, N$) and a time reversed version of the original ($y(t) = x(N - t + 1)$), with any difference being attributable to nonlinearity. They found that their time asymmetry technique produced strong discrimination with the chaotic Rossler equations, even with 50% additive Gaussian noise. No matter how efficient the statistic, it is important to note that asymmetry is only a sufficient indicator of nonlinearity, it is a not a necessary condition, so some nonlinear dynamics may display little asymmetry.

Although surrogate testing is extremely useful, results should be interpreted with caution. Chaos does imply nonlinearity as measured by appropriate surrogate tests, but nonlinearity does not necessarily imply chaos. Nonstationary linear processes and nonlinear stochastic processes may also cause a positive finding, as may non-monotonic measurement functions. For example, Schreiber and Schmitz (2000) showed that a stationary second-order linear autoregressive process observed through a measurement function made noninvertible by taking successive differences, a common procedure in time series analyses, produces a positive surrogate test result. Noninvertible measurement functions may arise in other cases, such as the measurement of signal power (i.e., squared amplitude).

Spurious findings of nonlinearity may also occur if a time series is unevenly sampled or contains missing values (Schmitz & Schreiber, 1999). The surrogate generation method of Schreiber (1998) may be applied to these cases. This method can create surrogate series for any null hypothesis that can be expressed as a cost function (computed by the TISEAN program `randomize`). Although this method is quite general, it uses a simulated annealing algorithm, and so can be very computationally expensive. Stam, Pijn, and Pritchard (1998) also show that AAFT surrogate testing may produce false positive results where strong periodicities are present. They demonstrate that this problem can be fixed if the series is truncated to a length that is an integer multiple of the dominant period. However, this also usually requires that a more computationally expensive discrete Fourier transform be used for surrogate generation, rather than the fast Fourier transform usually employed by AAFT algorithms. Schreiber's (1998) cost function–based surrogate algorithm can also address problems caused by periodic components, again at an increase in computational cost.

Even with the available prediction and surrogate-testing methods, noise remains extremely problematic for the characterization of nonlinear dynamics. Derivative plots are particularly useful for characterizing dynamics, but noise makes them difficult to interpret, so some sort of noise filtering is required. Traditional approaches to noise removal, such as frequency domain filters, cannot be applied to chaotic signals because they are, like noise, broadband. Instead, geometric filters (Grassberger et al., 1993), which smooth local neighborhoods of the attractor, can be used. In each neighborhood, the signal is reconstructed using the first few principal components of a linearization, usually achieved by singular value decomposition to ensure numerical stability. Because chaotic signals mainly project onto the dominant components, whereas noise is distributed evenly across all components, the reconstructed signal has an improved signal-to-noise ratio. Corrections are usually also made for the effects of curvature and sensitive dependence (Schreiber, 1999) and the filter may be iterated.

Like surrogate testing, local projective geometric filters should be used cautiously and, in particular, the filter should not be iterated too many times (Mees & Judd, 1993). When deterministic dynamic structure is present, it may be distorted by over filtering. For example, Mees and Judd found that a circular attractor contracted substantially after 20 iterations. Over filtering can create apparent structure from noise because of the finite length of measured time series. Although in infinite samples noise is evenly distributed in phase space, inhomogeneities can occur in finite samples that are expanded by the filtering process. Hegger et al. (1999) provide an example where pure Gaussian noise produced a structured delay plot after 10 iterations. They advise that little more than three iterations should be required if chaos is present. As was the case for prediction, larger noise levels can be dealt with by using larger neighborhoods, and by retaining fewer principal components.

Nonlinear Dynamical Analysis of Noisy Data

In this section, we apply techniques described in the last section to time series corrupted by noise. Three noisy versions of the Lorenz series in Figure 5.1a were created by adding independent zero mean Gaussian deviates with a standard deviation (SD) of 6.2 (i.e., the SD of the series in Figure 5.1a from $t = 2,000$ to 10,000), 8.7 and 12.4, creating stationary series with signal-to-noise ratios of 1:1 (50% noise, data file n1.dat), 1:2 (66.6% noise, data file n2.dat), and 1:4 (80% noise, data file n4.dat). Linear analyses of the noisy series produced results consistent with the addition of white noise. The spectrum flattened, and the number of significant partial autocorrelation coefficients, and hence the order of estimated AR models, reduced as noise increased. No systematic patterning was evident in residuals from the AR models of the noisy series. Many of the NDA techniques discussed so far were not useful when applied to the noisy series; derivative plots were unstructured at all delays, no clear first minimum could be found in the mutual information, and false nearest neighbor plots gave no clear indication of a minimum embedding dimension.

Testing for Stationarity

The first task, and perhaps the most difficult and crucial task in the dynamical analysis of behavioral data, is establishing stationarity. Stationarity is an important requirement for NDA, and when it does not hold, or cannot be shown to hold, one must be skeptical about the validity of results. Unfortunately, stationarity checks are inherently vulnerable to noise because they are based on the comparison of sub-sequences of a series, and so sample size is reduced. As illustrated by Figure 5.1a, weak stationarity checks (on the mean, variance, and auto-covariance) are not very discriminating of nonstationarity in the Lorenz series. What is required is a method of checking whether the full joint probability distribution, rather than just its lower order moments, is constant.

Schreiber's (1997) cross-prediction method of detecting nonstationarity is an attractive candidate, because neighborhood aggregation can make prediction noise tolerant. This method determines the degree to which information about one segment of a time series allows prediction of another segment of a time series. If the time series is stationary and each segment sufficiently long, equal cross prediction should be possible, because each segment will be governed by the same joint probability distribution.

Figure 5.5a illustrates the mean squared error for one-step prediction for the first 10,000 values of the noiseless Lorenz series broken into 10 consecutive segments of length 1000. The algorithm used locally constant prediction in delay-17 five-dimensional coordinates, averaged over a minimum of 30 neighbors, to generate one-step-ahead predictions. Larger embedding

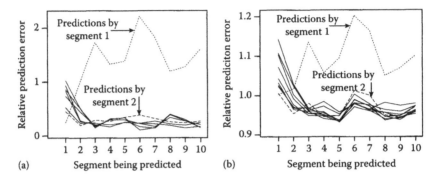

FIGURE 5.5
One-step cross prediction errors obtained using 10 sequential 1000 element segments (separate lines) to obtain predictions for all segments (abscissa) for (a) the series in Figure 5.1a using a five-dimensional delay-17 embedding and a minimum neighborhood size of 30 (nstat_z lorenz1.dat -#10 -m5 -d17 -k30 -o) and (b) the series in Figure 5.1a plus Gaussian noise with double the SD of the Lorenz series using a 10-dimensional delay-one embedding and a minimum neighborhood size of 30 (nstat_z lorenz1.dat -#10 -m10 -d1 -k40 -o).

dimensions and neighborhood sizes produced a similar pattern of results, as did smaller delays. The cross prediction method correctly detects the initial nonstationarity as the series settles into the attractor and stationary behavior thereafter. The result is most striking for the first 1000 values with the next 1000 showing only a slight effect. These results support the earlier decision to remove the first 2000 values from further analyses as a cautious response, although in practice, the evidence might only support removing the first 1000.

Figure 5.5b demonstrates that the cross prediction method still works when the series is corrupted with up to 80% noise. Prediction was quite poor overall because of high noise levels but the first segment is clearly singled out and even the second segment differs slightly. The parameters used were those that produced the best cross-prediction (i.e., prediction of other segments). Although this required some trial and error, the search was tractable as only two parameters were important: the embedding dimension and the neighborhood size. For the 80% noise series, averages over a minimum neighborhood size of 40 in 10-dimensional coordinates were best. In the 50% noise case, the predict default neighborhood size of 30 was best at either 9 or 10 dimensions. For uniformity, 10 dimensions will be used for all noise levels. In all cases, a delay of one was clearly superior to all other delays, indicating that longer delays are not required when noise is high. Further analyses of the noisy Lorenz series were carried out using the 8000 values from $t = 2,001$ to 10,000.

These results demonstrate the utility of Schreiber's (1997) cross-prediction method with high noise levels and relatively short subsequences of 1000. In principle, more sophisticated prediction methods, or methods that are more appropriate given knowledge of the underlying dynamics, can be incorporated as they become available. The degree of averaging can be adjusted to suit the level of noise, and Schreiber claims that this approach can work with subsequences as short as 300–400. He presents cross-prediction results in a similar way to recurrence plots, using grey scale to indicate the level of error, but in the present case the graphs in Figure 5.5 were clearer.

An advantage of prediction methods is that parameters such as delay and embedding dimension can be chosen based on what gives the best predictions. Selection of parameters via minimizing cross-prediction error is particularly appropriate as it avoids over-fitting (Browne, 2000), which can be an important problem because of the flexibility of local prediction techniques. Typically, it is wise to use a large embedding dimension with prediction methods so that all available information can be utilized. However, there is an increased computational cost associated with the higher-dimensional representation. For the Lorenz data, the 10-dimensional delay-one coordinates that were optimal for cross-prediction were usually appropriate for the following analyses. Some adjustment was necessary, but in general, this involved changing only a single parameter, making the process of finding appropriate settings tractable.

Testing for Nonlinearity

Surrogate tests aim to determine whether a series contains nonlinear structure, and so is a viable candidate for NDA. For each stationary noisy series, 99 surrogate series were created using Schreiber and Schmitz's (2000) iteratively refined AAFT algorithm (e.g., for 50% noise, surrogates n1.dat -n99 -o, which produces surrogates with names n1.dat_surr_#, where # = 001...099). As discussed by Theiler, Linsay, and Rubin (1993), a mismatch between the beginning and end of the series poses a problem for surrogate-generation schemes that match linear autocorrelations in the original and surrogate series using Fourier methods. For finite series, Fourier amplitudes correspond exactly to the autocorrelation function only if the series is one period of a repeating sequence. Where this does not hold, tests of nonlinearity may produce false positives. Endpoint mismatch can be corrected by choosing a subsequence of the original series that matches end points as closely as possible (Ehlers, Havstad, Prichard, & Theiler, 1998).

TISEAN provides the program end to end, which chooses a subsequence with a length that is a multiple of 2, 3, or 5 (as required by a fast Fourier transform), to minimize both mismatch and phase slippage between the beginning and the end of a series. For the Lorenz data, only the 80% noise series required truncation, as reported by the output of endtoend n4.dat, by removing the first 193 points, resulting in a series of length 7776 (n4s.dat), 2.8% shorter than the original series. An alternative approach, useful where a matching subsequence cannot be found or the truncation required is too great, is to avoid Fourier-based methods and directly match the second-order autocorrelation structure using Schreiber's (1998) method.

Tests used the generally most discriminating measures of nonlinearity examined by Schreiber and Schmitz (1997): one-step prediction and cubic time asymmetry. The multiple input file–processing capability of the TISEAN `predict` and `timerev` programs was used to process the original and surrogate files in one pass.[3] Cubic time asymmetry was not discriminative at any noise level and for differences at any delay, performing even worse than was reported by Schreiber and Schmitz for the Lorenz series. One possible reason is that Schreiber and Schmitz used noise created by phase randomization so that the noisy series had the same spectrum as the original series. The Gaussian noise used here has a different spectrum and produces some large deviates that result in increased sampling variability for the cubic time-reversal index.

Locally constant prediction, in contrast, performed well, detecting the nonlinearity in the original sequence at a 97% level of confidence even for the 80% noise series, as shown in Figure 5.6. Figure 5.6 shows root mean

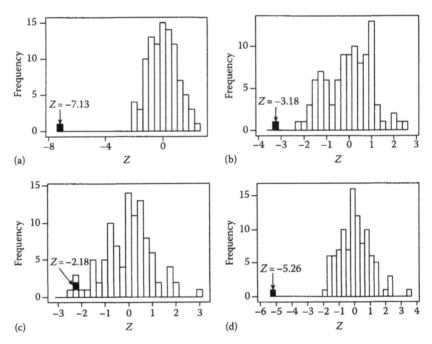

FIGURE 5.6

Histograms of standardized root mean squared one-step prediction error (Z) in a 10-dimensional delay-one embedding for original (filled bar) and surrogate (unfilled bars) series (a) 50% Noise and 1.6 SD neighborhood radius (`predict n1.dat n4.dat_surr_0?? -m10 -d1 -v1.6 -o`), (b) 66.6% Noise and a 1.6 SD neighborhood radius (`predict n2.dat n4.dat_surr_0?? -m10 -d1 -v1.6 -o`), and (c) 80% Noise and a 2.6 SD neighborhood radius (`predict n4.dat n4.dat_surr_0?? -m10 -d1 -v2.6 -o`). (d) 14-Step prediction error for the 80% noise series with a 1.6 SD neighborhood radius (`predict n4.dat n4.dat_surr_0?? -m10 -d1 -v1.6 -o -s14`). The predict program outputs RMSE; standardization used the mean and SD of the surrogate data.

square prediction error (RMSE) standardized using the mean and SD of the surrogate distribution. Z scores for the original series were -7.13, -3.18, and -2.18 at 50%, 66.6%, and 80% noise series. Some experimentation was required with the extent of neighborhood averaging used by the TISEAN predict program. This program allows the radius of the neighborhood to be specified either in absolute units or as a fraction of the series SD. The later method (specified by the $-v$ option) was used, as it automatically scales for the differing SDs of the noisy series.

To the nearest 0.1, a neighborhood size of 1.6 SD produced the smallest RMSE for the original series, and for the 50% (RMSE = 6.879) and 66.6% (RMSE = 9.478) noise series. A delay-one 10-dimensional embedding was also optimal, in agreement with the cross-prediction findings, despite the larger neighborhoods used here. The same setting also produced the best prediction for the 80% noise series (RMSE = 13.058) but discrimination was poor, with the original series being placed only at the 18th percentile. Larger neighborhoods produced progressively better discrimination for the 80% noise series, with a radius of 2.6 producing the best discrimination, at the third percentile as shown in Figure 5.6c, but slightly poorer prediction (RMSE = 13.316). For the other two series a broad range of neighborhood sizes resulted in prediction error for the original series at the first percentile.

Stam et al.'s (1998) method, comparing nonlinear prediction for the original and time-reversed series, was also applied to the noisy Lorenz data. In all cases the time-reversed data had a larger prediction error: 6.889, 9.488, and 13.038 for 50%, 66.6%, and 80% noise with 1.6 SD neighborhoods, and 13.322 for 80% noise with a 2.6 SD neighborhood. However, the differences were not nearly as large as reported by Stam et al. for the Rossler equations, or the average differences for the surrogate prediction tests. This method also suffers from the drawback that it does not provide a confidence level.

In an attempt to improve on the locally constant prediction used by the predict program, surrogate tests were also constructed using local linear model prediction error as calculated by the TISEAN onestep program. Most parameters were the same as for predict, but the best discrimination was produced with small minimum neighborhood around 35, where the original series is at the fourth percentile and $Z = -2.03$. As with locally constant prediction, error decreased with neighborhood size. The onestep program outputs prediction error divided by the series SD, which decreased from 0.984 for neighborhoods of 35–0.94 for neighborhoods of 400. As these results were barely on par with locally constant prediction, and their computation was at least an order of magnitude slower, predict seems preferable.

Although not examined by Schreiber and Schmitz (1997), prediction error for more than one step was calculated to determine if better discrimination could be achieved. The results, shown in Figure 5.7, were costly to compute, requiring an overnight run for each series. They show that discrimination between the surrogates and the original series was substantially improved for longer prediction lags. Figure 5.7b shows that, although a neighborhood of 2.6 SD produced the best discrimination for lag one predictions, much better discrimination was produced at longer lags by a neighborhood of 1.6 SD, which is also optimal for minimizing overall prediction error. Figure 5.6d shows the error distribution for lag 14 predictions, which produced the best discrimination across all lags, demonstrating the potential improvement in discrimination afforded by an examination of multiple prediction lags.

Noise Filtering

The analyses performed so far are important for the application of NDA to data with unknown properties because they validate two fundamental

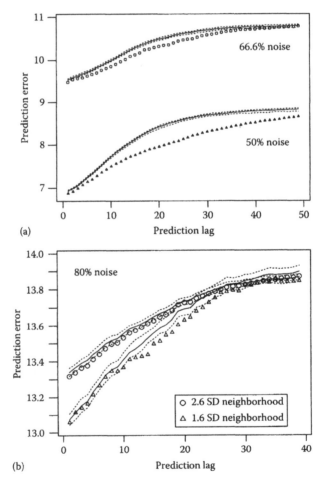

FIGURE 5.7
Root mean square locally constant prediction error as a function of prediction lag for (a) 50% and 66.6% noise series with a 1.6 SD neighborhood radius, (b) 80% noise series with a 1.6 SD radius and a 2.6 SD radius.

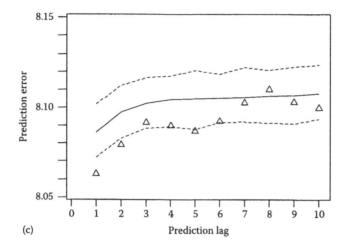

(c)

FIGURE 5.7 (continued)
(c) coarsely sampled 80% noise series, with a 1.6 SD radius. The symbols are prediction errors for the original series. The solid line is the mean prediction error for the 99 surrogate series. The dotted lines are a 95% confidence interval for surrogate prediction, estimated by the 3rd and 97th order statistics of the surrogate distribution.

assumptions of most NDA techniques: stationarity and the presence of non-linear dynamics. Passing both tests helps to ensure that the data are sufficient in terms of both quantity and quality for NDA and also validates the application of geometric filters to remove noise. The high noise levels in these examples defeated the direct application of quantitative indicators, such as dimensionality or Lyapunov coefficients, and qualitative indicators, such as derivative plots. However, with appropriate filtering, much of the structure in the original time series can be recovered (e.g., Figure 5.1b), derivative plots are structured (e.g., Figure 5.2c and d), and even D2 estimation reveals something close to a scaling region (e.g., Figure 5.4c), although it underestimates the Lorenz attractor's dimensionality. Before examining these results, however, it is useful to examine the behavior of the filter itself.

Figure 5.8a shows the increase in the correlation between the original and filtered series as a function of the iteration of Grassberger et al.'s (1993) algorithm (e.g., `ghkss nl.dat -m10 -dl -q2 -k400 -i9 -o`, the `-i9` and `-o` parameters create files containing the filtered series for each iteration $n = $ 1–9, with extensions `.opt.n`). The algorithm performs orthogonal projections onto a q-dimensional manifold in a neighborhood of minimum size k for each data point. Correlations are shown for three values of q, 1, 2, and 3, along with two minimum neighborhood sizes, 400 and 800, corresponding to neighborhoods containing 5% and 10% of the series, respectively. The delay-one 10-dimensional coordinates adopted in previous analyses also produced the highest correlations, but larger neighborhoods were used for filtering than for cross-prediction for two reasons. First, larger neighborhoods can be used because the entire series of length 8000 is available, whereas only 1000 observations were available for cross-prediction. Second, larger neighborhoods produced better correlations between the filtered and original series, although the improvement with increasing size reduced for larger neighborhoods and less noise, so a difference between size 400 and 800 is only evident for the 80% noise series in Figure 5.8a.

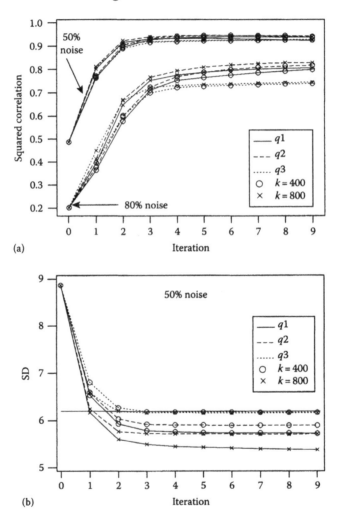

FIGURE 5.8

(a) Squared correlation between the original Lorenz series from $t = 2,001$ to 10,000 and filtered series, and (b) the SD of the filtered series, as a function of filter iterations retaining one to three singular values ($q1$–$q3$). For the 50% noise series, minimum neighborhood size was either 5% ($k400$) or 10% ($k800$) of the series length, whereas for the 80% noise series, it was 10% ($k800$) or 20% ($k1600$) of the series length. The SD of the original Lorenz series (6.2) is indicated by a solid flat line in (b) and (c).

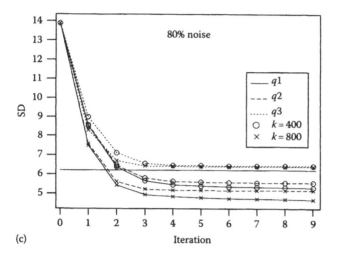

FIGURE 5.8 (continued)

Generally, increasing filter iterations and neighborhood size, and lowering the dimensionality of the projections, produced smoother derivative plots (e.g., Figure 5.2c and d). Schreiber (1999) advises that even locally constant projections (–q0) may be used, but here derivative plots revealed some clear artifacts induced by filtering for locally constant and one-dimensional projections (–q1). The best results visually were produced by two-dimensional projections (–q2), which may be because this closely matches the dimensionality of the Lorenz series (2.05). However, the two-dimensional projection could cause some shrinkage of the underlying dynamics (cf. Mees & Judd, 1993), as revealed by plots of the filtered series SD in Figure 5.8b and c, and comparison of Figure 5.1a and b. Series filtered with three-dimensional projections, in contrast, almost exactly matched the SD of the underlying dynamics at higher iterations, but did not correlate as highly with the original series as lower-dimensional projections. These findings suggest that the asymptotic SDs of the filtered series for a range of projection dimensions may be used to determine what proportion of variance in a noisy series is due to nonlinear dynamics, and perhaps to provide a rough estimate of their dimensionality.

In applications, the information in Figure 5.8a is not available, but some guidance in parameter setting can be gained by examining SDs and derivative plots for the filtered series. Casdagli (1991) and Mpitsos (1994) have argued that the best evidence for low-dimensional chaos is a complex but structured delay plot. As shown in Figure 5.2c, after six iterations using two-dimensional projections, the derivative plot of the filtered 50% noise series is quite structured, and contains smooth trajectories. The corresponding derivative plot using three-dimensional projections (Figure 5.2d) is less clear and the trajectories less smooth, but important aspects of the dynamics are still evident, such as quicker changes near the center of each state and slower changes between states and at extremes. Derivative plots for the higher noise series were correspondingly less structured, but appropriate filter setting still revealed the qualitative aspects of the dynamics. It is wise to always examine filtered series using derivative or delay plots, as some settings can produce artifacts, such as under filtering of the beginning and end of the series.

Discussion

The approach taken in this paper should be generally useful, but some of the specific conclusions are no doubt limited to the Lorenz dynamics and additive Gaussian noise. In practice, it is likely that an iterative approach will be required to refine theory and measurement techniques. As well as different dynamics, different noise models may be required, such as colored noise or noise in dynamical parameters. Parameter noise can be particularly problematic as nonlinear systems can be extremely sensitive to some changes in parameters but virtually invariant with other parameter changes. Casdagli (1997) extended the idea of embedding to parameterized families of dynamical systems, allowing nonstationarity due to slowly varying parameters, and developed methods based on recurrence plots for reconstructing slowly varying parameter changes. Hegger, Kantz, Matassini, and Schreiber (2000) advocate "over-embedding" of such systems with p varying parameters in an $m > 2(d + p)$-dimensional space.

An interesting possibility suggested by the results on stationarity testing is that separate blocks of trials in behavioral experiments or segments of physiological recordings can be concatenated in order to obtain a series of sufficient length for NDA. Often, avoidance of nonstationarity due to fatigue requires that measurement be broken up into blocks with rest periods between them. Longstaff and Heath (1999) report cases where separate segments of stationary chaotic time series can be concatenated without confounding NDA. In practice, it will not be known if each segment is a sample from the same stationary series, so a method of checking is needed. Conversely, it is often useful to detect change points, as they may signal underlying changes or the need for intervention. Cross-prediction stationarity checks can help to perform both functions. Although concatenation is attractive, and possibly necessary if NDA is to have impact in many areas of life sciences, a number of questions spring to mind. Should embedding points that span segments be eliminated? Should segments be trimmed before concatenation to enforce constraints such as periodic continuity? To what degree does concatenation distort low-frequency structure as its period approaches the length of subsequences? Hopefully future research will help to answer these questions.

The results of testing for nonlinearity were encouraging, although quite a deal of parameter tuning was required to produce sufficiently powerful tests. One relatively unexplored possibility opened up by these results is that higher power may be available at longer prediction lags. Although the results given in Figure 5.6d are promising, a post hoc approach like choosing the lag with the greatest discrimination will inflate Type 1 error. Should the use of longer prediction lags prove generally useful, a test that integrates over lags will be required, along with simultaneous confidence intervals, rather than the point-wise intervals shown in Figure 5.7.

For all prediction methods applied to noisy series, a delay of one was found to produce the best results, contrasting sharply with the noiseless case where longer delays were preferred. One possible implication is that dynamics will be difficult to identify in noisy series unless they are finely sampled, as in Figure 5.1. Coarse sampling may occur in practice if the dynamics evolve on a much faster time scale than the measured behavior. A preliminary exploration of this issue, using a series of length 10,000 created by sampling every 100th value from the first million values of the Lorenz *x* series, found that the prediction-based surrogate test was still able to detect nonlinearity even with 80% noise. As shown in Figure 5.7c, significant indications of nonlinear dynamics were obtained at only the first two lags, contrasting with the results for a finer sampling. These results are encouraging for applications where the coarseness of sampling is not within experimental control.

Although geometric filtering was very effective in removing noise and recovering the underlying nonlinear dynamics, clearly there is a good deal of art in producing "clean" series and derivative plots, with an attendant risk of inflated Type 1 error. The neighborhood size and projection dimension of geometrical filters act in much the same way as kernel bandwidth and regression order do to control smoothness in nonparametric regression functions (e.g., Wand & Jones, 1995). The two techniques are conceptually related, except that nonparametric regression operates in the time domain for a time series, whereas geometric filtering acts in the embedding space. The job of the analyst will be made much easier if automatic neighborhood size selection methods can be developed, in much the same way that automatic bandwidth selection methods are now available for nonparametric regression. Fortunately, both global and local approaches to geometric filtering are very active areas of research (e.g., Roweis & Lawrence, 2000; Tenenbaum, de Silva, & Langford, 2000).

The filtering results reported here are encouraging because they enable graphical identification of qualitative features of the dynamics underlying a noisy series. However, filtering can both fail to remove sufficient noise and introduce systematic distortions that affect estimates of quantitative dynamical indices, such as D2. Figure 5.4c shows the best local slope plot that was found for the filtered 50% noise Lorenz series. Something approaching a scaling region occurred around a distance of 5–6, but the slope estimates did not converge and overestimate the Lorenz attractor's dimension. Another apparent scaling region occurred around a distance of 8–9 but it is very narrow and substantially underestimates the dimensionality of the Lorenz attractor. A possible reason for these problems is the use of a two-dimensional projection, which results in shrinkage of the underlying dynamics. Figure 5.4d shows the results when filtering used three-dimensional projections (i.e., the series shown in Figure 5.2d). No proper scaling region is revealed, although the region around 10 is close and underestimates dimensionality only slightly.

If the present results are any guide, absolute estimates of D2 from filtered series are not likely to be very accurate, although they may suffice for relative comparisons. Caution should be exercised in the estimation of quantitative indices from filtered series, because, as with any smoothing technique, filtering can introduce systematic bias. Either bias corrections must be developed, or alternative noise tolerant indices used that can be applied to unfiltered series. One graphical technique associated with D2 estimation, the space-time plot, was informative without filtering (e.g., Figure 5.3b). The same Theiler window as for the original series was indicated, and evidence consistent with deterministic dynamics provided: constant rather than increasing contours for longer time separations. Note, however, that increasing contours do not necessarily rule out deterministic dynamics, as they may be caused by a combination of deterministic and stochastic dynamics. In general, none of the techniques reviewed here can definitively differentiate nonlinear stochastic and chaotic processes. Cencini, Falcioni, Kantz, Olbrich, and Vulpiani (2000) discuss this issue and a possible analytic approach.

In summary, the results presented here suggest that methods based on delay-space averaging and prediction provide the most powerful means to address problems caused by measurement noise. Although it will probably never be possible to *prove* chaos in a noisy measured time series, robust and powerful algorithms are now available to test for and quantify nonlinear structures rather than simply assuming it. Prediction-based methods are attractive not only for their noise tolerance, but also if the aim of a dynamic theory is to explain all systematic temporal variation, as they provide direct measures of the available structure that are relatively theory free, in the sense of not requiring knowledge of the form or parameters of determining equations. Geometrical filtering methods, which are also based on delay-space averaging, appear to provide the best means of graphically characterizing nonlinear dynamics in noisy time series.

Acknowledgments

Initial work on this project was completed by the first author while writing a review of Heath (2000) (Heathcote, 2002) and supervising Dr. Alice Kelly's PhD project. We would like to thank the School of Behavioural Sciences,

University of Newcastle, and an Australian Research Council Large Grant for funding support and the TISEAN project (Hegger et al., 1999) for making their software freely available.

Notes

1. Stationarity up to order k dictates invariance only of joint moments up to order k (Rao & Gabr, 1984). For a linear Gaussian process, $k = 2$ stationarity implies "strong" stationarity, invariance of the full joint probability distribution, as moments greater than two are zero.

2. Recurrence quantification analysis (RQA, Trulla, Giuliani, Zbilut, & Webber, 1996; Webber & Zbilut, 1994) provides a number of indices to quantify recurrence plots. Thomasson, Hoeppner, Webber, and Zbilut (2001) claimed that RQA "does not require assumptions about stationarity, length or noise" (p. 94). However, the RQA indices are formed by aggregating local measures across a time series and aggregation does require stationarity if a rigorous meaning is to be attached to the aggregate values. Both series length and noise are important in obtaining precise estimates, and an embedding may not even be possible due to noise, so noise and length are relevant to RQA. It is also difficult to decide what distance should be chosen to define recurrence, a critical parameter that has a very strong effect on the values of the RQA indices.

3. `timerev n1.dat n1.dat_surr_0?? -d1 > nltr.dat` and `predict n1.dat n1.dat_surr_0?? -m10 -d1-v1.6 -o > n1pred.dat`. The "??" characters are DOS single character wildcards, -d specifies the delay, and the redirection of standard output (>) stores the measure of nonlinearity in a file. For the `predict` command, the –m option specifies the embedding dimension, and –o writes predicted values to a separate output file with _pred appended to the name of each input file.

References

Barahona, M. & Poon, C.-S. (1996). Detection of nonlinear dynamics in short, noisy time series. *Nature, 381,* 215–217.

Box, G. E. P. & Jenkins, G. M. (1976). *Time series analysis: Forecasting and control.* San Francisco: Holden Day.

Brown, S. & Heathcote, A. (2003). Averaging learning curves across and within participants. *Behaviour Research Methods, Instruments & Computers, 35,* 11–21.

Browne, M. W. (2000). Cross-validation methods. *Journal of Mathematical Psychology, 44,* 108–132.

Cryer, J. D. (1986). *Time series analysis.* Boston: PWS-Kent.

Casdagli, M. (1991). Chaos and deterministic versus stochastic non-linear modelling. *Journal of the Royal Statistical Society, Series B, 54,* 303–328.

Casdagli, M. (1997). Recurrence plots revisited. *Physica D, 108*, 12–44.

Casdagli, M., Eubank, S., Farmer, J. D., & Gibson, J. (1991). State space reconstruction in the presence of noise. *Physica D, 51*, 52–98.

Cencini, M., Falcioni, M., Kantz, M., Olbrich, E., & Vulpiani, A. (2000). Chaos or noise: Difficulties of a distinction. *Physics Review E, 62*, 427–437.

Cleveland, W. S. (1979). Robust locally weighted regression and smoothing scatterplots. *Journal of the American Statistical Association, 74*, 829–836.

Davison, A. C. & Hinkley, D. V. (1997). *Bootstrap methods and their application.* Cambridge: Cambridge University Press.

Diks, C., van Houwelingen, J. C., Takens, F., & DeGoede, J. (1995). Reversibility as a criterion for discriminating time series. *Physics Letters A, 201*, 221–228.

Eckmann, J. P., Oliffson-Kamphorst, S., & Ruelle, D. (1987). Recurrence plots of dynamical systems. *Europhysics Letters, 4*, 973.

Ehlers, C. L., Havstad, J., Prichard, D, & Theiler, J. (1998). Low doses of ethanol reduce evidence for nonlinear structure in brain activity. *Journal of Neuroscience, 18*, 7474–7486.

Fraser, A. M. & Swinney, H. L. (1986). Independent coordinates for strange attractors from mutual information. *Physical Review A, 33*, 1134–1140.

Gilden, D. L. (1997). Fluctuations in the time required for elementary decisions. *Psychological Science, 8*, 296–301.

Grassberger, P. (1987). Evidence for climatic attractors: Grassberger replies. *Nature, 326*, 524.

Grassberger, P., Hegger, R., Kantz, H., Schaffrath, C., & Schreiber, T. (1993). On noise reduction methods for chaotic data. *Chaos, 3*, 127–141.

Grassberger, P. & Procaccia, I. (1983). Characterization of strange attractors. *Physical Review Letters, 50*, 189–208.

Gregson, R. A. M. (1988). *Nonlinear psychophysical dynamics.* Hillsdale, NJ: Lawrence Erlbaum Associates.

Gregson, R. A. M. (1992). *n-Dimensional nonlinear psychophysics.* Hillsdale, NJ: Lawrence Erlbaum Associates.

Gregson, R. A. M. (1995). *Cascades and fields in perceptual psychophysics.* Singapore: World Scientific.

Guastello, S. J. (1995). *Chaos, catastrophe, and human affairs: Applications of nonlinear dynamics to work, organizations, and social evolution.* Hillsdale, NJ: Lawrence Erlbaum Associates.

Heath, R. A. (2000). *Nonlinear dynamics: Techniques and applications in psychology.* Mahwah, NJ: Earlbaum.

Heathcote, A. (2002). Book review: An introduction to the art; nonlinear dynamics: Techniques and applications in psychology by R. A. Heath. *Journal of Mathematical Psychology, 46*, 609–628.

Heathcote, A. & Brown, S. (2004). Beyond curve fitting? Comment on Liu, Mayer-Kress and Newell (2003). *Journal of Motor Behavior, 36*, 225–232.

Heathcote, A., Brown, S., & Mewhort, D. J. K. (2000). Repealing the power law: The case for an exponential law of practice. *Psychonomic Bulletin and Review, 7*, 185–207.

Hegger, R., Kantz, H., Matassini, L., & Schreiber, T. (2000). Coping with nonstationarity by over-embedding. *Physical Review Letters, 84*, 4092–4095.

Hegger, R., Kantz, H., & Schreiber, T. (1999). Practical implementation of nonlinear time series methods: The TISEAN package. *Chaos, 9*, 413–435.

Kelly, A., Heathcote, A., Heath, R. A., & Longstaff, M. (2001). Response time dynamics: Evidence for linear and low-dimensional nonlinear structure in human choice sequences. *Quarterly Journal of Experimental Psychology, 54,* 805–840.

Kelso, J. A. S. (1995). *Dynamic patterns: The self-organization of brain and behaviour,* Cambridge, MA: MIT Press.

Kennel, M. B., Brown, R., & Abarbanel, H. D. I. (1992). Determining embedding dimension for phase-space reconstruction using a geometrical construction. *Physical Review A, 45,* 3403–3411.

Koebbe, M. & Meyer-Kress, G. (1991). Use of recurrence plots in the analysis of time-series data. In M. Casdagli & S. Eubank (Eds.) *Nonlinear modelling and forecasting* (pp. 361–376). Reading, MA: Addison-Wesley.

Longstaff, M. & Heath, R. A. (1999). A nonlinear analysis of the temporal characteristics of handwriting. *Human Movement Science, 18,* 485–524.

Lorenz, E. N. (1969). Atmospheric predictability as revealed by naturally occurring analogues. *Journal of Atmospheric Science, 26,* 636.

Luce, R. D. (1995). Four tensions concerning mathematical modelling in psychology. *Annual Review of Psychology, 46,* 1–26.

Mees, A. I. & Judd, K. (1993). Dangers of geometric filtering. *Physica D, 68,* 427–436.

Mpitsos, G. J. (1994). The chaos user's tool kit. *Integrative Physiological and Behavioural Science, 29,* 307–310.

Newell, K. M., Liu, Y.-T., & Mayer-Kress, G. (2001). Time scales in motor learning and development. *Psychological Review, 108,* 57–82.

Nusse, H. E. & Yorke, J. A. (1997). *Dynamics: Numerical explorations* (2nd ed.). New York: Springer-Verlag.

Osborne, A. R. & Provenzale, A. (1989). Finite correlation dimension for stochastic systems with power-law spectra. *Physica D, 35,* 357–381.

Provenzale, A., Smith, L. A., Vio, R., & Murante, G. (1992). Distinguishing between low-dimensional dynamics and randomness in measured time series. *Physica D, 58,* 31–49.

Rao, T. S. & Gabr, M. M. (1984). *An introduction to bispectral analysis and bilinear time series models.* New York, Springer-Verlag.

Ratcliff, R. (1979). Group reaction time distributions and an analysis of distribution statistics. *Psychological Bulletin, 86,* 446–461.

Roweis, S. T. & Lawrence, K. S. (2000). Nonlinear dimensionality reduction by locally linear embedding. *Science, 290,* 2323–2326.

Sauer, T., Yorke, J., & Casdagli, M. (1991). Embedology. *Journal of Statistical Physics, 65,* 579–616.

Schmitz, A. & Schreiber, T. (1999). Testing for nonlinearity in unevenly sampled time series. *Physical Review E, 59,* 4044–4047.

Schreiber, T. (1997). Detecting and analysing nonstationarity in a time series using nonlinear cross predictions. *Physical Review Letters, 78,* 843–846.

Schreiber, T. (1998). Constrained randomisation of time series data. *Physical Review Letters, 80,* 2105–2108.

Schreiber, T. (1999). Interdisciplinary application of nonlinear time series methods. *Physics Reports, 308,* 1–64.

Schreiber, T. & Schmitz, A. (1997). Discrimination power of measures for nonlinearity in a time series. *Physical Review E, 55,* 5443–5447.

Schreiber, T. & Schmitz, A. (2000). Surrogate time series. *Physica D, 142,* 346–382.

Stam, C. J., Pijn, J. P. M., & Pritchard, W. S. (1998). Reliable detection of nonlinearity in experimental time series with strong periodic components. *Physica D, 112,* 361–380.

Sugihara, G. & May, R. (1990). Nonlinear forecasting as a way of distinguishing chaos from measurement error in time series. *Nature, 344,* 734–741.

Takens, F. (1981). *Detecting strange attractors in turbulence: Lecture notes in mathematics, 898.* New York: Springer.

Tenenbaum, J. B., de Silva, V., & Langford, J. C. (2000). A global geometric framework for nonlinear dimensionality reduction. *Science, 290,* 2319–2323.

Theiler, J. (1990). Estimating fractal dimension. *Journal of the Optical Society of America A, 7,* 1055–1073.

Theiler, J. & Eubank, S. (1993). Don't bleach chaotic data. *Chaos, 3,* 335–341.

Theiler, J., Eubank, S., Longtin, A., Galdrikian, B., & Farmer, J. D. (1992). Testing for nonlinearity in time series: The method of surrogate data. *Physica D, 58,* 77–94.

Theiler, J., Linsay, P. S., & Rubin, D. M. (1993). Detecting nonlinearity in data with long coherence times. In A. S. Weigend and N. A. Gershenfeld (Eds.), *Time series prediction: Forecasting the future and understanding the past.* Reading, MA: Addison-Wesley.

Theiler, J. & Rapp, P. E. (1996). Re-examination of the evidence for low-dimensional, nonlinear structure in the human electroencephalogram. *Electroencephalographic Clinical Neurophysiology, 98,* 213–222.

Thomasson, N., Hoeppner, T. J., Webber, C. L., & Zbilut, J. P. (2001). Recurrence quantification in epileptic EEGs. *Physics Letters A, 279,* 94–101.

Tong, H. (1990). *Non-linear time series, a dynamical system perspective.* Oxford: Oxford University Press.

Trulla, L. L., Giuliani, A., Zbilut, J. P., & Webber, C. L. (1996). Recurrence quantification analysis of the logistic equation with transients. *Physical Letters A, 223,* 255–260.

Wand, M. P. & Jones, M. C. (1995). *Kernel smoothing.* London: Chapman & Hall.

Webber, C. L. & Zbilut, J. P. (1994). Dynamical assessment of physiological systems and states using recurrence plot strategies. *Journal of Applied Physiology, 76,* 965–973.

Weigend, A. S. & Gershenfeld, N. A. (1993). *Time series prediction: Forecasting the future and understanding the past.* Reading, MA: Addison-Wesley.

6

The Effects of the Irregular Sample and Missing Data in Time Series Analysis*

David M. Kreindler and Charles J. Lumsden

CONTENTS

Human self-report time series data are typically marked by irregularities in sampling rates; furthermore, these irregularities are typically natural outcomes of the data generation process. Relatively little has been published to assist the analysis of irregularly sampled data. We report the results of a series of computational experiments on synthetic data sets designed to assess the utility of techniques for handling irregular time series data. The behavior of a conservative quasiperiodic, a dissipative chaotic, and a self-organized critical dynamics were sampled regularly in time, and the regular sampling was disrupted by data point removal or by stochastic shifts in time. Missing data segments were then patched by means of segment concatenation, by segment filling with average data values, or by local interpolation in phase space. We compared results of nonlinear analytical tools, such as autocorrelations and correlation dimensions, using complete and patched sets, as well as power spectra with Lomb periodograms of the decimated sets. Local interpolation in phase space was particularly successful at preserving key

* The original article was edited for format and first published as Kreindler, D. M., & Lumsden, C. J. (2006). The effects of the irregular sample and missing data in time series. *Nonlinear Dynamics, Psychology, and Life Sciences, 10,* 187–214. Reprinted by permission of the Society for Chaos Theory in Psychology & Life Sciences.

features of the original data, but required potentially impractical quantities of intact data as a primer. While the other patching methods are not limited by the need for intact data, they distort results relative to the intact series. We conclude that irregularly sampled data sets with as much as 15% missing data can potentially be resampled or repaired for analysis with techniques that assume regular sampling without introducing substantial errors.

How best to deal with missing data is an important question for investigators interested in applying a time series analysis (TSA) to human behavior. In the study of human behavioral dynamics—in particular, in studies based on self-report—missed data points are the rule rather than the exception. Practically speaking, people are incapable of generating self-reports on an exact schedule over any substantial length of time. When working with human subjects in longitudinal studies, a variety of reasons—including fatigue, other bodily needs, loss of motivation, carelessness, or unwillingness to tolerate the excessive degrees of intrusion that may be necessary to ensure consistency in data reporting—will typically result in imperfect data reporting rates, temporal inaccuracy, or temporal variability.

Most TSA methods currently in widespread use in the behavioral science community are based on the assumption of regularly sampled data. Missed samples, or samples that are sampled quasi-regularly (i.e., approximately regularly, but with some variation around the expected sampling time) will introduce error terms into the calculations that will propagate through the results. The properties of these errors have not been extensively reported. Thus, relatively little has been published addressing the impact of missing data on the results of TSA. Moreover, few specific guidelines exist addressing how best to deal with missing data. In particular, the study of systems whose underlying dynamics are not yet determined, but are hypothesized to be nonlinear, present particular problems. In cases when sensitive dependence on initial conditions is present, synthesized or "corrected" data are likely to introduce errors, the significance of which can be difficult to assess.

The general options facing an investigator with time series in which data are missing or sampled irregularly have been well summarized by Sprott (2001). If there are gaps in the data and the inter-gap sequence lengths suffice, the TSA can be done on each inter-gap sequence and the results combined. For the relatively short data series that typify behavioral self-report studies (generally in the range of several thousand data points at most), this "divide-and-conquer" approach is, unfortunately, not often practical. Alternatively, the gaps can be filled in some manner, for example, by packing them with zeros or with the series mean value (predicted to give poor outcomes in general) (Dimri & Prakash, 2001; Press, Teukolsky, Vetterling, & Flannery, 1993; Sprott, 2001), or by ignoring the gaps and taking the data as properly sampled series of reduced length, or by using a nonlinear predictor method to fill in the gaps. If the gaps result from stochastic irregularities in the sampling time rather than from missed reporting events, some mix of nonlinear interpolation and curve fitting might produce a useful data series sampled on uniform time steps if the gaps are not too big. However, in the absence of studies that report the statistical properties of missed or irregular samples relevant to human behavior, it is impossible to translate such general insights into practical criteria applicable to time series self-report protocols.

The purpose of this study is to use the method of surrogate data modeling to assess the efficacy of the time sampling repair options identified by Sprott and others. The sampling irregularities generated by our models replicate those we have recently established for mood self-report data in long-term protocols (with durations ranging from 18 months to 16 years of self-report data). To the best of our knowledge, our analysis provides the first evidence that standard methods of nonlinear interpolation can give robust estimates of power spectrum behavior and correlation dimension when substantial proportions of the data are missing, with the maximum permissible amount of missing data depending on the dynamic characteristics of the underlying system. In addition, our results suggest that irregularities due to inaccurate (rather than missed) times of self-report, and due to patching of gaps in the data with series mean values have an effect on the analysis equivalent to the superposition of broadband noise on the pristine series.

We are also able to confirm that, for the calculation of correlation dimensions, filling gaps with series means or zeroes is on the whole a bad idea (Sprott, 2001) compared to nonlinear interpolation across the gaps (if gap size is small enough to permit this step) or to constructing a time series of shorter length by ignoring the gaps and shrinking them to zero. In contrast, ignoring the gaps and shrinking them to zero introduces distortions in the frequency domain due to the disruption of temporal relationships between data, for example, manifesting as errors in the power spectrum. Time series based on self-selected report times, in which we find up to 85% of data is missing compared to an equivalent regular report schedule (Kreindler, Levitt, & Lumsden, 2004b), do not respond to any of these palliative steps unless the underlying system is dynamically self-similar and, so, may require different techniques. Our findings apply to test series prepared from the output of models with widely different behavior, including conservative quasi-periodic, dissipative chaotic, and self-organized critical (putatively, "edge-of-chaos"; Bak, 1996; Turcotte, 2001) dynamics, and may therefore be of general relevance to behavioral data in which the temporal sampling is less than perfectly regular.

Methods

Time Series Length

In TSA applications in the natural sciences, time series lengths of 10^5–10^6 are commonplace (e.g., Bak, Christensen, Danon, & Scanlon, 2002; Christensen, Danon, Scanlon, & Bak, 2002; Sprott, 2003); in contrast, time series of psychiatric and psychological phenomena typically contain on the order of 10^3 points (e.g., Gottschalk, Bauer, & Whybrow, 1995; Kreindler, Levitt, & Lumsden, 2004a; Kreindler & Lumsden, 2002). We therefore produced computer-generated data sets with lengths in the range 10^3–10^4 points: 10^4 where we found it necessary to enable accurate resolution of the nonlinear dynamical properties of the series, but 10^3 where possible to better correspond to published studies. Results were plotted using Gnuplot v4.0 (http://www.gnuplot.info, Feb 2005).

Dynamics

To explore the generality of our sample repair methods, we considered outputs from a conservative, a dissipative chaotic, and a self-organized critical model.

Conservative. This was a driven harmonic oscillator with coefficients of the driving functions selected to yield an equation of motion composed of incommensurate sinusoid frequencies in proportions 13:47:93, as shown in Figure 6.1a, in which the slowest-frequency component was allowed to cycle through approximately eight complete cycles over 13,000 points. This oscillator's dynamics are equivalent to a four-dimensional autonomous phase flow with trajectories confined to a 3-torus in that space. These frequencies were selected to provide multiple near-harmonic and nonharmonic ratios across a range of frequencies. A selection of alternative ratios was considered; the results (data not shown) did not differ substantially from those reported here.

Dissipative chaotic. This was the first of 13,000 coordinates of a Lorenz dynamics time series, as shown in Figure 6.1b, generated using a set of MATLAB® routines (Wan, 2005) with initial values $(x_0, y_0, z_0) = (1, 0, 0)$ and the control parameters set at typical values ($r = 28$; $b = 8/3$; $\sigma = 10$); the routines used Runge–Kutta fourth-order integration (Press et al., 1993) with time increment $dt = 0.01$. The initial point $(1, 0, 0)$ relaxed to the attractor in about 60 time steps, and the inclusion or exclusion of this short initial transient had no detectable effect on the subsequent analysis (data not shown). The Lorenz chaos has been extensively studied and has a well-characterized attractor with a broadband power spectrum in the x and z coordinates (Farmer, Crutchfield, Frochling, Packard, & Shaw, 1980) and a correlation dimension of 2.068 ± 0.086 (Sprott, 2003).

Self-organized criticality (SOC). This was a set of total grain counts generated using a standard Bak sandpile model (Bak, 1996), as shown in Figure 6.1c. Our Bak sandpile was a 64×64 cell, two-dimensional grid, with one grain (perturbation excitation) added per time step and the resulting total grain load counted (Sprott, 2003) for 13,000 steps. On the basis of extensive prior studies (Kreindler & Lumsden, 1998), we preloaded the pile with $64^2 \times (h/2)$ grains, where $h = 4$ is the critical height of the pile, and then allowed 1000 time steps for the pile to come to a steady state prior to beginning the recording

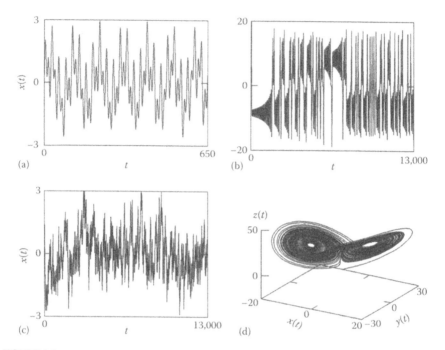

FIGURE 6.1

Time series plots of our conservative, dissipative chaotic, and self-organized critical model time series. (a) sin(13*t*) + sin(47*t*) + sin(93*t*); only the first 650 points of the series have been plotted for clarity. (b) The initial 13,000 values of the *x* component of the Lorenz series plotted against time, *t*, using parameter values (*r*, *b*, σ) = (28, 8/3, 10) and starting values (x_0, y_0, z_0) = (1, 0, 0). (c) 13,000 values of total grain count over time in a 64 × 64 Bak sandpile that has been uniformly randomly preloaded with 64 × 64 × 2 grains and allowed 1000 steps to evolve to a critical state. (d) The initial 13,000 values of the Lorenz series, (*x*(*t*), *y*(*t*), *z*(*t*)), plotted using the same parameter and initial values as in (b).

of grain counts. The resulting time series was rescaled to generate a series with mean 0 and variance 1.

The dynamical properties of SOC systems have not as yet been studied as extensively as nonlinear dynamical chaos. It has recently been hypothesized, however, that SOC systems in their steady state have evolved not to a chaotic attractor but have instead self-organized to the "edge of chaos" (Bak, 1996), in which adjacent trajectories diverge in the power law, rather than the Lyapunov exponential, fashion over time (Turcotte, 2001). SOC systems yield $1/f^\alpha$ power spectra over multiple orders of magnitude; however, the interactions between system components spanning all scales of time and distance that define an SOC system (Bak, 1996; Jensen, 1998) result in long-range temporal and spatial correlations, and might not be properly characterized as purely $1/f^\alpha$ colored noise.

The inclusion of an SOC-type dynamics in our study was of particular interest to us. We have proposed elsewhere that the dynamics of both normal human mood variation and mood change in rapidly cycling bipolar disorder (American Psychiatric Association, 1994) may be self-organized critical (Kreindler & Lumsden, 1999; Kreindler & Lumsden, 2002) rather than low-dimensional chaotic (Gottschalk et al., 1995). Further evaluation of this hypothesis, however, requires methods for TSA on mood data series containing temporal imperfections of the kind investigated in this report.

The models for the sinusoid and sandpile dynamics were coded in the C++ programming language using the gcc 3.4.3 compiler (The Free Software Foundation, http://www.gnu.org/software/gcc/, February 2005) under Fedora Core 3 Linux (http://fedora.redhat.com, 2004); the Lorenz model dynamics were generated in MATLAB v12 (The MathWorks, Inc. (2000), Natick, MA, USA) under Microsoft Windows 2000 Pro SP4 (Microsoft Corp., Redmond, WA, USA) on a Pentium 4M-based system running at 1.7 GHz with 768 Mb of RAM. These regularly sampled series were then input to procedures that simulated the effect of imperfect data reporting over time.

Irregular sampling of the simulated time series. There are two kinds of sampling irregularities that arise in self-report data sets: reports that arrive at times other than the expected ones due to *variation* of reporting times around the expected reporting time, and complete loss of reports due to *failure* to report at an expected reporting time. Each of these kinds of irregularities was simulated separately.

To simulate the irregularity resulting from loss of proportion of reports, two sampling protocols were developed in the C programming language and applied to the output of the dynamical models. Because of their erosive effect on the regular order of the originating time series, we refer to these routines as time series "decimation."

Decimation protocol 1. We recently completed a study (Kreindler et al., 2004a, 2004b) in which handheld computer (HHC)-based techniques (Kreindler, Levitt, Woolridge, & Lumsden, 2003) were used to collect mood self-report information every 12 h ± 10 min in a community-based sample over an 18 month interval—approximately 1095 self-reports per participant—from ($n = 20$) healthy adults and ($n = 20$) adults with a DSM-IV diagnosis of rapidly cycling bipolar disorder (American Psychiatric Association, 1994). Despite the use of the HHCs to provide audio prompts to remind participants of scheduled reporting times, our reporting rate across all subjects averaged 84%, with no significant difference between patients and healthy controls. This reporting rate, and its corresponding missing data count of approximately 15%, serves as one motivating case for the present study.

To simulate the effects of missed data reports of this type, the time series data sets from each dynamical model were processed to varying degrees by removing series elements using a uniform random distribution to decide which points were eliminated. We considered decimated series from each dynamical model in which 0%, 1%, 2%, 3%, 5%, 10%, 15%, 20%, 25%, 30%, 35%, 40%, 45%, 50%, 60%, 70%, 80%, 90%, and 95% of the data were removed from the originating time series.

Decimation protocol 2. We have previously reported an initial study of a 16 year mood record that is quite unlike the HHC-based assessment described above. This 16 year mood record consists entirely of self-reports generated at subject-selected times. A diary entry was produced whenever a subjectively significant mood change was noted to have occurred and not at fixed times (Kreindler & Lumsden, 2002). We found that when considered over the entire 16 year interval of the diary record, the magnitude of the time intervals (the "gaps") between successive reports demonstrates power-law scaling best fit with a power-law exponent of –2.7. The self-reports, when entered into the diary, were often generated at 1200 or 0000 h. Taking 12 h as a baseline sampling interval, and neglecting the 220 reports that occurred more frequently than this, we derived a sampling template of 1,849 events occurring over 12,270 sampling periods in 16 years, for an effective event rate of 15% reporting. We therefore created a decimation procedure that used a power law with an exponent of –2.7 to determine the interval between sampled points; this procedure was implemented as a C language routine using the gcc v.3.4.3 compiler. Using this routine, each sampled point was followed by a gap of size of S elements, where S is the minimum value such that $R < \left(\sum_1^S x^{-2.7} \right) / \left(\sum_1^{168} x^{-2.7} \right)$, and R is a value drawn from the uniform random distribution between 0 and 1. We considered decimated series from each dynamical model in which 0%, 1%, 2%, 3%, 5%, 10%, 15%, 20%, 25%, 30%, 35%, 40%, 45%, 50%, 60%, 70%, 80%, 90%, and 95% of the data were removed from the originating time series.

The impact of *variation* of reporting times around the expected reporting time was explored using a modified version of our decimation protocol 1, discussed further below.

Patching the Decimated Time Series

We assessed four different methods of treating the gaps in the regularly sampled model data produced by our time series decimation protocols:

"As-is". Some standard methods of frequency domain analysis do not require regular time sampling (see below). We used the decimated series in this step without further processing.

Gap closure. The elements of the decimated series were treated as temporally consecutive, thereby not introducing artificial values at the cost of distorting the temporal relationships among the decimated series' elements.

Mean filling. The eliminated elements were replaced with series mean values (Press et al., 1993; Sprott, 2001), so as to preserve the remaining elements' temporal relationships at the cost of introducing erroneous values.

Local interpolation. A more complex alternative to gap closure or mean filling is to replace eliminated elements with estimates of the missing values. Rather than using linear interpolation on the raw data set (e.g., Dimri & Prakash (2001)), we used local interpolation in the phase space to generate replacement values. The TISEAN nonlinear TIme SEries ANalysis package (Hegger, Kantz, & Schreiber, 1999; Hegger, Kantz, & Schreiber, 2000) currently enjoys widespread use for this and related applications in nonlinear TSA. TISEAN provides a routine, nstep(), that applies Taylor series methods to the local dynamics of the phase flow to predict series values (Hegger et al., 1999). One difficulty of using nstep() is that a "primer" of good-quality data is needed to establish the dynamics in the phase space before further prediction can be done. For nstep(), any unbroken subset of length $d \times M_{max}$ (where d is the delay and M_{max}, the maximal embedding dimension desired in the attractor reconstruction) will provide an adequate primer. While in practice it may be difficult to obtain a primer, for the purpose of this initial study we assumed that such a primer was available and explored the properties of the reconstructed data set.

To assess the usefulness of an nstep()-based local interpolation routine for patching, a series of experiments were conducted with our three model systems. The sinusoid series's trivial dynamics were well predicted by nstep() (data not shown). We next used the first 13,000 points of the three-dimensional Lorenz series (with all parameters and initial points set to the values noted above) and the sandpile series (also as above). The first 6500 points of both series were left intact. The subsequent 6500 points were decimated to varying degrees, using the decimation protocols 1 and 2 as described above. nstep() was then repeatedly used to fill the resulting gaps so as to reconstruct the full 13,000 point set: for each gap (x_n, y_n, z_n) for the Lorenz series, or (x_n) for the sandpile series, $n = (j,...,k)$, $j \leq k$, nstep() was used to predict $(k - j + 1)$ points, using the first $(n - 1)$ points as its basis for prediction. For the patched sinusoid and SOC series, the phase space delay was set relative to the first zero, t_0, of the autocorrelation function, as described below. In the case of the Lorenz model, we instead calculated t_0 using the square of the x component, as described below. We then calculated correlation dimensions (see below) for each resulting patched series, using the patched portion of the series (i.e., the final 6500 points) only.

Time Series Analysis

Frequency domain. Accurate estimation of the power spectrum using the fast Fourier transform (FFT) (Press et al., 1993) requires data sampled at regular time intervals. Lomb (1976), however, described a powerful method (implemented as an algorithm in Press et al. (1993)) that does not need regular data spacing in time. As opposed to the conventional power spectrum's Fourier analysis, Lomb's method carries out linear least squares fitting of a sinusoid to the data set, eliminating the requirement for regular sampling. The resulting "Lomb periodogram" resembles a conventional power spectrum in that it quantifies the frequencies' relative contribution as a function of frequency. We therefore compared the frequency domain behavior of each of our dynamical models as predicted by both the Lomb and the standard power spectrum methods at each level of data decimation and for each method of data gap closure. We also compared the resulting frequency domain plots to the behavior of a time series of equal length that modeled extreme temporal irregularity due to variation in expected reporting times, as described next.

We compared the spectral behaviors established in the previous paragraph with a time series characterized by extreme temporal irregularity—comparable to a subject who submits the number of reports requested, albeit not necessarily at the requested times (e.g., Hyland, Kenyon, Allen, & Howarth, 1993). Using a variant of our decimation protocol 1, we simulated sparse random-time sampling by randomly selecting 13,000 points from a 100,000-point compound sinusoid time series generated using the same model over the same interval with more frequent sampling, retaining each point P_i if $R_i < 1/13$ (where R_i was a random value drawn from a uniform distribution on the interval), $i = \{1, 2, ..., 13,000\}$, and dropping P_i otherwise. We then computed Lomb periodograms for both the intact and sampled time series.

Autocorrelation function and correlation dimension. We calculated the autocorrelation function and the correlation dimension for each of our patched decimated series using the TISEAN routines autocor() and d2(), and compared these to the behavior of the regularly sampled, undecimated series for each dynamics. The calculation of the correlation dimension depends on the selection of an appropriate delay, d. Too small values of d will cause spurious results due to erroneous inclusion of points proximate in time, while inappropriately large values of d may result in insufficient numbers of points for reconstruction in the phase space at higher values of M, where M is the embedding dimension, once $d \times M$ is on the order of the size of the data set. While the selection of values for d that will yield best results depends on the nature of the specific system being studied, the first zero, t_0, of the autocorrelation function often provides a useful starting point (Kantz & Schreiber, 1999). For each series, we calculated t_0 and compared the d2() output across a range of possible delay values, d, ranging between 0 and $10t_0$. The adequacy of the resulting d estimates was verified by the inspection of the space–time separation plots (Cohen & Servan-Schreiber, 1992) for adequate decorrelation of the dynamics and of the correlation dimension plots for power-law scaling behavior (Grassberger & Procaccia, 1983) over the largest range of inter-neighbor distances for the values surveyed. The d2() output behavior of the trigonometric and sandpile dynamics was on the whole not very sensitive to variations in d, and, unless otherwise noted, for our analysis of the decimated time series based on these models, we used $d = t_0$ in evaluating the correlation dimension. The Lorenz model was an exception to this: we found that a delay of approximately $0.16d$—the value resulting from calculating t_0 of the square of the x variable of the Lorenz model—rather than d itself best resolved the d2() behavior of the time series constructed from the Lorenz model with the parameters we used, and gave a dimension estimate of 1.99 ± 0.01 for the 3D Lorenz attractor in close agreement with earlier reports (Grassberger & Procaccia, 1983) that used a delay of 0.25d. (The 0.25d delay is known to correspond to the time required for a significant decorrelation of the phase flow of a single "wing" of Lorenz's butterfly-shaped attractor surface (Kantz & Schreiber, 1999), while d is more a time for the phase point's tour of the Lorenz attractor as a whole.)

Results

Effects of Missing Points and Temporal Inaccuracy

Mean filling. Uniform random decimation (decimation protocol 1) with mean filling of the gaps produced results in the frequency domain comparable to the addition of broadband white noise to the pristine signal. The more prominent spectral features were preserved but the signal-to-noise ratio escalated as the degree of decimation increased. Figure 6.2 illustrates the trend for the Lorenz data. The sinusoid and sandpile data gave similar results. The intrusion of the white noise is understandable because, in effect, the time series is being rewritten with randomly spaced clusters of null values. A sine function fit (as in either Lomb's method or with an FFT-based power spectrum) will then unmask a broad range of frequencies in which no frequency is preferred.

Power-law decimation (decimation protocol 2) with mean-fill patching of the simulated time series gave a similar effect, but with progressively larger amounts of colored noise with slope −2 added to the spectrum, rather than white noise.

Gap closure. Patching using gap closure altered the spacing of series elements. As shown in Figure 6.3, while this did not result in a noticeable addition of a broadband noise component, it did have the effect of shifting the frequencies of spectral features relative to their original frequency, as a result of distortion of temporal relationships between series elements.

Temporal uncertainty. Random-time sampling of the sinusoid time series revealed that, as shown in Figure 6.4, while prominent spectral features are preserved, the effect of this kind of temporal irregularity also is analogous to the addition of white noise to the signals from a system sampled at regular time intervals, as shown in Figure 6.2.

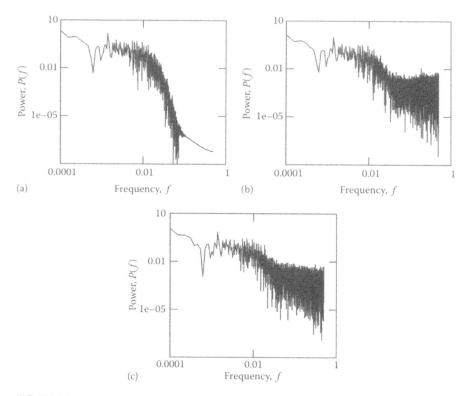

FIGURE 6.2
Mean filling: (a) Lomb periodogram of the intact Lorenz series, plotting the series' power, P, as a function of frequency, f. (b) Lomb periodogram of the Lorenz series following 20% uniform random decimation (protocol 1) and mean filling, equivalent to the addition of broadband white noise. (c) Lomb periodogram of the Lorenz series following 20% power-law decimation (protocol 2) and mean filling, equivalent to the addition of broadband colored noise, with slope –2.

Autocorrelation function. Figure 6.5 illustrates that the first zero of the autocorrelation functions, t_0, for our data varied with both the degree of decimation and the specific patching strategy used. In our Lorenz model, decimation rates greater than 5% patched with gap closure resulted in steady decreases in t_0 as the decimation rate increased. In contrast, mean filling resulted in t_0 remaining essentially constant up to decimation rates of 90%. Results for the sinusoid series were essentially similar. In contrast, decreases in t_0 roughly proportionate to the decimation rate were observed with the sandpile series when either patching method was applied. There was little difference noted between decimation protocols 1 and 2 (with either gap filling or mean replacement) at similar rates of decimation. In our sandpile model, both mean filling and gap closure resulted in declines in t_0 roughly proportionately to the decimation rate.

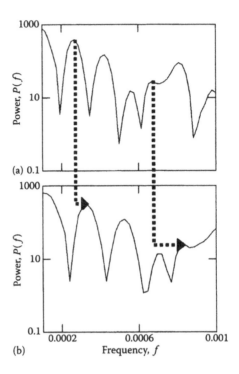

(a)

(b)

FIGURE 6.3
Gap closure: Lomb periodograms of (a) the intact Lorenz series and (b) the Lorenz series decimated at a rate of 20% using uniform random (protocol 1) decimation and patched with gap
closure. For clarity, only a portion of the low-frequency end of the spectrum is plotted, using
linear rather than logarithmic frequency, f. The dashed arrows connecting corresponding
spectral features in (a) and (b) highlight the shift of spectral features toward higher frequencies caused by patching with gap closure.

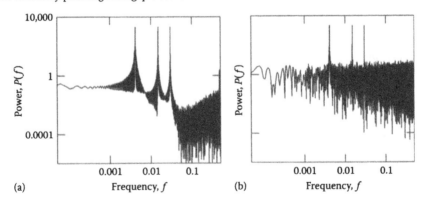

(a)

(b)

FIGURE 6.4
Temporal uncertainty: Lomb periodograms of (a) the compound sinusoid series, sampled at
13,000 equally spaced intervals, are compared with (b) 13,000 samples of the compound sinusoid series drawn randomly from 100,000 equally spaced samples over the same interval as
used with (a). The broadband noise apparent in (b) is absent in (a).

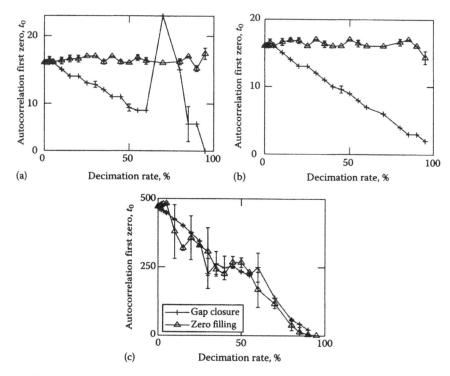

FIGURE 6.5

First zero of the autocorrelation function, t_0: Plot of t_0 versus the decimation rate using uniform random decimation (protocol 1). The means and standard deviations are indicated, calculated with $n = 10$ decimated sets for each decimation level, using (a) the sinusoid series, (b) the square of the x component of the Lorenz series, and (c) the sandpile series.

Correlation Dimension

Sinusoidal model. The inspection of the sinusoidal model's $C2(r)$-versus-r plot, where $C2$ is the correlation integral calculated at radius r, reveals four distinct regions for embedding dimension $M > 4$, as shown in Figure 6.6: the $r < 0.05$ region, for which $C2$ scales as r^M; the $0.05 < r < 0.6$ region, in which r is sufficiently small that $C2$ scales as $r^{1.3}$, consistent—given the limited length of the series—with its appearing one-dimensional, locally; the $0.6 < r < 2.5$ region, in which $C2(r) \sim r^{3.4}$, approximating the expected result of $C2(r) \sim r^3$; and the asymptotic region for $r > 2.5$.

With decimation protocol 1 (uniform random decimation) and gap-closure patching, progressive increases in the level of decimation result in loss of the scaling regions and emergence of $C2 \sim r^M$ scaling across all r, as would be expected with an increasing stochastic component of the signal: as the level of decimation increases, the gap closure operating on the underlying compound sinusoid would be expected to progressively better approximate a stochastic rather than deterministic pattern; attractor reconstruction using

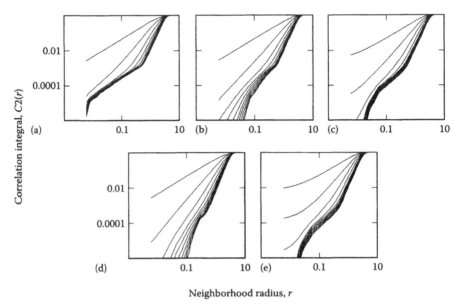

FIGURE 6.6
Sinusoid C2(r) versus r: Plot of the correlation integral, C2, as a function of the radius, r, over which C2 is calculated. (a) Intact series; (b) 5% uniform random (UR) decimation with gap-closure (GC) patching; (c) 5% UR decimation with mean-fill (MF) patching; (d) 5% power-law (PL) decimation with GC; and (e) 5% PL decimation with MF patching.

this noise should fill whatever space is used for reconstruction, resulting in a $C2 \sim r^M$ scaling pattern. In particular, increases in decimation rates above 3% result in loss of convergence of C2 for $0.6 < r < 2.5$. In contrast, all decimations with mean-fill patching lead to two kinds of changes: an initial progressive steepening of the C2(r)-versus-r slopes as the decimation rate increased, then progressive flattening of all curves due to the increasing proportion of identical, series-mean values as the system becomes progressively more point-attractor-like.

The use of decimation protocol 2 (power-law decimation) and gap-closure patching truncated the $C2 \sim r^1$ mid-range section and progressively steepened the slope at larger values of r for all decimation rates. The combination of decimation protocol 2 and mean filling resulted in better preservation of the $C2 \sim r^1$ and $C2 \sim r^3$ scaling regions as well as the asymptotic region until a decimation rate of 5%, above which point-attractor features increasingly predominate.

Lorenz model. The Lorenz model's C2(r)-versus-r plot, as shown in Figure 6.7, calculated with the intact data set and a d of 0.16, exhibits scaling for $r < 10$ where $C2 \sim r^2$, in good agreement with typical values of the correlation dimension (i.e., 2.068...). For values of $r > 10$, there is an asymptotic region. Gap-closure patching following decimation using protocol 1 resulted in

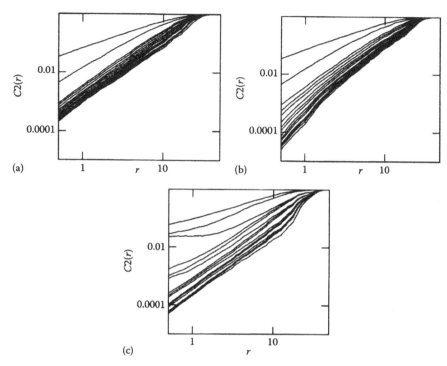

FIGURE 6.7
Lorenz $C2(r)$ versus r: Plot of the correlation integral, $C2$, as a function of the radius, r, over which $C2$ is calculated. (a) Intact series; (b) 15% UR decimation with CG patching; and (c) 15% UR decimation with MF patching.

good preservation of the r^2 scaling region up to decimation rates of 15%. Above this, the $C2(r)$-versus-r plot displayed increasing signs of stochastic ($C2 \sim r^M$) scaling as the decimation rate increased further. In contrast, while mean-fill patching resulted in a progressive emergence of point attractor features (i.e., $C2 \sim r^0$ scaling) for increasing values of r, r^2 scaling was preserved over a broad range of r, particularly at higher values of M, up to decimation rates of 15%. Decimation using protocol 2 resulted in similar findings (data not shown).

SOC model. As shown in Figure 6.8, our sandpile model's $C2(r)$-versus-r plot exhibited three distinct regions of scaling: $C2 \sim r^{1.22}$ for $r < 0.5$, $C2 \sim r^{0.6M}$ for $r > 1$, and an asymptotic region for $r > 2$. Gap-closure patching using protocol 1 resulted in only minimal changes in features of the $C2(r)$-versus-r plot for $r > 0.5$ across decimation rates from 0% to 60%. This is hypothesized to be a direct consequence of the self-similar properties of the sandpile series. In contrast, mean-fill patching resulted in increases in the slopes of the higher-M curves for all $r < 2$ as the decimation rate increased from 0% to 35%, followed by progressive flattening of all curves for all r as the decimation rate rose further and the proportion of identical mean values in the

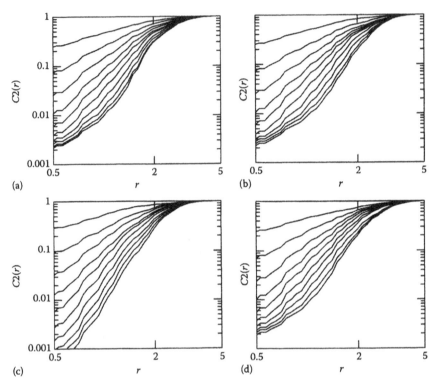

FIGURE 6.8
Sandpile $C2(r)$ versus r: Plot of the correlation integral, $C2$, as a function of the radius, r, over which $C2$ is calculated. (a) Intact series; (b) 15% UR decimation with CG patching; (c) 60% UR decimation with CG patching; and (d) 15% UR decimation with MF patching.

series increased. Decimation using protocol 2 resulted in similar findings (data not shown).

Patching with nstep(). Patching the dissipative chaotic Lorenz model with nstep() yielded impressive but different results for data sets decimated with either the uniform random decimation protocol 1 or the power-law decimation protocol 2. We found that nstep() was successful at patching the series for decimation rates of up to a threshold rate of around 40% in the case of protocol 1, as shown in Figure 6.9, and around 15% for protocol 2 (data not shown), with the maximum distances between patched and original points less than 0.1 and 0.3 units, respectively. $C2(r)$-versus-r plots generated using the patched data were essentially identical up to these threshold rates. Because nstep() predicts one missing point at a time, using all earlier points to reconstruct the attractor, longer stretches of missing data will propagate and magnify errors more than shorter segments. As a result, decimation protocol 1, which generated data sets with multiple

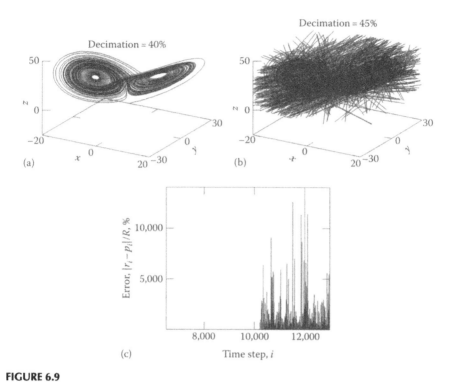

FIGURE 6.9

nstep() patching with the Lorenz series: (a) Plot of the Lorenz series after 40% UR decimation and patching with nstep(); (b) same as (a), but with 45% UR decimation. (c) Plot of the percentage error as a function of the time step, i. The percentage error is calculated as $|r_i - p_i|/R$, where $\{r_i\}$ are the points in the original Lorenz series; $\{p_i\}$, the original Lorenz series following 45% UR decimation and patching with nstep(); and R, the greatest dimension of the Lorenz attractor, estimated as $R = 87$. The error is initially negligible, but then abruptly begins to exceed R by thousands of percent for $i > 10{,}250$, consistent with an abrupt breakdown in the ability of nstep() to adequately patch the series at this decimation level.

isolated or short stretches of missing data, was on the whole better handled by nstep() than was protocol 2, which more frequently introduced extended segments of missing points. Beyond these thresholds, the patching routine intermittently generated a series of patch values that departed the attractor on a diverging trajectory; divergence occurred during both short and long gaps. The performance of the patching routine at threshold decimation rates implies that breakdowns in patching at a point r_n are due to a combination of local dynamics on the attractor at r_n, cumulative errors in the reconstruction of the attractor for r_i, $i < n$, and the size of the gap at r_n.

As shown in Figure 6.10, when applied to the self-similar SOC sandpile model, nstep() patched gaps in the time series robustly with either of the decimation protocols. In particular, while there was some increase in the distances between elements of the original and patched series up to a decimation rate

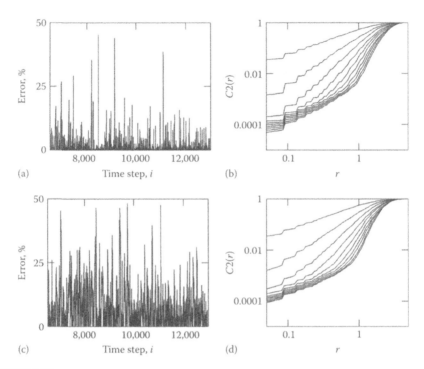

FIGURE 6.10

nstep() patching with the sandpile series: Percentage error of the nstep()-patched series for UR-decimated sandpile series decimated at rates of (a) 20% and (c) 90%. The percent error was calculated similarly as Figure 6.9c, but with $R = 6$. In contrast to the Lorenz case, with the sandpile series, the distances remain bounded, limited to no more than three times the series' variance. $C2(r)$-versus-r plots, created using these patched series, are shown in (b) and (d), respectively. Compared with Figure 6.8a, a $C2(r)$-versus-r plot generated using the intact series reveals essentially no differences.

of 20%, patched values remained within three times the series' variance of the original series values at all times for decimation rates of up to 90%. No significant differences were noted on the resulting $C2(r)$-versus-r plots.

As summarized in Table 6.1, the quality of the reconstructed time series was affected by the patching method, by the underlying statistical pattern of the missing data, and by the substrate system's dynamics. nstep()-based patching was able to reconstruct decimated sets at equal or (in all but one case) greater decimation rates than either of the other two patching methods that we investigated; however, in the real world, the need for an intact primer of adequate length may preclude its use. With gap closure and mean filling, the quality of the reconstructed sets was adequate up to a threshold level, and then gradually deteriorated as the decimation rate increased further. While this was true as well when nstep()-based reconstruction was applied to the SOC model, in the case of the Lorenz model, the quality of reconstruction abruptly deteriorated once a threshold rate was surpassed.

TABLE 6.1

Maximum Patchable Decimation Rates for
Preserving the Principal Features of the
$C2(r)$-versus-r Diagrams

	Protocol 1 (UR) (%)	Protocol 2 (PL) (%)
Sinusoidal model		
Mean fill	0	0
Gap closure	3	0
Lorenz model		
Mean fill	15	15
Gap closure	15	15
nstep()	40[a]	15[a]
SOC model		
Mean fill	0	0
Gap closure	60	60
nstep()	90	90

Decimation rates up to which each patching protocol is capable of UR = uniform random decimation and PL = power-law decimation.

[a] Abrupt rather than gradual deterioration in
the quality of the reconstruction once the tab-
ulated decimation rate is exceeded.

Discussion

The collection of real-world behavioral data in quantities and quality adequate to apply conventional TSA techniques can be problematic, especially
if there are trade-offs between the quality of the data, the intrusiveness of
the sampling protocol, and the degree of naturalism of the data. For example, one recent study (Silipo, Deco, Vergassola, & Bartsch, 1998) reported a
100% subject response rate with a very small ($n = 4$) number of subjects using
twice-daily reporting over 512 days. However, it did so at the expense of temporal precision (12 ± 1 h); furthermore, no information was reported on participant retention or dropout rates. In our own work with volunteer subjects
on regular schedules over 18 month periods, response rates have typically
averaged about 84%. Furthermore, as we noted earlier, there are significant
protocols in which subjects self-select their report times rather than attempt
to comply with a rigid schedule for data submission. Thus, taken relative to
an idealized standard of the dynamical system sampled with perfect regularity, "missing" time series data are a fact of life in behavioral science, and
are likely to be of increasing significance as attention turns from short-term,

highly controlled studies to long-term assessments of subjects living in situ in their family and community.

Our findings suggest that the investigator faced with less than pristinely regular data does not need to abandon the potential benefits of nonlinear TSA. What is required, first and foremost, is a careful assessment of the statistical properties of the temporal distribution of the sampled data, after which decisions can be made as to the patching or reconstruction of the time series in terms of data values spaced regularly in time. Some methods, such as the Lomb periodogram, do not require regular data spacing in time and may be used with the original series data. Many others, such as the standard autocorrelation function, power spectrum, and standard correlation dimension procedures, run on equally spaced time steps. It is almost essential to test their merit for each intended TSA application by using surrogate data, which has been generated at regular time steps by nonlinear models of the relevant class or classes, and then decimated to mimic the temporal irregularities of the observed time series.

In the surrogate data time series reported here, we find that the quality of the reconstruction is affected by the type of dynamical system, by the selection of the time or the frequency domain as the arena for analysis, by the temporal distribution of the data, by the amount of missing data, and by the reconstruction method. Despite this complexity, the basal impact of temporal irregularity is an effect equivalent to adding broadband noise (white or colored) to the pristine signal. With properly chosen patching or reconstruction steps, the key structure is preserved above the noise in power spectra and correlation integrals for decimation rates approaching or exceeding 15% of missing data, as shown in Table 6.1. On this basis, for example, data series of the kind possible with current long-term *in situ* self-report protocols (Kreindler & Lumsden, 2002; Kreindler et al., 2004b) are amenable to nonlinear TSA using existing methods.

As discussed in the "Methods" section, nonlinear prediction routines such as nstep() use attractor reconstruction via embedding, resulting in the need for a time series without missing reports of length $d \times M$. In most real-world cases of experimental data, investigators have no knowledge about the geometry of a hypothesized underlying attractor, resulting in the need to rely on general rules for the choice of d. There are various opinions about the optimal size of the embedding delay, d: whereas using t_0 is more conservative, using the first value of the autocorrelation at which $1/e$ decay has occurred will result in a lower value and, therefore, in a shorter minimum series length. In the case of relatively brief time series, the choice of a shorter or longer value of d may limit the maximal value of M for which results are calculable. Conversely, choosing a value for d that is excessively long—for example, $t_0 = 1024$ using the x component rather than $t_0 = 16$ using the *square* of the x component in the case of the Lorenz series—can yield spurious results. To further complicate matters, our results show that, in some classes of series, series decimation can have an effect on the value of t_0. A patching method, such as our nstep()-based method, which preserves the series length and has the least effect on t_0, may therefore be an important element of a multistep reconstruction procedure. One additional point is that, regardless of what can or cannot be calculated, work by Tsonis and Elsner (1989) and Ruelle (1990) linking the size of the data set with the maximal embedding dimension, M, for which results can be considered meaningful, provides important constraints for interpreting these kinds of calculations on short time series.

Finally, as our 16 year data set demonstrates, perhaps "missing" data is a misnomer: rather, "irregular" sampling should be the preferred description in certain cases, since it does not imply that something needs to be filled in to replace points that may not exist in the first place; however, far fewer tools exist at present for examining and characterizing the time domain behavior of irregular series. Studies exploring links between missing versus irregular data will be a vital step in understanding both the application of TSA techniques to real behavioral data and the quantitative aspects of subjective data in general.

Because of their widespread use in the community, our study focused on the performance of selected standard tools as the quality of temporal sampling varied. Such performance data are essential to the rational application of current methods to complex problems. The subsequent challenge—that of further improving the mathematical tools themselves—is sure to attract more and more attention as data series based on unavoidably imperfect time sampling continue to gain importance in behavioral analysis.

Acknowledgments

We thank Bell University Labs at the University of Toronto for computational resource support in aid of the work reported here. David M. Kreindler thanks the Department of Psychiatry, University of Toronto, for research time support.

References

American Psychiatric Association. (1994). *Diagnostic and statistical manual of mental disorders* (4th ed.). Washington, DC: American Psychiatric Association.

Bak, P. (1996). *How nature works: The science of self-organized criticality.* New York: Copernicus/Springer-Verlag Inc.

Bak, P., Christensen, K., Danon, L., & Scanlon, T. (2002). Unified scaling law for earthquakes. *Physical Review Letters, 88,* 178501.

Christensen, K., Danon, L., Scanlon, T., & Bak, P. (2002). Unified scaling law for earthquakes. *Proceedings of the National Academy of Sciences USA, 99*(Suppl. 1), 2509–2513.

Cohen, J. D., & Servan-Schreiber, D. (1992). A neural network model of catecholamine modulation of behavior. *Psychiatric Annals, 22*, 125–129.

Dimri, V. P., & Prakash, M. R. (2001). Scaling of power spectrum of extinction events in the fossil record. *Earth and Planetary Science Letters, 186*, 363–370.

Farmer, D., Crutchfield, J., Frochling, H., Packard, N., & Shaw, R. (1980). Power spectra and mixing properties of strange attractors. *Annals of the New York Academy of Sciences, 357*, 453–472.

Gottschalk, A., Bauer, M. S., & Whybrow, P. C. (1995). Evidence of chaotic mood variation in bipolar disorder. *Archives of General Psychiatry, 52*, 947–959.

Grassberger, P., & Procaccia, I. (1983). Characterization of strange attractors. *Physical Review Letters, 50*, 346–349.

Hegger, R., Kantz, H., & Schreiber, T. (1999). Practical implementation of nonlinear time series methods: The TISEAN package. *Chaos, 9*, 413.

Hegger, R., Kantz, H., & Schreiber, T. (2000). *TISEAN 2.1 Nonlinear Time Series Analysis* [Web Page]. Retrieved February 16, 2005, from http://www.mpipks-dresden. mpg.de/~tisean

Hyland, M. E., Kenyon, C. A. P., Allen, R., & Howarth, P. (1993). Diary keeping in asthma: Comparison of written and electronic methods. *British Medical Journal, 306*, 487.

Jensen, H. J. (1998). *Self-organized criticality: Emergent complex behavior in physical and biological systems* (Cambridge Lecture Notes in Physics No. #10). Cambridge, U.K.: Cambridge University Press.

Kantz, H., & Schreiber, T. (1999). *Nonlinear time series analysis* (Cambridge Nonlinear Science Series.) New York: Cambridge University Press.

Kreindler, D., Levitt, A., & Lumsden, C. (2004a, July). *Correlation dimension estimates of extended mood time series in health and bipolar disorder using wireless handheld data collection.* Presentation to the 14th annual international conference of the Society for Chaos Theory in Psychology and the Life Sciences, Milwaukee, WI.

Kreindler, D., Levitt, A., & Lumsden, C. (2004b, October). *Collecting extended and accurate mood self-report time series data using wireless handheld telemetry in bipolar disorder.* Presentation to the 54th Annual Meeting of the Canadian Psychiatric Association, Montreal, PQ.

Kreindler, D., & Lumsden, C. (1998, August). *The mathematics of mood variation: Mood scaling via self-organized criticality.* Presentation to the 8th annual international conference of the Society for Chaos Theory in Psychology and the Life Sciences, Boston, MA.

Kreindler, D., & Lumsden, C. (1999). *The mathematics of mood variation: Mood scaling via self-organized criticality.* Poster presentation to the American Psychiatric Association Annual Convention, Washington, DC.

Kreindler, D., & Lumsden, C. (2002, August). *Self-organized criticality in bipolar disorder: A case study.* Presentation to the 12th annual international conference of the Society for Chaos Theory in Psychology and the Life Sciences, Portland, OR.

Kreindler, D., Levitt, A., Woolridge, N., & Lumsden, C. J. (2003). Portable mood mapping: the validity and reliability of analog scale displays for mood assessment via hand-held computer. *Psychiatry Research, 120*, 165–177.

Lomb, N. R. (1976). Least-squares analysis of unequally spaced data. *Astrophysics and Space Science, 39*, 447–462.

Press, W. H., Teukolsky, S. A., Vetterling, W. T., & Flannery, B. P. (1993). *Numerical recipes in C: The art of scientific computing* (2nd ed.). Cambridge, MA: Cambridge University Press.

Ruelle, D. (1990). The Claude Bernard Lecture, 1989. Deterministic chaos: The science and the fiction. *Proceedings of the Royal Society of London. Series A, Mathematical and Physical Sciences, 427*, 241–248.

Silipo, R., Deco, G., Vergassola, R., & Bartsch, H. (1998). Dynamics extraction in multivariate biomedical time series. *Biological Cybernetics, 79*, 15–27.

Sprott, J. C. (2001). *Multifractals: Chaos and time-series analysis.* 11/28/00 Lecture #13 in Physics 505, University of Wisconsin Madison [Web Page]. Retrieved October 2004, from http://sprott.physics.wisc.edu/phys505/lect13.htm

Sprott, J. C. (2003). *Chaos and time-series analysis.* New York: Oxford University Press.

Tsonis, A. A., & Elsner, J. B. (1989). Chaos, strange attractors, and weather. *Bulletin of the American Meteorological Society, 70*, 14–23.

Turcotte, D. L. (2001). Self-organized criticality: Does it have anything to with criticality and is it useful? *Nonlinear Processes in Geophysics, 8*, 193–196.

Wan, E. A. *Time Series Data* [Web Page]. Retrieved February 1, 2005, from http://www.cse.ogi. edu/~ericwan/data.html

7

A Dynamical Analysis via the Extended-Return-Map*

Jay-Shake Li, Joachim Krauth, and Joseph P. Huston

CONTENTS

* The original article was edited for format and published as: Li, J.-S., Krauth, J., & Huston, J. P. (2006). Operant behavior of rats under fixed-interval reinforcement schedules: A dynamical analysis via the extended return map. *Nonlinear Dynamics, Psychology, and Life Sciences, 10,* 215–240. Reprinted with permission of the Society for Chaos Theory in Psychology & Life Sciences.

The extended-return-map (ERM) was employed to analyze the inter-response time data of operant experiments using fixed-interval (FI) schedules and food reinforcement. After intensive training over numerous sessions, rats gradually developed several types of temporal patterns of lever-pressing (LP) behaviors, which were visualized through different patterns of data point distributions in the ERM. Analyses with randomly shuffled data sets confirmed that these patterns depended on the sequential order of the inter-response time data, indicating that they reflected dynamics of the behavior. A procedure was developed to quantify the difference between patterns in the ERPs, thus enabling the comparison between sessions and animals. Simulations suggested that, in addition to the two-state break-and-burst responding, both multiple switches of behavioral states during the inter-reinforcement periods and the acceleration of LP rate should be taken into consideration for the dynamics found in the data.

It has long been known that FI schedules lead to a distinct temporal pattern of behavior in which rate of responding increases between the deliveries of successive reinforcements. (A brief description of Skinner-box experiments and FI schedules of reinforcement can be found in Appendix 7.A.) This pattern has been described as a scalloped curve (Ferster & Skinner, 1957; Skinner, 1938), or in some studies as two-state break-and-burst responding (Dews, 1978; McGill, 1963; Schneider, 1969; Shull, 1991). There have been several theoretical attempts to model the FI-schedule performance (Gibbon, 1977; Hoyert, 1992; Killeen & Fetterman, 1988; Machado, 1997; Machado & Cevik, 1998). For example, Machado (1997) built a model of FI responding, which was capable of learning the scalloped curve of responses. In the work of Shull, Gaynor, and Grimes (2001), the idea of "two-state" process was the starting point. They suggested that variable-interval schedule performance was composed of periods of engagement in responding alternating with periods of disengagement. However, the regularity of such behaviors under FI-schedule control is constantly interrupted by variability in the length of the pause, the degree of acceleration, the total number of responses emitted in the inter-reinforcement period, etc. The picture of operant behavior found under the control of such schedules is, thus, a mixture of periodicity and irregularity. Variability can be seen not only between different individuals, but also within the same animal between different sessions, or even during different inter-reinforcement periods within a session.

In recent years, nonlinear dynamical systems approaches have been widely used to study complicated phenomena and have achieved remarkable success in many fields. The fundamental difference between the nonlinear dynamical systems and the traditional approach regards the causes of variability. In the traditional, linear approach, researchers tend to look for additional variables or external disturbances to explain variability in the data. On the contrary, the research on the nonlinear dynamical systems leads to the finding that simple deterministic systems with nonlinear characteristics are capable of producing one-dimensional time-series data with turbulent appearances. Examining the traditional statistical properties is not sufficient to distinguish the behavior of a nonlinear system's dynamics from that produced by a purely stochastic process (e.g., May, 1976). The graphical tools used in the nonlinear dynamical systems approach, on the other hand, can generate distinct patterns in a multidimensional map that reveals deterministic characteristics of the originally one-dimensional time-series data (Kantz & Schreiber, 1997).

One of these tools, the return-map, is designed for a special type of time-series data, namely the interevent times, and has been successfully applied in the analysis of many physical systems (e.g., Shaw, 1984). It is constructed by plotting the interevent time at any moment against its predecessor. If there are deterministic relationships between interevent times, the patterns shown in the map will reflect the mathematical properties of these relationships. The return-map has also been applied with limited success in the analysis of the inter-response time data in free operant experiments (Palya, 1992; Weiss, 1970). There seemed to be no direct dependency between neighboring inter-response times. Li and Huston (2002) modified the return-map and introduced a new method, the ERM, to construct a two-dimensional diagram out of one-dimensional inter-response time data. In an ERM, averages of several inter-response times were plotted against averages of their predecessors. If there are deterministic relationships between neighboring moving averages, their mathematical properties might be reflected in the ERM patterns. The authors trained rats with different intermittent schedules of reinforcement. Distinct lattice patterns appeared in the ERMs after animals had underwent several sessions of training. The ERM also enabled the tracing of the development of schedule control of behavior over multiple training sessions. Furthermore, simulation studies suggested that an abrupt switch between behavioral states was an essential part in the dynamics of operant behavior controlled by FI schedules.

However, there are still some questions regarding the application of the ERM that remain unanswered. First, the ERM itself is only a qualitative description of the dynamics of operant behavior. There is a need of a measure to quantify the differences between two ERMs. Such a measure can enable the comparison between behaviors of different animals and the comparison between behaviors of the same animal in different sessions. Second, it is still unclear whether or not the ERM can provide information about the animal's behavior that differs from those brought about by traditional variables, such as the rate of responding. Finally, more studies are needed to understand the link between patterns in the ERMs and the actual behaviors observed in a Skinner-box.

In the present study, we presented the results of a more thorough study of FI responding using the ERM. A procedure was introduced to quantify the differences between two ERMs, thus enabling a quantitative comparison between data from different sessions and data produced by different animals. To illustrate how different FI dynamics can reveal the ERM patterns seen in the experimental data, we also applied the ERM to data generated by computer simulations. The starting point of our studies was the "two-state-conception." That is, simulation programs alternate between LP and non-lever-pressing (non-LP) states. Later, additional properties, namely the acceleration of LP rate and multiple switches of behavioral states during the inter-reinforcement period, were implemented into simulation programs. They were found to be responsible for the more complicated dynamics of operant behavior under the control of FI schedules.

Data Analysis

Extended-Return-Map

The data gathered here is the time interval between two adjacent lever presses, the so-called inter-response time data. The procedure to construct the ERM was described previously (Li & Huston, 2002). Briefly, let T_n ($n = 1, ..., N$) be the nth inter-response time of a Skinner-box experiment, where N is the total number of responses in a session. First, a new time-series U_m, which consisted of moving averages of inter-response time data, was calculated:

$$U_m = \frac{1}{f} \sum_{i=0}^{f-1} T_{m+i}. \tag{7.1}$$

The parameter f defined the width of the moving-average-window. Then, a set of two-dimensional points, ERM, was defined by

$$ERM \equiv \{(U_j, U_{j+f}) \mid j = 1, ..., N - 2f + 1\}. \tag{7.2}$$

The patterns in the ERM changed with f. If $f = 1$, the ERM reduced to the original return-map. A larger value of f led to the formation of more clusters of points. However, if the value of f was too large, clusters came too close to each other and began to overlap. That interfered with the observation of the fine structures of ERM patterns. We have tried different choice of f (results not shown here) and found empirically that it was advantageous to define the parameter f with the integer closest to one half of the averaged number of responses per reinforcement in that session. Throughout the present work, the choice of the parameter f followed this rule.

Quantification of Differences between Two ERMs

Under the control of FI schedules, many rats produced patterns in the ERM similar to those shown in Figure 7.1a. It was found that the position of the cluster of points marked by "*a*" roughly equaled "inter-reinforcement period/*f*" (IRP/*f*). The inter-reinforcement period was defined by the experimental setup. We could make use of this property and defined a lattice structure to cover the ERM. Each square in the lattice structure had the same width *w*, and its value was defined according to

$$w = \frac{2 \cdot ((IRP/f) - U_{min})}{(2 \cdot W - 1)}, \qquad (7.3)$$

where U_{min} was the smallest moving average of inter-response times calculated in Equation 7.1. The parameter W was an integer and it specified the size of the squares. Its value influenced the ability of the procedure to differentiate two ERMs. For lack of theoretical rationale to select the value of W, heuristic arguments based on intensive research on numerous ERMs of different sessions and rats led to the choice of $W = 5$ in the present study.

Beginning from the left-bottom corner, let S_{pq} be the square in the *p*th column and the *q*th row of the lattice structure; the position of S_{pq} was defined by

$$S_{pq} = \{(x_p, y_q, x_p + w, y_q + w) \mid x_p = U_{min} + pw, y_q = U_{min} + qw; p, q = 0, 1, 2 \ldots\}. \qquad (7.4)$$

An example of such a lattice structure is shown in Figure 7.1b.

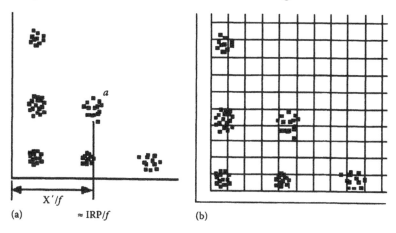

(a)   X'/*f*   ≈ IRP/*f*   (b)

FIGURE 7.1
(a) A scheme of the lattice-like ERM pattern. (b) A lattice structure defined over the ERM pattern in (a). The relative point density in each square defines an ERM vector.

Finally, the relative density of data points ρ_{pq} in the square S_{pq} was calculated according to

$$\rho_{pq} = \frac{\Theta(S_{pq})}{\sqrt{\sum_{k,l} \Theta^2(S_{kl})}}. \tag{7.5}$$

The variable $\Theta(S_{pq})$ was the number of data points in the square S_{pq}. A data point (U_j, U_{j+f}) belonged to the square S_{pq}, if it satisfied: $x_p \leq U_j < x_p + w$ and $y_q \leq U_{j+f} < y_q + w$. The series $\{\rho_{pq}\}$ could be regarded as a pq dimensional vector (ERM vector). Here p and q were the number of columns and rows of the lattice structure, respectively. The sum of squared differences of vector elements (ERM-SS) and the vectorial distance between two ERMs, A and B, were defined by

$$\text{ERM-SS} = \sum_{pq} (\rho_{pg}^A - \rho_{pg}^B)^2 \tag{7.6}$$

and

$$|\Delta\text{ERM-V}| = \sqrt{\text{ERM-SS}}, \tag{7.7}$$

respectively.

Part 1: Experimental Study

Method and Procedure

The experiment was done on six males, experimentally naive white Wistar rats. They were 2 months old and weighed between 220 and 260 g at the beginning of the experiment. Since the present work was aimed at introducing the analyzing tool ERM, we omitted other details of the experiment and outlined here only the necessary information for the understanding of the procedure. After 5 days (15 min/day) of manual shaping, followed by a 2 day pause and 1-day (90 min) training under continuous reinforcement, the reinforcement schedule was changed abruptly to FI. The FI training continued in the next 26 days. Three inter-reinforcement periods, 20, 40, and 60 s, were in use. They were designated as FI 20s, FI 40s, and FI 60s throughout the text. Each animal was assigned one of the three inter-reinforcement periods. During the 26 days of FI training, the assignment did not change. Each animal received one 90-min session daily.

Results

Typical ERM Patterns of FI Responding

The upper panel in Figure 7.2 shows the ERMs of rats 3 and 4 as examples for two typical ERM patterns of operant responding in session 26. The ERM of rat 3 shows several clusters that constitute a more-or-less lattice-like structure. In addition, there is a weak diagonal connection (indicated by an empty arrowhead) between clusters, and the cluster indicated by an empty arrow slightly elongates along the Y-axis. The ERM of rat 4 shows, in addition to a noticeable cluster in the center (indicated by a solid arrow), a dense distribution of data points in the bottom-left corner, which forms roughly a triangle. Another extinct ERM pattern of FI responding that was often seen

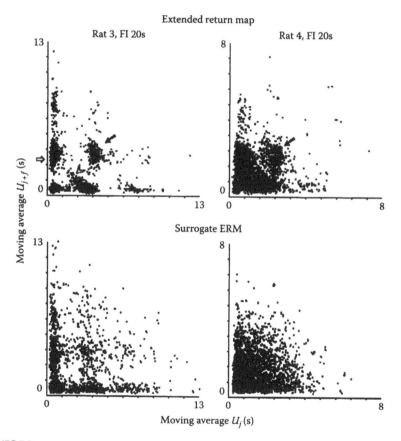

FIGURE 7.2

Upper panel: The ERMs of rats 3 and 4 data from session 26. Solid arrows indicate the central cluster. An empty arrow indicates the elongation of a cluster along Y-axis in the ERM of rat 3. The arrow head points to a weak diagonal connection between clusters. *Lower panel*: The ERMs of surrogate data sets, which were generated by randomly reorganizing the sequential order of IRT data used in the ERMs in upper panel.

(especially in earlier sessions of FI training) shows clusters elongated along the X-axis as well as the Y-axis. As a result, the pattern looks like the capital letter "L" plus a noticeable cluster in the center. The ERM of rat 4 in session 6 (Figure 7.4) is an example for it. The ERMs of the other animals' responses in session 26 had either lattice- or triangle-like structure (results not shown). No consistent correlation between the types of ERM patterns and the lengths of the FI schedules could be found.

As mentioned previously, the parameter f of the ERMs was defined by rounding, to its closest integer, one half of the number of lever presses per inter-reinforcement period in the session. Under this condition, the position of the center cluster in the map (indicated by solid arrows in the upper panel of Figure 7.2) followed a well-defined rule, namely, its X- and Y-coordinates were approximately equal to the inter-reinforcement period divided by the parameter f. By using this feature, we had adjusted the maximal scale of every ERM accordingly, so that for all ERMs this center cluster was located approximately in the same position of the diagram. This adjustment was extended to those rats which show ERM patterns other than the lattice structures; for example, the ERMs of the surrogate data sets or the ERMs from earlier sessions.

ERM Patterns of Surrogate Data

Now we wished to determine whether our findings in the ERMs really reflect the system's dynamics, or whether they are simply caused by the analyzing procedure. We tested the results with surrogate data sets which were generated by randomly reorganizing the sequential order of the same inter-response time. We applied the pseudo random number generator "Mersenne Twister" developed by Matsumoto and Nishimura (1998) in the generation of surrogate data sets. The surrogate data sets had the same statistical properties as the original ones, such as means, variances, and power spectrum, while the dynamical information, that is, the deterministic relationship of the system's present state with its history, was destroyed. If the patterns in the ERMs from surrogate data sets differed from those of the original ones, then the differences must be due to the fact that the dynamical information presented in the original data had been removed. The ERMs of surrogate data sets are shown in the lower panel of Figure 7.2. Inspection of the surrogate ERMs reveals that the distinct lattice as well as triangle structures disappeared or, at least, became more diffuse after the reorganizing process.

Correlation Dimension of the Surrogate ERM Patterns

In addition to this qualitative study, we conducted a quantitative comparison between the experimental and the surrogate ERMs. We repeated the randomization process 10,000 times and calculated the correlation dimension for each of the surrogate ERM and, thus, acquired a distribution of the surrogate correlation dimensions. The correlation dimension is a kind of fractal dimension that reflects geometrical features of data sets. The definition of the correlation dimension and the process to calculate it from experimental data were introduced independently by Grassberger (1983) and by Hentschel and Procaccia (1983). The size of the area outside the experimental value of correlation dimension in the distributions of surrogate correlation dimensions reveals the probability of obtaining the experimental inter-response time data simply by chance. A similar procedure was also used in our previous work (Li & Huston, 2002). The results shown in Table 7.1

TABLE 7.1

The Size of the Area[a] Outside the Experimental Value in the Distribution of Surrogate Correlation Dimensions

Rats	Area Size
1	0.0037
2	0.0001
3	0.0389
4	0.0009
6	0.0005
9	0.0438

[a] Total area normalized to unity.

suggest that it is very unlikely that the experimental sequences of inter-response time data were acquired simply by chance, since all experimental values were located far from the central region of the distributions of the surrogate correlation dimensions.

Development of the Temporal Dynamics of FI Responding

The process of acquisition of these specific behavioral dynamics could also be assessed using ERMs. The ERMs of sessions 1, 2, 3, 6, 9, 12, 15, 18, 21, 24, 25, and 26 of rats 3 and 4 are shown in Figures 7.3 and 7.4 as examples. Sessions 1, 2, 3, 24, 25, and 26 were chosen because they represent the beginning as well as the end of the learning process. The other sessions were chosen in constant distance of sessions. Initially, all animals showed similar behavioral dynamics, as indicated by the similar patterns in ERMs of session 1. Some also shared similar intermediate stages. Gradually each rat developed its individual dynamics of FI responding. A careful examination on the ERMs of sessions 24–26 in the end-phase of the learning process reveals that even the fine structures seen in the ERM patterns were preserved from session to session, which is indicative of the stabilization of the behavioral dynamics of the animals under the given experimental conditions.

Learning Curve of the FI Responding

The change of animal's behavior with sessions of FI training can be regarded as a kind of learning curve. In most of the traditional analyses, the rate of responses was used to measure the change of behavior over sessions. This observable reflects the quantitative aspect of behavioral changes. The ERM-SS can also be used to plot a learning curve. Since the ERM reflects the temporal patterns of the behavior, the ERM-SS learning curve can be regarded as a measure of behavioral changes from the qualitative aspect.

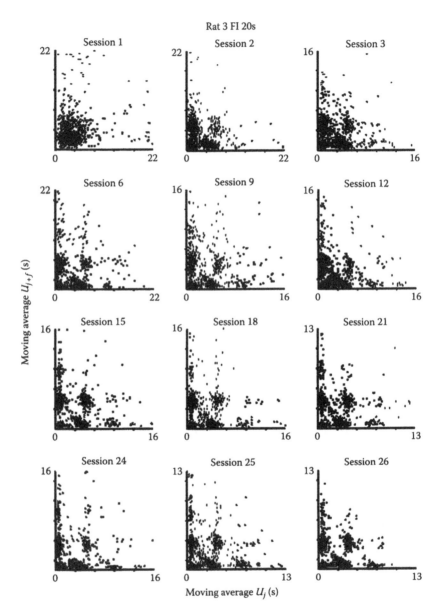

FIGURE 7.3
The ERMs of sessions 1, 2, 3, 6, 9, 12, 15, 18, 21, 24, 25, and 26 for rat 3. The developing process of the behavioral dynamics under the control of FI schedules can be seen.

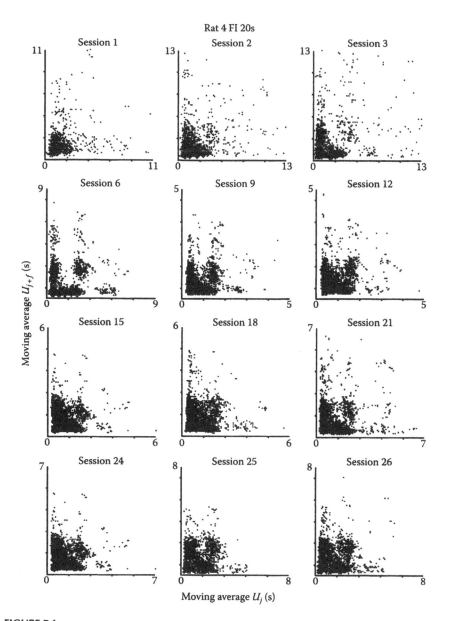

FIGURE 7.4

The ERMs of sessions 1, 2, 3, 6, 9, 12, 15, 18, 21, 24, 25, and 26 for rat 4. The developing process of the behavioral dynamics under the control of FI schedules can be seen.

In Figure 7.5a, the learning curves constructed using both observables are shown. The result of each animal was plotted separately. The learning curves of the traditional observable (dotted lines) were constructed using squared difference of response rates between earlier sessions and session 26. We took the squared difference because the ERM-SS is also a squared quantity. Besides, a squared difference, like the ERM-SS, is always positive. The ERM-SS used for plotting the learning curve was calculated using one ERM from the earlier sessions and the other ERM from session 26. Relative values were used for the ordinate. The results indicate that the ERM-SS learning curves indeed approached asymptotically a final level, which implied that the behavioral patterns approached asymptotically to a final stable state.

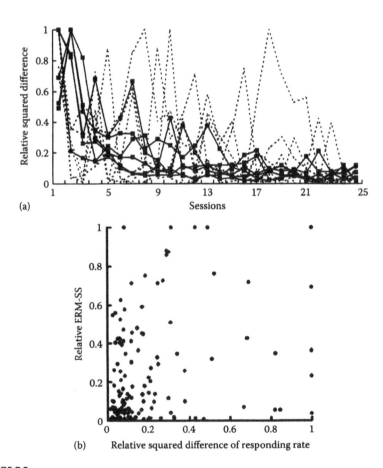

FIGURE 7.5
(a) Learning curves constructed using ERM-SS (solid lines) and squared differences of responding rate (dashed lines). Ordinate: Differences of earlier sessions to session 26. Relative values were used. Abscissa: sessions. (b) Scatter plot of relative ERM-SS by relative squared differences of responding rate. Data of six rats were pooled.

The response-rate learning curves of some animals, on the contrary, showed strong fluctuations even after numerous sessions of FI training.

We also plotted a scatter diagram (Figure 7.5b) using the relative ERM-SS as Y-coordinates and the relative squared difference of response rates as X-coordinates. Pooled data are used in this plot. Furthermore, the Pearson product–moment correlation coefficient between the relative ERM-SS and the relative squared difference of response rates was calculated for each animal separately, and the results are shown in Table 7.2. Both results suggest that the ERM-SS and the squared difference of responding rate had only low-to-moderate correlation.

TABLE 7.2

Pearson Product–Moment Correlation Coefficients between the ERM-SS and the Squared Difference of Responding Rate

Rats	R
1	0.25
2	0.43
3	0.60
4	0.38
6	0.52
9	0.34

Part 2: Simulation Study

Simulation Programs

In our previous work, we presented simulation studies, which indicated that an abrupt switching between behavioral states during the inter-reinforcement period was essential for the appearance of lattice patterns in the ERM (Li & Huston, 2002). Employing this idea as a starting point, we now present some basic patterns of FI-responding dynamics that can explain the formation of the lattice-, triangle-, and "L"-like structures in the ERMs shown above. Results of simulations are presented to supplement the analytical explanations. A detailed formulation of the simulation programs can be found in Appendix 7.B, and the parameters used are presented in Table 7.3.

Simulation Results

Lattice Structure

We begin our discussion with a regular alternation between the non-LP and the LP sections, as shown in Figure 7.6a. For simplicity, we also initially presume that the rate of LP within the LP sections is approximately constant, and the length of non-LP sections is much longer than the averaged inter-response time within LP sections. Under these conditions, and using a parameter f smaller than one half of the averaged number of lever presses per reinforcement, we obtained three clusters of points in the ERM, as shown

TABLE 7.3

Parameters Used in Computer Simulations

Programs		1		2		3	
		Figure 7.6d	Figure 7.6e	Figure 7.7c	Figure 7.8c	Figure 7.9c	Figure 7.9d
Length of non-LP state	O_{nlp}	1200	1200	400	600	600	400
	R_{nlp}	1	800	800	300	500	900
	M_{nlp}	—	—	—	—	600	400
Acceleration state	O_{acc}	—	—	—	500	—	—
	R_{acc}	—	—	—	300	—	—
	O_{ad}	—	—	—	200	—	—
Length of LP state	O_{lp}	—	—	—	—	0	0
	R_{lp}	—	—	—	—	400	200
	M_{lp}	—	—	—	—	800	1000
Probability of LP	m	0.03	0.03	0.03	0.03	0.03	0.03
	a	0.1	0.1	0.1	0.1	0.1	0.1
	S_o	0.0002	0.0002	0.0002	0.0002	0.0002	0.0002
FI (s)		20	20	20	20	20	20

in Figure 7.6c (clusters a, b, and c), and in the results of simulation 1 shown in Figure 7.6d. The points in the cluster "*a*" had both X- and Y-coordinates within the LP sections (windows X_a and Y_a). If either the X- or the Y-window crossed a non-LP section, clusters "*b*" (with windows X_b, Y_b) and "*c*" (with windows X_c, Y_c) appeared.

Now we must further consider the situation that some data points had Y- or X-windows across two or more non-LP sections (Y_d and X_f in Figure 7.6b), or that both X- and Y-windows crossed at least one non-LP section ($X_e Y_e$ in Figure 7.6b). Since the length of non-LP sections was usually much longer than the averaged inter-response time in LP sections, the resulting magnitude of one of the coordinates (X or Y) in the former cases was about twice that of the situations Y_b and X_c discussed above. In the later case, the magnitude of both X- and Y-coordinates (X_e and Y_e) was about the same as that of the situations Y_b and X_c. As a consequence, additional clusters "*d*," "*e*," and "*f*" appeared in the ERM (Figure 7.6c). They formed, together with the clusters "*a*," "*b*," "*c*," a lattice pattern. The result of computer simulation of this situation is shown in Figure 7.6e. Note that some clusters elongated slightly along X- or Y-axes. The cause of this phenomenon is explained below.

L-Type Structure

The distance between clusters depended mainly on the length of non-LP sections. In most of the inter-response time data, the length of non-LP varied

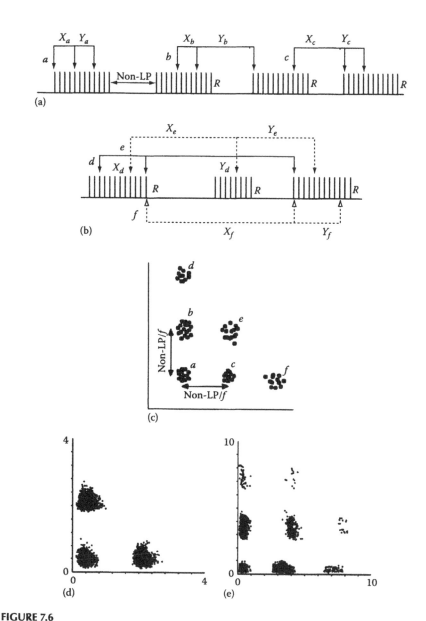

FIGURE 7.6

(a) Simple alternations of non-LP and LP behavioral states. (b) One of the two averaging windows crossed two non-LP states (solid line, Y_d, and dashed line, X_f), or both averaging windows crossed one non-LP state (dotted line, X_e and Y_e). (c) A schematic lattice-like ERM pattern resulted from the forward movement of averaging windows in (a) and (b). (d), (e) Results of computer simulations for situations in (a) and (b).

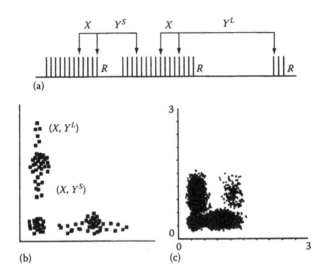

FIGURE 7.7
(a) Variation of the lengths of non-LP behavioral states, XY^L and XY^S. (b) Schematic ERM pattern. Elongation of clusters in the ERM can be observed. (c) Result of computer simulation.

from inter-reinforcement period to inter-reinforcement period, as indicated by the windows Y^S and Y^L in Figure 7.7a, where Y^S covered a shorter and Y^L a longer non-LP section. This variation resulted in an elongation of clusters, as shown in Figure 7.7b. In extreme cases, all three clusters were interconnected and an "L"-like pattern appeared. This type of ERM pattern was often found in earlier sessions during the training, for example, the ERM of rat 4 in session 6 (Figure 7.4). The simulation result for this situation is shown in Figure 7c. Note that there is a weak cluster in the center of the map. It is analog to the cluster "e" in Figure 7.6c, and its appearance can be explained using the arguments discussed previously.

Triangle Structures

We now consider the situation in which the rate of LP was not constant, but accelerated in the beginning of the LP sections, as shown in Figure 7.8a. This resulted in smaller non-LP sections followed by some relatively longer inter-response times in the LP sections. When the averaging windows X_g and Y_g moved forward, the resulting Y-coordinates of data points in the ERM increased gradually, while the X-coordinates were small and remained approximately constant. This corresponded to the cluster "g" elongated along the Y-axis shown in Figure 7.8b. Next we consider what happened when the averaging windows X_h and Y_h moved forward. In this case, the X-coordinates of data points in the ERM increased, while the Y-coordinates

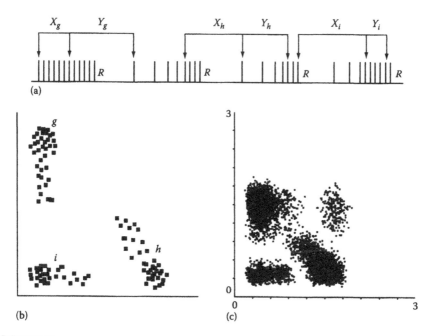

FIGURE 7.8

(a) An acceleration state is inserted between non-LP and LP states. (b) Schematic ERM pattern. Elongation of clusters along the Y-axis (cluster "g"), diagonal line (cluster "h"), and X-axis (cluster "i") appears. (c) Result of computer simulation.

decreased. The result was the cluster "h" elongated along the diagonal direction (Figure 7.8b). Finally, we consider the case of windows X_i and Y_i, where the X-coordinates decreased gradually and the Y-coordinates were small and remained approximately constant, when the averaging windows moved forward. This led to the cluster "i" elongated along the X-axis (Figure 7.8b). In summary, there appeared a triangle pattern in the ERM. The simulation of this situation is shown in Figure 7.8c. Note that in Figure 7.8c, there is a weak cluster in the center of the ERM, which can be explained in the same way as the cluster "e" in Figure 7.6c.

The triangle structure in the ERMs can also be explained by an alternative behavioral pattern shown in Figure 7.9a, in which multiple switches between non-LP and LP states during the inter-reinforcement period were hypothesized. We first consider the moving of windows X_g and Y_g in Figure 7.9a. Similar to the situation "g" in Figure 7.8a, Y-coordinates of data points in the ERM increased gradually when the windows moved from the range marked with solid line and filled arrowhead to that marked with dashed line and empty arrowhead. At the same time, the X-coordinates were small and remained approximately constant. Under this condition, the cluster "g" in the ERM appeared (Figure 7.9b). Similar consideration for the moving of

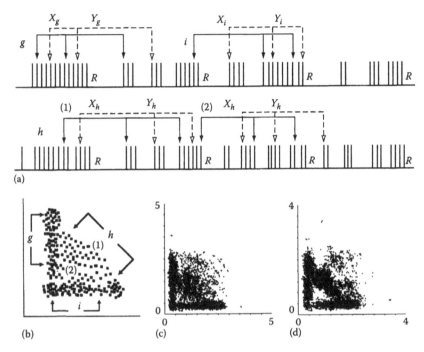

FIGURE 7.9
(a) Multiple switches of behavioral states during the inter-reinforcement periods. Solid lines and filled arrowheads represent the original positions of averaging windows. Dashed lines and empty arrowheads represent the new position after crossing a non-LP state. (b) Schematic ERM pattern. Diffused points in cluster "h" appeared. (c), (d) Results of computer simulations with different parameter sets. Filled triangle (c) and empty triangle (d) can be found.

windows X_i and Y_i from the range marked with solid line and filled arrowhead to that marked with dashed line and empty arrowhead revealed the cluster "i" in Figure 7.9b. The situation for the forward moving of windows X_h and Y_h is more complicated. Here both the averaging windows X_h and Y_h included some non-LP states. The resulting data points had comparable X- and Y-coordinates. As a result, they were located between clusters "g" and "i." Depending on the location of the averaging windows and on the distribution of non-LP states, we might have situation (1), in which X-coordinates increased while Y-coordinates decreased, or situation (2), in which X-coordinates decreased while Y-coordinates increased. The resulting points can be uniformly distributed in the area between X- and Y-axes, as in the cluster "$h(1)$" and "$h(2)$" in Figure 7.9b, and in the simulation results shown in Figure 7.9c. Here, the ERM pattern looked like a solid triangle. By setting different values for the simulation parameters, it was also possible to generate data points concentrated more along the diagonal direction as shown in Figure 7.9d.

Discussion

The results of the present study confirmed our previous finding that the ERMs can serve as an analyzing tool for the study of dynamics of operant behavior controlled by FI schedules (Li & Huston, 2002). In addition to the lattice-like structures found previously, some other distinct patterns, such as the triangle- and "L"-like structures, could be seen. This might be due to the using of more animals, and/or prolonged training under the FI schedules. The study with surrogate data sets suggested that these distinct ERM patterns depended on the sequential order of the inter-response time data; hence, they reflected dynamics of the FI responding. A procedure was suggested to convert the point-density distribution of an ERM into a vector, which could be used for the calculation of the ERM-SS and the vectorial distance. Both measures described quantitatively the difference between two ERMs. Theoretically, the ERM vector also enabled the calculation of mean vector of an experimental group.

The characteristic dynamics of FI responding gradually developed over sessions of training under the FI schedules. Similar results have been found previously (Li & Huston, 2002). With the aid of the newly defined ERM-SS, a learning curve of the temporal patterns of FI responding could be constructed. It indicated an asymptotical approach of the behavioral patterns toward a final stable state. This curve was somewhat different from that constructed using a traditional variable, the rate of responses. Furthermore, the relative ERM-SS and the squared difference of responding rate showed quite low correlation, suggesting that the two variables delivered different information about animals' behavior.

There could be several explanations for the varieties of ERM patterns found. For one, different ERM patterns might reflect subgroups within the animals. Since the subjects of the present study, the Wistar rats, are genetically heterogeneous, it is not surprising that different subtypes of behaviors were found under similar experimental conditions. It will be interesting to see whether or not it is possible to select an inbreed lineage of rats which shows, for example, exclusively FI responding with triangle-like ERM patterns. Such an inbreed lineage of rats may be useful in the study of certain behavioral disorders, such as the attention-deficit/hyperactivity disorder (ADHD), since it has been found that the FI responding of children with ADHD differed significantly from that of the normal children (Sagvolden, Aase, Zeiner, & Berger, 1998). The unusual behavior of ADHD children might be expressed in terms of particular ERM patterns.

On the other hand, different ERM patterns could be indicative of coexisting stable states of behavior under the same experimental conditions. Upon the abrupt change of reinforcement schedules, the animals must adapt their behavior accordingly. Even under the control of the same reinforcement schedule, and under the presumption that the individual differences were small, animals could be expected to develop different types of stable behaviors according to a mechanism analog to the process of bifurcation. Intensive works on the theoretical model of FI responding are required to study this process. Previously, Machado (1997) has built a model of FI responding which was capable of learning the scalloped curve of responses. Another model proposed by Hoyert (1992) could simulate the real-time dynamics of FI responding. However, it is not known they could account for the emergence of different types of behavioral dynamics under the same FI schedule. It will be interesting to examine the dynamical properties of these models with the ERM, and to see whether or not they can simulate different types of ERM structures seen in the present study.

The ERM patterns discerned in our data could be explained with the simplified and schematic dynamic patterns discussed in the text. We started with the "two-state-responding-conception" similar to that studied by several other groups (Schneider, 1969; Dews, 1978; Shull et al., 2001). Simple variation of the lengths of the non-LP state could lead to the formation of the lattice pattern. However, it is not sufficient to explain more complicated structures, such as the triangle pattern. Our simulations suggested that an additional acceleration state or multiple switches of behavioral states during the inter-reinforcement period might have played an important role. The experimental data might ultimately best be subsumed by several of such basic schemes. Alternatively, they could serve as a starting point for the building of a more comprehensive model to explain the dynamics of FI-schedule-controlled behavior.

The application of ERMs is not restricted to the study of operant behaviors, such as that observed in a Skinner-box experiment. In principle, any time-series data generated by mechanisms similar to a point-process can be analyzed using this tool. Whether or not it will reveal useful information depends on the properties of the data. The present study indicated that if events in the data set tend to appear in quasi-periodical bursts separated by longer pauses, then ERMs can deliver information about the fine structures hidden in the temporal organization. In addition to the inter-response time of a Skinner-box experiment, the firing of single neurons also produces interspike intervals with similar temporal structure. Thus, the analysis of neuronal firing seems to be a suitable subject of study for the application of ERMs.

Acknowledgment

This study was supported by a grant (DFG Hu 306/15-3) from the German Science Foundation and by a grant (NSC 93-2413-H-194-007) from the National Science Council of the Republic of China.

Appendix 7.A Skinner-Box Experiments and FI Schedules

A Skinner-box typically contains one or more levers which an animal can press, and one or more places in which reinforcers like food can be delivered. The contingency between the desired behavior, LP, and the delivery of foods increases its occurring frequency. In principle, it is possible for a rat to learn the correct response by trial and error. However, the rat is extremely unlikely to press the lever by chance, so a "shaping" procedure is often used. In the beginning, instead of rewarding the rat for producing exactly the LP behavior, it is rewarded whenever it performs a behavior which approximates to LP. The closeness of the approximation to the desired behavior required in order for the rat to get a pellet is gradually increased so that eventually it is only reinforced for pressing the lever.

In the initial definition of reinforcement, every occurrence of the desired behavior is rewarded. However, in nature environments, the same behavior seldom always leads to exactly the same consequence. Thus, it is more realistic to study the effects of intermittent schedules of reinforcement. The FI schedule is one example of intermittent schedules of reinforcement. Under this experimental condition, reinforcers will be delivered only after the desired behavior occurs after the passage of a fixed period of time, the so-called inter-reinforcement period.

Appendix 7.B Simulation Programs

The first simulation was implemented with a single switch of behavior from the non-LP to LP state. The non-LP state began with the previous reinforcement, and its length L_{nlp} was decided by the following equation:

$$L_{nlp} = O_{nlp} + \text{random}(R_{nlp}). \tag{7.8}$$

Here "random ( )" is the pseudo random number generator "Mersenne Twister" that can generate random integers within the range $(0-R_{nlp})$ with uniform probability. Parameters O_{nlp} and R_{nlp} determined the magnitude and variance of L_{nlp}. The time resolution in the simulation was 0.01 s.

After the end of the non-LP state, one lever press was generated immediately and the LP state began. The probability $P(t_{lp})$ of further lever presses was a sigmoid function described by the following equations:

$$P(t_{lp}) = m \times S(t_{lp}), \tag{7.9}$$

$$S(t_{\text{lp}}) = S(t_{\text{lp}} - 1) + a \times [1 - S(t_{\text{lp}} - 1)], \tag{7.10}$$

$$S(0) = S_O. \tag{7.11}$$

The discrete integer t_{lp} counted the time after previous lever press. Again the time resolution was 0.01 s. Parameter m scaled the sigmoid function $S(t_{\text{lp}})$ ($S \in [0\ldots1]$) to the actual probability of lever presses $P(t_{\text{lp}})$. Initial value and the increasing rate of S were determined by parameters S_O and a.

The second simulation was the same as the first one, except that there was an accelerating state implemented between the non-LP and LP states. At the end of the non-LP state, one lever press was generated immediately and the acceleration state began. The length of acceleration states was determined by the following equation:

$$L_{\text{acc}} = O_{\text{acc}} + \text{random}(R_{\text{acc}}). \tag{7.12}$$

During this state, the probability of further lever presses was also described by Equations 7.9 through 7.11. However, unlike the LP state, an acceleration element A_{ad} was added to the time counter t to compute the length of inter-response times. A_{ad} was calculated through the following equation:

$$A_{\text{ad}} = O_{\text{ad}}\left[1 - \frac{t_{\text{ad}}}{L_{\text{acc}}}\right], \tag{7.13}$$

where t_{ad} was an integer which counted the time after the beginning of the acceleration state. The factor $[1 - (t_{\text{ad}}/L_{\text{acc}})]$ was the quotient of time left in the acceleration state and decreased linearly. The parameter O_{ad} determined the magnitude of the additional factor. The acceleration state ended when t_{ad} exceeded L_{acc}. Unlike the end of a non-LP state, no lever press was generated. The computation of LP states, which came after the acceleration state, was the same as in the first simulation.

In the third simulation, multiple switches between behavioral states were implemented. After reinforcement, the first non-LP state began. The lengths of non-LP and LP states (L_{nlp} and L_{lp}) were determined by Equations 7.14 and 7.15, respectively:

$$L_{\text{nlp}} = O_{\text{nlp}} - M_{\text{nlp}}\left(\frac{t_{\text{rf}}}{\text{FI}}\right) + \text{random}(R_{\text{nlp}}), \tag{7.14}$$

$$L_{\text{lp}} = O_{\text{lp}} - M_{\text{lp}}\left(\frac{t_{\text{rf}}}{\text{FI}}\right) + \text{random}(R_{\text{lp}}), \tag{7.15}$$

where an integer t_{rf} counted the time after reinforcements with time resolution 0.01 s and FI was the length of the inter-reinforcement period. Parameters

O_{nlp}, R_{nlp}, O_{lp}, and R_{lp} determined the magnitude and variance of the length, and parameters M_{nlp} and M_{lp} determined the slope of the linearly decreasing or increasing lengths of non-LP and LP states. Similar to the first simulation, one lever press was generated immediately after each non-LP state ended. During LP states, the probability of LP was determined by Equations 7.9 through 7.11.

References

Dews, P. B. (1978). Studies on responding under fixed-interval schedules of reinforcement: II The scalloped pattern of the cumulative record. *Journal of the Experimental Analysis of Behavior, 29*, 67–75.

Ferster, C. B., & Skinner, B. F. (1957). *Schedules of reinforcement.* New York: Appleton-Century-Crofts.

Gibbon, J. (1977). Scalar expectancy theory and Weber's law in animal timing. *Psychological Review, 84*, 279–325.

Grassberger, P. (1983). Generalized dimensions of strange attractors. *Physics Letters, 97A*, 227–230.

Hentschel, H. G. E., & Procaccia, I. (1983). The infinite number of generalized dimensions of fractals and strange attractors. *Physica, 8D*, 435–444.

Hoyert, M. S. (1992). Order and chaos in fixed interval schedules of reinforcement. *Journal of the Experimental Analysis of Behavior, 57*, 339–363.

Kantz, H., & Schreiber, T. (1997). *Nonlinear time series analysis.* New York: Cambridge University Press.

Killeen, P., & Fetterman, G. (1988). A behavioral theory of timing. *Psychological Review, 95*, 274–295.

Li, J.-S., & Huston, J. P. (2002). Non-linear dynamics of operant behavior: A new approach via the extended return map. *Reviews in the Neurosciences, 13*, 31–57.

Machado, A. (1997). Learning the temporal dynamics of behavior, *Psychological Review, 104*, 241–265.

Machado, A., & Cevik, M. (1998). Acquisition and extinction under periodic reinforcement, *Behavioural Processes, 44*, 237–262.

Matsumoto, M., & Nishimura, T. (1998). Mersenne Twister: A 623-dimensionally equidistributed uniform pseudorandom number generator. *ACM Transactions on Modeling and Computer Simulation: Special Issue on Uniform Random Number Generation, 8*, 3–30.

May, R. M. (1976). Simple mathematical models with very complicated dynamics. *Nature, 261*, 459–467.

McGill, W. J. (1963). Stochastic latency mechanisms. In R. D. Luce, R. R. Bush., & E. Galanter (Eds.), *Handbook of mathematical psychology*, Vol. 1 (pp. 309–360). New York: Wiley.

Palya, W. L. (1992). Dynamics in the fine structure of schedule–controlled behavior. *Journal of the Experimental Analysis of Behavior, 57*, 267–287.

Sagvolden, T., Aase, H., Zeiner, P., & Berger, D. (1998). Altered reinforcement mechanism in attention-deficit/hyperactivity disorder. *Behavioural Brain Research, 94*, 61–71.

Schneider, A. B. (1969). A two-state analysis of fixed-interval responding in the pigeon. *Journal of the Experimental Analysis of Behavior, 12*, 677–687.

Shaw, R. S. (1984). *The dripping faucet as a model chaotic system.* Santa Cruz, CA: Aerial Press.

Shull, R. L. (1991). Mathematical description of operant behavior: An introduction. In I. H. Iversen & K. A. Lattal (Eds.), *Techniques in the behavioral and neural sciences: Vol. 6. Experimental analysis of behavior, Part 2* (pp. 243–282). Amsterdam: Elsevier.

Shull, R. L., Gaynor, S. T., & Grimes, J. A. (2001). Response rate viewed as engagement bouts: effects of relative reinforcement and schedule type. *Journal of the Experimental Analysis of Behavior, 75*, 247–274.

Skinner, B. F. (1938). *The behavior of organisms.* New York: Appleton-Century-Crofts.

Weiss, B. (1970). The fine structure of operant behavior during transition states. In W. N. Schoenfeld (Eds.), *The theory of reinforcement schedules* (pp. 277–311). New York: Appleton-Century-Crofts.

8

Adjusting Behavioral Methods When Applying Nonlinear Dynamical Measures to Stimulus Rates*

Barbara Bruhns Frey

CONTENTS

* The original article was edited for format and originally published as: Frey, B. B. (2006). Adjusting behavioral methods when applying nonlinear dynamical measures to stimulus rates. *Nonlinear Dynamics, Psychology, and Life Sciences, 10,* 241–273. Reprinted by permission of the Society for Chaos Theory in Psychology & Life Sciences.

The nonlinear dynamical perspective addresses an area of behavior that tends to be ignored: trial-to-trial variability under seemingly identical conditions. Data exhibiting sequence effects are excellent candidates for a nonlinear dynamical analysis, because of their dependence on previous trial events. When applying nonlinear dynamical tools to behavioral data, the assumptions underlying the tools may require changes in experimental methodology. In this chapter, the assumptions of trial timing and their potential impact on nonlinear dynamical analysis are explored. This is done with the intent of initiating a discussion on the appropriate control of trial timing. To examine these issues, each participant's response times (RTs) on simple two-choice tasks were treated as a time series and submitted along with two comparisons to two nonlinear dynamical measures. Two experimental variables, stimulus rate and stimulus–response mapping, were manipulated in order to examine the issue of controlling trial timing via stimulus rate. Significant differences between the observed and comparison time series were found for estimated dimensionalities (*m*). Differences in dimensionality (*m*) estimates were also found between the experimental variables. Of the two methods for controlling stimulus rate, response–stimulus interval and interstimulus interval, the latter is recommended when applying nonlinear dynamical measures.

This chapter examines RTs from a nonlinear dynamical perspective. Advances in computer technology and nonlinear dynamical tools enable investigation into behavioral nonlinear dynamical structure. Nonlinear dynamical tools are sensitive to characteristics that traditional linear methods are unable to detect. Yet, few researchers have applied these nonlinear dynamical tools to behavioral data, and assumptions underlying the tools may require changes in experimental methodology. In this chapter, the assumptions of trial timing and their potential impact on nonlinear dynamical analysis are explored. This is done with the intent of initiating a discussion on the appropriate control of trial timing.

Time series analysis and nonlinear dynamical measures have been applied to behavioral data, in particular RT data (Clayton & Frey, 1996; 1997; Gilden, 1997; Gilden, Thornton, & Mallon, 1995; Heath, 2000; Kelly, Heathcote, Heath, & Longstaff, 2001; Ward, 1994, 1996, 2002). These studies have uncovered some characteristics that *hint* of a nonlinear dynamical account. Kelso (1995), Thelen and Smith (1994), Port and Van Gelder (1995), Heath (2000), and Ward (2002) provide an excellent summarization of the behavioral findings.

Typically in RT experiments, participants are given a warning signal and after a short delay a stimulus appears. The participant's task is to classify the stimulus as quickly as possible while trying to maintain high accuracy. For example, in a two-choice J–K task, the participant presses one key to indicate the presence of a "J" and another key for the presence of a "K." Following the response, there is a short delay before the next trial. This delay is termed the response–stimulus interval (RSI).

In many experiments examining RT, the trial timing is not always clearly defined. Some early studies provide only the total time of the experiment. Others use ambiguous terms such as time lag. Examination of the methods suggests most studies control or manipulate RSI. RSI serves as an index of the rate at which the stimuli are presented. As the stimulus rate is increased, less time is available to prepare for the next trial.

Another measure of the stimulus rate is interstimulus interval (ISI), which is the time between the presentations of stimuli. ISI is related to RSI by the equation $ISI = RT + RSI$. Since RT is a varying dependent measure, only one of the two measures can be made a constant at a given time. Figure 8.1 illustrates the timing of trial events using the two measures of stimulus rate.

Probability theory led to the traditional view of RT as a random variable X (Klein, 1997). From this perspective, observations $(X_1, ..., X_n)$, where X_i is an observation and n equals the number of observations, are independent and reflect some aspect of the underlying system, such as processing involved in generating the RT (e.g., decision making, retrieval, response generation). From this view, RTs are assumed to be composed of deterministic and stochastic components (Luce, 1986). For example, the deterministic component can be reflected as a measure of central tendency for all observations collected under a particular trial condition (i.e., the same state). The random contribution can be modeled via a stochastic process (e.g., $w(0, \sigma^2)$ where w corresponds to a probabilistic function with a mean of 0 and a variance of σ^2). If the processes underlying one response are independent of the next (i.e., assume processing is reset

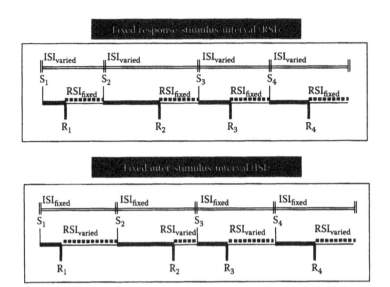

FIGURE 8.1

The top diagram indicates the trial events for a fixed RSI. The bottom diagram depicts the trial events for a fixed ISI. The horizontal lines correspond to time advancing from left to right. R corresponds to a response made to a specific stimulus (S).

each trial) and variance in RT across equivalent conditions result from random contributions, then the method for controlling stimulus rate should not matter.

Numerous findings such as priming and sequence effects suggest that RTs do depend on previous trials (Luce, 1986; Kirby, 1980). Rather than the exception, trial-to-trial dependencies are ubiquitous, widely observed across various dependent measures and paradigms. Since RTs depend on prior trials, the method of controlling stimulus rate might matter, particularly when applying nonlinear dynamical measures. Nonlinear dynamical measures generally exploit three basic characteristics: dependence on past history (dynamics), low dimensionality (i.e., limited number of degrees of freedom or active variables), and sensitivity to initial conditions.

Many of the nonlinear dynamical tools are used to analyze discrete time series. These time series are assumed to be obtained by taking measurements at equal interval from an underlying continuous process. The consequence of taking discrete measurements from a continuous system is a loss of information and possibly a distortion of the underlying continuous system. As an illustration, consider a simple sine wave. A sampling rate that is equal to the period of the sine wave provides a consistent estimate (i.e., no variability) of the dependent variable. If the sine wave is randomly sampled, the estimate of the dependent variable would range from −1 to 1 and there would be considerable variability compared to the first scenario. The dependent measure collected with these two methods of sampling would lead to very different outcomes. Failing to sample at equal intervals can introduce false dynamics and can result in an overestimate of the nonlinear dynamical characteristics (Kantz & Schreiber, 1997, 2002).

From the perspective of cognition, RT is viewed as a measure of information processing. When measuring the processes underlying RT, it is assumed the underlying processes are continuous and RT is a discrete measure of those processes. Although the experimental manipulations focus on between trial sequence effects, it is assumed that the underlying processing includes both within-trial and between-trial components.

Achieving a uniform sample when examining RT is a problem. One solution is to view the onset of the trial events and the time to respond to the events (RT) as a phase relationship. By fixing the ISI so the stimuli are presented at equal intervals, we establish a uniform presentation rate of the stimuli. RT becomes a point estimate of the relative phase with respect to a uniform sampling rate. This view of RT is similar to the method that Kelso (1995) used to measure the phase difference in finger movements relative to a fixed interval signal.

Another problem that occurs when collecting behavioral time series is disruptions in the time series. Disruptions can occur when a participant takes a break after a series of trials or between sessions. Many nonlinear dynamical measures require long time series and are not designed to interpolate between disruptions in the data or to correct for missing observations in the time series. Ignoring the disruptions in the time series by concatenating data can also result in false dynamical patterns (Kantz & Schreiber, 1997, 2002). The tasks used in the following experiments were selected because participants are able to accurately and rapidly respond to the stimuli. Participants are able to complete a large number of trials within a single session.

Two experiments are conducted to investigate the application of a nonlinear dynamical approach to two-choice tasks. In the experiments, the trial-to-trial response latencies are treated as consecutive samples in a time series. To reduce the risk of falsely classifying a time series, each observed time series is compared with two surrogate control conditions and submitted to two tests. The control comparisons ensure that the tests can discriminate the observed time series from time series generated under two alternative hypotheses.

The two experiments test for nonlinear dynamical structure across two measures of stimulus rate (i.e., RSI and ISI). Experiment 1 manipulates RSI and task difficulty through stimulus–response mapping. Stimulus–response mapping refers to the spatial relationship between the stimulus and the response. For example, with direct mapping (i.e., congruent) and a stimulus presented on the left side, the participant's task is to press the response key on the left. When the mapping is crossed (i.e., incongruent) and the stimulus occurs on the left, the response key on the right should be pressed. By manipulating RSI and stimulus–response mapping, a variety of sequence patterns occur (e.g., Kirby, 1980). The variety of outcomes and explanations of those outcomes provide a rich data set to examine from a nonlinear dynamical perspective. In conjunction with the manipulation of stimulus–response mapping, Experiment 2 examines sampling rate via ISI. The lack of evidence for nonlinear structure in prior investigations may be due to the method of controlling stimulus rate.

Experiment 1A: RT Dynamics across Stimulus–Response Mapping and RSI

Experiment 1 is an extension of Vervaeck and Boer's (1980) work examining sequence effects. *Sequence effect* is the term used to describe the dependence of behavior on the past history of stimuli and responses. Sequence effects are observed across a wide range of paradigms and dependent measures (for a review, see Kirby, 1980; Kornblum, 1973; Luce, 1986). Numerous experimental variables (e.g., practice, stimulus–response mapping, participant expectancies, probability of an alternation, number of alternatives, stimulus and response probabilities, stimulus rate) affect both the magnitude and the direction of sequence effects. The ubiquity of sequence effects means the results have the potential to be generally applicable. Sequence effects exhibit characteristics that are consistent with a dynamical account. The complexity of the effects related to stimulus rate suggests that there may be nonlinear influences.

When a response is enhanced due to the repetition of an event (e.g., stimulus "J" followed by another "J"), the resulting enhancement is called a repetition effect. Conversely, an alternation effect is enhanced performance following an alternation of events (e.g., stimulus "J" followed by stimulus "K"). The order of a sequence pattern describes the amount of *past* history used to partition the patterns. First-order patterns correspond to the simple repetition and alternation effects.

Participants generally exhibit a repetition effect at short RSI (e.g., 50 ms; e.g., Kirby, 1976; Vervaeck & Boer, 1980). When RSI is increased, for example, to 1000 ms, participants performing a similar task typically produce an alternation or an abated repetition effect (e.g., Kirby, 1972, 1976). Suggesting behavioral asymmetry with stimulus rate, even with biases toward alternations via instructions or stimulus probabilities, alternation effects are seldom observed at very short RSIs (e.g., Kirby, 1980; Kornblum, 1973; Luce, 1986).

At long RSIs, there appears to be a process that is under the control of the participant (e.g., expectancies). A number of studies have demonstrated that expectancies influence the pattern of sequence effects (e.g., Kirby, 1976; Vervaeck & Boer, 1980; Kornblum, 1969). For example, after a minimal delay, participants exhibit the ability to choose to enhance either repeating or alternating responses.

In Experiment 1, there are two levels of stimulus–response mapping (direct, crossed) and two RSIs (short, long). The analysis is used to examine the trial-to-trial dependencies in more detail from a nonlinear dynamical perspective. The experimental manipulations allow for testing if the mean differences between experimental groups are reflected in nonlinear measures (i.e., levels of RSI or stimulus–response mapping).

Method

Participants

Twenty-four Vanderbilt University undergraduates participated in partial fulfillment of an introductory psychology course requirement. Twenty-three participants were right handed. All participants had normal or corrected-to-normal vision and participated for 1.5 h or less.

Design

A mixed-subject $2 \times 2 \times (2 \times 2)$ design was used. The two between-subject variables were RSI (short, long) and stimulus–response mapping (direct, crossed). The within-subject variables were stimulus type (LEFT, RIGHT) and first-order sequence type (repetition, alternation).

Stimuli, Stimulus Location, and RSI

The probability of a stimulus (i.e., LEFT or RIGHT) was approximately equal across sequence type and pattern. The stimulus location (left or right of the fixation) depended on the stimulus–response mapping condition. To discourage guessing based on a fixed stimulus interval, the RSI varied slightly within each condition. The delays within the RSI condition were randomly selected with equal probability. The three levels of the short RSI were 15, 45, and 75 ms. The three levels of the long RSI were 750, 900, and 1050 ms.

Stimulus–Response Mapping (Direct, Crossed)

Stimulus–response mapping was manipulated by changing the assignment between the hand used to respond (left, right) and the stimulus. In the direct condition, the correct response to the stimulus (e.g., LEFT) matched the response-hand (i.e., key pressed by the left hand). In the crossed condition, the stimulus (e.g., LEFT) and the response were incongruent (e.g., when stimulus LEFT was presented, the correct response was a right-hand key press).

To increase the difficulty in response selection between the two groups, the location of the stimulus was also mapped in a direct or crossed manner to the response location (The Simon Effect; Simon & Rudell, 1967; Wickens & Hollands, 2000). For the direct condition, the word "LEFT" appeared to the left of the fixation and the word "RIGHT" appeared to the right of the fixation. This was reversed for the crossed condition (e.g., the word "LEFT" appeared to the *right* of the fixation). In the crossed condition, the correct response to the stimulus RIGHT, presented on the left of the fixation point, was a left-key press. Left-responses and right-responses were made by pressing the "z" and the "/" keys with the left and right hand index fingers, respectively.

Different-hand response keys were chosen because they are slightly faster than same-hand responses (Annett & Annett, 1979) and exhibit less trial-to-trial variability (Helmuth & Ivry, 1996).

Procedure

At the beginning of the experiment, the task was explained to the participants. Participants were instructed to respond as quickly as possible and to maintain accuracy above 95%. The experiment consisted of 18 warm-up trials with accuracy feedback on each trial. Warm-up trials were followed by 1550 experimental trials with no accuracy feedback or breaks to rest.

After the participant initiated the experiment by pressing the 'z' key, a fixation point was displayed in the center of the screen for 100 ms. Following the fixation, the test stimulus (i.e., LEFT or RIGHT) was displayed to either the left or right of the fixation depending on the stimulus–response mapping. Participants responded by pressing the appropriate key. After the RSI elapsed, a new trial began.

Materials

A series of stimulus files were generated. For each stimulus file, a computer algorithm shuffled an equal number of both stimuli until there were approximately an equal number (i.e., 6) of every possible sequence of length eight (i.e., $S_{(n-7)}S_{(n-6)}S_{(n-5)}S_{(n-4)}S_{(n-3)}S_{(n-2)}S_{(n-1)}S_n$; number of combinations equals $2^8 = 256$, where S corresponds to the stimulus (LEFT or RIGHT) and n corresponds to the trial number). The same set of stimulus files was used for each between-subject condition.

Apparatus

Microcomputer Experimental Laboratory (MEL; Schneider, 1988) software on IBM-compatible PCs was used to control stimulus presentations, to collect responses, and to record response latencies for all experiments. MEL software was used in these studies because it is designed to provide millisecond (ms) precision in presenting and collecting stimuli. The measurement accuracy, generally under 10 ms, for RT depends on the type of keyboard and the key that is pressed. The participants viewed the computer monitor from approximately 45 cm.

Results

For each experiment, the first part of the result section summarizes the analysis of the mean proportion correct and RT. Incorrect response latencies and RTs less than 30 ms and more extreme than two standard deviations from the mean were excluded from the mean RT analyses (trimming is based on individual means and standard deviations). This section examines the data from a traditional sequence effects perspective and investigates whether the literature is confirmed.

Recent findings of fractal properties in behavioral data such as RT by Gilden and others (Clayton & Frey, 1997; Gilden, 1997; Gilden et al., 1995; Pressing & Jolley-Rogers, 1997; Ward, 2002) argue against using statistical measures such as the mean and variance (Liebovitch, 1998). The reason is that for fractals as more data are included the sample mean and variance do *not* converge on population parameters. For fractals, the sample mean and variance do not exist (Liebovitch, 1998). An alternative is to study how the time series evolves.

The latter half of the result sections examine the utility of nonlinear dynamical methods in studying sequence effects. In addition to examining whether the nonlinear measures reflect the differences in the means, this section examines whether the nonlinear dynamical measures provide new insights into sequence effects.

Summary of the Analysis of the Means

Participants maintained high accuracy. Average proportion correct was 0.97. Differences in proportion correct concurred with response latencies and were generally inversely related (i.e., no evidence for speed accuracy trade off existed). RTs increased as the RSI increased (compare 327–412 ms for short and long RSI, respectively; $F(1, 20) = 20.16$, $p < .05$). Consistent with the literature (Kirby, 1980), response latencies were slower in the crossed (402 ms, hard task) condition compared to the direct (337 ms, easy task) stimulus–response mapping condition, $F(1, 20) = 11.78$, $p < .05$. The interaction between RSI and stimulus-response mapping was not significant ($p > .10$).

Responses to repeated stimuli (repetitions) were 14 ms faster than responses to alternating stimuli, $F(1, 20) = 6.78$, $p < .05$ (compare 363–377 ms for repetitions and alternations, respectively). The interaction between sequence type and RSI was significant, $F(1, 20) = 4.88$, $p < .05$. During the short-RSI, responses to repetitions had a 26 ms advantage over alternations (compare 314–340 ms for repetitions and alternations, respectively). As more time was available, participants responded equivalently to both repetitions and alternations (compare 411–413 ms for repetitions and alternations, respectively). This finding is consistent with the literature (Kirby, 1976, 1980; Kornblum, 1973; Luce, 1986; Vervaeck & Boer, 1980).

The interaction between sequence type and stimulus-response mapping was significant, $F(1, 20) = 4.93$, $p < .05$. Participants in the easier direct stimulus–response mapping condition were able to respond quickly to both repetitions and alternations (compare 336–338 ms for repetitions and alternations, respectively). As task difficulty increased via a crossed stimulus–response mapping, participants responded 26 ms faster to repetitions than alternations (compare 389–415 ms for repetitions and alternations, respectively).

Responses to the stimulus RIGHT (365.2 ms) were 8.9 ms faster than the stimulus LEFT (374.1 ms), $F(1, 20) = 5.78$, $p < .05$. The interaction between type of stimulus and stimulus–response mapping was marginally significant, $F(1, 20) = 3.20$, $p < .10$. Because most of the participants were right handed and participants tend to exhibit a dominant-hand bias in simple RT tasks, it is not surprising that under the direct stimulus–response mapping, responses to the stimulus RIGHT (329.4 ms) were faster than to the stimulus LEFT (345.0 ms). Under the crossed stimulus–response mapping condition, the advantage for stimulus RIGHT over stimulus LEFT dissipated (compare 401.0–403.2 ms, respectively). The main effect and the interaction with type of stimulus suggest an effective manipulation of stimulus–response mapping.

Table 8.1 presents the marginal three-way interaction between sequence type, RSI, and stimulus–response mapping, $F(1, 20) = 3.71$, $p < .10$. The differences in the magnitude of the repetition effect replicate previous studies. As the RSI is increased, the advantage of repetitions dissipates. Likewise, given two tasks that differ only in difficulty, the more difficult task exhibits a larger repetition effect.

Abbreviated labels have been applied to the groups. For RSI, the letters L and S correspond to the long and short condition, respectively. For stimulus–response mapping, D corresponds to the direct condition and C corresponds to the crossed. When participants had plenty of time to perform a simple task, an abated repetition effect (i.e., 0.6 ms, LD group) was observed. The groups with either increased task difficulty (i.e., 3.6 ms, LC group) or limited time (i.e., 3.5 ms, SD group) exhibited similar intermediate repetition effects. As expected, the largest repetition effect (i.e., 47.8 ms, SC group) was observed in the condition with the least amount of time and the most difficult task. One issue is whether the mean differences across the four groups reflect changes in the underlying dynamical processing. If there are changes in the underlying dynamical processing, then the largest differences in the nonlinear measures should be observed between the LD and SC groups.

TABLE 8.1

Response Times (ms) for Sequence Type (Repetition, Alternation), RSI (Short, Long) and Stimulus-Response Mapping (SRM; Direct, Crossed)

RSI by Stimulus–Response Mapping[a]	Repetition (R)	Alternation (A)	Repetition Effect Difference (A − R)
Short RSI Direct SRM (SD)	308.4	311.9	3.5
Short RSI Crossed SRM (SC)	320.4	368.2	47.8
Long RSI Direct SRM (LD)	364.0	364.6	0.6
Long RSI Crossed SRM (LC)	458.1	461.7	3.6

[a] The letters S and L combined with D and C are used to form four descriptors of each group (i.e., SD, SC, LD, and LC).

Nonlinear Dynamical Analysis

Description of Time Series

Each participant's consecutive response latencies were treated as a time series (no observations were removed and practice trials were not included in the time series). Participants' time series are referred to as the observed time series. Each observed time series was difference lag one and the grand mean was removed. Then, two control comparison time series, shuffled and ARIMA, were generated from each observed time series. The control conditions tested whether surrogate data generated under two different linear models can be rejected (e.g., Scheier & Tschacher, 1996, 2002; Theiler, Eubank, Longtin, Galdrikian, & Farmer, 1992).

The first comparison time series is generated by shuffling, 10 times, the original time series that is suspected of being dynamical. Consequently, any trial-to-trial structure is destroyed (i.e., dependencies from trial-to-trial), while the distributional properties (e.g., standard deviation) are maintained. The shuffled and original time series differ only in the temporal order of the observations (i.e., shuffling destroys the trial-to-trial dynamics). The first comparison between the original and shuffled time series tests for dynamic structure. In addition, the shuffled time series minimizes the risk of falsely classifying the original. The shuffled time series tested the assumption that trial-to-trial variance is temporally uncorrelated gaussian noise (i.e., the hypothesis of IID).

The second control comparison models trial-to-trial variability as linearly autocorrelated Gaussian noise. In this second condition, the trial-to-trial structure of the original time series is fit by an autoregressive integrated moving average model, i.e., ARIMA(p, d, q). Unlike the shuffled comparison, an ARIMA model can form linear correlations between trials. An ARIMA model is a more plausible linear account of the trial-to-trial dependencies than the first control comparison. An ARIMA model assumes that temporal dependencies result from overlapping independent contributions. An ARIMA model is composed of several components (i.e., autoregressive, moving average, and integrated). ARIMA(p, d, q) models are constructed such that the variability results from p overlapping independent processing times (autoregressive) and q overlapping random contributions (moving average).

For the autoregressive (AR) component, the dependencies result from the carry over of one of the previous trial's deterministic component, such as processing time (e.g., X_{t-1}). In the moving average (MA) contribution, the dependencies are caused by the carry over of random contributions (e.g., Z_t). The last component, Integration (I), is achieved by taking the difference, lag d (e.g., $d = 1$, $X_n - X_{n-1}$), of the time series. Some types of nonstationarity can be corrected by differencing the time series.

To generate the ARIMA comparison time series, this chapter utilized two methods, smallest canonical correlation method (SCAN) and extended

TABLE 8.2

ARIMA $(p, 1, q)$ Models for Each Participant by Condition[a]

	Short RSI	Long RSI
Direct SRM[b]	(0,1,1), (0,1,1), (0,1,1),	(0,1,1), (0,1,1), (0,1,1),
	(0,1,1), (0,1,3), (0,1,4)	(0,1,1), (0,1,4), (0,1,4)
Crossed SRM	(0,1,1), (0,1,1), (0,1,2),	(0,1,1), (0,1,1), (0,1,1),
	(0,1,3), (0,1,3), (0,1,5)	(0,1,1), (0,1,4), (2,1,3)

[a] p corresponds to the order of the AR(p) component, 1 to differencing lag one and q corresponds to the order of the MA(q) component.

[b] SRM corresponds to stimulus-response mapping.

sample autocorrelation function (ESACF), to identify the order of the ARIMA parameters. The SCAN and ESACF methods are both described and available via SAS (SAS Institute Inc., 1999). Both methods work with stationary and nonstationary time series. Ideally, the two methods converge on the same model. However, this was not always the case. When the results do not converge, the models were estimated by using the region of overlap to choose the order based on parsimony (i.e., least number of parameters). Table 8.2 presents the order of the ARIMA models for Experiment 1.

The three types of time series (observed, shuffled, and ARIMA) per individual were submitted to two nonlinear dynamical measures, the BDS test and the (CI) correlation integral estimate of dimensionality (m). To test for differences in type of time series as well as the between-subject variables, the estimate of dimensionality (m) for the three types of time series for each participant are treated as a within-subject variable and are submitted to a repeated measures analysis of variance (ANOVA). In general, ANOVAs are robust to violations of the normality assumption. Throughout the experiments, if there was a significant violation of sphericity assumption, the significance level of the repeated measures ANOVA was adjusted based on Greenhouse–Geisser (G–G) epsilon ($\hat{\varepsilon}$) and the relevant epsilon value is presented after the obtained F-value. When there is a violation of the sphericity assumption and the significance level is adjusted (G–G), the power of the test is reduced. Depending on the number of participants (e.g., $n > 30$), MANOVA is a more powerful alternative (Maxwell & Delaney, 2000).

There were three planned comparisons for the estimate of dimensionality (m). The first comparison tested for a difference between the observed time series and the ARIMA time series. The second compared the observed and shuffled time series. The last contrasted the observed time series estimates of the LD and SC groups (the groups expected to have the largest mean differences). For all planned comparisons, the stepdown Bonferroni procedure was used to control familywise Type I error rates. If Mauchly's sphericity test indicated a violation of the homogeneity assumption ($p < .05$), error terms specific to the contrasts were used.

BDS Test

The BDS statistic (Brock, Hseih, & LeBaron, 1991) utilizes the CI to test whether trial-to-trial variability is IID. Rejection of the hypothesis of IID suggests that the trial-to-trial variability exhibits structure that is inconsistent with a stochastic process. The CI measures how the phase space is filled as the embedding dimension increases. A random time series should fill the phase space uniformly for all embedding dimensions. A chaotic time series, or one with temporal structure, does not fill the phase space evenly. In a chaotic time series, the phase space is filled only in the attractor region.

The BDS test is calculated by taking a weighted difference between the CI calculated for the observed time series and the expected CI for a random process. The BDS statistic divided by the standard deviation of the statistic is normally distributed and results in a z-score (i.e., z-score = BDS/SD) that expresses the probability of structure in the time series. The BDS test is obtainable as part of the *t*-series library in R/S + (Hornik, 2003). Brock et al. (1991) also make the test available.

Brock et al. (1991) recommend using an epsilon (ϵ) that is between 0.5 and 2.0 times the standard deviation of the time series. In addition, the embedding dimensions (m) should not *exceed* the length of the time series divided by 200 (i.e., $m < N/200$, where N equals the length of the time series). Following Heath's (2000) convention, BDS z-scores are averaged across a range of embedding dimensions. The maximal embedding dimension is determined via the equation $m < N/200$. The lower value of the embedding dimension is set to 4.

Given that variability generated from a stochastic process (i.e., white noise) is assumed to be IID, rejection of the null hypothesis suggests that there is structure in the time series that needs further investigation. Table 8.3 summarizes the number of significant BDS z-scores across RSI and stimulus–response mapping. The hypothesis of IID was rejected for all 24 observed time series. Because the ARIMA time series have trial-to-trial structure, it is not surprising that the hypothesis of IID was rejected for 19 of the 24 ARIMA time series. As anticipated (because the shuffling destroyed trial-to-trial dependencies that exist in the observed time series) the null hypothesis of IID was not rejected for the shuffled time series.

TABLE 8.3

The BDS Statistic (Brock et al., 1991) Tests for Trial-to-Trial Dependencies[a]

RSM	Short RSI			Long RSI		
	Observed	Shuffled	ARIMA	Observed	Shuffled	ARIMA
Direct	6	0	5	6	0	5
Crossed	6	0	5	6	0	4

[a] The table lists the number of z-scores out of 6 per cell that are greater than critical-z, p < .01. z-Scores are averaged across embedding dimensions 4–8 and the four levels of Epsilon. For each time series, levels of Epsilon correspond to standard deviation times 0.5, 1.0, 1.5, and 2.0.

The ARIMA time series describe at least some of the dynamical structure of the observed time series. One question is how well the ARIMA models account for the structure in the observed time series. The goodness of fit is tested by submitting the residuals from the ARIMA model to the BDS test. Because the z-scores for 20 of the 24 residual time series were greater than the critical z-score, the hypothesis of IID can be rejected for most of the residual ARIMA time series. Thus, most of the linear ARIMA models leave a portion of the trial-to-trial dependencies in the observed time series unexplained. The linear ARIMA account does not provide an adequate model of the trial-to-trial dependencies in the observed time series. Although it does not discriminate between nonlinear stochastic and chaotic time series, the BDS test is a powerful test for nonlinearity. Rejection of the null hypothesis for the observed time series advocates further nonlinear analyses.

Dimensionality (m) Estimates

The dimensionality of a time series is the minimum number of dynamical variables needed to predict or model that time series. In this chapter, dimensionality (m) is estimated using the Grassberger and Prococcia (1983) CI. The CI reflects the cumulative frequency of the Euclidean distance between points across increasing distance criteria (i.e., radius, ε). As the radius increases, more distances between points are included. For the time series in this chapter, the CI are estimated using Kantz and Schreiber (1997, 2002) algorithm (also available via the TISEAN application, Hegger and Kantz, 1999).

Typically, the linear scaling region on a log–log plot of CI $C_m(\varepsilon)$ across increasing radii (ε) for a particular embedding dimension (m) is used to estimate the dimensionality (m). For each time series, the linear scaling region was obtained with an algorithm that searched the scaling region from small to large radii to find the largest linear scaling region by using a minimum R^2 (i.e., .99) and length of scaling region criteria (30 points). If no region was found, the minimum R^2 was reduced by .01 and the procedure was repeated until a region was found meeting the criteria. The slope of the linear scaling region was used as an estimate dimensionality (m) for a particular embedding dimension (m).

The dimensionality of the time series is derived using the point at which the estimated dimensionality (m) stops increasing as the embedding dimension increases (i.e., the saturation point). A random time series should fill the phase space uniformly across embedding dimensions (m). The expected CI for a random time series across embeddings can be denoted as $C_m(\varepsilon) = (C_m(\varepsilon))^m$. Thus, the expected value of a random time series on a log–log is the embedding dimension, m. If the time series does not fill the phase space uniformly, the slope should be less than the embedding dimension, m.

With very large samples, the CI dimensionality (m) estimates can be used to calculate an accurate point estimate. With small samples, this estimate is typically underestimated. One way to test for an underestimate is to compare the obtained time series to a shuffled version of the obtained time series. If the estimate dimensionality (m) of the shuffled time series is below the expected value, then underestimation is very likely. Also, poor implementation of the measures can result in underestimation (e.g., including the curved portion of the CI function in the slope estimate). Due to the limited time series length and possible contamination via exogenous noise (e.g., irregular sample), the results from the application of CI dimensionality (m) measure should not be viewed as a precise estimate. Rather, it should be viewed as a possible means for discriminating between difference classes of time series and for detecting a directional change in the estimates resulting from experimental manipulations. For this reason, throughout this chapter, dimensionality (m) will be used to indicate that the estimate corresponds to the slope of the CI for embedding dimension (m) and the term dimensionality will be used when discussing the possible dimensionality of a time series.

The estimated dimensionalities (m) were submitted to a 2 × 2 (3 × 3) ANOVA. The two between-subject variables were stimulus–response mapping and RSI. Embedding dimensions (5, 6, and 7) and the three types of time series (observed, shuffled, and ARIMA) served as the within variables. Neither main effect for stimulus–response mapping or RSI was significant. The main effect for embedding dimension was significant, $F(2, 40) = 413.32$, G–G $\hat{\varepsilon} = 0.8299$, $p < .05$. As embedding dimension increased, the estimated dimensionality (m) increased (i.e., 4.57, 5.29, and 5.95 for embedding dimensions 5, 6, and 7, respectively).

The main effect for type of time series was significant, $F(2, 40) = 4.96$, G–G $\hat{\varepsilon} = 0.8662$, $p < .05$. The observed time series estimate (5.02) was significantly less than the ARIMA time series estimate (5.48), $F(1, 40) = 58.56$, $p < .017$. The observed time series estimate did not differ from the shuffled (5.30) time series estimate, $p > .05$. The comparison between the LD and SC groups observed time series was not significant, $p > .05$.

The three-way interaction between type of time series, embedding dimension and stimulus–response mapping was significant, $F(4, 80) = 3.49$, G–G $\hat{\varepsilon} = 0.9360$, $p < .05$ (Table 8.4). The estimated dimensionality (m) (i.e., CI slope) increased for all three time series as embedding dimension rose. Lack of a saturation point suggests that the dimensionality for all three time series was greater than 5. Comparing across stimulus–response mapping for the observed time series, the estimate for the direct task was slightly lower than the crossed task. This suggests that dimensionality may decrease as a task becomes easier.

The estimates in Table 8.4 hint that dimensionality may differ with task difficulty. Since all four tasks were fairly easy to perform, the minimal differences in the estimated dimensionalities (m) might be the consequence of limited differentiation between the tasks. Experiment 1B was conducted to

TABLE 8.4

Dimensionality (*m*) Estimate Calculated from the Correlation Integral (CI)
for Embedding Dimension by Stimulus-Response Mapping (SRM) and Type
of Time Series

Time Series	SRM	Embed Dim 5	Embed Dim 6	Embed Dim 7
Observed	Direct	4.25	4.82	5.54
Observed	Crossed	4.38	5.19	5.94
Shuffled	Direct	4.56	5.42	6.09
Shuffled	Crossed	4.71	5.31	5.72
ARIMA	Direct	4.71	5.44	5.96
ARIMA	Crossed	4.79	5.56	6.43

increase the task difficulty and to test if the increased difficulty was reflected
in larger dimensionality (*m*) estimates.

Experiment 1B: RT Dynamics across Task Difficulty Using RSI

Experiment 1B replicates the two most extreme groups in Experiment 1A
using a task that is slightly more difficult. The difficult task corresponds to
the short RSI condition paired with the crossed stimulus–response mapping
condition (SC group). The pairing of the long RSI condition with the direct
stimulus–response mapping condition should produce the easiest pairing
(LD group).

The primary procedural difference between Experiments 1A and 1B is the
stimulus location. In Experiment 1A, the stimulus location is dependent on
the stimulus–response mapping. Consequently, participants could use the
compatible mapping between the stimulus location and response to reduce
the difficulty of the crossed condition. Notes from participant debriefings
confirm that a number of the participants used location rather than stimulus
to make their decisions. Eliminating the direct relation between the stimulus location and response should increase task difficulty (Kornblum & Lee,
1995). In Experiment 1B, the stimulus location (left or right) varies randomly
from trial-to-trial and is independent of the stimulus–response mapping.

Method

Participants

Twenty-two Vanderbilt University undergraduates participated in partial
fulfillment of an introductory psychology course requirement. Twenty participants were right handed. All participants had normal or corrected-to-
normal vision and participated for 1.5 h or less.

Design

A mixed-subject 2 × (2 × 2) design was used. Volunteers participated in one of two groups (LD or SC). The LD group was assigned the long (L) RSI and the direct (D) stimulus–response mapping condition. The SC group was given a short (S) RSI and crossed (C) stimulus–response mapping. The within-subject variables were type of stimulus (LEFT, RIGHT) and first-order sequence type (repetition or alternation).

Procedure, Materials, and Apparatus

The stimulus location (left or right) was randomly selected with equal probability. Long and short RSI conditions were identical to Experiment 1A. All other aspects were identical to Experiment 1A.

Results

The two levels of the between variable, group, were LD and SC. Two participants, one in each group, were removed from the analyses because of poor performance (i.e., 51% and 87% accuracy for LD and SC participants, respectively). The remaining participants successfully maintained high accuracy levels.

Summary of the Analysis of the Means

The average accuracy levels for the LD and SC groups were 97% and 96%, respectively. The response latencies were trimmed using the same criteria and methods as in Experiment 1A. Although latencies were generally faster under long RSIs and direct stimulus–response mapping conditions (compare 516.1–581.5 ms for the LD and SC groups), this difference was not significant ($p > .05$).

A 42 ms first-order repetition effect was found. Responses to repetitions (528 ms) are faster than responses to alternations (570 ms), $F(1, 18) = 41.71$, $p < .05$. There is a significant interaction between sequence type and group, $F(1, 18) = 5.05$, $p < .05$. This interaction is expressed as a greater difference between repetitions and alternations for the SC group (compare 502 and 530 ms for LD repetitions and alternations, respectively, to 553 and 610 ms for SC repetitions and alternations, respectively). Consistent with the literature, the repetition effect (i.e., AE − RE) increased from 28 to 57 ms as the task difficulty increased and as RSI decreased.

Nonlinear Dynamical Measures

The procedures for generating the three times series and the nonlinear dynamical measures are described in Experiment 1a. The three time series for each participant were submitted to the battery of nonlinear tests. The order of the ARIMA models for each group is listed in Table 8.5.

TABLE 8.5

ARIMA (p, 1, q) Models for Each Participant by Condition[a]

	Short RSI	Long RSI
Direct SRM		(0,1,1), (0,1,1), (0,1,1), (0,1,1), (0,1,1),
		(0,1,2), (0,1,2), (0,1,3), (1,1,2), (3,1,4)
Crossed SRM	(0,1,1), (0,1,1), (0,1,1), (0,1,1), (0,1,1),	
	(0,1,1), (0,1,1), (0,1,1), (0,1,2), (0,1,3),	

[a] p corresponds to the order of the AR(p) component, 1 to differencing lag one, and q corresponds to the order of the MA(q) component. SRM corresponds to stimulus-response mapping.

Three planned contrasts were conducted for the dimensionality (m) estimates. The first comparison tested for a difference between the observed and the ARIMA time series. The second compared the observed and shuffled time series. The last contrast tested whether the observed time series differ between the LD and the SC groups.

BDS Test

The BDS z-scores for groups across type of time series are listed in Table 8.6. The table lists the number of z-scores out of a total of 10 per cell that are greater than critical-z, $p < .01$. z-Scores are averaged across embedding dimensions 4–8 and the four levels of Epsilon. For each time series, levels of Epsilon correspond to standard deviation times 0.5, 1.0, 1.5, and 2.0. SRM corresponds to stimulus–response mapping. The hypothesis of IID was rejected for all observed time series and for 19 of 20 ARIMA time series. As expected, the null hypothesis of IID cannot be rejected for the shuffled time series. The BDS test was employed as a goodness of fit test; the z-scores of 14 of the 20 ARIMA residual time series indicated that the ARIMA model left a significant amount of the trial-to-trial dynamics unaccounted for.

The BDS test confirmed that the observed time series contained trial-to-trial structure and that many of the ARIMA models did not provide a satisfactory account of the trial-to-trial dependencies. These findings support additional investigation into the possibility of nonlinear dynamical characteristics.

TABLE 8.6

The BDS Statistic (Brock et al., 1991) Tests for Trial-to-Trial Dependencies

	Type of Time Series		
Group	Observed	Shuffled	ARIMA
LD (Long RSI, Direct SRM)	10	0	9
SC (Short RSI, Crossed SRM)	10	0	10

Dimensionality (*m*) Estimates

Dimensionality (*m*) estimates were submitted to a 2 × (3 × 3) ANOVA. The variables were group (LD and SC), embedding dimensions (5, 6, 7) and time series (observed, shuffled, and ARIMA). The observed time series estimates did not differ from the comparison time series estimates. The estimated dimensionality (*m*) increased with each embedding dimension, $F(2, 36) = 6.92$, G–G $\hat{\varepsilon} = 0.7473$, $p < .05$. The means for embedding dimensions 5, 6, and 7 were 4.76, 5.52, and 6.07, respectively.

A significant interaction between embedding dimension and group was found, $F(2, 36) = 213.60$, G–G $\hat{\varepsilon} = 0.7473$, $p < .05$. For the LD group, the means for embedding dimensions 5, 6, and 7 were 4.73, 5.41, and 6.19, respectively. For the SC group, the means for embedding dimensions 5, 6, and 7 were 4.79, 5.63, and 5.95, respectively. Due to the lack of significant differences between the time series and closeness of the estimates, this interaction is hard to interpret. No other effects, interactions, or comparisons were significant ($p > .05$).

Both the BDS and the estimated dimensionality (*m*) findings suggest that there may be nonlinear dynamical structure in the observed time series. However, the evidence is minimal. Since the dimensionality (*m*) estimates for observed time series did not differ from the shuffled time series, some of the significant differences in the experimental variables may be due to biases caused possibly by an insufficient sample size. As discussed in the next experiment, some of these findings and lack of findings may be the result of false-dynamics induced via the experimental methods. Experiment 2 explores the application of nonlinear dynamical methods to behavioral data by examining the issue of achieving a uniform sample via a fixed ISI. Experiment 2 may aid in revealing nonlinear dynamical structure.

Experiment 2: RT Dynamics across Stimulus–Response Mapping and ISI

Experiment 1, along with other research (e.g., Clayton & Frey, 1996, 1997; Ward, 1994; Cooney & Troyer, 1994), provides mixed results in support of a nonlinear dynamical account of response latencies. The nonlinear dynamical measures exhibited some limited differences from control conditions. Nonlinear dynamical differences between RT times series are not generally found across experimental variables. None of these studies used a fixed ISI. The nonlinear estimates may have been distorted by the method of controlling stimulus rate.

Using a simple four-choice task, Kelly et al. (2001) extended previous cognitive findings by employing a uniform ISI. In response to one of four horizontally displayed lights, participants indicated the stimulus by pressing the key below the displayed light. Kelly et al. submitted the observed time series along with control time series to nonlinear noise reduction techniques (see TISEAN application, Hegger and Kantz, 1999). Their results supported a nonlinear dynamical account.

Kelly et al.'s findings suggest that a fixed ISI may be necessary to detect nonlinear dynamical structure. Although Kelly et al. obtained evidence in support of a nonlinear dynamical account, it is possible that their conclusions resulted from estimate biases caused by false structure. By splicing the time series at regular interventions, after the breaks taken every 4 min, unintentionally dynamical structure could be reflected in the nonlinear dynamical measures. Likewise, false dynamics could be introduced by splicing the time series when a response was missed.

Experiment 2 is a replication of Experiment 1 with the modification of a fixed ISI. It extends the work of Kelly et al. by removing breaks from the experiment. Sampling rate may be a critical variable when applying behavioral time series to nonlinear dynamical measures. Presenting the stimulus at a fixed interval and measuring RTs in relation to the stimuli achieves a uniform sampling rate (Kelso, 1995). This uniform sampling rate may be necessary to assess the true nonlinear dynamical nature of the underlying processes.

Method

Participants

Thirty-two Vanderbilt University undergraduates participated in the study. The participants received partial fulfillment of an introductory psychology course requirement. Twenty-eight participants were right handed. All participants had normal or corrected-to-normal vision and participated for 1.5 h or less.

Design

This experiment used a simple mixed-subject design as Experiment 1. The two between-subject variables were ISI (short 900 ms, long 2000 ms) and stimulus–response mapping (direct, crossed). The within-subject variable was first-order sequence type (repetition or alternation).

Stimuli and Stimulus Location

The probability of a stimulus (LEFT, RIGHT) was approximately equal across sequence type and pattern. The stimulus occurred equally to the left or right of the fixation and varied independently from the stimulus-response mapping condition.

Stimulus–Response Mapping

Stimulus–response mapping was either direct or crossed. In the direct condition, the correct response to the stimulus (e.g., LEFT) matched the response finger (i.e., left key pressed). In the crossed condition, the stimulus (e.g., LEFT) and the response were incongruent (e.g., right key pressed).

Procedure

Except for changing the fixation, limiting the duration of the stimulus and controlling ISI, Experiment 2 used procedures similar to Experiment 1B. At the beginning of the experiment, the task was explained to the volunteers who were instructed to respond as quickly as possible and to maintain accuracy above 95%. In addition, a strong emphasis was placed on keeping the number of missed items to a minimum. Participants were given feedback on 108 practice trials. If a participant's accuracy was below 60% on the last 50 practice trials, they were given another set of 108 practice trials. These warm-up trials were followed by 1550 experimental trials in which no accuracy feedback was given.

After the participant initiated the experiment by pressing the "z" key, a fixation (i.e., ***** + *****) was displayed in the center of the screen for 100 ms. Following the fixation, the test stimulus (i.e., LEFT or RIGHT) replaced the asterisks on the left or right of the plus symbol for 500 ms. Then, the stimulus and asterisks were erased and a plus symbol was displayed in the center of the screen for the duration of the ISI. A response could be entered any time after the test item had been displayed and before the next test item was displayed (i.e., total time: 900 or 2000 ms). If the response was not made during the allotted time, the RT was set to zero. Participants indicated their response by pressing the appropriate key on the computer keyboard. Left- and right-responses were made by pressing the "z" and the "/" keys with the left- or right-hand index finger, respectively. After the ISI elapsed, a new trial began with the onset of the fixation.

Materials and Apparatus

The stimulus files from Experiment 1 were used. These files were modified slightly so that the stimulus positions were ordered so that the probability of each stimulus and position was approximately equal across stimulus sequences of length eight (i.e., $p(S_{(n-7)}S_{(n-6)}S_{(n-5)}S_{(n-4)}S_{(n-3)}S_{(n-2)}S_{(n-1)}S_n) = 1/2^8 = 1/256$). The same stimulus files were used in each between-subject condition. All other aspects of the experiment were identical to Experiment 1.

Results

Summary of the Analysis of the Means

The switch to a fixed ISI in Experiment 2 resulted in a task in which participants had difficulty maintaining high accuracy. Consequently, the criterion for excluding a participant from analysis was loosened to an overall accuracy of 78%. Four participants were excluded from the analysis. Two of these participants were in the crossed stimulus–response mapping using a short-ISI (SC group). One participant was dropped from the crossed stimulus–response mapping using a long ISI (LC group) and one participant was dropped from the direct stimulus–response mapping using short-ISI (SD group). Data from 28 participants were used in the analysis. There were seven participants in each group (stimulus–response mapping × ISI).

The proportion correct was larger for the direct (0.910) stimulus–response mapping as compared to the crossed (0.863) stimulus–response mapping, $F(1, 24) = 4.35$, $p < .05$. This suggested that manipulating the stimulus–response mapping made the task more difficult. However, this difference was not observed in response latencies. There were a number of interactions with stimulus–response mapping and these are discussed in the subsequent sections.

The main effect for ISI was not significant for either RT or proportion correct ($p > .10$). The lack of a main effect was surprising considering the RTs in Experiment 1 were faster for the short ISI (e.g., 327 and 412 ms for short and long, respectively). However, the switch from a fixed RSI to a fixed ISI resulted in a harder task. This was reflected in the lower overall proportion correct of 0.886 as compared to 0.97 in Experiment 1. In Experiment 1A, RTs increased from 337 to 402 ms in the direct and crossed RSI conditions, respectively. An average RT of 453.9 ms in Experiment 2 suggests that the change to a fixed ISI increased the difficulty level for the short ISI such that RTs were equivalent to the long ISI condition.

A 18.7 ms first-order repetition effect was observed for RTs, $F(1, 24) = 15.75$, $p < .05$. Responses to repetitions were faster than alternations (i.e., 444.6 ms versus 463.3 ms) and were more accurate than alternations (i.e., 0.899 versus 0.874; $F(1, 24) = 6.48$, $p < .05$).

For proportion correct, a significant interaction between sequence type and stimulus–response mapping, $F(1, 24) = 4.41$, $p < .05$, suggested that the crossed condition was more difficult than the direct condition for the alternations. In addition, the difference between repetitions and alternations was greater in the crossed condition.

Nonlinear Dynamical Measures

The nonlinear dynamical analyses followed the same procedures described in Experiment 1A. The three types of time series (i.e., observed, shuffled, and ARIMA) were submitted to a battery of nonlinear dynamical measures. Table 8.7 presents the order of the ARIMA models in each between-subject condition. For each of the measures, three planned comparisons were performed, which were similar to those performed in Experiment 1.

TABLE 8.7

ARIMA (p, 1, q) Models for Each Participant by Condition[a]

	Short 900 ms ISI	Long 2000 ms ISI
Direct SRM	(0,1,2), (0,1,2), (0,1,2), (0,1,3), (0,1,3), (0,1,3), (0,1,4)	(0,1,1), (0,1,1), (0,1,1), (0,1,2), (0,1,2), (0,1,2), (0,1,3)
Crossed SRM	(0,1,2), (0,1,2), (0,1,2), (0,1,2), (0,1,3), (0,1,3), (1,1,1)	(0,1,1), (0,1,1), (0,1,1), (0,1,1), (0,1,1), (0,1,1), (0,1,1)

[a] p corresponds to the order of the AR(p) component, 1 to differencing lag one and q corresponds to the order of the MA(q) component. SRM corresponds to stimulus–response mapping.

BDS Test

The z-scores for the BDS statistic lead to the rejection of the hypothesis of IID for all of the observed time series and all but one of the ARIMA time series (see 900 ISI direct stimulus–response mapping condition). Table 8.8 lists the number of z-scores that are greater than the greater than critical-z, $p < .01$ for both ISI and stimulus-response mapping. z-Scores are averaged across embedding dimensions 4–8 and four levels of Epsilon. The four-levels of Epsilon for each time series correspond to standard deviation times 0.5, 1.0, 1.5, and 2.0. The hypothesis of IID was not rejected for the shuffled time series.

The z-scores for 24 of the 28 residual time series were greater than the critical z-score. This suggests that the ARIMA models did not provide an adequate account of the observed trial-to-trial dependencies. Consistent with Experiment 1, the results for the BDS test point toward additional nonlinear dynamical analysis.

Dimensionality (m) Estimates

The estimates for each participant were submitted to a $2 \times 2 \times (3 \times 3)$ ANOVA with ISI (900 ms, 2000 ms) and stimulus–response mapping (direct, crossed) as the between-subject variables. Embedding dimension (5, 6, and 7) and type of time series (observed, shuffled, and ARIMA) were the within-subject variables.

A main effect was found for type of time series, $F(2, 48) = 9.20$, G–G $\hat{\varepsilon} = 0.9874$, $p < .05$. The estimates for the observed time series (5.10) were less than for both the shuffled (5.35), $F(1, 48) = 7.71$, $p < .025$, and the ARIMA time series

TABLE 8.8

The BDS Statistic (Brock et al., 1991) Tests for Trial-to-Trial Dependencies

Stimulus-Response Mapping	ISI 900 ms			ISI 2000 ms		
	Observed	Shuffled	ARIMA	Observed	Shuffled	ARIMA
Direct	7	0	6	7	0	7
Crossed	7	0	7	7	0	7

(5.48), $F(1, 48) = 17.81, p < .017$. The comparison between the LD and SC groups observed time series was not significant, $p > .05$.

As the embedding dimension increased the dimensionality (m) estimates increased, $F(2, 48) = 575.77$, G–G $\hat{\varepsilon} = 0.7255$, $p < .05$ (see Ave., Table 8.9). Table 8.9 depicts the marginal interaction between type of time series and embedding dimension, $F(4, 96) = 2.25$, G–G $\hat{\varepsilon} = 0.7541$, $p < .10$. The increase in dimensionality (m) across embedding dimension was roughly parallel for both the shuffled and ARIMA time series. In comparison, the estimates for observed time series appeared to grow less and less with each embedding dimension. The observed time series may have a limited dimensionality (e.g., between 5 and 8).

A significant three-way interaction between embedding dimension, stimulus–response mapping and ISI was found, $F(2, 48) = 4.28$, G–G $\hat{\varepsilon} = 0.7255$, $p < .05$. This interaction, depicted in Table 8.10, shows that the SC group estimates were higher than estimates for the other three groups.

The marginal four-way interaction between type of time series, embedding dimension, stimulus-response mapping and ISI was significant, $F(4, 96) = 2.20$, G–G $\hat{\varepsilon} = 0.7541$, $p < .10$. The dimensionality for three of the observed time series groups (i.e., SD group, LD group, and LC group) might differ from the remaining time series (see Figure 8.2). The findings suggest that the observed time series for the three groups have lower dimensionality (m) than the comparison conditions and the observed time series for the SC group. Furthermore, the interactions involving the between-subject variables, ISI and stimulus-response mapping, suggest that dimensionality may increase with task difficulty.

TABLE 8.9

Mean Estimated-Dimensionality (m) for the Type of Time Series (Observed, Shuffled, and ARIMA) by Embedding Dimension (5, 6, and 7) Interaction along with the Means

Time Series	Embed Dim 5	Embed Dim 6	Embed Dim 7	Average
Observed	4.50	5.11	5.70	5.10
ARIMA	4.74	5.49	6.19	5.47
Shuffled	4.65	5.34	6.05	5.35
Average	4.63	5.31	5.98	

TABLE 8.10

Mean Estimated-Dimensionality (m) for the Three-Way Inter-Action between Embedding Dimension (5, 6, and 7), Stimulus-Response Mapping (SRM; Direct, Crossed) and ISI (900 ms, 2000 ms)

ISI	SRM	Embed Dim 5	Embed Dim 6	Embed Dim 7
900 ms	Direct	4.59	5.26	5.82
2000 ms	Direct	4.57	5.29	5.99
900 ms	Crossed	4.73	5.46	6.24
2000 ms	Crossed	4.64	5.24	5.88

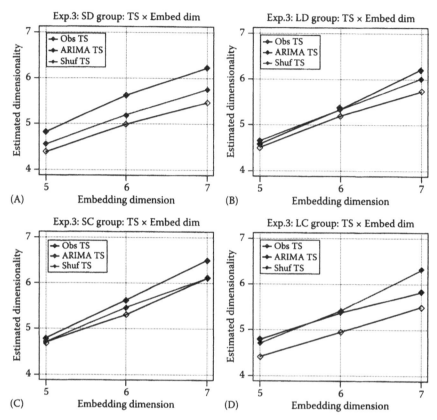

FIGURE 8.2
Estimated dimensionality (*m*) across embedding dimension across type of time series (TS). Obs TS, ARIMA TS, and Shuf TS correspond to the observed, ARIMA and shuffled time series, respectively. Segments A, B, C, and D correspond to the four groups resulting from crossing ISI (900 (S) ms, 2000 (L) ms) with stimulus–response mapping (direct (D), crossed (C)). Thus, the SD group corresponds to the short (S) ISI combined with the direct (D) stimulus–response mapping.

Discussion

BDS Test

Across all of the experiments, the hypothesis of IID was rejected via the BDS test for the observed time series and most of the ARIMA time series. In addition, the goodness-of-fit tests resulted in the rejection of the hypothesis of IID for the ARIMA residuals. Using the BDS as a goodness of fit test, the unaccounted for structure in the ARIMA residuals indicates that the ARIMA does not provide an adequate account for the dynamical behavior in the observed time series. These findings support rejecting an account that assumes independent trial events and strongly indicated that further nonlinear analysis should be pursued. Although the BDS test does not provide information about the type of nonlinear structure and the differences in nonlinear structure between experimental conditions, it is a powerful test of nonlinearity.

An important question is whether the experimental manipulations are reflected in the nonlinear dynamical measures. Although the BDS test is excellent at detecting nonlinear structure, it is not a useful tool for detecting differences in nonlinear structure between experimental conditions. For this reason, the estimated dimensionality (m) along with mean RTs are discussed in more detail concerning differences due to the experimental manipulations.

Estimated Dimensionality (m)

Across all of the experiments, the estimated dimensionality (m) did not saturate with increased embedding dimension. Although RT differed across the groups in Experiment 1, the experimental manipulations using a fixed RSI were only slightly reflected in the dimensionality (m) estimates.

Despite the limited differences in mean RT in Experiment 2, the dimensionality (m) estimates differed across the two between-subject variables and the observed time series had lower dimensionality (m) estimates compared to both comparison time series. With easier tasks, the estimated dimensionalities (m) appear to be decreased. This pattern maps to the sequence effect pattern of increased strategy utilization with an easier task. The group differences in the dimensionality (m) estimates might be reflecting changes in participant strategies. By examining the two methods of controlling stimulus rate, RSI and ISI, the findings in this study suggest that some of the lack of findings in the application of nonlinear dynamical methods might result from using RSI.

Limitations

From a behavioral perspective, 1550 trials is quite a large number of responses to make in one session without breaks. Consequently, there is a potential for fatigue, deterioration in performance with time on task. Two types of fatigue may occur: (a) physiological fatigue and (b) mental fatigue. Typically, there is a sudden drop in performance with fatigue and response variability increases (Guastello, 1995). Across the experiments, the accuracy rates were fairly high (greater than 95% and 85% for Experiments 1 and 2, respectively). The task, including responding with different hands, was developed so that the participants could accurately perform a large number of trials in one session. Nonstationary time series is a possible consequence of fatigue and participants performing such a large number of trials in a single session. An alternative is to allow breaks and treat the time series as a series of spliced steady states.

Although 1550 trials seem like a reasonable number of observations for a behavioral study, this is not necessarily the case for nonlinear dynamical analysis. The minimum required length of a time series depends on the nonlinear dynamical measure and characteristics of the time series. Consider estimating the dimensionality using the Grassberger and Procaccia (1983) CI , if the dimensionality estimate is around 4, then a minimum of 3,111,700 trials[1] may be necessary in order to obtain a point estimate of the dimensionality with 5% precision (Smith, 1988). Eckmann and Ruelle (1992) make less conservative estimates of approximately 400 data points.[2] Estimating the minimum number of data points is a debated issue. In addition, these estimates of the minimum number of data points assume a noise free data set. As discussed earlier, the estimated dimensionality (m) should not be viewed as a precise estimate. Instead, it serves as a tool to discriminate between classes of time series and possibly detecting directional changes resulting from experimental manipulations.

Conclusion

The mixed results when nonlinear measures are applied to RT may be the consequence of the methods used for data collection. The methodological examination of stimulus rate indicates that the nonlinear dynamical measures are sensitive to variability in the onset of each stimulus. When conducting nonlinear dynamical analysis, ISI should be manipulated rather than RSI. Dimensionality (m) estimates appear to decrease with reduced task difficulty and increased stimulus rate. This finding points to the potential applicability to modeling behavioral data.

Acknowledgments

I want to thank my husband, Lewis Frey, and my advisor, Keith Clayton, for all their support and guidance.

Notes

1. $N_{min} \geq (42)^M = (R(2 - Q)/(2(1 - Q)))^M$; where N_{min} = minimum number of data points, $Q = 0.95$ precision of estimate, $R = 4$ and M = the greatest integer less than the dimensionality.
2. $2 \log(N) > d \log(1/p)$; where N = minimum number of data points, d = dimensionality and p = precision = 0.05.

References

Annett, M., & Annett, J. (1979). Individual differences in right and left reaction time. *British Journal of Psychology, 70*, 393–404.

Brock, W. A., Hseih, D. A., & LeBaron, B. (1991). *Nonlinear dynamics, chaos, and instability: Statistical theory and economic evidence.* Cambridge, MA: MIT Press.

Clayton, K., & Frey, B. (1996). Inter- and intra-trial dynamics in memory and choice. *In* W. Sulis & A. Combs (Eds.), *Nonlinear dynamics in human behavior* (pp. 90–106). Singapore: World Scientific.

Clayton, K., & Frey, B. (1997). Studies of mental 'noise.' *Nonlinear Dynamics, Psychology and Life Sciences, 1*, 173–180.

Cooney, J. B., & Troyer, R. (1994). A dynamic model of reaction time in a short-term memory task. *Journal of Experimental Child Psychology, 58*, 200–226.

Eckmann, J. P., & Ruelle, D. (1992). Fundamental limitations for estimating dimensions and Lyapunov exponents in dynamical systems. *Physica D, 56*, 185–187.

Gilden, D. L. (1997). Fluctuations in the time required for elementary decisions. *Psychological Science, 8*, 296–301.

Gilden, D. L., Thornton, T., & Mallon, M. W. (1995). 1/f noise in human cognition. *Science, 267*, 1837–1839.

Grassberger, P., & Prococcia, I. (1983). Characterization of strange attractors. *Physical Review Letters. 50*, 346–349.

Guastello, S. J. (1995). *Chaos, catastrophe, and human affairs.* Mahwah, NJ: Lawrence Erlbaum Associates.

Heath, R. A. (2000). *Nonlinear dynamics: Techniques and applications in psychology.* Mahwah, NJ: Lawrence Erlbaum Associates.

Hegger, R., & Kantz, H. (1999). Practical implementation of nonlinear time series methods: The TISEAN package. *Chaos, 9(2)*, 413–435.

Helmuth L. L., & Ivry, R. B. (1996). When two hands are better than one: Reduced timing variability during bimanual movements. *Journal of Experimental Psychology: Human Perception & Performance, 22*, 278–293.

Hornik, K. (2003). The R FAQ. ISBN 3-901167-51-X. From http://www.r-project.org/

Kantz, H., & Schreiber, T. (1997). *Nonlinear time series analysis.* New York: Cambridge University Press.

Kantz, H., & Schreiber, T. (2002). *Nonlinear time series analysis* (2nd ed.). New York: Cambridge University Press.

Kelly, A., Heathcote, A., Heath, R., & Longstaff, M. (2001). Response time dynamics: Evidence for linear and low-dimensional nonlinear structure in human choice sequences. *Quarterly Journal of Experimental Psychology A, 54*, 805–840.

Kelso, J. A. S. (1995). *Dynamic patterns: The self-organization of brain and behavior.* Cambridge, MA: MIT Press.

Kirby, N. (1972). Sequential effects in serial reaction time. *Journal of Experimental Psychology, 96*, 32–36.

Kirby, N. (1976). Sequential effects in two-choice reaction times: Automatic facilitation or subjective expectancy? *Journal of Experimental Psychology: Human Perception and Performance, 2*, 567–577.

Kirby, N. (1980). Sequential effects in choice RT. In A. T. Welford (Ed.), *Reaction times* (pp 129–172). London, U.K.: Academic Press.

Klein, J. L. (1997). *Statistical visions in time: A history of time series analysis* (pp. 1662–1938). New York: Cambridge University Press.

Kornblum, S. (1969). Sequential determinants of information processing in serial and discrete choice reaction time. *Psychological Review, 76*, 113–131.

Kornblum, S. (1973). Sequential effects in choice RT: A tutorial review. In S. Kornblum (Ed.), *Attention & Performance IV* (pp. 259–288). New York: Academic Press.

Kornblum, S., & Lee, J. (1995). Stimulus-response compatibility with relevant and irrelevant stimulus dimensions that do and do not overlap with the response. *Journal of Experimental Psychology: Human Perception and Performance, 21*, 855–875.

Liebovitch, L. S. (1998). *Fractals and chaos: Simplified for the life sciences.* New York: Oxford University Press.

Luce, R. D. (1986). *Response times: Their role in inferring elementary mental organization* (pp 254–272, 306–318, 399–421). New York: Oxford University Press.

Maxwell, S. E., & Delaney, H. D. (2000). *Designing experiments and analyzing data: A model comparison perspective.* Mahwah, NJ: Lawrence Erlbaum Associates.

Port, R. F., & Van Gelder, T. (1995). *Mind as motion: Explorations in the dynamics of cognition.* Cambridge, MA: MIT Press.

Pressing, J., & Jolley-Rogers, G. (1997). Spectral properties of human cognition and skill. *Biological Cybernetics, 76*, 339–347.

SAS Institute Inc. (1999). *SAS OnlineDoc®, Version 8.* Cary, NC: SAS Institute Inc. Website for ARIMA, SCAN, and ESACF: http://gsbapp2.uchicago.edu/sas/sashtml/ets/chap7/sect21.htm#idxari0110

Scheier, C., & Tschacher, W. (1996). Appropriate algorithms for nonlinear time series analysis in psychology. In W. Sulis & A. Combs (Eds.), *Nonlinear dynamics in human behavior* (pp. 27–43). Singapore: World Scientific.

Schneider, W. (1988). Micro experimental laboratory: An integrated system for IBM PC compatibles. *Behavior Research Methods, Instruments, and Computers, 20*, 206–217.

Simon, J. R., & Rudell, A. P. (1967). Auditory S–R compatibility: The effect of an irrelevant cue on information processing. *Journal of Applied Psychology, 51*, 300–304.

Smith, L.A. (1988). Intrinsic limits on dimension calculations. *Physics Letters A, 133*, 283–288.

Theiler, J., Eubank, S., Longtin, A., Galdrikian, B., & Farmer, J. D. (1992). Testing for nonlinearity in time series: The method of surrogate data. *Physica D, 58*, 77–94.

Thelen, E., & Smith, L. B. (1994). *A dynamic systems approach to the development of cognition and action.* Cambridge, MA: MIT Press.

Vervaeck, K. R., & Boer, L. C. (1980). Sequential effects in two-choice reaction time: Subjective expectancy and automatic after-effects at short response-stimulus intervals. *Acta Psychologica, 44*, 175–190.

Ward, L. (1994, June). Hypothesis testing, nonlinear forecasting, and the search for chaos in psyhophysics. Paper presented to the *Fourth Annual Conference of the Society for Chaos Theory in Psychology*, Baltimore, MD, June 24–27.

Ward, L. (1996). Chaos in psychophysics? Hypothesis testing and nonlinear forecasting approaches. In W. Sulis & A. Combs (Eds.), *Nonlinear dynamics in human behavior* (pp. 77–89). Singapore: World Scientific.

Ward, L. (2002). *Dynamical cognitive science.* Cambridge, MA: MIT Press.

Wickens, D., & Hollands, J. G. (2000). *Engineering psychology and human performance* (3rd ed.). Englewood Cliffs, NJ: Prentice-Hall.

9

Entropy

Stephen J. Guastello

CONTENTS

This chapter organizes a cluster of interesting relationships among entropy, information, dimensionality, and complexity. It is easiest to start with complexity: The concept was introduced to the nonlinear dynamical system (NDS) lexicon to emphasize that it was virtually impossible to account for all the possible interactions among agents in a social system and iterate their outcomes over time, even though there was a substantial set of deterministic rules governing their behavior. For this purpose, agent-based models and simulations of similar design were introduced; these fall outside the boundaries of this book's objectives. A related concern with complexity, however, surrounds the question, "How complex are the dynamics of a particular process?" This question has seen many answers (Biggiero, 2001), but several plausible answers surround the fractal dimension, which can characterize the scope of the movements of a changing system, and the values of information and entropy metrics. There are also connections between information and entropy functions and dimensionality that are convenient and useful.

A single and precise definition of entropy is not forthcoming because of the shifts in its meaning that occurred over the last 150 years. In its first historical epoch, the concept of entropy was introduced by Clausius in 1863 in conjunction with relationships among pressure, volume of a container, and temperature along with broader problems in heat transfer. Entropy soon became a concept of *heat loss* or *unavailable energy* (Ben-Naim, 2008).

The second phase of meaning was associated with Boltzmann's atomistic interpretation of entropy, which concerned the movement of gas molecules throughout a container space. Although it is not possible to follow the movement of every molecule, it is possible to give an average value for the molecules in a particular condition. In one extreme, we have the condition of absolute zero temperature where molecules of a gas stand still; in the other extreme, we have Brownian motion, where molecules are bouncing randomly and completely filling the container's space. This paradigm led to the concept of entropy as being one of *order* versus *disorder*.

According to the second law of thermodynamics, which dates back to the late nineteenth century, entropy in the form of heat loss is inevitable in any system, so that the eventual outcome is "heat death." In this scenario, the system (if not the entire universe) will gravitate toward an equilibrium point where, in essence, nothing more will happen once the heat energy has been expended, and the system is completely disordered.

Two physical principles from the late twentieth century appear to contradict the heat death scenario. One is the notion of the expanding universe, which has the effect of keeping us away from even attaining the equilibrium point. Economists (e.g., Keynes, 1936/1965) have relied on a similar notion of an expanding economy to explain why particular economic systems can continue without collapse in spite of apparent limits to growth. Of course, no one has determined whether cosmic expansion translates into meaningful social or economic events on earth.

The other phenomenon, self-organization (Nicolis & Prigogine, 1989), is more germane: Rather than dissipating energy, the system reorganizes itself—without any assistance from outside agents—to produce a more efficient use of its energy. The spin-glass phenomenon serves as one example. If molten glass is subjected to high-intensity mixing, the molecules do not homogenize. Rather, molecules group together depending on the similarities of their electron spins; this principle has been studied in the context of dynamical systems more than once (Kauffman, 1993; Sulis, 2008). In more broadly defined conditions, the mixing of agents produces hierarchical structures that serve to reduce entropy or disorder and keep the system apparently stable. The hierarchical structures often take the form of driver–slave relationships among the subsystems, such that the behavioral dynamics of one subsystem drive the dynamics of another (Haken, 1984, 1988). The key point is that the system increases its internal order spontaneously, and reduces its rate of lost potential energy.

Several mechanisms of self-organization have been advanced in recent decades. The common theme among them is the newfound order results from the flow of information among the subsystems. Furthermore, it is possible to observe important transitions in systems' behavior by tracking the level of entropy across time or experimental conditions.

Shannon Entropy

Shannon's (1948) concepts of information and entropy originated with a problem connected to the engineering of long-distance telephone systems. As such they were removed from thermodynamics problems and constituted the third epoch in the history of entropy. In this theory, a *bit* of information is the amount of information that is required to determine a dichotomous outcome; one can readily see how the concept of the bit became fundamental to computer science (Gray, 1990). For a multichotomous outcome involving N equiprobable states, $N - 1$ bits are needed. If a system were truly random and produced an arbitrarily long string of observations, one would require as many bits of information to determine the series as there were original data points.

The computations assume that unequal a priori probabilities are the norm, however. The Shannon entropy (H_s) for the set of categories, S, with probabilities of occurrence p_i, is

$$H_s = \sum_{i=1}^{r} p_i \left[\ln \left(\frac{1}{p_i} \right) \right] \tag{9.1}$$

Information, I, was defined as H_s, which is simple enough, and the two entities add up to H_{max}, which is in turn a function of the number of possible states:

$$H_{max} = H_s + I = \ln(S) \tag{9.2}$$

Although the relationship is straightforward, H_{max}, and thus the rest of Equation 9.2 can be more elusive than appearances indicate. For some problems, some states are theoretically possible but never actually occur within the lifespan of the series of data observations that make up a time series, and such states become for all intents and purposes nonexistent. For instance, if the system contained two states A and B, but the concern was for patterns of pairs of states, there would be four resulting states AA, AB, BB, and BA. If the concern was for triplets of states, however, there would be eight possibilities (AAA, AAB, ABA, ABB, BAA, BBA, BAB, and BBB), and H_{max} would be larger so long as none of the triplets failed to appear in the time series. The missing states would not be known until after the fact, however; hence, H_{max} becomes elusive. For this reason, theorists have taken to ignoring the relationship in Equation 9.2, and treating 9.1 as "information entropy."

The Shannon entropy function evolved as an important concept in information technology, contemporary thermodynamics, symbolic dynamics, and as a means for comparing time series and studying the properties of NDSs. Ben-Naim (2008) went so far as to advocate disbanding the use of entropy concepts within physics to be consistent with the evolving information theory concepts. Schmitz (2007) advocated a similar renovation of thinking in economics, ecology, and biology. The effectiveness of the information and entropy constructs in economics is subject to some debate, as principles derived from physics sometimes collide with known principles in economics (Mantegna & Stanley, 2000; Rosser, 2008). Meanwhile, information theory constructs were introduced to psychology decades ago (Quastler, 1955) as a contribution to decision theories that were developing at that time, but with little connection to thermodynamic metaphors. In short, if principles of physics serve as good metaphors for understanding phenomena in the behavioral sciences, the isomorphisms reside in the mathematical and statistical relationships and not in the physics per se.

Symbolic dynamics involves patterns of states of systems over time. Although its central concepts are covered in Chapters 19 through 22, it will be necessary to make reference to patterns of states over time while tackling the more immediate objective of using entropy-based concepts to characterize time series. The fractal dimension and information functions have both been used to quantify the complexity of a time series. In the case of the latter, the complexity of a time series is the amount of information required to describe it (Sarkar & Barat, 2008).

Shannon Entropy and Complexity

Before embarking further, consider two applications of Shannon entropy to express notions of complexity. In one, the psychological construct of the self was alleged to play an important role in the buffering against the effects of stress to the extent that the self was more complex (Linville, 1987). The complexity of the self could be the result of a complex developmental process, which is known to be nonlinear and dynamic in nature (Lunkenheimer & Dishion, 2009; Marks-Tarlow, 1999; van Geert, 2009). Linville's experimental procedure required participants to read a stack of index cards that showed self-concept ideas, and then sort them into categories, making as many categories as they felt were necessary to accommodate all the cards. As one might expect, some participants used more categories than others did, and the distributions of cards into categories were unequal. H_s was computed for each participant's categorization; it rendered the differing numbers of states and distributions into a single number that could be used to compare in a relatively standard experimental design and analysis.

In another situation, Tschacher, Scheier, and Grawe (1998) hypothesized that psychotherapy, if successful, would bring about an ordering effect among some key concerns of the patients. The patients completed a brief questionnaire containing 33 items during each of at least 40 therapy sessions. The researchers compared the first 20 sessions with the last 20 sessions. The 33 items were reduced to a smaller number of dimensions or categories through the use of O-type factor analysis, which was performed separately for the starting and ending group of sessions. H_s was computed for each block of sessions.

Tschacher et al.'s hypothesis was actually rooted in the notion of a ratio between order and disorder, rather than the absolute values of H_s. They noted, as we did above, that H_{max} was not forthcoming, although the goal was to take a ratio between the two amounts. To get around this limitation, they adopted an ordering statistic developed by Banerjee, Sibbald, and Maze (1990):

$$\Omega = 1 - \left(\frac{H_{act}}{\Pi s_i^2} \right) \tag{9.3}$$

where H_{act} was defined as H_s, and potential entropy (or information) was defined as the cross-product of the variances (s_i^2) of all states.

Information Dimension

The applications of Shannon entropy considered so far involved a single set of discrete states and associated probabilities, where the states appear in a one-dimensional time series. The concept can be expanded to cover continuously valued variables and more than one dimension. Recall from Chapter 4 that the dimensionality of a trajectory associated with an attractor is calculated by first imposing boxes (circles) on the two-dimensional array, where the boxes have a side (or radius) r. Of course some of the boxes remain empty, and the two-dimensional embedding is larger than the resulting fractal dimension.

If we were to take each of the boxes, call them "states," and calculate the probability of the system visiting each of the states, it would be possible to calculate H_s for the system. If r were allowed to decrease, there would be more states, and more information would be needed to describe the states. The information dimension, D_I, describes the speed at which more information would be needed as r decreases (Sprott, 2003):

$$D_I = \lim_{r \to \infty} \left(\frac{\sum_i p_i \log p_i}{\log r} \right) \tag{9.4}$$

For practical purposes, a single value of r would probably be used to calculate D_I from experimental data. The accuracy of the estimate of the "true" dimension would depend on the size of r nonetheless.

Cross-Entropy

Cross-entropy is sometimes known as Boltzmann entropy or the Kullbeck–Liebler information metric (Gregson, 1995):

$$H_B = -\sum_{i=1}^{h} p_i \ln\left(\frac{p_i}{m_i}\right) \tag{9.5}$$

where
  p_i is once again the probability associated with state i
  h is the total number of states

The nuance is that p_i is relative to m_i, a predicted value of p_i, emanating from a theoretical model M. Importantly, M can be a complex nonlinear dynamical process with bifurcations and discontinuities that are usually intractable through linear prediction methods (p. 22).

Nonlinear psychophysical processes have produced many situations where Equation 9.5 has been valuable. Other classes of applications involve multidimensional stimuli with trajectories that can be graphed. The level of convergence and divergence among the trajectories constitutes the entropy concept with one trajectory functioning as M in Equation 9.5 with respect to another (Neemuchwala, Hero, & Carson, 2005).

Mutual Entropy

The mutual entropy function bears a strong resemblance to a correlation. Unlike cross-entropy, which is predicated on comparing predicted and actual probabilities, mutual entropy is based on a more direct concordance between two variables, X and Y, having i and j states, respectively. The entropy associated with the joint states X_iY_j is defined as the entropy associated with X plus the entropy associated with Y, minus the entropy associated with their joint combinations:

$$I(X;Y) \equiv H(X) + H(Y) - H(X,Y) \tag{9.6}$$

which is calculated as (Ben-Naim, 2008)

$$I(X;Y) = \sum p_{ij}\log_2\left(\frac{p_{ij}}{p_i p_j}\right) \tag{9.7}$$

In light of its similarity to the Shannon function, mutual entropy is perhaps more commonly known as mutual information.

In one contemporary application, Lai, Mayer-Kress, and Newell (2008) studied a psychomotor problem involving aiming ability where participants tried to hit a target with a graphic pen. The theoretical problem concerned movement velocities, which are known to change in the course of a short movement. The two variables were movement time and percentage of the movement trajectory to the target. These two variables were carved into categories of values. The mutual entropy or information between position and velocity was then compared across experimental conditions.

An aiming study could just as well address a related but different problem of aiming accuracy. In that case, the two measures would be the proximity to the target's X and Y coordinates. The experimental apparatus would need to be rigged to position the target at different places on the screen, and the difference between the response position and the target position would be calibrated as being off by so many millimeters up-down or left-right. The measurements would then be broken into categories. Mutual entropy or information could then be calculated and compared across experimental conditions, whatever they would happen to be.

In another category of examples, Lunkenheimer and Dishion (2009) reported a study where they were studying parent–child interactions. Units of interactions were observed and calibrated on a continuous scale that divided conceptually into four categories: 1 = negative, 2 = directive, 3 = neutral, and 4 = positive. Table 9.1 is a stylized* representation of the data from one of their dyads. The original data were actually a time series; the data

TABLE 9.1

A 4 x 4 Array of Frequencies and Probabilities for Calculating Mutual Entropy (Information)

X	Y 1	2	3	4	Row Total	Row Pr
1(freq)	1	8	0	4	13	0.0878
$p(x,y)$	.0068	.0541		.0270		
2(freq)	3	30	0	6	39	.2635
$p(x,y)$	.0203	.2027		.0405		
3(freq)	9	40	0	10	59	.3986
$p(x,y)$	.0608	.2703		.0676		
4(freq)	10	25	0	2	37	.2500
$p(x,y)$	.0676	.1689		.0135		
Column total	23	103	0	22	148	
Column Pr	.1554	.6959	0	.1486		100

* This is *not* a reanalysis of the actual data, but a close representation of the patterns that appeared in a graphic made from the data.

in the rows of the table designated as "(freq)" were the relative frequencies at which each cell was visited during the sequence of events. If one were to apply Equation 9.7 to the table of frequencies, the result would be a mutual entropy of 0.321, which is only a small relationship between X and Y.

Note that the empty cells given in the column for $Y = 3$ would not produce a calculation that contributed to the summation; they would be ignored. The peculiar positioning of the empty cells did occur, however; they were retained in the table for use later on in this chapter.

If all the points appeared in the same cell, the argument to the $\log_2$ operator would be 1.0, and the log would be 0, and mutual entropy would be 0 as well. If the points were scattered over the entire 4×4 grid, there would be no correlation there either. If all the points were lined up on the diagonal however, mutual entropy would be 2.000.

If one consults enough sources on entropy statistics, it is often possible to find the same formula written with the base-10 log, base-2 log, and natural log. Sometimes there is a computational convenience involved, although the meaning of information is more closely tied to the latter two choices. In principle, the choice of log base does not make a great deal of difference, so long as all the conditions being compared are calculated the same way.

Topological Entropy

The next two concepts, the Lyapunov exponent and topological entropy, rely on some convenient relationships among NDS constructs. Whereas Shannon distinguished between information that an outside observer required to predict or describe the system's states and entropy as whatever states and probabilities were not accounted for by available information, Prigogine (Prigogine & Stengers, 1984; Nicolis & Prigogine, 1989) removed the role of the observer altogether. If one were to consider chaotic systems, which generate random-appearing sequences from simple functions containing affine transformations, a small amount of information explains a lot of activity. Chaotic systems in motion thus *produce* information. Thus, information and entropy became one and the same entity, and captured by the concept of topological entropy (Crutchfield, 1994; Nicolis & Prigogine, 1989; Ott, Sauer, & Yorke, 1994).

Lyapunov Exponent

A Lyapunov exponent is a measure of stability for a dynamical system, and has been useful for summarizing the level of chaos or turbulence in the system (Ruelle, 1991). It is calculated (Kaplan & Glass, 1995, p. 334–335; Puu, 2000, p. 157) by taking pairs of initial conditions y_1 and y_2 and their iterations one step ahead in time, which would be y_2 and y_3. If the ratio of absolute values of differences

$$L \cdot \frac{|y_3 - y_2|}{|y_2 - y_1|} \tag{9.8}$$

is less than 1.0, the series is contracting. If the value of L is greater than 1.0, the function is expanding and sensitive dependence is present. The Lyapunov exponent, λ, is thus

$$\lambda \cdot \ln[L] \tag{9.9}$$

For an ensemble of trajectories in a dynamical field, exponents are computed for all values of y. If the largest value of λ is positive, and the sum of λ_i is negative, then the series is chaotic in the sense that the trajectories are unpredictable, bounded, and sensitive to initial conditions.

The calculation of Equation 9.9 is made on the entire time series and averaged over N values, where N is the last entry in the time series:

$$\lambda = \left(\frac{1}{N}\right) \sum_{N=1}^{N} |\ln(L)| \tag{9.10}$$

As N increases to infinity, Equation 9.10 generalizes as (Wiggins, 1988, p. 54)

$$y = e^{\lambda t} \tag{9.11}$$

A positive value of λ indicates an expanding function, which is to say, chaos. A negative λ indicates a contracting process, which could be a fixed point if it is strongly negative. A $\lambda = 0$ indicates a limit cycle attractor; a slightly positive λ indicates a quasiperiodic function, and a slightly negative λ indicates a dampened oscillator.

An algorithm for calculating the Lyapunov spectrum that was first introduced by Wolf et al. (1985) was eventually replaced by Equation 9.12 (Rosenstein, Collins, & DeLuca, 1993), which does a better job of separating exponential expansion due to deterministic chaos from expansions produced by noise:

$$L_k = \frac{1}{2(N-k-m+1)} \sum_{n=m}^{N-k} \log \sum_{j=0}^{m-1} (X_{2 \cdot k} - X_{1 \cdot k})^2 \tag{9.12}$$

where
  N is the length of the time series
  X is the time series variable observed k steps apart
  m is the embedding dimension (Sprott, 2003, p. 253)

The inner summation is over all available pairs of observations within a dimension; the outer summation is over all subcalculations across dimensions. The largest Lyapunov exponent, $\lambda = dL_k/dk$.

The Lyapunov exponent can also be calculated statistically through nonlinear regression, thereby separating the deterministic structure from noise in a different way that allows point prediction as well. That story continues in Chapter 13. For present purposes, it is noteworthy that λ can be regarded as a measure of complexity by itself, or converted to a fractal dimension, $D_F = e^\lambda$ (Ott et al., 1994).

Kolmogorov–Sinai Entropy

Topological entropy, also known as Kolmogorov–Sinai (K–S) entropy, extends the idea behind Shannon entropy to a multidimensional array with continuous data represented as discrete states in q dimensions (Ott et al., 1994, p. 39; Heath, 2000, p. 193):

$$H_{KS} = \frac{1}{1-q} \ln \sum_{i=1}^{p} p_i^q \tag{9.13}$$

Under limiting conditions, H_{KS} is equal to the largest Lyapunov exponent. H_{KS} is the same as H_s when $q = 2$. It is technically undefined for a single categorical variable, but for all intents and purposes the analysis defaults to H_s again for a single dimension (Ott et al., 2004). Topological entropy appears again in Chapter 21 with symbolic dynamics.

Deflection Models

Although some concerns were raised in Chapter 3 about the effect of detrending on the correlation dimension, some entropy-based metrics make use of a similar procedure or a similar concept. This section introduces diffusion entropy, approximate entropy, sample entropy, and detrended fluctuation analysis.

Diffusion Entropy

The central idea is that a time series is really a stationary point with deflections in either direction (Scarfetta, Hamilton, & Grigolini, 2001). The notion of diffusion harkens back to the classical thermodynamic idea of heat transfer. The probability density function of a variable x over time results from deflections that are distributed as an inverse power law that is a function of time,

$$\text{pdf}(z,t) = \left(\frac{1}{t^B}\right) F\left(\frac{z}{t^B}\right) \tag{9.14}$$

and quantified by Shannon entropy:

$$H_s = -Ip(z,t) \ln\left[p(z,t)\right]dz \tag{9.15}$$

The variable x has been replaced by z, which is the probability density associated with the value of x. Within one increment of time, several x values will be observed. The mean of these values is transformed to z. The probabilities become known after a long-enough time series has transpired. The calculation $H_s = p^*\ln(p)$ is straightforward. For two increments of time, there would be two means that become $p^*\ln(p)$ that are added together. The process continues for three increments and so on; with enough time, the entire distribution would be represented. One can then plot H_s as a function of $\ln(t)$ (Sarkar & Barat, 2008). The slope of the line is the fractal dimension B, as in Equation 9.14 and other examples of power laws.

Approximate Entropy

Recall that the classic calculations of the correlation dimension and the Lyapunov exponent are susceptive to the influence of noise, and some algorithms are better than others for getting around the influence of noise. The approximate entropy index, ApEn (Pincus, 1991), was devised to separate order from disorder, where order resulted from a deterministic process. Lower values mean more order. In principle, points in an ordered system that start out close together should stay close together as the time series evolves.

ApEn evolved in a series of steps that began with the Grassberger–Procaccia (1983) algorithm for the correlation dimension and K–S entropy, followed by a series of steps that resulted in Eckmann–Ruelle (E–R; 1985) entropy:

$$\Phi^m(r) = (N - m + 1) - 1 \sum_{i=1}^{N-m+1} \log C_i^m(r) \tag{9.16}$$

In Equation 9.16, N is the length of the time series, which is divided into m overlapping segments; C is the correlation dimension; and r is the radius used to compute C. $\Phi^m(r)$ is the average log-odds of an observation falling within a given range of another.

Also important is that ApEn can be calculated with fidelity on $N = 100$. Its calculation simplifies to

$$\text{ApEn}(m,r,N) = \Phi^m(r) - \Phi^{m+1}(r) \tag{9.17}$$

The log-odds can be ascertained directly instead of working via the correlation dimension. This particular approach became distinguished as *sample entropy*. If $m = 2$, two adjacent points are compared, and $m + 1$ means that we would compare points that are two steps away. $m = 3$ is a relatively common choice also (Heath, 2000; Richman & Moorman, 2000). The odds in question would be the proportion of adjacent points that are within r of each other. r is a small proportion of the standard deviation; Heath suggests $.01 - .25$ SD. Psychologists might like to use the standard error of measurement.

ApEn has seen numerous applications in the behavior sciences. In the last 5 years alone, the topics have included psychomotor coordination and control (Cavanaugh et al., 2006; Hong, Brown, & Newell, 2008; Hong, Lee, & Newell, 2007; Hong & Newell, 2008; Wijnants, Bosman, Hasselman, Cox, & Van Orden, 2009), EEG activity (Jausovec & Jausovec, 2007; Jausovec, Jausovec, & Gerlic, 2006; Papadelis et al., 2007a; 2007b), daily mood variations and related self-reports (Burton, Heath, Weller, & Sharpe, 2010; Glenn et al., 2006; Pincus, 2006; Pincus, Schmidt, Palladino-Negro, & Rubinow, 2008; Rao, Rao, & Yeregani, 2006), heart rate data (Goncalves, Bernardes, Rocha, & Ayres-de-Campos, 2007; Sarà et al., 2008; Sara & Pistoia, 2010; Sarkar & Barat, 2008; Yeragani, Krishnan, Engels, & Gretebeck, 2005), and arterial oxygen levels related to sleep apnea (del Campo, Hornero, Zamarron, Abasolo, & Alvarez, 2006). ApEn can be calculated through the R statistical package (R Development Team, 2008).

Detrended Fluctuation Analysis

Detrended fluctuation analysis (Peng, Havlin, Stanley, & Goldberger, 1995) examines entropic deviations from local means in a time series, and recovers a fractal dimension by way of a power-law scaling relationship. Unlike diffusion entropy, detrended fluctuation analysis assumes that the time series could be nonstationary. A time series consisting of N observations of a variable y is broken up into k equal segments of length n. The sum of squared deviations from the segment means M is then calculated:

$$F(y) = r \frac{1}{N} \sum_{k=1}^{k} \sum_{n=1}^{n} (y_k - M_k)^2 \qquad (9.18)$$

The process is then repeated for increasing segment lengths. Next, $\log[F(y)]$ is plotted as a function of $\log(n)$. The slope of the resulting line is the fractal dimension. The technique has seen numerous applications with physiological data, such as heartbeat and gait (Goldberger et al., 2002; Deffeyes et al., 2009).

State Space Grids

The final technique for this chapter combines some of the principles in the entropy statistics with phase space diagrams. The objective of state space grids (Lewis, Lamey, & Douglas, 1999) is to find an attractor in a two-variable phase space. Consider again the data in Table 9.1. It was possible to compute its mutual entropy, which said something about the association between the two variables, but the question of attractors is not whether the order of states is consistent across the two variables, but whether special combinations of the two variables exist that constitute an attractor. Furthermore, the curious separation of frequency density in the example suggests that perhaps two attractors could be present there and not just one.

The analysis begins with a table containing frequencies shown for each cell in the matrix. The frequency in each cell is then compared against an expected value where the number of observations is equal. The observed and expected values are subtracted, squared, and summed over cells:

$$F(xy) = \sum \left(F_{obs} - F_{exp} \right)^2 \tag{9.19}$$

All cells are used at first. Then the process is repeated with the least frequent cells removed. Then it is repeated again with the second-least frequent cells removed, and so on. Finally, $F(xy)$ is plotted as a function of analytic step. There should be a precipitous drop at the point where the attractor's cells are separated from the others.

In the example, the starting total frequency was 148 observations and 16 cells. The expected value is 9.250 observations per cell. On the second step, all cells with frequencies of 0 are removed. The total is still 148 observations with 12 remaining cells; the expected value is 12.333 observations per cell. On the third step, the cell with a frequency of 1 is removed; the total drops to 147 with 11 cells and an expected value of 13.364 observations.

The plot of $F(xy)$ by steps appears in Figure 9.1. The steepest decent is from step 8 to step 9, at which cells with frequencies of 9 or less have been dropped out. One could reasonably demark cell frequencies of 10 or more as part of an attractor, and ignore the other cells as noise or chatter. A block of four contiguous cells would remain as one attractor, and another single cell of 10 observations would be a separate attractor. On the other hand, the sharp decline continues almost as sharply to step 10, where cells with frequencies of 10 or less are removed, and stops. The conclusion would be that there is one attractor consisting of three contiguous cells.

Two other calculations are possible with state space grids. Both require using the original time series. *Influence* is the proportion of times a point that is not on an attractor (excluding points that are already on a different attractor) moves onto the attractor in the next time interval. If a point leaves an attractor,

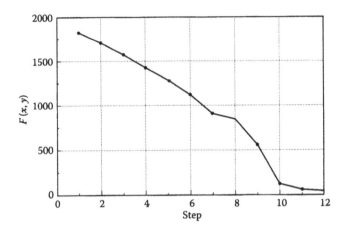

FIGURE 9.1
Plot of frequency variance by analytic step as low-density cells are eliminated.

stability is the length of time that it takes for the point to return to the attractor. Influence and stability can be compared across experimental conditions.

State space grids are not widely used yet, but they have taken root with problems in developmental psychology where the behavior of dyads is involved (Granic & Hollenstein, 2006; Hollenstein, 2007; Hollenstein, Granic, Stoolmiller, & Snyder, 2004; Lunkenheimer & Dishion, 2009; Martin, Fabes, Hanish, & Hollenstein, 2005), or in clinical psychology where two variables such as anxiety and depression often fluctuate simultaneously (Katerndahl & Wang, 2007).

References

Banerjee, S., Sibbald, P. R., & Maze, J. (1990). Quantifying the dynamics of order and organization in biological systems. *Journal of Theoretical Biology, 143*, 91–111.

Ben-Naim, A. (2008). *A farewell to entropy: Statistical thermodynamics based on information.* Singapore: World Scientific.

Biggiero, L. (2001). Sources of complexity in human systems. *Nonlinear Dynamics, Psychology, and Life Sciences, 5*, 3–20.

Burton, C., Heath, R. A., Weller, D., & Sharpe, M. (2010). Evidence of reduced complexity in self-report data from patients with medically unexplained symptoms. *Nonlinear Dynamics, Psychology, and Life Sciences, 14*, 15–26.

Cavanaugh, J. T., Guskiewicz, K. M., Giuliani, C., Marshall, S., Mercer, V. S., & Stergiou, N. (2006). Recovery of postural control after cerebral concussion: New insights using approximate entropy. *Journal of Athletic Training, 41*, 305–313.

Crutchfield, J. P. (1994). The calculi of emergence. Computation, dynamics, and induction. *Physica D, 75*, 11–54.

Deffeyes, J. E., Kochi, N. K., Harbourne, R. T., Kyvelidou, A., Stuberg, W. A., & Stergiou, N. (2009). Nonlinear detrended fluctuation analysis of sitting center-of-pressure data as an early measure of motor development pathology in infants. *Nonlinear Dynamics, Psychology, and Life Sciences, 13*, 351–368.

del Campo, F., Hornero, R., Zamarron, C., Abasolo, D. E., & Alvarez, D. (2006). Oxygen saturation regularity analysis in the diagnosis of obstructive sleep apnea. *Artificial Intelligence in Medicine, 37*, 111–118.

Goldberger, A. L., Amaral, L. A. N., Hausdorff, J. M., Ivanov, P. C., Peng, C. K., & Stanley, H. E. (2002). Fractal dynamics in physiology: Alterations with disease and aging. *Proceedings of the National Academy of Sciences, 99*, 2466–2472.

Goncalves, H., Bernardes, J., Rocha, A. P., & Ayres-de-Campos, D. (2007). Linear and nonlinear analysis of heart rate patterns associated with fetal behavioral states in the antepartum period. *Early Human Development, 83*, 585–591.

Glenn, T., Whybrow, P. C., Rasgon, N., Grof, P., Alda, M., Baethge, C., & Bauer, M. (2006). Approximate entropy of self-reported mood prior to episodes in bipolar disorder. *Bipolar Disorders, 8*, 424–429.

Granic, I. & Hollenstin, T. (2006). A survey of dynamic systems methods for developmental psychopathology. In D. Cicchetti & D. J. Cohen (Eds.), *Developmental psychopathology. Vol. 1: Theory and method* (pp. 889–930). Hoboken, NJ: Wiley.

Grassberger, P., & Prococcia, I. (1983). Characterization of strange attractors. *Physical Review Letters, 50*, 346–349.

Gray, R. M. (1990). *Entropy and information theory.* New York: Springer-Verlag.

Gregson, R. A. M. (1995). *Cascades and fields in perceptual psychophysics.* Singapore: World Scientific.

Haken, H. (1984). *The science of structure: Synergetics.* New York: Van Nostrand Reinhold.

Haken, H. (1988). *Information and self-organization: A macroscopic approach to self-organization.* New York: Springer-Verlag.

Hollenstein, T. (2007). State space grids: Analyzing dynamics across development. *International Journal of Behavioral Development, 31*, 384–396.

Hollenstein, T., Granic, I., Stoolmiller, M., & Snyder, J. (2004). Rigidity in parent-child interactions and the development of externalizing and internalizing behavior in early childhood. *Journal of Abnormal Child Psychology, 32*, 595–607.

Hong, S. L., Brown, A. J., & Newell, K. M. (2008). Compensatory properties of visual information in the control of isometric force. *Perception & Psychophysics, 70*, 306–313.

Hong, S. L., Lee, M.-H., & Newell, K. M. (2007). Magnitude and structure of isometric force variability: Mechanical and neurophysiological influences. *Motor Control, 11*, 119–135.

Hong, S. L. & Newell, K. M. (2008). Visual information gain and the regulation of constant force levels. *Experimental Brain Research, 189*, 61–69.

Heath, R. A. (2000). *Nonlinear dynamics: Techniques and applications in psychology.* Mahwah, NJ: Erlbaum.

Jausovec, N., & Jausovec, K. (2007). Personality, gender and brain oscillations. *International Journal of Psychophysiology, 66*, 215–224.

Jausovec, N., Jausovec, K., & Gerlic, I. (2006). The influence of Mozart's music on brain activity in the process of learning. *Clinical Neurophysiology, 117*, 2703–2714.

Kaplan, D., & Glass, L. (1995). *Understanding nonlinear dynamics*. New York: Springer-Verlag.

Katerndahl, D., & Wang, C.-P. (2007). Dynamic covariation of symptoms of anxiety and depression among newly-diagnosed patients with major depressive episode, panic disorder, and controls. *Nonlinear Dynamics, Psychology, and Life Sciences, 11*, 349–366.

Kauffman, S. A. (1993). *Origins of order: Self-organization and selection in evolution*. New York: Oxford.

Keynes, J. M. (1965). *General theory of employment, interest, and money* (2nd ed.). New York: Harcourt Brace. Originally 1936.

Lai, S.-C., Mayer-Kress, G., & Newell, K. N. (2008). Mutual information in the evolution of trajectories in discrete aiming movements. *Nonlinear Dynamics, Psychology, and Life Sciences, 12*, 241–260.

Lewis, M. D., Lamey, A. V., & Douglas L. (1999). New dynamic system method for the analysis of early socioemotional development. *Developmental Science, 2*, 457–475.

Linville, P. W. (1987). Self-complexity as a cognitive buffer against stress-related illness and depression. *Journal of Personality and Social Psychology, 52*, 663–676.

Lunkenheimer, E. S., & Dishion, T. J. (2009). Developmental psychopathology: Maladaptive and adaptive attractors in children's close relationships. In S. J. Guastello, M. Koopmans, & D. Pincus (Eds.), *Chaos and complexity in psychology: The theory of nonlinear dynamical systems* (pp. 282–306). New York: Cambridge University Press.

Martin, C. L., Fabes, R. A., Hanish, L. D., & Hollenstein, T. (2005). Social dynamics in the preschool. *Developmental Review, 25*, 299–327.

Neemuchwala, H., Hero, A., & Carson, P. (2005). Image matching using alpha-entropy measures and entropic graphs. *Signal Processing, 85*, 277–296.

Nicolis, G., & Prigogine, I. (1989). *Exploring complexity*. New York: Freeman.

Mantegna, R. N., & Stanley, H. E. (2000). *An introduction to econophysics: Correlation and complexity in finance*. Cambridge, U.K.: Cambridge University Press.

Marks-Tarlow, T. (1999). The self as a dynamical system. *Nonlinear Dynamics, Psychology, and Life Sciences, 3*, 311–346.

Ott, E., Sauer, T., & Yorke, J. A. (Eds.). (1994). *Coping with chaos*. New York: Wiley.

Papadelis, C., Chen, Z., Kourtidou-Papadeli, C., Bamidis, P. D., Chouvarda, I. et al. (2007a). Monitoring sleepiness with on-board electrophysiological recordings for preventing sleep-deprived traffic accidents. *Clinical Neurophysiology, 118*, 1906–1932.

Papadelis, C., Kourtidou-Papadeli, C., Bamidis, P. D., Maglaveras, N., & Pappas, K. (2007b). The effect of hypobaric hypoxia on multichannel EEG signal complexity. *Clinical Neurophysiology, 118*, 31–52.

Peng, C. K., Havlin, S., Stanley, H. E., & Goldberger, A. L. (1995). Quantification of scaling exponents and crossover phenomena in nonstationary heartbeat time series. *Chaos, 5*, 82–87.

Pincus, S. M. (1991). Approximate entropy as a measure of system complexity. *Proceedings of the National Academic of Sciences USA, 88*, 2297–2301.

Pincus, S. M. (2006). Approximate entropy as a measure of irregularity for psychiatric serial metrics. *Bipolar Disorders, 8*, 430–440.

Pincus, S. M., Schmidt, P. J., Palladino-Negro, P., & Rubinow, D. R. (2008). Differentiation of women with premenstrual dysphoric disorder, recurrent brief depression, and healthy controls by daily mood rating dynamics. *Journal of Psychiatric Research, 42*, 337–347.

Prigogine, I., & Stengers, I. (1984). *Order out of chaos: Man's new dialog with nature.* New York: Bantam.

Puu, T. (2000). *Attractors, bifurcations, and chaos: Nonlinear phenomena in economics.* New York: Springer-Verlag.

Quastler, H. (1955). *Information in psychology.* Glencoe, IL: Free Press.

Rao, V. S. H., Rao, C. R., & Yeragani, V. K. (2006). A novel technique to evaluate fluctuations of mood: Implications for evaluating course and treatment effects in bipolar/affective disorders. *Bipolar Disorders, 8*, 453–466.

R Development Team (2008). *R: A language and environment for statistical computing.* [Computer software]. Vienna, Austria: R Foundation for Statistical Computing.

Richman, J. S., & Moorman, J. R. (2000). Physiological time series analysis using approximate entropy and sample entropy. *American Journal of Physiology: Heart and Circulatory Physiology, 278*, 2039–2049.

Rosenstein, M. T., Collins, J. J., & DeLuca, C. J. (1993). A practical method for calculating largest Lyapunov exponents from small data sets. *Physica D, 65*, 117–134.

Rosser, J. B., Jr. (2008). Debating the role of econophysics. *Nonlinear Dynamics, Psychology, and Life Sciences, 12*, 311–323.

Ruelle, D. (1991). *Chance and chaos.* Princeton, NJ: Princeton University Press.

Sarà, M., & Pistoia, F. (2010). Complexity loss in physiological time series of patients in a vegetative state. *Nonlinear Dynamics, Psychology, and Life Sciences, 14*, 1–14.

Sarà, M., Sebastiano, F., Sacco, S., Pistoia, F., Onorati, P. et al. (2008). Heart rate nonlinear dynamics in patients with persistent vegetative state: A preliminary report. *Brain Injury, 22*, 33–37.

Sarkar, A., & Barat, P. (2008). Effect of meditation on scaling behavior and complexity of human heartbeat variability. *Fractals, 16*, 199–208.

Scafetta, N., Hamilton, P., & Grigolini, P. (2001). The thermodynamics of social processes: The teen birth phenomenon. *Fractals, 9*, 193–208.

Schmitz, J. E. J. (2007). *The second law of life: Energy, technology, and the future of the earth as we know it.* Norwich, NY: William Andrew Publishing.

Shannon, C. E. (1948). A mathematical theory of communication. *Bell System Technical Journal, 27*, 379–423.

Sprott, J. C. (2003). *Chaos and time series analysis.* New York: Oxford.

Sulis, W. (2008). Stochastic phase decoupling in dynamical networks. *Nonlinear Dynamics, Psychology, and Life Sciences, 12*, 327–358.

Tschacher, W., Scheier, C., & Grawe, K. (1998). Order and pattern formation in psychotherapy. *Nonlinear Dynamics, Psychology, and Life Sciences, 2*, 195–216.

van Geert, P. (2009). Nonlinear complex dynamical systems in developmental psychology. In S. J. Guastello, M. Koopmans, & D. Pincus (Eds.), *Chaos and complexity in psychology: The theory of nonlinear dynamical systems* (pp. 242–281). New York: Cambridge University Press.

Wiggins, S. (1988). *Global bifurcations and chaos.* New York: Springer-Verlag.

Wijnants, M. L., Bosman, A. M. T., Hasselman, F., Cox, R. F. A., & Van Orden, G. C. (2009). 1/f scaling in movement time changes with practice in precision aiming. *Nonlinear Dynamics, Psychology, and Life Sciences, 13,* 79–98.

Wolf, A., Swift, J. B., Swinney, H. L., & Vastano, J. A. (1985). Determining Lyapunov exponents from a time series. *Physica D, 16,* 285–317.

Yeragani, V. K., Krishnan, S., Engels, H. J., & Gretebeck, R. (2005). Effects of caffeine on linear and nonlinear measures of heart rate variability before and after exercise. *Depression and Anxiety, 21,* 130–134.

10

Analysis of Recurrence: Overview and Application to Eye-Movement Behavior

Deborah J. Aks

CONTENTS

The rationale for using nonlinear dynamical systems (NDS) tools is often tied to evidence showing temporal dependencies and nonlinearities in human behavior. As we sample broader ranges of our test stimuli, what we find, rather than proportional scaling, are behavioral changes resembling power laws and containing sequential dependencies; for a review of studies, see Guastello, Koopmans, and Pincus (2009); Holden, Van Orden, and Turvey (2009); Shelhamer (2007); and Ward (2002). Unfortunately, these properties can violate assumptions of popular general linear modeling (GLM) statistics. Thus, to learn whether recurring patterns may be hidden amidst behavioral noise, we need to use analyses designed for this purpose. In Aks (2009), I summarized some NDS techniques and showed applications to eye movements in visual search. Various analyses showed long-range $1/f$ correlations across the sequence of eye fixations (when horizontal and vertical eye movements were analyzed separately). The results suggested that the generating mechanism can be understood further through alternative analyses, for which recurrence quantification analysis (RQA) and recurrence plots (RP) are well suited. With their high sensitivity to data order, RQA can

help bypass limits of GLM analyses, and assumptions of sequential independence. We can now identify and quantify dynamics such as whether behaviors recur over time (%RECUR) as part of a string of consecutive behaviors
(%DET), the extent to which the repeating sequence is stable (%LAM), and
whether these dynamics can be represented more simply in a 3D state space
(SS). As Webber and Zbilut (2005) aptly describes, "RQA functions like a
microscope, snooping out higher-dimensional subtleties in [...] dynamics
that are not obvious (to) first-dimensional representation" (p. 82).

State-Space Analysis and Time-Delay Reconstruction Using RQA and RP

RQA and RP, developed by Zbilut and Webber (1992), Webber and Zbilut
(1994, 2007), and Marwan (2003), are methods of nonlinear data analysis used
to investigate how behavioral patterns emerge and repeat (approximately)
over time. RQA quantifies the frequency and duration of recurrences in a
3D representation of a behavior unfolding over time, or in its *"ss trajectory."*
Poincare contributed significantly to the topological conceptualization of SS
as a flow of trajectories and how it can be used to study a system's dynamics
(see Shelhamer, 2007, for an interesting history of SS reconstruction).

Can we infer characteristics of the original system using only the history of
its output? Takens' (1981) theorem argues so: "Observing changes in a just a
single variable over time has sufficient information to represent the system's
dynamic as a 3D topological structure."[1] The basis for RQA analytic recovery
of rules governing behavioral dynamics lies in the assumption that the effect
of all (unobserved) variables is reflected in the observed output. Detecting
these dynamics is accomplished by *time-delay reconstruction* (TDR), and the feat
of dimensional reduction and expansion. For a detailed background reading
on RQA and TDR, see Webber and Zbilut (1994, 2005), Marwan, Romano, Thiel,
and Kurths (2007), Pellecchia and Shockley (2005), and Shelhamer (2007).[2]

TDR enables us to visualize the spatial structure of the dynamical system
by taking the 1D time series and expanding it into a higher-dimensional
space where the dynamics of the underlying generator may be revealed. The
goal is to capture system states each time we have an observation of that
system's behavior. Using TDR, or *delayed coordinate embedding*, all states of
the system are plotted against a time-lagged version of itself: $X(t)$ against
$X(t + delay)$. We expand the 1D signal into a multidimensional space, and
substitute each observation in the original $X(t)$ with vector $Y_i = \{x_i, x_{i-L},$
$x_{j-2L}, ..., x_{i-(m-1)L}\}$, where i is the time index, m is the embedding dimension,
and L is the time delay or lag. As a result, we have a series of vectors: $Y = \{y_1,$
$y_2, y_3, ..., y_{N-(m-1)L}\}$, with N data points in the original series. These can then
be represented as a 3D *SS representation*.

Figures 10.1 through 10.3 show the steps leading to a TDR using a simple periodic signal generated by a sinusoidal equation iterated over time (τ):

$$y(\tau) = A \sin(\omega\tau + \phi) \tag{10.1}$$

This function produces regular oscillations in the time series when its three parameters are fixed: *amplitude* (A), or magnitude of change in the behavior; *frequency* (ω), or the # of oscillations in a specified period of time; and *phase* (φ), or the starting point in the cycle at $t = 0$. The oscillating behavior is a classic "limit cycle." There is an elliptical or circular flow (in the SS) depending on L, and the topological viewpoint that is revealed: Just 200 data points and a lag of $L = 8$ are sufficient to show the complete cycle of the sine wave (as though we are viewing the circular SS from an orthogonal view). When $L = 1$, the same function produces an elliptical orbit in the SS (as though viewing the circle from its side).

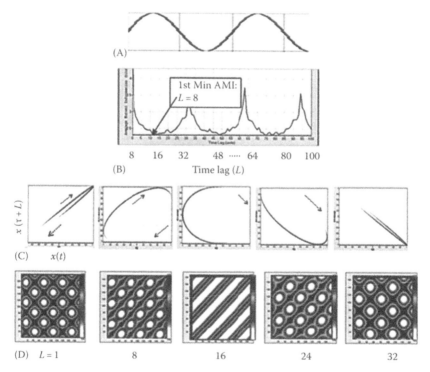

FIGURE 10.1

Illustration of how to estimate RQA parameters using a time series of sinusoid behavior (A); information bits (I) at each lag (L) are calculated by MI. The first minimum I is at $L = 8$ (B); corresponding SSs (C), and RPs at $m = 2$. Black points indicate high recurrence and points close together. RP patterns use initially 200 data points (D). When $L = 16$, periodicity is most dominant in diagonal lines (%DET), which capture the full 64-point cycle of the sine wave. As L deviates from 16, the dot structure becomes prominent, reflecting far points in SS, and when dots are aligned vertically, these reflect transitions in SS trajectory (%LAM).

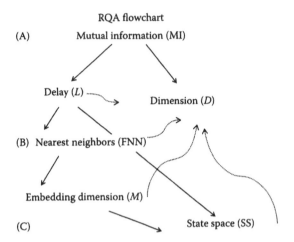

FIGURE 10.2
RQA steps used to obtain appropriate parameters for constructing SS. (A) Calculate AMI to get initial estimate of L, which is used in determining FNN. (B) Calculate FNN to estimate appropriate embedding dimension for SS trajectory. (C) Using appropriate parameters from these calculations may reveal a coherent structure in SS, reflecting determinism. AMI, FNN, and alternative methods, such as autocorrelation and correlation integrals, should converge on an estimate of actual dimension of the system.

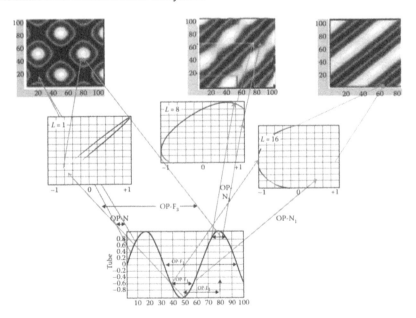

FIGURE 10.3
Relationship between points in time series (bottom sine wave), and three SS representations (middle closed-loop attractor), and RPs (top) at three different delays ($L = 1$, left; $L = 8$, middle; and $L = 48$, right). Illustrated behavioral states differ in phase, In or Out of phase (IP or OP), and proximity, Near (N) or Far (F).

Since real behavior is rarely as regular as a computationally generated sine wave, we examine what happens when we introduce some variability. Figure 10.4 shows white noise added to the sinusoid. A similar behavior may be produced by the human eye moving back and forth while tracking a point moving in space. Although with human behavior, variability is not nearly as uniform, and is often a form of colored noise, as is described extensively in the literature (e.g., Holden et al., 2009). Using the sinusoid function (Equation 10.1), I generated and tracked a dot moving back and forth, from the left to the right side of a computer display. I also tracked other oscillating patterns, such as a point moving periodically in a vertical and a circular path. I tested these simple tasks, with the goal of understanding scan paths used in more

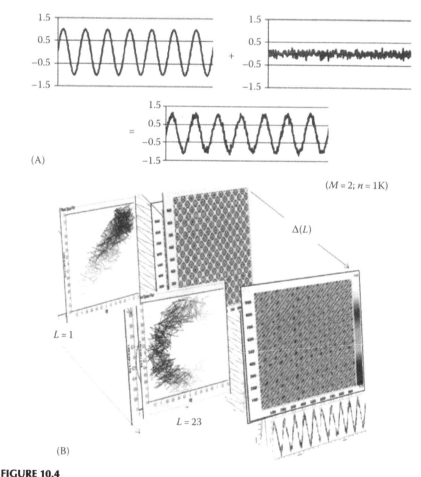

FIGURE 10.4
Combining sinusoid with additive noise (A) produces closed-loop SS trajectories with elliptical ($L = 1$) or circular ($L = 16$) trajectories when data series contains 1 K data points and $M = 2$. Fuzzy boundaries in SS trajectories and isolated points in RPs reflect local random transitions superimposed on the periodic behavior (B).

complicated search and object-tracking tasks. Similar to Aks (2009), correlated patterns occur in all fixation series, for both horizontal and vertical eye movements. Furthermore, the obvious oscillatory dynamics in these simple tasks is instructive in determining whether similar dynamics may reside in more complex behaviors.

While tracking a red dot oscillating across the display screen, I recorded my eye movements using a video-based eye-tracking system (SR-Research Eyelink1000; Osgoode, Ontario). Figure 10.5 shows my scan paths superimposed on the actual motion paths. Notice how scanning is far more variable than the smooth path of the target, and how faster motion produces discontinuities in the scan paths. Yet, interestingly, such eye movements used in *"real motion"* tracking are called *smooth pursuit* eye movements (e.g., Findlay & Gilchrist, 2003).

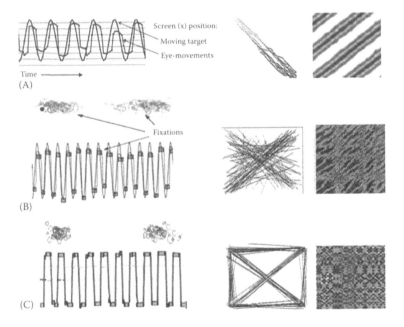

FIGURE 10.5
Smooth pursuit (horizontal) eye-movements in real (A and B) and apparent (C) motion tracking tasks. In real motion tracking, subjects tracked a *single* dot moving repeatedly back and forth from the left to right side of a computer display for 25S. Eye-movements resemble paths of the moving object when sampled at fixed rate (e.g., 100 Hz), but are far more variable and tend to undershoot the target (A). SS and RPs show clearly the oscillating behavior similar to a noisy sinusoid with pronounced periodicities (%Diag) and transitions (%LAM). The fixation sequence for the same real motion task focuses on the discrete and more variability fixations. In apparent motion tracking: *two* targets alternate (at same rate of 1 Hz), and scanning is far more discrete as illustrated by a gap in the central region (C). The fixation series tends to be more regular with eyes landing reliably closer to target locations. SS trajectories show clear oscillating paths in the criss-cross patterns corresponding to SS and RPs with higher LAM% and Diag% than in real motion trials.

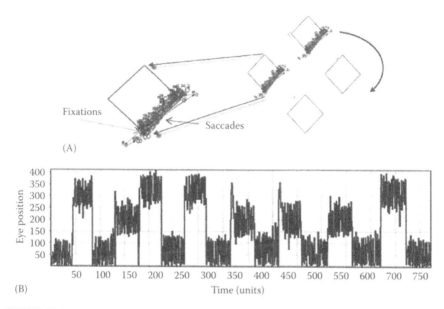

FIGURE 10.6

Subjects repeatedly scan one edge of a diamond back and forth until instructed to move (clockwise) to another diamond edge (A). Left diamond is a magnified version of 1 of the 4 diamonds making up the full display on right. Superimposed on some edges are a plot of saccadic eye movements and fixations. Scanning from corner to corner of each diamond edges produces a dense periodic pattern resembling a rapid oscillator. Clockwise shifts to new diamond edges produce a coarse and slow periodic pattern (B). Units are in msec.

In contrast, the form of eye movements defined (in part) by discrete and ballistic properties are the *saccades*. Figure 10.6 shows examples produced from tracking *two* dots alternating on the left and right sides of the computer display. Because these appear as a single dot moving in space, they are classified as *apparent motion*. Finally, Figures 10.6 and 10.7 show oscillating eye-movements in an edge-scanning task described in Aks (2009). Next, I step through an RQA analysis of scan paths obtained from tracking both real and apparent motion.

First, consider the forces that drive oscillating eye movements, and how these may be represented in the SS. Familiarity with SS attractors, such as fixed points, limit cycles, and repellers, is useful here. For background reading, see Guastello and Liebovitch (2009), and Shelhamer (2007). The simplest SS description for the repeated back and forth movements, or periodic oscillations produced by Equation 10.1, is well represented by a *"limit cycle,"* where a single point serves as an attractor, and the ordered sequence of states revolves around this attractor. The repetition of these cycles over time produces the resulting behavioral patterns that we are studying.

But is the same true of human eye movements? Consider how the dots, in our tracking tasks, exert a pulling (and pushing) force on our gaze. The pull of each dot (driven by the task and the subject's will to track the object) can be represented as a "flow" pattern in the SS, with behavioral states *attracted*

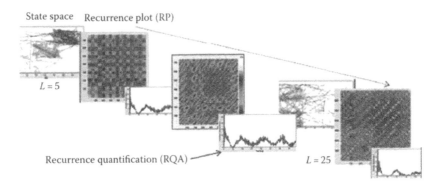

FIGURE 10.7
SS diagrams and RPs from the edges scanning task show transitional and stable states (e.g., when $L = 5$, there are three distinct attractor regions in SS corresponding to eye shifts to new diamond edges, and as well the three peaks in the RQA plots. RP boundaries indicate transitional states for slow ($L = 5$) and rapid shifts ($L = 25$). Diagonal patterns ($L > 5$) correspond to stable periodic behaviors, and SS trajectories moving along parallel paths over an extended period of time.

toward and (in the two-target case of apparent motion) *repelled* away from each dot. Rather than describing the alternating forces as a product of a single attractor (as in the case of real motion tracking), or the sine function (Equation 10.1 with A, ω, or φ parameter constraints), the two targets (in the apparent motion task) may generate the same cycling behavior from two points that alternate attracting and repelling the states of the SS trajectory.

Distinguishing between mechanisms underlying similar behavior can be challenging for a variety of reasons, including the need to determine optimal values for TDR and SS modeling. As we shall see, the ambiguity and trade-offs in the selection of appropriate parameters makes it important to confirm results with alternative NDS assessments. For example, in the case of competing forces acting on a behavior, their separate influences often are obscured so that one may appear to have prevailed in stabilizing the behavior. The seemingly stable state may in fact be on the verge of transitioning to a new state. Or, in other words, the system may be in a *critical* state (Bak, Tang, & Wiesenfeld, 1987). Telltale properties of critical states may be identified with alternative NDS analyses, such as spectral or various scaling methods where we are likely to find $1/f$ or other long-range correlations, intermittency, and behavior that are distributed as a power law. Later, we will rehash the importance of critical behavior and how its complexity affords flexibility for a system to switch between different states.

Here, we focus on RQA to help distinguish behavioral patterns by quantifying the reliability of different flow patterns and their likelihood of recurrence. Both local and global structures can be recovered, and, unlike many other methods, RQA can be used on short and nonstationary data series. Thus, fatigue and practice effects ubiquitous in much of human behavior can be analyzed separately from other dynamics. RQA assumes a property well-suited to human behavior—that the tested behavior is multidimensional. And by projecting the behavior onto a multidimensional space, the resulting SS representation can reveal dynamics driving the complex behavior. So, even though we (as observers) may have access only to low-dimensional output (e.g., the series of scalar observations, such as the sequence of eye fixations), we can recover even those dynamics with a complex and obscured set of sources.

RQA Methods

The software used for RQA analyses is *Visual Recurrence Analysis* (*VRA; v.4.6*) by Kononov (2005). I also used *Chaos Data Analyzer* (CDA) by Sprott and Rowlands (1995) to confirm SS topologies and parameter estimates. In addition, I used colored noise, and sinusoid data sets included with CDA. Additional RQA softwares, which can be found online, include *RQA* from C. Webber, *RQA 8.1*: http://homepages.luc.edu/~cwebber, and *Cross Recurrence Plots-Toolbox* (CRP) by N. Marwan, *CRP 4.5*: http://www.agnld.uni-potsdam. de/~marwan/toolbox.php.

Main Steps of RQA

A diagram showing the main RQA steps needed to produce valid SS representations is shown in Figure 10.2. Unless stated otherwise, the methods described here are based on Webber and Zbilut's (2005), including their seven required parameters: embedding dimension (M), delay (L), window range (W), norm (N), distance matrix (DM), radius (r), and line (Ln). These control matrix distances, rescaling, RP density, and other thresholds to detect system dynamics, and are described immediately following these key steps of RQA:

1. Select some variable x as an index of a behavior to track how it changes over time.
2. Replace each observation in the original signal X_t with a time-lagged version of itself: $Y_i = \{x_i, x_{i-L}, x_{j-2L}, \ldots, x_{i-(m-1)L}\} \rightarrow Y_i = \{y_1, y_2, y_3, \ldots, y_{(n-(m-1)L)}\}$.
3. Calculate distances between all pairs of vectors to form a DM consisting of the relative proximity of vectors in the reconstructed space. Max, Min, or the intermediate Euclidean distance may be used.
4. Expand the 1D time series into a multidimensional space, where the dynamics of the generating mechanism is predicted to reside. There are various ways to estimate the projected *embedding dimension* (M) including informed theory.

   Although it is important to consider theoretical factors that may be driving a behavior, this is not required of RQA. In the case of the sinusoid, we know the generating equation has three parameters—amplitude, phase, and direction—which can be represented in 3D space. However, collapsing these onto a lower-dimensional space is sufficient to reveal its SS limit cycle. In Figure 10.6, the two fixation points correspond to attractors for my alternating eye movements, and may correspond to two dimensions of the system. Internal mechanisms driving the eye movements add additional dimensions. Even without knowledge of causal factors, RQA can recover those that dominate the dynamics of the behavior. Because all of the dimensions driving the behavior are typically not known, *average mutual information* (AMI) and *false nearest neighbors* (FNN) may be used. These are described in "Recurrence parameters" section.

5. The resulting TDR shows a state's changing trajectory over time in the SS representation, and whether a sequence of behavior recurs with itself and its neighbors, and how many neighbors share information.

6. We then characterize recurring patterns by taking the time indices of each point on the SS, and represent these as coordinates (i, j) in the RP. A reference point is selected on an SS attractor, with a circular radius centered on this point. If the point on the attractor falls within this region, then a dot is placed at coordinates (i, j) on the RP. This process is repeated for all points until all have been exhausted. An $N \times N$ plot is formed with N as the number of points classified as part of the same attractor. The resulting RP contains a variety of information about the systems dynamics, including deviations from random behavior; periodicities; transitions; and local, global, and other forms of recurrences.

Recurrence Parameters

Mutual information (MI) measures the shared information (or mutual dependence) between variables, and is used to determine an optimal time delay (L) for the SS reconstruction (Fraser & Swinney, 1986). Given system state $X(t)$, an optimal L maximizes new information at $X(t + L)$. The average MI (or AMI) measures how many (average) bits of information $I(T)$ can be predicted at $X(t + T)$. Initial states have maximum bits $(X(t + 0) = X(t))$, and as T increases, $I(T)$ tends to decrease, but often fluctuates at later points in time. Thus, the first minimum of $I(T)$ is used to estimate L in the SS reconstruction. For the sinusoid, $I(T)$ drops to 0 when $L = 8$, as shown in Figure 10.2B.

FNN can be used to select M for uncovering the system's topology, and to learn whether neighbors of a given point in the time series are mapped correctly onto neighbors in the projected (M) space. FNNs are the points that have been projected erroneously into neighborhoods of other points to which they do not belong. Using the following key steps (from Kennel,

Brown, & Abarbanel; 1992), FNN selects M by estimating the minimal temporal separation of valid neighbors (M_{min}):

1. The nearest neighbor of every point in a given dimension is found: *For each vector* $X = (x_1, x_2, x_3,..., x_n)$ in the time series, find its nearest neighbor $Y = (y_1, y_2, y_3, ..., y_n)$ in a multidimensional space.

2. Check if points continue to be close in incrementally higher dimensions by computing *distance* (R) between vectors X and Y plots: $R = |x_{n+1} - y_{n+1}| \cdot y_{n+1}$. The %FNN drops to zero when an appropriate M has been reached.

3. Compare R with a *prediction error* generated by a network to determine whether points are correctly classified as neighbors on a common dimension. When its attractor is completely unfolded in a multidimensional space (similar to viewing the SS head on from an orthogonal view), R should be sufficiently small to detect if the nearest neighbor just found is false. We do this by comparing R with the errors made by the predictor. If the error is less than R, $|x_{n+1} - y_{n+1}| < |x_{n+1} - x_n|$, the nearest neighbor is "false."

Some cautionary notes are in order: (a) Like other parameter estimates, FNN requires setting threshold values to some somewhat arbitrary values. (b) The FNN method can be sensitive to the sampling rate of the time series.

Dimension (D) is the number of independent variables participating in the system at any given instant. The *Grassberger–Procaccia dimension* (GPD; Grassberger & Procaccia, 1983) is one measure of D, and can be estimated by observing *how the number of recurrences scales with increases in radius* (r). GPD can also be used as an upper limit for embedding dimension: $M_{max} > 2 \times$ GPD + 1 (Webber & Ziblut, 2005).

Embedding dimension (M) is estimated by the FNN analysis. M is increased incrementally until additional nearest neighbors no longer alter or add information to the dynamics (Kennel et al., 1992). This method works well for simple (stationary and low-noise) systems: M is approximated by the plateau in the FNN. For nonstationary, high-noise systems (as in the case of most human behaviors), the dimension tends to be inflated. A simple guideline is given, combining the suggestions of Webber and Zbilut (2005), Kononov (2005), and Shelhamer (2007): $D < M \leq 2D + 1$ with dimension (D) as the number of operating variables, or degrees of freedom, in the dynamical system under study. In most cases, however, D is unknown and can only be estimated by FNN or other methods. An alternative is a brute force approach performing *root mean square* (RMS) as M or as L changes. For biological and psychological systems, Webber and Zbilut (2005) recommend setting $M = 10$ or 20, but not higher, as artifacts masquerading as patterns of recurrence can result.

Time delay or lag (L) is the time separation of the components in the reconstructed vector of the system state where interaction across points is minimized, and the widest profile of the SS may be seen. L is estimated by the first minimum in the linear autocorrelation, or the nonlinear (average) MI function (Fraser & Swinney, 1986). A safe starting point is to use $L = 1$ to include all points, as is done for discrete behaviors (or maps). Webber and Zbilut (2005) even argue there is no need to find optimal delay, because the features of a system tend to be stable across different values of L, and living signals often do not have an optimum. L should be large enough to capture the main features of SS so that each reconstructed vector is sufficiently independent to carry new information about the trajectory. Furthermore, if L is too long, the coordinates may become too independent to the point of being random with respect to each other.

Threshold Parameters

Radius (*r*) is the distance threshold, $(r) < |x(i) - x(j)|$, surrounding a reference point on an attractor, and is used to transform the DM into a recurrence matrix (RM). Parameter r should be small enough to be sensitive to changes in the dynamics, but large enough to capture the dynamics. For the oscillatory eye movement (in Figure 10.5), r is based on original display screen units of 800×600 pixels.

Window range (*W*) is the range over which recurrence is captured in a certain M. To evaluate small-scale recurrences, set W to a small value, and for large-scale recurrences, set W at a large value with the following constraint: with an initial point (P_{init}) and an end point (P_{end}), $W < [(P_{end} - P_{init}) + 1]$. When $M > 1$ for N points, $M - 1 > P_{end}$, then, $P_{end} < N - M + 1$. For the sine wave in Figures 10.1 and 10.3, $W = 64$ captures the complete cycle of the wave.

Norm (*Nm*) defines geometrically the size and shape of the neighborhood surrounding each reference point. Like W, Nm can evaluate local (*MIN Nm*), global (*MAX Nm*), or intermediate (Euclidean) scales of recurrence. Unless noted otherwise, all analyses presented here use Euclidean Nm.

Line (*ln*) is used to quantify features of a time series without affecting the recurrence in RM. *ln* is typically set to the conservative estimate of two points (i.e., the shortest distance forming a line). If the length of the recurrence feature is shorter than *ln*, that feature is rejected. All analyses presented here use $ln = 2$.

Visualizing the Dynamics

Although it is straightforward to visualize the recurrence of states in 2D or 3D, most real systems, including those that drive human behavior, are higher-dimensional SSs that are difficult, if not impossible, to visualize. Eckmann, Kamphorst, and Ruelle (1987); and Packard, Crutchfield, Farmer, & Shaw (1980) provide a solution through projection onto 2D or 3D subspaces. Thus, the multidimensional SS trajectory can be evaluated through a 2D representation of its recurrences. The recurrence of a state at time i and at a different time j is depicted within a 2D matrix either with black and white dots, or dots varying in gray scales, or colors. In the black and white versions of RPs, black dots mark a recurrence. In all RPs, the x and y axes represent time indices i and j, with a point placed at (i, j). Points along the main (LOI, line of identity) diagonal represent *exact recurrence* or points recurrent with themselves. Diagonals off the LOI indicate periodicities in the data, and are quantified by the *percentage of determinism* (%DET). With human data, %DET is at best *approximate recurrences* and captures statistically similar patterns. To detect subtle patterns, the range or threshold for recurrence can be adjusted. Relaxing thresholds permits inclusion of more points on each iteration that are counted as recurrent. Thus, larger thresholds tend to produce wider diagonals and capture wider regions of recurrence.

Three quantitative measures of recurrence distinguish spurious from deterministic patterns: The first, *recurrence rate*, or the *percentage of recurrence* (%RECUR), measures the total density of recurring points in an RP, and thus includes both deterministic and random points. Because %RECUR is over-inclusive, it is important to find appropriate parameters (i.e., L and M) that minimize %RECUR, but maximize two measures of the deterministic recurrence: (a) %DET quantifies the periodicity, or system states with a similar time evolution by estimating the portion of %RECUR making up diagonal structures in the RPs. (b) Laminarity (%LAM) is an estimate of the percentage of all recurrence points that form vertical (and horizontal) lines. These signify stable or slowly changing states, and their edges indicate when the behavior is transitioning to another state. Additional measures are described below.

RP Signatures of Dynamical Structure

RPs together with windowed calculations of RQA (above) enable a time-dependent analysis of recurring patterns in a data series. Because of its sensitivity to the correlated structure, RQA is useful for detecting and identifying correlated patterns, along with transitions, or synchronization across behavioral states. In continuous gray-scale or colored RPs, distances between recurring points are represented. In these RPs (derived from the RM), the color's (relative) wavelength, or intensity of the gray level of each point (i, j) is proportional to the distance $|x(i) - x(j)|$. In the gray-scale RPs shown here, black points indicate high recurrence and white indicate low recurrence. RPs mostly contain dots, diagonal lines (parallel to LOI), and vertical (and horizontal) lines.

The symmetric property of the RPs renders the main diagonal as the "line of identity", and vertical lines equivalent to horizontal lines. When these appear as a band of patterns, they usually indicate a stable, or slowly changing states, with the edges marking the transitions between states. Shorter and narrower *vertical lines* can also arise from stationary states, or a path recurrent for several points looping around an isolated point in the SS. *Bands of longer vertical lines* correspond to segments of a trajectory remaining in the same SS region for some time (see %LAM below).

Diagonal lines represent segments of the SS trajectory that run parallel for some time, and include points that recur as part of a sequence of behavior. Those parallel to the LOI indicate a slowly moving trajectory, a long-lasting periodicity, a recurrent path or one that is part of a larger attractor with paths following each other (Shelhamer, 2007). Long-ranging periodicity is represented by *long diagonal lines* (uniformly spaced and equal in length to the LOI (i.e., %DET). Brief or transient periodicity is indicated by *short diagonal lines*, where SS trajectories rapidly diverge after coming close together at intermittent points in time. A reversal in a trajectory is represented by *lines orthogonal to LOI*. Here, nearby paths move in opposite directions with time.

Finally, a nondeterministic, or random, system will produce a *homogeneous pattern* over the RP. Points distributed uniformly indicate white noise, or a stationary process. There is no clustering or diagonals (unless r is set too high). "Patchy" patterns are a signature of colored noise. Brown noise produces broader clusters than pink noise. Fuzzy diagonal patterns can emerge for certain delays of colored noise indicating some deterministic structure in the system. We shall see similar patterns in eye movements used in tracking and search tasks. Table 10.1 summarizes additional patterns in RPs signifying different forms of recurrence.

Quantifying Recurrence

Although the qualitative information obtained from RPs is informative, quantifying the patterns is the real power behind RQA. Using small windows shifting along the LOI of the RP, we calculate the point density of the RPs, and their diagonal and vertical line structures, to determine the magnitude and frequency of time-dependent behavior making up the data series. Detailed descriptions of RQA can be found in the work of Zbilut and Webber (2007), and Shelhamer (2007). Briefly, the main quantitative measures of recurrence include three noted earlier, (a) %RECUR, (b) %DET, and (c) %LAM, and related measures, (d) MAXLn, (e) *entropy* (ENT), and (f) *trapping time* (TT).

As noted, %RECUR measures the total percentage of recurring points in the RP including both random and deterministic structures, and *%DET* is the percentage of recurring points forming diagonal lines and the periodicity of the behavior under study. Consecutive recurring points predicted by the longest diagonal line (*MAXLn*) measure a form of dynamical stability *inversely* related to system complexity, and how much the systems trajectories diverge

TABLE 10.1

Typical Patterns in RPs and Interpreting the Dynamics

RP Pattern	SS Dynamics
1. Homogeneity	1. Process is stationary; stable behavior or uniform white noise.
2. Fading to upper-left and lower-right corners of RP	2. Trend or a drift; nonstationary data.
3. Disruptions (white bands)	3. Nonstationary data; some states are rare or far from the normal; transitions may have occurred.
4. Periodic/quasiperiodic patterns	4. Cycles in the process; the time distance between periodic patterns (e.g., lines) corresponds to the period; different distances between long diagonal lines reveal quasiperiodic processes.
5. Single isolated points	5. Strong fluctuation in the process; if only single isolated points occur, the process may be an uncorrelated random.
6. Diagonal lines: a. Parallel to the LOI	6a. The evolution of states is similar at different epochs; the process may be deterministic.
b. Orthogonal to the LOI	6b. The evolution of states is similar at different times but they move in opposite directions (in SS); sometimes this is an indication for insufficient embedding.
7. Vertical and horizontal lines/clusters	7. Laminar states do not change or change slowly for some time. At the edge of the laminar bands, or when lines are narrow, transitions occur. When bands contain narrow lines, this indicates rapid transitions within a pattern, whereas edges of the bands indicate transitions between superposed pattern.
8. Long-bowed line structures	8. The evolution of states is similar at different epochs, but with different velocity; the dynamics of the system could be changing.

over time (i.e., the Lyapunov exponent). ENT measures this complexity as the loss of redundant information of the recurrence structure, and thus serves as a measure of a system's predictability. Small values of $MAXLn$ indicate more divergence in the SS trajectories, greater complexity (ENT), and behavior that is less predictable over time. Laminarity (%LAM) is the percentage of all recurrence points that form vertical (and horizontal) lines. These signify stable or slowly changing states, and their edges indicate when the behavior is transitioning to another state. While we would expect a behavior to be stable under a fixed set of conditions, a high %LAM is also typical of intermittent behaviors. TT is the (mean) length of vertical lines, and measures the average time for which the system is trapped (or changing only very slowly) within one state.

Illustration of Simple Oscillating Systems

We use behaviors from three systems to step through RQA. The first is the very simple sinusoidal system (Figure 10.1), the second uses the same sine wave plus noise (Figure 10.4), and the third approximates these with oscillatory eye movements generated by tracking an alternating stimulus (Figure 10.5). Figure 10.1 shows SSs and RPs of oscillating behavior, and how adjusting delay (L) produces different views of the SS, and different RP patterns. In a data set consisting of 1024 or more data points, setting $L = 8$ (the first minimum of AMI) is sufficient to uncover the dominant periodicity of the sine wave. When $L = 16$, the full (64-point) cycle of the sinusoid is revealed as a circular SS, and the periodic structure is most clearly depicted as diagonal lines in the RPs (Figures 10.1 and 10.3). As L deviates from 16 ($1 < L > 32$), the SS shape compresses into an elliptical shape, and RP diagonals become discontinuous, emphasizing the transition in path direction.

Gray levels in RP show the distances between recurring points, and the spacing between diagonal lines shows the interval and range of periodicity. If we sample different values of L within (and beyond) the full cycle of the sine wave, we can observe the evolution of the SS shape from an ellipse ($1 < L < 16$), to a circle ($L = 16$), back to an ellipse but reversed from the initial directions ($16 < L < 32$). The most compressed ellipses occur at $L = 1$ and $L = 32$, otherwise they become wider as delay approaches $L = 16$. When $L > 32$, this pattern repeats with $L = 48$, producing the full-cycle circular SS trajectory of $L = 16$, and its clear RP diagonal structure.

Figure 10.3 shows the relationship between RPs, SS states, and how these map back on to the original time series. These different representations provide unique information about changes to amplitude, and the proximity of the behaviors over time (in the time series), as well as the direction and relative phase of the different states (in the SS). For example, points that are close together and in-phase in the sinusoid time series appear close on the same SS trajectory and close together with similar color or brightness on the RP diagonal. But when points are out of phase in the TS, both close and far points tend to be far away on a common SS trajectory. As more variables contribute to the behavior, these relationships become more difficult to decipher in the SS representation, but the RPs and the quantification of recurrence help interpret the complex flow patterns appearing in the SS images.

Figure 10.5 shows what happens when (white) noise is added to the sine wave, and how SS and RPs still show a clear periodic structure. But now the regions are wider and "fuzzy" from the added noise. Also, short segments in longer delays (e.g., $L = 23$) show short-range periodicities in RPs superimposed on long-range oscillation: these are the haphazard paths traversing erratically along the "fuzzy" circular boundary of the SS trajectory. Rather than white noise, human variability tends to be a colored form of noise. Thus, rather than the uniform RPs patterns produced by simply adding white noise to the sinusoid behavior (in Figure 10.4), brown and pink noise produce distinctive heterogeneous and clustered patterns across many delays. Notice the uniform distribution of the white noise RPs (accounting for the simple additive effect when combined with the sinusoid). By contrast, (brown and pink) colored noise produced distinctive heterogeneous and clustered patterns across many delays ($L = 1$–10 at $M = 2$ were tested here). Brown noise produces large clusters that arise from the close proximity of successive states in the SS and points in the TS. Pink noise produces smaller clusters, often a sign of intermittent behavior.

Discussion

Recurring patterns in human behavior are ubiquitous, but their study is only in its infancy. RQA can detect and identify recurring patterns. In this chapter, we introduced RQA using a computer-generated sine wave series, and one with white noise added, and then applied RQA to these and various object-tracking and scanning tasks. It is noteworthy that oscillations emerge not only, where we would expect to see them, but also in more complex search and tracking tasks, where eye movements appear random. Even when systematic patterns appear, as is usually the case in the real world, and in most well-controlled eye-tracking experiments, when we examine the sequence of eye movements over time, we find not only oscillatory patterns, but also complex behavior with long-range correlations (e.g., Aks, 2009). The pervasiveness of these combined properties suggests a universal mechanism linking oscillations and "critical" behavior. Perhaps one may drive another. For example, combining different oscillatory mechanisms can produce complex yet effective behavior when performing a wide range of tasks.

There are many theoretical proposals that oscillatory mechanisms underlie human behavior with some arguing that attention samples in a periodic fashion (Van Rullen, Carlson, & Cavanagh, 2007). Similarly, our eyes may sample the environment in a periodic albeit noisy fashion. Such a mechanism would require only minimal resources, since it would likely be an automatic and effortless act.

What might underlie the pervasiveness of complex oscillatory behavior? System coupling is one possibility, and is supported by the pattern of results found here, as well as various findings from neuroscience. Examples of separable systems that interact in such a way include distant neural ensembles (Fries, 2001), and independent and modular computational units. Complex systems are defined in part by interconnectedness and criticality, and (often) within neuroscience, as patterns of synchrony.

Notes

1. Shalizi (2009) notes that Packard, Crutchfield, Farmer, and Shaw (1980) first published what is widely known as Taken's theorem, but credit David Ruelle as the originator of the solution to use time delays to reconstruct the SS of a system.
2. We can also define a bivariate extension, the cross RP $CR(i, j) = \Theta(\varepsilon - \| x(i) - y(j)\|)$ to evaluate how well two different systems are coupled over time (Marwan & Kurths, 2003). Richardson, Dale, and Kirkham (2007) use the CR analysis of gaze movements of two individuals engaged in conversation and the role of coupling between the speaker and the listener. They further show that shared background knowledge improves coordination.
3. The dynamical stability that MAXLn measures should not be confused with TND, which measures whether the mean data value changes over time from, for example, factors such as fatigue and practice. In these cases, TND deviates from 0 to reflect "drift" in the means. TND = 0 indicates an absence of drift and a stationary system whose mean is stable over time.

References

Aks, D. J. (2009). Temporal and spatial patterns in perceptual behavior: Implications for dynamical structure. In S. J. Guastello, M. Koopmans, & D. Pincus. (Eds.), *Chaos and complexity in psychology: Theory of nonlinear dynamical systems* (pp. 132–176). New York: Cambridge University Press.

Bak, P., Tang, C., & Wiesenfeld, K. (1987). Self-organized criticality: An explanation of the $1/f$ noise. *Physical Review Letters, 59,* 381–384.

Eckmann, J. P., Kamphorst, S. O., & Ruelle, D. (1987). Recurrence plots of dynamical systems. *Europhysics Letters, 4,* 973–977.

Findlay, J. M., & Gilchrist, I. D. (2003). *Active vision: The psychology of looking and seeing.* Oxford: Oxford University Press.

Fraser, A. M., & Swinney, H. L. (1986). Independent coordinates for strange attractors from mutual information. *Physics Review A, 33,* 1134–1140.

Fries, P. (2001). A mechanism for cognitive dynamics: Communication through neuronal coherence. *Trends in Cognitive Science 9,* 474–490.

Guastello, S. J., & Liebovitch, L. S. (2009). Introduction to nonlinear dynamics and complexity. In S. J. Guastello, M. Koopmans, and D. Pincus. (Eds.), *Chaos and complexity in psychology: Theory of nonlinear dynamical systems* (pp. 1–36). New York: Cambridge University Press.

Guastello, S. J., Koopmans, M., & Pincus, D. (2009). (Eds.), *Chaos and complexity in psychology: Theory of nonlinear dynamical systems.* New York: Cambridge University Press.

Grassberger, P., & Procaccia, I. (1983). Measuring the strangeness of strange attractors. *Physica D, 9,* 189–208.

Holden, J., Van Orden, G., & Turvey, M. T. (2009). Dispersion of response times reveals cognitive dynamics. *Psychological Review, 116*, 318–342.

Kennel, M., Brown, R., & Abarbanel, H. (1992). Determining embedding dimension for phase-space reconstruction using a geometrical construction. *Physics Review A, 45*, 3403–3411.

Kononov, E. (2005). *Visual recurrence analysis (VRA)* V.4.9. Retrieved April 8, 2010 from http://nonlinear.110mb.com/vra

Marwan, N. (2003). Encounters with neighbours: Current developments of concepts based on recurrence plots and their applications. Doctoral Dissertation, University of Potsdam, ISBN 3-00-012347-4. urn:nbn:de:kobv:517-0000856. Retrieved April 1, 2010 from http://www.recurrence-plot.tk/furtherreading.php

Marwan, N., & Kurths, J. (2003). Nonlinear analysis of bivariate data with cross recurrence plots. *Physics Letters, 330*, 214–223.

Marwan, N., Romano, M. C., Thiel, M., & Kurths, J. (2007). Recurrence plots for the analysis of complex systems. *Physics Reports, 438*, 237–329.

Packard, N. H., Crutchfield, J. P., Farmer, J. D., & Shaw, R. S. (1980). Geometry from a time series. *Physics Review Letters, 45*, 712–716.

Pellecchia, C. L., & Shockley, K. (2005). Applications of recurrence quantification analysis: Influence of cognitive activity on potential fluctuations. In M. A. Riley & G. C. Van Orden (Eds.), *Tutorials in contemporary nonlinear methods for the behavioral sciences* (pp. 95–141) Washington, DC: National Science Foundation. Retrieved from http://www.nsf.gov/sbe/bcs/pac/nmbs/nmbs.jsp

Richardson, D. C., Dale, R., & Kirkham, N. Z. (2007). The art of conversation is coordination: Common ground and the coupling of eye movements during dialogue, *Psychological Science, 18*, 407–413.

Shalizi, C. (2009, August). Recurrence times of stochastic processes (also hitting, waiting, and first-passage times). *Notebooks*. Retrieved April 1, 2010 from http://cscs.umich.edu/~crshalizi/notebooks/recurrence-times.html

Shelhamer, M. (2007). *Nonlinear dynamics in physiology: A state-space approach.* Singapore: World Scientific.

Sprott, J. C., & Rowlands, G. (1995). *Chaos data analyzer.* Raleigh, NC: Physics Academic Software (American Institute of Physics).

Takens, F. (1981). Detecting strange attractors in turbulence. In D. A. Rand & L. S. Young (Eds.), *Dynamical systems and turbulence* (pp. 366–381). New York: Springer-Verlag.

Van Rullen, R., Carlson, T., & Cavanagh, P. (2007). Dividing attention between multiple targets: simultaneous or sequential allocation? [Abstract]. *Journal of Vision, 7*, 642, 642a. Retrieved April 1, 2010 from http://journalofvision.org/7/9/642/, doi:10.1167/7.9.642

Ward, L. M. (2002). *Dynamical cognitive science.* Cambridge, MA: MIT Press.

Webber, C. L. Jr., & Zbilut, J. P. (1994). Dynamical assessment of physiological systems and states using recurrence plot strategies. *Journal of Applied Physiology, 76*, 965–973.

Webber, C. L. Jr., & Zbilut, J. P. (2005). Recurrence quantification analysis of nonlinear dynamical systems. In M. A. Riley & G. C. Van Orden (Eds.), *Tutorials in contemporary nonlinear methods for the behavioral sciences* (pp. 26–94). Retrieved March 1, 2005, from http://www.nsf.gov/sbe/pac/nmbs/nmbs.jsp

Webber, C. L. Jr., & Zbilut, J. P. (2007). Recurrence quantifications: Feature extractions from recurrence plots. *International Journal of Bifurcation and Chaos, 17,* 3467–3475.

Zbilut, J. P., & Webber, C. L., Jr. (1992). Embeddings and delays as derived from quantification of recurrence plots. *Physics Letters A, 171,* 199–203.

Zbilut, J. P., & Webber, C. L., Jr. (2007). Recurrence quantification analysis: Introduction and historical context. *International Journal of Bifurcation and Chaos, 10,* 3477–3481.

11

Discontinuities and Catastrophes with Polynomial Regression

Stephen J. Guastello

CONTENTS

This chapter considers two broad classes of techniques for identifying discontinuities that can be accomplished using the general linear model. The first, spline
regression, is an analytic technique that could be useful when generic nonlinearities are involved, but no particular nonlinear dynamical systems (NDS) processes is assumed. The second, catastrophe theory, consists of a limited group
of formal models that contain specific configurations of attractors, bifurcations,
and saddles that underlie discontinuous change processes. Catastrophe models
also have important uses in the analysis of self-organizing processes.

Spline Regression

Simple Splines

The idea behind spline regression is that a series of observations contains
identifiable segments where the local properties of the model or data are different. One might take a function that has one bend in the curve and analyze
the two data segments separately using two linear models instead of one
model for the whole data series. This particular strategy is known as piecewise linear regression, and can be as simple as it looks. Its limitation is that
if one were to take a curve that is too complex and break it into many locally
linear regions, having many locally linear models is tantamount to having
no model at all. One would also need a rule for determining the switching points from one linear model to another if there are too many switching
points. Another problem with the piecewise approach is that the sample size
for each of the pieces is smaller than the total N, which could make finding a
local regression weight difficult if the pieces are too small.

Spline regression shares the same objective for trend modeling, but uses
all the data in one regression analysis. The analysis exercise below was suggested by Darlington (1990). Figure 11.1 is plot of the U.S. census at 10-year
intervals from 1790 to 2000. There is a bend in the curve at 1890 when the
immigration rate spiked. There are several approaches to modeling this nonlinear trend with the general linear model: taking the log of time, population, or both; polynomial regression, or spline regression.

A spline is created by making a dummy-coded (0–1) variable and then multiplying it by the independent variable, which is time in this case. The resulting
variable is used as a second independent variable in multiple linear regression.

The results of the possible analyses appear in Table 11.1. Time was recoded
to simple integers 1–22. The r^2 for the linear model is used as the benchmark.
For the possible log combinations, log of the criterion variable, census, produced the best improvement. Spline regression Model 1 used the dichotomous variable only. Model 2 contained a multiplicative variable defined
above. Neither made a significant improvement over the linear model.
The best results were obtained for polynomial regression. The polynomial

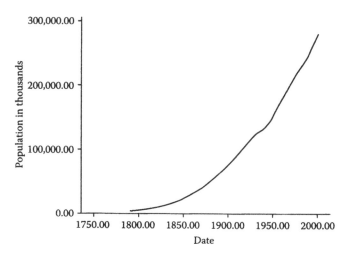

FIGURE 11.1
U.S. Census data 1790–2000.

TABLE 11.1

Spline and Other Regression Techniques for U.S. Census Data

Model	r^2/R^2	β	t
Linear	0.92		
Log time	0.65		
Log census	0.97		
Log census and time	0.94		
Spline Model 1			
Time		1.05	8.34*
Spline (0–1)	0.92	−0.11	−0.87
Spline Model 2			
Time		0.69	3.71*
Spline × time	0.93	0.29	1.57
Polynomial			
Linear		−0.18	−5.88*
Quadratic	0.99	1.18	37.53*

*$p < .001$.

model would be adopted as the best description of the census data. Analyses for other data sets could turn out differently, of course.

SETAR Models

Self-exciting threshold autoregression (SETAR) models (Tong, 1990) are a variety of piecewise linear regression designed to find trends in the peaks of highly volatile time series, such as electrocardiograms, electroencephalograms, electromyagrams, or recordings of peak daily anxiety levels for psychological patients under treatment for anxiety disorders. Typically there is a high frequency of small changes from one observation to the next, with larger changes being rarer. The premise is that the large changes are more meaningful than the small ones. The large changes are interpreted as having crossed a threshold, similar to the firing of a neuron.

A SETAR analysis starts with two time series of differences in the variate X, where X takes on a value above or below a threshold, T:

$$\Delta X_t = \beta_0 + \beta_1 \Delta X_{(t-1)} + \cdots + \beta_n \Delta X_{(t-n)} \quad \text{if } X < T, \tag{11.1}$$

$$\Delta X_t = \gamma_0 + \gamma_1 \Delta X_{(t-1)} + \cdots + \gamma_n \Delta X_{(t-n)} \quad \text{if } X > T, \tag{11.2}$$

where
  t is time
  β_i and γ_i are two sets of regression weights

In Equations 11.1 and 11.2, ΔX_t is predicted from a linear combination of previous ΔX, with as many lags as needed to account for ΔX_t.

Warren, Sprott, and Hawkins (2002) used the SETAR procedure to analyze anxiety levels and intensities of obsessive ruminations for psychological patients who kept logs of their experiences for 275 days. The regression models themselves were not especially remarkable, except to say that only one lag was needed and that they predicted outcomes better than a simple linear autoregressive model. The insight from the analysis came from examining the placement or location of the threshold. They began with $T = 1.5$ as a plausible value. The regression equations could be easily solved to produce equilibria at 0.07 and 0.57, respectively, for rumination data using Equations 11.1 and 11.2. Thus, if the patient exerted control to stay under T, rumination levels would rise only very slowly. If the patient tried to control T to 0.0, which is below the lower equilibrium value, the time series would cross a bifurcation point into a periodic or chaotic regime with large fluctuations. Noise in the system would exacerbate the fluctuations.

Catastrophe Theory

The central proposition of catastrophe theory is the classification theorem (Thom, 1975), which states (with qualifications) that, given a maximum of four control parameters, all discontinuous changes of events can be modeled by one of seven elementary topological forms. The forms are hierarchical and vary in the complexity of the behavior spectra they encompass. The models describe change between (or among) qualitatively distinct forms for behavior; they do not imply any notion of desirable or undesirable outcome. There is a continuity underlying each discontinuity that contributes to the explanatory power of the theory.

A spectrum of behavior contains steady states (attractors), changes in behavior that are shaped by underlying bifurcations, repellors, and saddle points. The changes in behavior are governed by one to four control parameters, depending on the complexity of the model. In NDS, a control parameter denotes an independent variable that has a particular function in the change process. A researcher would identify one or more psychological (or other substantive) measurements that would correspond to a particular function.

The elementary catastrophe models fall into two groups: the cuspoids and the umbilics. The elementary cuspoids involve one dependent measure. They have potential functions in three to six dimensions and response surfaces in two to five dimensions. They are the fold, cusp, swallowtail, and butterfly. Their names reflect fanciful interpretations of what parts of their geometry resemble. The elementary umbilics involve two dependent outcomes measures, three or four control parameters, and response surfaces in five or six codimensions. A codimension is the sum of dimensions associated with two or more order parameters (behavioral spectra) that are joined in a system. The umbilics are the wave crest (or hyperbolic umbilic), hair (or elliptic umbilic), and mushroom (or parabolic umbilic) models. Thom (1975), Zeeman (1977), Poston and Stewart (1978), Woodcock and Davis (1978), Gilmore (1981), Thompson (1982), and Arnold (1974) serve as general references for the structure of catastrophe models and provide numerous applications outside the social sciences.

Fold

The fold is the simplest model in the set. Its response surface is defined as the set of points where

$$\frac{df(y)}{dy} = y^2 - a, \qquad (11.3)$$

where
  y is the dependent measure
  a is the control parameter

The fold model (Figure 11.2) is the geometric building block of the elementary models. It describes a change in behavior from a stable state or attractor to an unstable state as a function of a. The relationship between a and y is a simple threshold model, with the important difference that, like other NDS models, change in behavior is function of both the control variable and the order parameter at a previous point in time or space. Behavior remains steady, even though a is changing. Once a hits a critical value, however, behavior changes abruptly. Because the change is in the direction of instability, the trajectory of y leaves the basin of the attractor and flies outward to no particular destination. The equilibrium values of y are determined by setting Equation 11.3 equal to 0. The negative root is the center of the stable attractor, and the positive value is the unstable center.

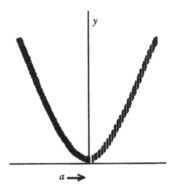

FIGURE 11.2
Fold catastrophe response surface.

The fold, like the other catastrophe models, has a potential function that depicts the model in static form:

$$f(y) = \frac{y^3}{3} - ay. \tag{11.4}$$

Catastrophe functions are not inherently static or dynamic; it is the application and data that give a model a static or dynamic character. Static models are inherent in probability density functions (pdfs) for the catastrophe models, which are covered in Chapter 13. Potential functions are integrals of the response surface equations. The concept is more interesting for the cusp and larger models, and it is revisited below.

Cusp

The cusp response surface is three-dimensional (Figure 11.3). It describes two stable states of behavior. Change between the two states is a function of two control parameters, asymmetry (*a*) and bifurcation (*b*). At low values of *b*, change

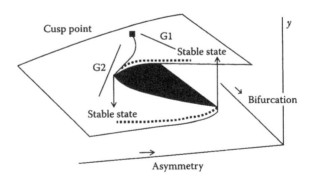

FIGURE 11.3
Cusp catastrophe response surface.

is smooth, and at high values of b it is potentially discontinuous, depending on the values of a. At high values of b and low values of a, changes occur around the lower mode and are relatively small in size. At middle values of a, changes occur between modes and are relatively large, assuming b is also large. At high values of a, changes occur around the upper mode and are again small.

The cusp response surface is the set of points where

$$\frac{df(y)}{dy} = y^3 - by - a. \tag{11.5}$$

Following the dotted line on Figure 11.3, change in behavior is denoted by the path of a control point over time. The point begins on the upper sheet denoting behavior of one type and is observed in that behavioral modality for a period of time. During that time, its coordinates on a and b are changing when suddenly it reaches a fold line and drops to the lower value of the behavior, which is qualitatively different from where it remains. Reversing direction, the point is observed in the lower mode until coordinates change to a critical pair of values, at which moment the point jumps back to the upper mode. There are two thresholds for behavior change, one ascending, and one descending. The shaded area of the surface is the region of inaccessibility in which very few points fall; it is produced by a repellor.

The cusp bifurcation set induces two diverging response gradients on the response surface, which are joined at a cusp point. The diverging gradients are labeled G1 and G2 in Figure 11.3. The cusp point is a saddle point and is known as the point of degenerate singularity; it is the most unstable point on the surface. Behavior at the cusp point could tip in the direction of either attractor as a result of a very small change in the control parameters. Analogous points exist in other catastrophe models also.

The attractor centers of the response surface can be found by setting Equation 11.5 equal to 0. The solution will produce three roots. The two negative roots represent attractor centers, and the positive root represents the repellor center. Critical values of y where instability takes place can be found by taking the first derivative of Equation 11.5, setting it to 0 and solving. The same analytic principles apply to the other models below. The regression equations for the catastrophe models can be used for this purpose.

Other features of catastrophe models include hysteresis, the delay rule, and the Maxwell convention. Hysteresis is where behavior changes back and forth, or up and down the manifold, which prevents the control point from retracing its path in both directions. The presence of hysteresis strongly suggests a cusp dynamic. The delay rule runs as follows: Once the control parameters are in place to promote a behavior change, there could be small time delay before the change occurs. This is analogous to the common observation of a "calm before the storm." The Maxwell convention addresses the situation where one gradient may be stronger than the other. The system is likely to remain in the stable state that is most probable.

Entropy and Self-Organization

The potential function for the cusp model is shown in Equation 11.6 and a graph of the function is shown in Figure 11.4:

$$f(y) = \frac{y^4}{4} - \frac{by^2}{2} - ay. \qquad (11.6)$$

Here the position of y on the vertical axis represents the level of potential energy or entropy in the system. Entropy is low when the control point is in a well, but high otherwise. In the upper left, the control points sit in a stable state behind the cusp point. As the bifurcation factor increases in strength (upper right), the control point becomes more instable and ready to fall into one of the wells in the lower left or lower right. The asymmetry parameter determines which well it will be.

The diagram is essentially showing a phase shift in the physical sense of ice turning to water and back again. The mathematical models are the same (Gilmore, 1981). Furthermore, the qualitatively distinct changes in the system that accompany a self-organizing process have been characterized as cusp catastrophes (Cerf, 2006; Guastello, 2002, 2005a; Haken, 1988, 1999). Other writers interested in the self-organizing dynamics have called attention to one of the two-well portions of the diagram, and sometimes leave the bifurcation parameter as a constant while working with the asymmetry parameter (Kelso, 1995; Vallacher & Nowak, 2009).

A random force occurs when a system is exposed to any entropy-inducing events, in which each element of the system (e.g., subjects in an experiment) has an equal probability of exposure (Agu, 1983). As elements increase in entropy, the probable location of each element in the space expands to the limits of its

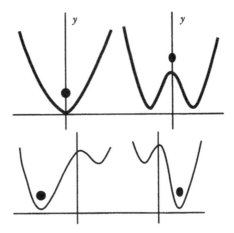

FIGURE 11.4
Phase shifts inherent in the potential function for the cusp catastrophe.

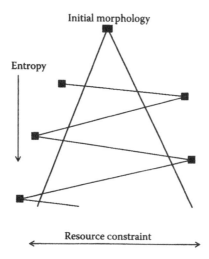

FIGURE 11.5
Relationship between entropy and the cusp bifurcation set.

confinement. Bifurcations serve to reduce entropy by partitioning the energized elements into neighborhoods of relative stability (Thompson, 1982).

Figure 11.5 represents a problem introduced by Crutchfield (2003) as part of an explanation for the evolution of animal or plant morphology. The increases in structural variation (entropy), which could favor one structure or another, act as a bifurcation factor. If an environmental constraint or condition favors one structure or another, a particular morphology will stabilize.

Swallowtail

The swallowtail catastrophe model describes a behavior spectrum consisting of two stable states plus two unstable areas. The surface, which is shown in Figure 11.6, is defined as the set of points where

$$\frac{df(y)}{dy} = y^4 - cy^2 - by - a. \tag{11.7}$$

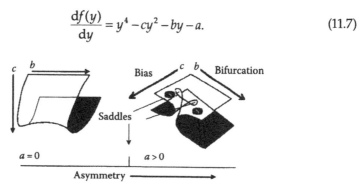

FIGURE 11.6
Swallowtail catastrophe response surface.

The new control parameter, *c*, and is variously known as the bias, swallow-tail, or second asymmetry parameter.

The surface is four-dimensional and requires three-dimensional sectioning for graphic rendition. Once again, the shaded area is the place where few points fall. When $a = 0$, behavior rests in an unstable state. Parameters *b* and *c* do not do much of anything here, except to predispose the control point for a particular gradient of change as *a* increases. When *a* increases, the control point moves through the shaded area until it lands on the portion of the surface shown at the right, which more interesting. When $a > 0$, behavior can change between two adjacent stable states (marked S), or fall through a twist and gap (saddles) in the surface, through the shaded regions back to the unstable state at the left.

Whereas the cusp has seen more applications than one can reasonably count at this point in history, and the fold is perhaps a little too dull to garner much attention, the swallowtail catastrophe model has seen only a few applications in the behavioral sciences. One notable suite of applications involves the emergence of leaders from a leaderless group of people working on a task (Guastello, 2007). Research with that model has relied on the static method of analysis that is covered in Chapter 13 rather than the dynamic difference equations that are described in this chapter.

Butterfly

The butterfly catastrophe model describes a spectrum of three qualitatively different behavioral outcomes and has lent itself to a modest range of applications (Guastello, 1995). Once again there are regions of inaccessibility between the sheets representing repellor forces or antimodes. The function is five-dimensional, and what appears in Figure 11.7 is the most interesting three-dimensional sectioning. The surface features a pocket, which allows the control point to slip through to the middle mode, or oscillate between the top and bottom modes. The butterfly response surface is defined as the set of points where

$$\frac{df(y)}{dy} = y^5 - dy^3 - cy^2 - by - a. \tag{11.8}$$

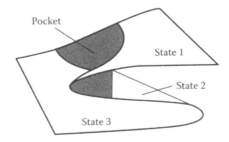

FIGURE 11.7
Response surface for the butterfly catastrophe model.

The new control parameter, *d*, is the butterfly or second bifurcation parameter. It governs the joint action of control parameters *b* and *c*, such that *b* and *c* can work together to permit behavior change across the three adjacent states, or alternatively, *b* and *c* can work interactively to permit behavior change between the extreme modalities only.

Umbilic Models

The wave crest (hyperbolic umbilic) catastrophe model describes a phenomenon that involves two order parameters (*x*, *y*) and three control parameters (*a*, *b*, and *c*). The two order parameters vary separately, each having the level of complexity of the fold model, which is quadratic with one stable and one unstable state. It got its name when mathematicians determined that the crest of every wave breaks at a 120° angle (Lu, 1976). It was later discovered, however, that wave crest geometry did not adequately describe wave forms. Although the wave crest has not seen much application in the behavioral sciences, its function served as a starting point for developing an important series of nonlinear models that is covered in Chapter 12.

The potential function for the wave crest model is

$$f(x, y) = x^3 + y^3 + ax + by + cxy. \tag{11.9}$$

The response surface is composed of the two partial first derivatives of the potential function:

$$\frac{df(x)}{dx} = 3x^2 + a + cy, \tag{11.10}$$

$$\frac{df(y)}{dy} = 3y^2 + b + cx. \tag{11.11}$$

The wave crest function contains two asymmetry parameters, one for each behavior. The two behaviors share a common bifurcation factor. Note that the bifurcation of *y* affects changes in *x* and vice versa.

The hair catastrophe model (elliptic umbilic) got its name from its bifurcation set, which tapers to a single thread that connects two swallowtail points. This model also contains two dependent measures and three control parameters. Both dependent measures function separately and each has one stable and one unstable state. The surface is composed of two partial first derivatives of the potential function:

$$\frac{df(y)}{dy} = 3y^2 - x^2 + a + 2cy, \tag{11.12}$$

$$\frac{df(x)}{dx} = 2xy + b + 2cx. \tag{11.13}$$

The hair model contains two asymmetry parameters, one for each behavior, and the two behaviors share a common bifurcation parameter. The quadratic function of x, however, affects changes in y. The model contains a modulus term, xy, which represents an interaction that takes place between the two dependent measures. Applications of this model in the social sciences are rare (Puu, 1991), although they are entrenched in optical physics (Berry, Nye, & Wright, 1979).

The mushroom catastrophe model (parabolic umbilic) is composed of two dependent measures and four control parameters. One dependent variable is characterized by two stable states because of its cubic component, and the other has one stable and one unstable state because of its quadratic component. There is an interaction between the two dependent measures and separate bifurcation and asymmetry terms for each dependent measure. The mushroom response surface is given by

$$\frac{df(y)}{dy} = 4y^3 + 2dy + x^2 + b, \tag{11.14}$$

$$\frac{df(x)}{dx} = 2xy + 2cx + a. \tag{11.15}$$

The mushroom catastrophe has found uses in the analysis of morbidity statistics in insurance data and in the analysis of creative problem-solving discussions (Guastello, 1995, 2002). In the latter application, the two dependent measures were separate components of the conversation flow. The component with the two stable states consisted of basic questions and responses among the discussion participants who were sometimes agreeing with each other and sometimes not. The component with the stable and unstable states consisted of especially creative contributions where the group suddenly headed in a new direction when they thought they had an idea that could solve the problem they were working on. The two bifurcation variables arose from situational variables that induced sudden increases in conversation levels. The two asymmetry variables were groups of personality variables that were connected to one dependent measure or another.

Nonelementary Models

According to the singularity rule, which derives from the classification theorem (Thom, 1975), there is only one possible response surface for a given bifurcation set. The singularity rule actually holds true for catastrophe models with five control parameters, as well as for those models with four or fewer controls. There are four catastrophe models with five control parameters. One model is a six-dimensional cuspoid. There are 3 seven-dimensional umbilics with five control parameters; they differ in the number of modulus terms they contain. For a glimpse at the complex geometry associated with some of those models, see Callahan (1980, 1982). They have not seen any uses in the behavioral sciences, however.

Catastrophe models of further complexity do not conform to the singularity rule. The simplest of those is the double cusp, which consists of two dependent measures and six control parameters. Each order parameter oscillates between two stable states of behavior. There are 4 eight-dimensional response surfaces associated with the double cusp. Mathematical work connected to the double cusp and other nonsingular catastrophe models can be found in Zeeman (1977) and Arnold (1974). The lack of a unique response surface would make data analysis intractable if an application could be defined. Researchers would be advised to break the geometry of the problem into smaller subspaces using elementary models, testing those, and then proceeding to larger models if needed.

Models and Classification

The taxonomy of models, together with the singularity rule, allows us to reduce the plethora of possible discontinuous change functions to a very small number of possibilities. Because topological models are rubberized rather than rigid, catastrophe models can withstand perturbations without changing their classification, so long as a tear in the surface is not introduced. This principle is known as diffeomorphism up to transversality.

By the same token, catastrophe models should be regarded as unbreakable packages. For instance, if there are two stable states and hysteresis between them, one can be confident that two control parameters are waiting to be identified. Thus, it is possible to infer global structure from some of the local dynamics. It also means that testing a model involves examining all its critical parts. Catastrophes are not generic nonlinearities that can be treated as polynomials in the sense of Equation 2.19. Its parts go together in a confluent way to account for the phenomenon at hand.

The complexity of catastrophe models is signified by the highest exponent for the behavioral variable in the surface equation, which would be the quadratic term for the fold, the cubic for the cusp, and so forth. The leading exponent denotes the number of control parameters in the model, each with its own unique function, and the complexity of the behavioral arrays they generate.

Causality

In the traditional concept of causality, if we deliberately do A today and obtain B tomorrow, and obtain the same pattern on a few replications just to be sure, then A causes B. Although some simplistic systems work that way, NDS theory shows that the causation of phenomena can be much more complex. In catastrophe models, a random force and a bifurcation act conjointly to produce a phenomenon. The concept of cause is now replaced by a combination of control parameters, bifurcation, entropy, and an autonomous process.

An autonomous process is the sequence of behaviors that emanate from the underlying structure of a system. A system with a particular bifurcation set will exhibit a distinct pattern of behavior given a sufficient random force. Entropy itself is not sufficient as other control parameters, notably asymmetry, affect the process. The nature of the attractors themselves is an important contributor to the end result. General NDS shows that there are several classic cases of attractors, each of which produces different patterns of behavior over time. Self-organized systems and coupled dynamics are essentially combinations of elementary processes.

Dynamic Difference Equations for Catastrophes

The method of dynamic difference equations can be accomplished through the multiple regression programs of standard statistical software such as the Statistical Package for the Social Sciences (SPSS). The technique evolved over a series of publications (Guastello, 1982a,b, 1987a, 1988, 1992a, 1995, 2002, 2005b). It can be used to evaluate hypotheses concerning all four cuspoids, and a variation using canonical correlation can be used with umbilic catastrophes. Dynamic difference equations have been used most often in cases where the order parameter has been measured for many people (or other entities) at two points in time. The technique can also be used for situations where one entity is measured many times in a time series analysis. For purposes of discussion, however, the most common variety is considered as the prototype.

The technique was intended for testing theory-driven hypotheses, rather than for an adventure in curve-fitting. Accurate models come from good theories and require several steps: (a) Identify the behavioral output variable, the number and type of stable system states. (b) Identify hysteresis, multiple threshold events, or other instabilities that might be present. (c) Select a model that accounts for the observed dynamics. When in doubt, try something simple. (d) Identify real-world variables that behave as asymmetry, bifurcation, swallowtail, butterfly, or other parameters. (e) Identify any surface gradients if possible. (f) Compare the proposed function with any known functions for the same phenomenon; this last step is part of the computational analysis that follows.

Cusp Model

The specification of a regression equation for a cusp begins with a definition of the cusp as a function over time:

$$dz = (z^3 - bz - a) \, dt, \tag{11.16}$$

where z is the behavioral outcome that has been corrected for location and scale as defined in Equation 2.11. The control variables, a and b, are also corrected for location and scale. dt is set equal to 1, meaning that all time intervals are of equal duration. Equation 11.16 becomes operational by replacing the left-hand side with change in behavior at two points in time, Δz, and inserting empirical weights among the terms:

$$\Delta z = \beta_0 + \beta_1 z_1^3 + \beta_2 z_1^2 + \beta_3 b z_1 + \beta_4 a, \qquad (11.17)$$

where z_1 is the behavioral outcome at the first of two points in time.

Equation 11.17 contains a quadratic term that is not included in the formal cusp model. If the data for one side of the cusp, e.g., shifting up compared to shifting down, is better developed in the data than the other side, the quadratic term will soak up variance associated with lopsided representation of the surface.[1]

A network of hypotheses is contained in Equation 11.17. First there is the overall statistical significance of the model, and size of the R^2, which describes how well the data approximate a true cusp model. The t-tests on the regression weights denote the contribution of each term in the model. In optimal situations, each term in the model accounts for a unique portion of criterion variance. The cubic power potential denotes and implies a bifurcation structure, and there should be a significant bifurcation effect present.

Under less than optimal conditions, bivariate correlations between the individual terms and the difference equation may be significant, but empirical weight may not all be significant in the multiple regression model. Here it would be useful to use a cross-validation strategy could confirm the conclusion. One might investigate collinearity among control variables and behavior measures, bivariate correlations between elements of the model and the difference scores.

Markov Assumption

The regression models for catastrophe surfaces assume that a simple one-stage Markov process is operating, meaning that the control point has no memory of where it came from beyond the prior instant. Violation of the assumption could result in flattening the surface (dampening catastrophic fluctuations) or expanding the dimensionality (retardation) of the surface by introducing additional convolutions.

In the earliest uses of the dynamic difference equations, the possibility of enhancing models by including additional polynomial terms at the end of the model after all the main parts were entered and tested was considered. The procedure was quickly dropped when it became apparent that the additional variance accounted for was relatively small at best. Most audiences at that time had difficulty assimilating the basic concept and procedure, so added complications were not helpful in communication. Others had difficulty distinguishing the notion of model with theoretical properties from an atheoretical polynomial modeling exercise; the use of additional polynomial terms would only serve to obfuscate the real distinction. In the long run, if one really suspects that a complex lag is operating, it would be better to design the study as a time series—many observations on one or few entities—and assess the presence and impact of lag by more obvious means.

Prototype Experimental Design

In a classical experiment, there are objects or people that we measure or manipulate in a theoretically interesting way. Those results are compared with those obtained from control groups where no such manipulation has taken place, or perhaps a reverse manipulation, depending on the nature of the theory. A similar principle applies to the testing of NDS hypotheses, only the contrast is made between the results of equations: the structural equation for the NDS model, and a model that is based on a comparison or default theory. Most of the time, the comparison models are two control equations that are structurally linear, in which case they would include the same qualitative variables a and b. The linear-difference model for the cusp is

$$\Delta y = \beta_0 + \beta_1 a + \beta_2 b. \tag{11.18}$$

If more than one experimental variable is hypothesized for either of the controls, additional terms containing those variables would be added to Equation 11.18. The linear pre–post model is essentially the same model as the linear-difference model, except that y_1 has been moved to the right side of the equation:

$$y_2 = \beta_0 + \beta_1 y_1 + \beta_2 a + \beta_3 b. \tag{11.19}$$

The dependent variable, y, needs no correction for location or scale because ordinary linear models are robust with respect to those transformations.

One of the first uses of the dynamic difference equation for the cusp compared and contrasted the cusp structure with a theoretical model that used a moderator term, ab. Thus, ab was added to the end of Equations 11.18 and 11.19. The term shows up in attitude studies (e.g., Clair, 1998) where the comparison theory of planned action regularly includes moderator relationships.[2] The reasoning behind the comparison models carries over to different types of nonlinear models in the next chapter. One might indeed find that a fair comparison is with a different nonlinear model rather than a linear model.

The R^2 coefficient for the nonlinear model should exceed the R^2 for the two linear alternatives. R^2 for Equation 11.19 is often larger than R^2 for Equation 11.18, but not always. There is no significance test for the difference in values of R^2 under the thinking that the nonlinear model need only be equal to the alternative to provide a reasonable competing explanation; all other things being equal, the nonlinear-theoretical model offers more of an explanation of a change process than saying that a weight combination of two or three variables adds up to proportional change. Furthermore, in the long run, effect size is more important than significance in the reasoning of meta-analysis. For two different batches of studies it was reported that, in cases where the researcher adopted the nonlinear hypothesis over the alternative, the ratio of coefficients was 2:1 in favor of the nonlinear hypothesis (Guastello, 1995, 2002).

Two types of hypothesis testing are involved. The comparison of R^2 coefficients models is a Bayesian type of hypothesis whereby predictions made from the model with the higher have higher odds of being correct. Neyman–Pearson hypothesis testing is used for determining whether regression weights are significantly different from 0.00. In some types of nonlinear models encountered in later chapters, Neyman–Pearson significance occasionally yields to computational significance. A model component is computationally significant if it improves the overall accuracy of the model, and perhaps alters the statistical significance of other components without being statistically significant itself.

SPSS Syntax

The data set for catastrophe models needs to be organized so that each observation contains the dependent measure (the variable that is hypothesized to show catastrophic behavior) at two points in time, and the values of the control value at Time 1. The user then needs to specify some COMPUTE statements before specifying the actual regression subprogram. COMPUTE statements are needed to transform dependent measures and other control variables with respect to location and scale.

$$\text{COMPUTE } z2 = (y2 - L)/ S. \tag{S11.1}$$

In statement S11.1, $y2$ is the dependent measure at Time 2, L is the value of location, and S is the value of scale. Actual numbers would replace L and S. The same syntax would be used for control variables and for changing $y_1 \rightarrow z_1$. Compute statements are also needed to define power terms such as z^3 (statement S11.2), a similar quadratic term, bifurcation interactive terms (S11.3), and the difference score (S11.4) that is used as the dependent measure in the catastrophe models.

```
COMPUTE zpow3 = z1**3.                          (S11.2)
COMPUTE bz = b*z1.                              (S11.3)
COMPUTE deltaz = z2 - z1.                       (S11.4)
```

The variable *b* listed in statement S11.3 is a variable that is hypothesized to function as a bifurcation term in a cusp. A particular application can have several variables called *b*. The variable called *a* in statement S11.5 is a variable that is hypothesized to function as an asymmetry variable.

Proceed to the main program statements after completing the COMPUTE statements. Use the regression subprogram.

```
Regression descriptives/ missing=pairwise       (S11.5)
/variables = dz zpow3 zpow2 bz a
/dependent = dz/ enter zpow3 zpow2 bz a.
```

Next, inspect the significance tests for each of the terms in the model. If the regression weight for zpow3 is not significant, drop zpow2 and try it again. In the event that there are several variables being tested as *a* and *b* variables, drop any variable for which the *p*-value on the regression weight is greater than 0.10. It is sometimes worthwhile to try the reverse hypothesis: that the variables first thought to behave as *b* are really behaving as *a*, and vice versa. To do so, define some more compute statements for the new bifurcation terms (S11.6) and run the regression as a two-step process (S11.7). In the event a variable contributes to both control parameters, one can reasonably conclude that the variable is behaving more like a gradient than like one of the control parameters in the pure sense.

```
COMPUTE az = a*z1.                              (S11.6)
REGRESSION descriptives/ missing=pairwise       (S11.7)
/variables = dz zpow3 zpow2 bz a b az
/dependent = dz /enter zpow3 zpow2 a /enter az b.
```

Statement S11.7 assumes that the variable that was first thought to behave as *b* was not significant, and that both zpow3 and zpow2 were acceptable.

Finally, the two linear alternative models would be defined as S11.8 and S11.9.

```
REGRESSION descriptives/ missing=pairwise       (S11.8)
/variables = dz b a
/dependent = dz /enter a b.
REGRESSION descriptives/ missing=pairwise       (S11.9)
/variables = z2 z1 b a
/dependent = z2 /enter z1 b a.
```

Order of Variable Entry

The third line of S11.5 enters all the variables simultaneously; the computer program then orders them in the most advantageous fashion. One could also make a case for entering the terms in order of importance, with the cubic entering first. It sometimes happens, however, that the order of variable entry is an important factor in whether all the parts of the model attain statistical significance. Darlington (1990) identified this phenomenon as complementarity. A complementarity effect is similar to a suppressor effect in that the role of a variable could change from nonsignificant to significant depending on what other variables in the model. The distinguishing feature is here is that the change in effect size, rather than the significance, is the primary concern.

If one were to reverse the control variables as described in S11.7, one is essentially using backward elimination: keeping all the variables that are good enough, throwing away those that are not, and then adding others that have a lower theoretical priority.

If the problem has many variables involved, it might be useful to reduce the number of variables before proceeding with the catastrophe model. Here one might conduct a stepwise linear regression between the research variables and the dependent measure, and use the surviving variables in the catastrophe model. The risk of this procedure is that a variable could have no linear relationship with the dependent variable, but a strong nonlinear relationship, which could be the bifurcation variable in the case of the cusp.

An alternative procedure would be to factor analyze the research variables in hope of reducing their number to only a very small number. Here the researcher inherits all the heuristics and ambiguities that usually go with a factor analysis; those issues fall outside the bounds of the present work, and are reviewed well in many standard texts on factor analysis.

Bus Accidents

The following example illustrates an analysis with the cusp that involves several variables that could contribute to the control parameters. It also makes use of two different specifications for the scale parameter.

The participants in the study (Guastello, 1991, 1992b) were 290 bus drivers in a major Midwestern city who completed a questionnaire on occupational hazards, stress exposures and consequences, safety management for their work environment, the number of accidents they had in their bus during the prior 3 years meeting a specified criterion of severity, and the number of accidents they had in their private automobiles meeting the same criterion. There were 256 complete surveys for this analysis.

In the theoretical model, there were two stable accident levels, a low level that was usually associated with personal driving and a high level that was usually associated with the job. Thus, the two points in time really consisted of personal versus occupational exposure. Measurements of subjective danger ratings and transit hazards were hypothesized as asymmetry variables.

Social and job stress, physical stress such as nonstandard work shifts, and anxiety were hypothesized as bifurcation variables.

The modeling results appear in Table 11.2 showing the *t*-tests on the regression weights, *F*-tests for the overall model, and R^2 and shrunken R^2 using Wherry's formula. It is a judgment call as to whether the shrinkage formula should be applied, but considering that the original study contained several more cusp models for different health consequences and models of different sizes, and different variables showed up in each one, the shrinkage formula helped with interpretation. There was not much shrinkage for the accident criterion.

The linear-difference model consisted of one variable, transit hazards. The nonsignificant variables were reintroduced into the model to allow a comparison on the basis of using all the variables rather than just the significant

TABLE 11.2

Cusp and Linear Models for Transit Accidents

Variable	t	F	R^2	Shrunken R^2
Linear difference model				
Transit hazards	6.29****	39.57****	0.14	.14
All variables		6.25****	0.20	0.17
Pre–post linear model				
Transit hazards	5.45****			
Car accidents	2.24**			
Anxiety	2.10**	20.82****	0.22	0.21
All variables		7.89****	0.26	0.23
Cusp Model 1				
$z_1{}^3$	5.27****			
$z_1{}^2$	−5.81****			
$z_1{}^*$ Anxiety	1.82			
$z_1{}^*$ Social/job stress	−2.34*			
$z_1{}^*$ Physical stress	0.29			
Age/experience	1.29			
Danger	0.93			
Environmental hazards	1.16			
Transit hazards	6.21****	24.75****	0.50	0.48
Cusp Model 2				
Transit hazards	7.32****			
$z_1{}^2$	−8.64****			
$z_1{}^*$ Social/job stress	−3.19***			
$z_1{}^3$	7.56***	237.05***	0.63	0.63

* p < .05.
** p < .10.
*** p < .01.
*** p < .001.

ones. The pre–post difference model contained three variables: transit hazards, car accidents (y_1), and anxiety.

Cusp Model 1 is shown with all the variables remaining at the last step. Significant effects were obtained for the cubic term and one bifurcation term (social and job stress); thus, there was sufficient evidence to conclude that a cusp was present. The asymmetry variable, transit hazards, completed the picture. Both the uncorrected and shrunken R^2 values for the cusp compared favorably with the linear counterparts. The cusp accounted for approximately twice as much variance as the next-best linear alternative.

Alternative Scale Formula

Cusp Model 1 in Table 11.2 was calculated using the ordinary standard deviation as the scale parameter, which is usually more than adequate. Occasionally, however, an alternative scale model that is based on statistical modes within the cusp model, rather than the overall mean, is a better choice. It might be a better choice in cases where the data mostly fall on the highly unfolded region of the response surface (with only minor representation of the low-bifurcation side of the surface) and the data points are clumped closely around their respective modes.

The alternative approach requires partitioning the frequency distribution of y into modes and antimodes. Then

$$\sigma_s = \sqrt{\frac{\sum_{y=1}\sum_{m=1}\left(y_{m-}^{m+} - m_i\right)^2}{N - M}} \tag{11.20}$$

(Guastello, 1992b, 1995, 2005b). The N is the total number of observations. M represents the number of statistical modes around which points are actually observed to vary. For instance, $M = 2$ for cusp applications where there are no points in the antimode. If there were points in the antimode, dispersion around the antimode should be included also; thus, $M = 3$. m_i is a modal value of y. Squared differences between y and m_i are summed over the range of y; the range extends from a lower boundary of $m-$ to an upper boundary of $m+$ within the neighborhood of m_i. The sums are summed over the two or three modal regions.

Cusp Model 2 in Table 11.2 is a reanalysis of the same data using the alternative measure of scale. This time the nonsignificant variables were removed. The same control variables were identified, but the R^2 increased. There was virtually no shrinkage in R^2.

Cusps for Program Evaluations

One of the early applications (Guastello 1982b) considered the use of the cusp catastrophe for training, therapy, and policy evaluation, which is a broad class of potential applications. The dependent measure would be a criterion that is measured "before" and "after" an intervention. The bifurcation variable could be the difference between a treatment group and a control group (dichotomous variable) or the amount of time exposed to the program or policy. The bifurcation variable could be one that reflects the treatment would work better for some people or in some situations rather than others, or in the case of a therapy, a proclivity toward recidivism. The asymmetry variable would be a disposition to perform better in training, such as an ability. Similarly, the asymmetry variable could be a proclivity toward succeeding in a therapy, such as an absence of comorbidity. In addition to the benefits of a more detailed theoretical explanation and a high percentage of variance accounted for, the cusp would indicate whether stable changes in behavior had been attained.

As an illustrative example, Byrne, Mazanov, and Gregson (2001) used the cusp to evaluate three programs designed to curtail adolescent smoking. The programs were focused on health, fitness, and self-confidence. A predisposition variable was tested as the asymmetry parameter, and social pressure was tested as the bifurcation parameter. The reports of cigarette consumption were taken at three points in time—before the program, immediately after, and 1 year after the program.

A summary of results appears in Table 11.3. Because the accuracy of the pre–post linear model exceeded that of the linear-difference model in all cases, only the pre–post results are shown. A full cusp with significant bifurcation and asymmetry terms was obtained at T2 for the health program, and its accuracy exceeds that of the linear model; the same results persisted at T3.

A full cusp was obtained for the fitness program at T2, although its accuracy was equivalent to that of the linear model. The gap between the cusp and linear models increased at T3, but the asymmetry variable was no longer making a significant contribution to prediction.

The cusp model for the self-confidence program did not fare well against the linear model at T2. Its results improved markedly at T3, although the overall predictive accuracy was weaker than what was observed for the other two programs.

TABLE 11.3

Shrunken R^2 for Cusp and Linear Models Obtained for Three Intervention Programs

Program	Model	R^2 at T2	R^2 at T3
Heath	Full cusp	0.45	0.45
	Pre–post	0.27	0.24
Fitness	Full cusp	0.31	0.32[a]
	Pre–post	0.31	0.17
Self-confidence	Cusp, no[a]	0.14	0.23
	Pre–post	0.21	0.09

[a] The asymmetry variable was lost at T3.

A reasonable interpretation of these results would be as follows: (a) The prediction of behavior afforded by the cusp for all three programs improved over time, whereas the accuracy of the linear models waned. This outcome is related in part to the cusp's capacity to predict recidivism as well as sudden change in the desired direction. (b) A stronger cusp models implies a more fully developed manifold (unfolded region of the surface), thus, affording more local stability at both attractor points. Programs with greater overall impact on their target audience should produce greater levels of local stability. (c) In this regard the best program was the health program, followed by the fitness program. (d) Social pressure, the bifurcation variable, has the impact of splitting the target audience into those who complied and those who did just the opposite. The contrarian response is known as a reactance effect in psychology.

Fold, Swallowtail, and Butterfly Models

Hypotheses concerning the fold, swallowtail, and butterfly models can be tested in the same manner as the cusp. Their characteristic regression equations are, respectively,

$$\Delta z = \beta_0 + \beta_1 z_1^2 + \beta_2 a, \tag{11.21}$$

$$\Delta z = \beta_0 + \beta_1 z_1^4 + \beta_2 z_1^3 + \beta_3 c z_1^2 + \beta_4 b z_1 + \beta_5 a, \tag{11.22}$$

$$\Delta z = \beta_0 + \beta_1 z_1^5 + \beta_2 z_1^4 + \beta_3 d z_1^3 + \beta_4 c z_1^2 + \beta_5 b z_1 + \beta_2 a. \tag{11.23}$$

R^2 coefficients for Equations 11.21 through 11.23 would be compared with those obtained from their respective linear-difference and linear pre–post equations.

The mainstay of the reasoning behind cusp analysis and interpretation carries over to the other catastrophe models, although there are some nuances that are associated with the complexity of the models themselves. The fold is simpler than the cusp, containing only a quadratic term and an asymmetry term. The correlation between Δz and is 0.49–0.55 for a sequence of random numbers pulled from a random number table; however (Guastello, 1992a), thus, the asymmetry variable is critical to establishing a fold model. As mentioned in Chapter 3, however, random numbers are not as random as people would like to think. They are produced by an algorithm that contains a self-avoidance feature. Thus, it would follow that such an algorithm *is* a deterministic dynamical process, and unless there is reason to believe that real data collected in a study came from such an algorithm, a fold and a random number sequence have little chance of being confused. In other words, the numbers do not know where they came from, but the researcher does know.

The asymmetry variable is less critical for a cusp because there are distinctive nonlinear parts that are associated with the bifurcation manifold. In this regard the bifurcation variable has a higher priority than the asymmetry variable.

Umbilic Catastrophe Models

The umbilic catastrophe models contain two dependent variables, and thus present a new analytic challenge. The ideas, nonetheless, parallel those already developed. In the case of the wave crest, the two differential functions result in a pair of polynomial difference equations:

$$\Delta u = \beta_0 + \beta_1 u_1^2 + \beta_2 a + \beta_3 cv, \tag{11.24}$$

$$\Delta v = \beta_0 + \beta_1 v_1^2 + \beta_2 b + \beta_3 cu, \tag{11.25}$$

where u and v are dependent measures x and y, respectively, after they have been corrected for location and scale. The control parameters are a, b, and c. The modulus term (constant 3) that appears in the deterministic formula is now absorbed into the empirical weight for the variable associated with it; the same manipulation has been made in the statistical models for the hair and mushroom catastrophes as well. The theoretical presence of the modulus terms suggests that some empirical weights are meant to be three or four times larger for some terms of the statistical umbilic catastrophe models. Not enough is known about the umbilic functions in social science to guarantee that such ratios in weight size would actually be observed. The known empirical cuspoid models that are discussed throughout the book frequently show sharply different weight sizes, and the largest weights are often associated with power polynomials.

There are two choices of procedure for testing an umbilic hypothesis such as a wave crest model. One is to treat the two functions as two separate models, each with its own pair of linear-difference control models

$$\Delta x = \beta_0 + \beta_1 a + \beta_2 c + \beta_3 y, \tag{11.26}$$

$$\Delta y = \beta_0 + \beta_1 b + \beta_3 c + \beta_4 x, \tag{11.27}$$

respectively, and its own pair of linear pre–post control models,

$$x_2 = \beta_0 + \beta_1 x_1 + \beta_2 a + \beta_3 c + \beta_4 y, \tag{11.28}$$

$$y_2 = \beta_0 + \beta_1 y_1 + \beta_2 b + \beta_3 c + \beta_4 x. \tag{11.29}$$

R^2 coefficients for the various equations can be compared in the usual fashion.

The second option is to add Equations 11.24 and 11.25 together to form a canonical regression model where the two partial derivatives can acquire empirical weights and their relative contributions can be assessed:

$$\gamma_1 \Delta u + \gamma_2 \Delta v = \beta_0 + \beta_1 u_1^2 + \beta_2 a + \beta_3 cv + \beta_4 v_1^2 + \beta_5 b + \beta_6 cu, \tag{11.30}$$

where γ_i are empirical canonical weights (Guastello, 1982a, 1987b).

Similarly, the hair and mushroom catastrophe models can be expressed as a pairs of regular polynomial regression functions, or as canonical models:

$$\gamma_1 \Delta u + \gamma_2 \Delta v = \beta_0 + \beta_1 u_1^2 + \beta_2 v_1^2 + \beta_3 u_1 v_1 + \beta_4 cv + \beta_5 cu + \beta_6 a + \beta_7 b, \tag{11.31}$$

$$\gamma_1 \Delta u + \gamma_2 \Delta v = \beta_0 + \beta_1 v_1^2 + \beta_2 u_1^2 + \beta_3 u_1 v_1 + \beta_4 cu + \beta_5 cu + \beta_6 dv + \beta_6 a + \beta_7 b. \tag{11.32}$$

For any of the umbilic structural models, canonical correlation analysis will produce two weighted functions of Δu and Δv that are optimally correlated with weighted functions of u, v, a, b, c, and d in the case of the mushroom. Hence, two canonical r^2 coefficients are produced. In the known uses of the mushroom (Guastello, 1987b, 1995), the structure of the model was determined by comparing the first (and larger) canonical r^2 for the mushroom against the R^2 for a cusp as the next-best alternative theoretical model. Significant weights were also obtained for all components of the mushroom models.

The well-recognized problem with any canonical correlation analysis is that the weighted functions of predictors and criteria (left- and right-hand sides of the equations) can be highly associated with each other, but at the same time they could have low direct relationships to the actual underlying measurements. A redundancy analysis would quantify the degree of relationship between the canonical variates and the original variables.

Critiques

The method of dynamic difference equations encountered a critique from Alexander, Herbert, DeShon, and Hanges (1992), and a reply (Guastello, 1992a). The crux of Alexander et al.'s argument resurfaced in Witkiewitz, van der Maas, Hufford, and Marlatt (2007): "[T]he Guastello approach capitalized on chance variation in the data and tended to overfit the catastrophe regardless of the true distribution of the data" (p. 383). Alexander et al. took a cusp function that came from an actual study (they did not state what method was used to ascertain the function) and generated many thousands of cases of Δy, a, b from the function. They created difference scores artificially by pairing each observation with another one that had a very small difference in a;

all differences in *a* were of equal size. Their attempt to compute a cusp model fared poorly. They obtained a higher R^2 when they added some Gaussian noise to *y*, but did not obtain a bifurcation factor. The fundamental flaw in Alexander et al.'s analysis resided in the definition of difference scores. Instead of creating $df(y)/dy$ or $df(y)/dt$, they created dy/da. If one takes the cusp potential function and differentiates with respect to *a*, one obtains a linear function of *y* and not the cusp response surface!

The reply to Alexander et al. (1992) showed that differences in random numbers produced a quadratic function, not a cusp. Furthermore, if one takes the same pairs of random numbers, and invents a bifurcation variable that corresponds perfectly to the large and small differences in pairs of random numbers, as *b* should do, and invents a similarly perfect measure of *a*, one obtains a very nice cusp (Guastello, 1992a). Thus, it is fair to conclude that (a) the polynomial method works as intended, (b) the critique was seriously flawed, and (c) good hypotheses for the control variables are critical for drawing any proper conclusions about the presence of a catastrophe function.

Witkiewitz et al. (2007) also wrote, "Hufford [et al., 2003] used the Guastello (1982[b]) polynomial regression approach because the alternative approaches required too large of a sample size, but the authors questioned the appropriateness of the polynomial regression equations as a test of the catastrophe model" (p. 383). The issue Hufford et al. raised was, "Given that the linear models are not nested within the cusp catastrophe model, we are unable to directly compare the three models" (p. 225) referring to the cusp, pre–post, and linear-difference models. They suggested that cusps be tested as hierarchical models with the linear model entered first and the cusp components entered last. As mentioned here previously, *the comparison between the R^2 coefficients is a comparison between two theoretical explanations for data.* The cusp is not a generic polynomial; the bifurcation factor is an important part of the explanation also. Nonlinear models often have linear components that are accounted for in part by the polynomial terms; thus, the hierarchical suggestion gives undue credit to the linear model for any overlapping variance and any benefits of complementarity effects. The asymmetry variable is also common to all three models.

Hufford et al. (2003) tested a cusp on two small data sets ($N = 41$ and 43) that were evaluations of an alcohol treatment program. The cusp hypothesis seemed promising because of its ability to capture recidivism. For one sample they obtained an $R^2 = 0.58$ for the cusp, significant cubic, quadratic, and asymmetry components, but no bifurcation component; R^2 for the best linear alternative was 0.19. For the second sample they obtained an $R^2 = 0.83$ for the cusp, with significant cubic and quadratic components, but neither control parameter attained significance; R^2 for the best linear alternative was 0.14. The reanalysis of the data in Witkiewitz et al. (2007) using a different computational approach to catastrophe analysis (which is considered in Chapter 13) resulted in the same basic conclusion. Thus, one should not blame the analysis for a weak theory. This point carries over to other types of NDS systems, analyses, and hypotheses. The attrition rate of over 70% in their samples could have been responsible for the ineffectiveness of their control parameters.

Cross-Validation

Any lingering doubts about whether the cusp or any other regression model capitalizes on chance can be resolved through any of several means for ascertaining population statistics from sample statistics; see Darlington (1990) for a critical comparison. Resampling techniques have advantages over split-sample techniques, particularly if the research sample is large enough for a full-sample analysis, but not so large as to permit the assumption that splitting the sample will cover the entire topological range both times. Resampling, or bootstrapping, got its start from the problem of finding a population regression coefficient from a sample when the sample was too small to sustain the usual tests of significance, but the r was large enough to be useful had it come from a larger sample.

There are two published reports with resampling techniques that illustrate the point. The earlier one (Guastello, 1995, pp. 245–251) recounted a cusp analysis for the onset of a set of medical disorders that were thought to originate from occupational stress exposures. A cusp model was obtained with all its parts, and its R^2 (0.70, $N = 238$) was better than the next-best linear model ($R^2 = 0.28$). The data set was then copied enough times to produce a total quantity of 15,590 cases. Ten samples of 200 cases were randomly drawn from the expanded database. The regression equation that was produced from the calibration sample was used to predict values of Δz from z_1, a's and b's in the ten samples. The mean r^2 was 0.62, and the median value was 0.64. Using the median, the amount of reduction in R^2 was the difference between $r = 0.84$ and 0.79. One can thus conclude that the model was robust with regard to chance effects.

The second example (Guastello, 2002, pp. 132–148) was an application of the cusp to the prediction of performance and turnover in U.S. Air Force recruits. A cusp model was obtained with all its parts, and its R^2 (0.30, $N = 5038$) was better than the next-best linear model ($R^2 = 0.21$) for predicting performance and dropping out at 30 months out of a 48-month contract. The calibration sample was randomly drawn from a data based containing 59,397 enlisted personnel. Ten more random samples of 1000 cases were drawn from the same database. The regression equation that was produced from the calibration sample was used to predict values of Δz from z_1, a's and b's in the 10 samples. The mean r^2 for the cusp was 0.26, and the median value was 0.27. The mean and median mean r^2 for the next-best linear model was 0.18. By all the usual heuristics, both the cusp and linear models were resilient under resampling, and the cusp was a better explanation of the phenomenon. Capitalization on chance can be ruled out as a contrary explanation.

Notes

1. The cusp geometry is composed of two oriented folds; thus, the use of the quadratic term to account for more of one than the other makes sense from a topological point of view. It is not simply a holdover from common polynomial regression.
2. The use of moderators in the comparison models for hypotheses involving the swallowtail or butterfly catastrophes becomes unwieldy because of all the possible interaction terms that could be involved. Thus, moderators should only be used if the default theory actually specifies moderator relationships.
3. Of related interest, Barrett, Caldwell, and Alexander (1985) advanced the position that dynamic performance in occupational settings did not exist. It was met with firm replies to the contrary (Austin, Humphreys, & Hulin, 1989; Hofmann, Jacobs, & Baretta, 1993).

References

Agu, M. (1983). A method for identification of linear or nonlinear systems with the use of externally applied random force. *Journal of Applied Physics, 54,* 1193–1197.

Alexander, R. A., Herbert, G. R., DeShon, R. P., & Hanges, P. J. (1992). An examination of least-squares regression modeling of catastrophe theory. *Psychological Bulletin, 111,* 366–374.

Arnold, V. I. (1974). Normal forms of functions in the neighborhoods of degenerate critical points. *Russian Mathematical Surveys, 29,* 10–50.

Austin, J. T., Humphreys, L. G., & Hulin, C. L. (1989). Another view of dynamic criteria: A critical analysis of Barrett, Caldwell, & Alexander. *Personnel Psychology, 42,* 583–596.

Barrett, G. V., Caldwell, M. S., & Alexander, R. A. (1985). The concept of dynamic criteria: A critical analysis. *Personnel Psychology, 38,* 41–56.

Berry, M. V., Nye, J. F., & Wright, F. J. (1979). The elliptic umbilic diffraction catastrophe. *Philosophical Transactions of the Royal Society A, 291,* 453–484.

Byrne, D. G., Mazanov, J., & Gregson, R. A. M. (2001). A cusp catastrophe analysis of changes to adolescent smoking behavior in response to smoking prevention programs. *Nonlinear Dynamics, Psychology, and Life Sciences, 5,* 115–138.

Callahan, J. (1980). Bifurcation geometry of E6. *Mathematical Modeling, 1,* 283–309.

Callahan, J. (1982). A geometric model of anorexia and its treatment. *Behavioral Science, 27,* 140–154.

Cerf, R. (2006). Catastrophe theory enables moves to be detected towards and away from self-organization: The example of epileptic seizure onset. *Biological Cybernetics, 94,* 459–468.

Clair, S. (1998). A cusp catastrophe model for adolescent alcohol use: An empirical test. *Nonlinear Dynamics, Psychology, and Life Sciences, 2,* 217–241.

Crutchfield, J. P. (2003). When evolution is revolution: Origins of innovation. In J. P. Crutchfield & P. Schuster (Eds.), *Evolutionary dynamics* (pp. 101–133). New York: Oxford University Press.

Darlington, R. B. (1990). *Regression and linear models.* New York: Wiley.

Gilmore, R. (1981). *Catastrophe theory for scientists and engineers.* New York: Wiley.

Guastello, S. J. (1982a). Moderator regression and the cusp catastrophe: Application of two-stage personnel selection, training, therapy and program evaluation. *Behavioral Science, 27,* 259–272.

Guastello, S. J. (1982b). Color matching and shift work: An industrial application of the cusp-difference equation. *Behavioral Science, 27,* 131–137.

Guastello, S. J. (1987a). A butterfly catastrophe model of motivation in organizations: Academic Performance. *Journal of Applied Psychology, 72,* 165–182.

Guastello, S. J. (1987b). Catastrophe modeling of the accident process: Risk dispersion for 10 industrial classes. *Social and Behavioral Sciences Documents, 17,* MS 2817.

Guastello, S. J. (1988). Catastrophe modeling of the accident process: Organizational subunit size. *Psychological Bulletin, 103,* 246–255.

Guastello, S. J. (1991). Psychosocial variables related to transit accidents: A catastrophe model. *Work & Stress, 5,* 17–28.

Guastello, S. J. (1992a). Clash of the paradigms: A critique of an examination of the polynomial regression technique for evaluating catastrophe theory hypotheses. *Psychological Bulletin, 111,* 375–379.

Guastello, S. J. (1992b). Accidents and stress-related health disorders: Forecasting with catastrophe theory. In J. C. Quick, J. J. Hurrell, & L. M. Murphy (Eds.), *Work and well-being: Assessments and interventions for occupational mental health* (pp. 262–269). Washington, DC: American Psychological Association.

Guastello, S. J. (1995). *Chaos, catastrophe, and human affairs: Applications of nonlinear dynamics to work, organizations, and social evolution.* Mahwah, NJ: Lawrence Erlbaum.

Guastello, S. J. (2002). *Managing emergent phenomena: Nonlinear dynamics in work organizations.* Mahwah, NJ: Lawrence Erlbaum Associates.

Guastello, S. J. (2005a). Statistical distributions and self-organizing phenomena: What conclusions should be drawn? *Nonlinear Dynamics, Psychology, and Life Sciences, 9,* 463–478.

Guastello, S. J. (2005b). Nonlinear methods for the social sciences. In S. Wheelan (Ed.), *The handbook of group research and practice* (pp. 251–268). Thousand Oaks, CA: Sage.

Guastello, S. J. (2007). Nonlinear dynamics and leadership emergence. *Leadership Quarterly, 18,* 357–369.

Haken, H. (1988). *Information and self-organization: A macroscopic approach to self-organization.* New York: Springer-Verlag.

Haken, H. (1999). Synergetics and some applications in psychology. In W. Tschacher & J.-P. Dauwalder (Eds.), *Dynamics, synergetics, and autonomous agents* (pp. 3–12). Singapore: World Scientific.

Hofmann, D. A., Jacobs, R., & Baratta, J. E. (1993). Dynamic criteria and the measurement of change. *Journal of Applied Psychology, 78,* 195–205.

Hufford, M. R., Witkiewitz, K., Shields, A. L., Kodya, S., & Caruso, J. C. (2003). Relapse as a nonlinear dynamic system: Application to patients with alcohol use disorders. *Journal of Abnormal Psychology, 112,* 219–227.

Kelso, J. A. S. (1995). *Dynamic patterns: The self-organization of brain and behavior.* Cambridge, MA: MIT Press.

Lu, Y.-C. (1976). *Singularity theory and an introduction to catastrophe theory.* New York: Springer-Verlag.

Poston, T., & Stewart, I. (1978). *Catastrophe theory and its applications.* London: Pitman.

Puu, T. (1991). Structural change in flow-based spatial economic models: A survey. *Socio-Spatial Dynamics, 2,* 1–18.

Thom, R. (1975). *Structural stability and morphogenesis.* New York: Benjamin-Addison-Wesley.

Thompson, J. M. T. (1982). *Instabilities and catastrophes in science and engineering.* New York: Wiley.

Tong, H. (1990). *Non-linear time series: A dynamical system approach.* New York: Oxford University Press.

Vallacher, R. R., & Nowak, A. (2009). The dynamics of human experience: Fundamentals of dynamical social psychology. In S. J. Guastello, M. Koopmans, & D. Pincus (Eds.), *Chaos and complexity in psychology* (pp. 370–401). New York: Cambridge University Press.

Warren, K., Sprott, J. C., & Hawkins, R. C. (2002). The spirit is willing: Nonlinear bifurcations and mental control. *Nonlinear Dynamics, Psychology, and Life Sciences, 6,* 55–70.

Witkiewitz, K., van de Maas, H. L. J., Hufford, M. R., & Marlatt, G. A. (2007). Nonnormality and divergence in posttreatment alcohol use: Reexamining the Project MATCH data "another way." *Journal of Abnormal Psychology, 116,* 378–394.

Woodcock, A. E. R., & Davis, M. (1978). *Catastrophe theory.* New York: Avon.

Zeeman, E. C. (1977). *Catastrophe theory: Selected papers 1972–1977.* Reading, MA: Addison-Wesley.

12

Nonlinear Regression and Structural Equations

Stephen J. Guastello

CONTENTS

Whereas multiple linear regression is a generalized technique for fitting lines, nonlinear regression (NLR[1]) is an analogous technique for fitting curves such as

$$y = e^x, \tag{12.1}$$

$$y = ax^u, \tag{12.2}$$

or

$$pr(y) = \frac{1}{[1 + \exp(-a - x)]} \qquad (12.3)$$

as possible examples. In these examples, y is a dependent or outcome variable, x is an independent variable, a is a constant, and u is an exponent that is thought to be fixed for the relationship between x and y, but otherwise unknown. Equation 12.3 describes a trend toward an asymptote where $0 < y < 1$, where pr(y) could be the proportion of the distance between the start and the asymptote or a cumulative probability, depending on how one generates the numbers for a given problem.

NLR was introduced nearly 50 years ago (Albert & Gardner, 1967; Hartley, 1961; Marquardt, 1963). It can assess virtually any differentiable function; the function can involve static or time series models, NDS models, or otherwise. Many of its uses involve exponential structures such as Equations 12.2 and 12.3 (Guastello, 1995, 2002; Huet, Bouvier, Gruet, & Jolivet, 1996; Ratkowski, 1983; Seber & Wild, 2003; Stortelder; 1998; Tong, 1990). NDS has had considerable uptake in the natural sciences and biology, but only limited use in the social sciences thus far. Uses in the social sciences are growing, however, with the introduction of NDS theory for defining models that researchers might want to test. NLR is available on the more comprehensive statistical packages such as Statistical Package for the Social Sciences (SPSS; Chicago IL). Its basic principles are described below, followed by examples that involve NDS hypotheses and a series of exponential models that have broad relevance to NDS.

The NLR Computational Procedure

Model Specification

For the first step in the process, the researcher specifies a model to be tested by placing regression weights, θ_i, in strategic locations. Thus, Equations 12.1 through 12.3 would be tested as

$$y = e^{\theta x}, \qquad (12.4)$$

$$y = ax^{\theta}, \qquad (12.5)$$

or

$$pr(y) = \frac{1}{[1 + \exp(\theta_1 + \theta_2 x)]}. \qquad (12.6)$$

In Equation 12.4, there is an indirect proportionality between y and the predictor variable x. In Equation 12.5, the power of x was unknown and needed to be estimated, so the regression weight replaced u. In Equation 12.6, the constant 1 is taken literally as 1.00, the constant a that needed to be estimated is specified as a regression weight, and a weight for proportionality is placed in front of x.[2] In nonconstrained NLR, which is the default, the final weights of θ_1 and θ_2 will carry a positive or negative sign, so the negative signs shown in Equation 12.3 are treated as subsets of possible additive numbers.[3]

Besides the range of possible equations that can be tested, we encounter another difference between multiple linear regression and NLR. In the general linear model, a final constant corresponding to the y-intercept is always assumed and is automatically calculated. In NLR, constants could occur anywhere, so the user must specify the placement of any constants. Thus, if one prefers to include a constant that looks like the ordinary regression constant in Equation 12.4, the model would need to be specified as

$$y = \exp(\theta_1 x) + \theta_2. \tag{12.7}$$

Equations 12.4, 12.6, and 12.7 are three models for determining single attractors in a time series. Equation 12.7 can denote a fixed point or chaotic attractor, depending on the sign of θ_1. Equation 12.6 assumes a fixed point which the researcher can specify, but does not distinguish the nature of the attractor. The connection between exponential structure and NDS is expanded later in the chapter.

In the general linear model, there is a one regression weight for each variable in the model plus one more for the y-intercept. There is no such correspondence in NLR. For instance, Equation 12.6 could be embellished as[4]

$$y = \theta_1 + \left[\frac{(\theta_1 - \theta_2)}{\left(1 + \exp\left(\theta_3 (x - \theta_4)\right)\right)} \right], \tag{12.8}$$

which is a model for comparing two asymptotic functions in the relationship between y and x (Huet et al., 1996). Note that θ_1 appears in the model twice.

Iteration

NLR proceeds by an iterative calculation process, known as a Newton–Raphson search. Suppose we have a model such as Equation 12.4 or 12.5. Let q_n be a value of a parameter at step n. The algorithm (a) inserts q_n into $f(y)$ and (b) inserts q_n into $df(y)$. (c) It calculates the parameter estimate on the next step as

$$q_{(n+1)} = q_n - \left[\frac{f(y)}{df(y)} \right].$$ (12.9)

The algorithm assesses whether the new q reduces the variance of the predicted values of y more than trivially. Iteration continues until the answer is NO.

The iteration process requires initial values of the parameters to start the program, which the researcher must supply. Sometimes there is a precedent from earlier research regarding what the initial values should be, but often there is not. Thus, the recommendation here is to start the procedure with equal values of θ_i, which should be relatively small in absolute value, such as 0.5, and let the results fall where they may.[5]

Parameters and R^2

The results of an NLR analysis take the form of a final set of parameter values, confidence intervals (CIs) on the parameter estimates, an analysis of variance (ANOVA) table that produces R^2, and a set of correlations among parameter estimates. The final set of parameter values has the same relevance and meaning as β_i in the general linear model. Unlike the general linear model, however, the CIs are calculated differently for each parameter. The calculation depends on the type of function the particular parameter has within the model. For an extended description of how the CIs are computed, see Seber and Wild (2003). Otherwise, it is sufficient to say that they work as any other CIs: if the range does not include 0.00, then the obtained value is significantly different from 0.00. The standard practice in NLR analysis only uses the 95% CI, or $p < .05$. The R^2 value for the overall model is important, but all parts of the theoretical model are needed to draw a conclusion about the efficacy of a model. The reasoning here differs from that which underlies the general linear model; each regression weight has a specific role to play, rather than simply as one more approximation of an ultimate linear model that connects independent variables with a dependent variable. Similarly, an overall F-test for the model can be calculated from the mean-square of the model and mean-square residual, but it is redundant with and less important than the tests on the individual weights.

Here, we encounter another difference between NLR and the general linear model. In bivariate linear regression, the size of the standardized regression weight *is* the correlation. In multiple linear regression, the sizes of the standardized regression weights are corrected for other variables in the model, but the corrected sizes become larger as R^2 becomes larger. No so with NLR. The nature of the calculations is such that the size of the parameters and R^2 are disjointed. The sizes of particular parameters are affected by the presence of other parameters in the model, however.

A related phenomenon in NLR is that the total number of parameters in a model does not automatically boost the R^2 as in multiple linear regression. Instead the introduction of dubious parameters or weak variables could challenge the program to find optimal weights, with the result of a reduction in R^2. Guastello, Pincus, and Gunderson (2006) reported some examples of the R^2 for a two-parameter nonlinear autoregressive model dropping when a third parameter was introduced. The significant effects within the three-parameter model were theoretically important; however, so their three-parameter model was retained and interpreted in spite of the drop in mean R^2 from .73 to .67.[6]

Parameters and Independent Variables

NLR analysis produces correlations among parameter estimates. This feature does not have a direct counterpart in the general linear model, although it serves approximately the same purposes as examining the bivariate correlations among independent variables. Once the final NLR weights have been determined, it is possible to determine how the parameter estimates would have to be altered to accommodate each observation in the data set. Thus, each observation in the data set can be said to have its own set of parameter values. As mentioned already, there is no reason why the number of parameters should be equal to the number of variables in the analysis.

Sometimes researchers find it useful to generate parameters for cases using a model that replaces an independent variable with a regression parameter. For instance an element in a model such as $\theta_1 by^2$, where b is a variable, would be represented as $\theta_1 y^2$. b is essentially treated as constant and absorbed into the estimates of θ_1. Then as a separate step, values of θ_1 are generated for each case and correlated (linear) with possible variables that could explain θ_1. This technique does facilitate shotgun analysis of possible candidates for b. Thus, any tests of hypotheses concerning b would be indirect rather than direct. Another limitation is that the R^2 coefficient for a model without b would probably be higher than it would be for any particular choice of b; R^2 attenuates because any candidate b would be imperfectly correlated with θ_1. Thus, indirect testing of variables is an option, but not one that is widely adopted.

Backward Elimination

The majority of NLR applications involve a theoretical model with critical parts and a small number of substantive variables that are also part of the theoretical relationship. It is possible to define models, however, with more parameters than are actually needed, in which case something akin to multicollinearity occurs among parameter estimates. High correlations among the parameter estimates could explain why some parameters attain statistical significance and others do not, should there be a problem. As we saw with the catastrophe models in Chapter 11, some elements of the model can be more important than others theoretically, so the reasonable approach would be to drop out the least important element or parameter, recalculate, and look for improvements.

Defining the Variance Function

NLR can be performed using either the least squares or maximum likelihood variance function. The least squares method is the default in SPSS NLR and is widely used. Research on the error aspects of the NLR statistical method has probably paid more attention to maximum likelihood (e.g., Stortelder, 1998), whereas research on the deterministic functions favors least squares (Tong, 1990, 1994). The least squares method assumes homoscedasticity around the curve. If this assumption is met reasonably well, there is no substantial difference in final parameter estimates between least squares and maximum likelihood (Huet et al., 1996).

The distributions of NLR parameters are IID if the system is stationary (Nicholls & Quinn, 1982). Any non-IID variance is either accounted for by the model itself or lost. If the data are expected to contain a seriously nonstationary component, a maximum likelihood error function would help to find the dominant model. Nicholls and Quinn recommended, however, that iteration processes with maximum likelihood be initialized with values that are close to the real values. Referring to Figure 3.3, the iteration process would pull the parameter estimates to fit one branch of the function or the other, and could possibly become trapped in the wrong function or not find either one. Close initial values might be obtained by an initial analysis with least squares. A potential drawback of maximum likelihood estimation is that SPSS requires a user-defined function that is going to vary with the model structure. Seber and Wild (2003) and Stortelder (1998) are good guides for possible choices of maximum likelihood functions. Additional parameter estimates could be required, which is another feature to consider.

The use of location and scale corrections as defined previously often have a substantial advantage in meeting the assumption for least squares. Thus, if one were to test Equations 12.4 through 12.7, y would be replaced by z as defined in Equation 2.11, and x would be similarly transformed.

The procedure of transforming variables before analysis is actually a simplification of what is sometimes done by using a complex definition of scale, and replacing y with an $f(y, y_0, \sigma_s)$ within the model itself. The transformation function typically carries additional parameters that need to be estimated. Although the number of parameters has only a loose relationship to R^2 at best, the heuristic of preferring fewer to more parameters in a model is still well-advised. More degrees of freedom that are expended at the calibration phase of model development would predict lower generalizability of the regression weights for point estimation of new observations from outside the original sample.

An alternative means for managing heteroscedasticity or nonstationarity is to find an explanation for it. For instance, if the data look like the problem in Figure 3.3, a bifurcation variable would be implied, which would explain the switch between the line and the curve. A spline might also solve the problem. If a dominant function can be identified, it might be useful to analyze the results for a second function; see the section "Compound models" later in this chapter.

Model Comparison

NDS models are usually developed in a context where an alternative model, usually based on linear assumptions, often exists. Thus, comparisons of NDS models with theoretical alternatives are recommended as a general strategy as it was in the context of catastrophe modeling in Chapter 11.

Inasmuch as NDS models are often autoregressive time series with one entity being measured at many points in time, two types of alternative models are possible. One is the linear autoregression model

$$y_2 = \beta_0 + \beta_1 y_1 + \beta_2 x. \tag{3.17}$$

The other evaluates y as a linear function of time:

$$y_2 = \beta_0 + \beta_1 t. \tag{12.10}$$

Nonlinear functions of time are also possible comparison models, depending on the extant theory in a given area; for an example, see Guastello, Nathan, and Johnson (2009). As mentioned earlier, Equation 3.17 capitalizes on local linearity in a time series, so the relative strength of an NDS model might not become apparent until the lag length is sufficiently large. Equation 12.10 makes an interesting contrast; NDS models are typically autoregressive where time is internalized, whereas time is treated as an exogenous variable in many non-NDS models.

In experiments where many entities are measured repeatedly in relatively short time series, for example, 12 groups of 8 observations each (Guastello & Guastello, 1998), the analysis could be accompanied by an ANOVA with repeated measures and other experimental factors involved. Time is represented as an external variable in a repeated measures analysis. A comparison effect equivalent to Equation 12.10 can be constructed by piecing together the component sums of squares:

$$\eta^2 = \frac{SS(Rm)}{[SS(Rm) + \Sigma SS(IRm) + SSresid]} \tag{12.11}$$

where
   Rm is the repeated measure on which the NDS time series is based
   $\Sigma SS(IRm)$ are sums of squares for interactions with R
   the last term is the sums of squares for the residual (within-subjects)
      component

SPSS Syntax

Data for a time series is organizing in columns, with each row representing separate steps in time. The following SPSS syntax statements will perform an analysis of a time series using the model shown in Equation 12.7. An example analysis follows. The first step is to transform the time series variable by location and scale, using Equation 3.11 or Syntax Statement 12.1; the transformed variable is named z2 to indicate that it is the observation at time 2. The second step is to create a lag variable of z that will pair each value of z2 with a value corresponding to that in the previous step. SPSS can create such a variable automatically using the syntax statement

$$\text{COMPUTE Z1} = \text{LAG(Z2, 1).} \tag{S12.1}$$

Other lag lengths can be created by substituting a different number for 1 inside the parentheses; give the resulting variable a different name if more than one lag length is to be used. Lag variables can also be created by copying and pasting the contents of one column of the spread sheet into another. See the example below for determining a viable lag length for a given problem.
   The NLR analysis then requires three statements:

$$\text{MODEL PROGRAM a = 0.5 b = 0.5.} \tag{S12.2}$$

$$\text{COMPUTE PRED = exp(a*z1) + b.} \tag{S12.3}$$

$$\text{NLR z2 with z1.} \tag{S12.4}$$

S12.2 specifies the characters in the subsequent statement that are regression weights, each followed by equals sign and its initial value. The Greek characters θ_1 and θ_2 are substituted by Roman characters a and b. S12.3 specifies the model. S12.4 executes the model and names the dependent variable, following by the keyword with, followed by the independent variables included in the model. In some editions of SPSS, NLR z2 is sufficient, followed by a period. Windows-based operation of the NLR program gives three windows, one for defining the parameters and initial values, one for typing in the equation (everything appearing after the equals sign in 12.3, without the period), and one to identify the dependent variable.

An autocorrelation spectrum for z2 can be executed with the syntax

```
ACF VARIABLES = z2
/NOLOG
/MXAUTO 16                              (S12.5)
/SERROR=IND
/PACF.
```

The first statement specifies the time series variable. The following subcommand specifies that log transformations of z2 are not requested. The second subcommand specifies the number of lags to be tested; this example tests the lag out to 16. The third subcommand specifies Gaussian error or white noise as the error term when calculating the standard error of the autocorrelation; this option could be useful depending on the problem. The last subcommand requests partial autocorrelations; here the autocorrelation at any lag length longer that 1 is corrected for the influence of lag at shorter lengths.

In the example that follows, log transformation was avoided because it did not represent a known alternative theoretical model. Lag lengths up to 4 were interesting. Standard errors of the autocorrelation and the partial correlations were not interesting.

Motivational Flow

The concept of motivational flow (Csikszentmihalyi, 1990) describes the state of consciousness of someone who is heavily immersed in their work such that they lost the sense of time and a good deal of awareness of extraneous events around them. Flow states occur when intrinsic motivation for the task is high, the task requires high levels of demand on a person's skills, and the task is challenging. Guastello, Johnson, and Rieke (1999) hypothesized that the level of flow experience is a chaotic process over long-enough periods of time. Participants in their study kept a log book for 1 week in which they recorded the levels of skill required and challenges for the activities they were doing; entries were recorded at 15 min intervals.

The data for this demonstration came from one of the log books. The measure of flow was produced by multiplying the ratings of skills and challenges (on a 0–9 scale). Skills and challenges were divided by their standard deviations before multiplying to produce flow. Flow was used as z2 in the autoregression model. The NLR was Equation 12.7; x and y were replaced by Flow at two points in time. Its relationship to chaos is explained later in the chapter; the regression weight for the exponent was important, however, because it signified the value of the largest Lyapunov exponent for the time series.

Any periods of time spent sleeping were dropped out of the time series, leaving a total of 428 observations (Figure 12.1). Data were missing for 63 observations, leaving 365 observations for the analysis.

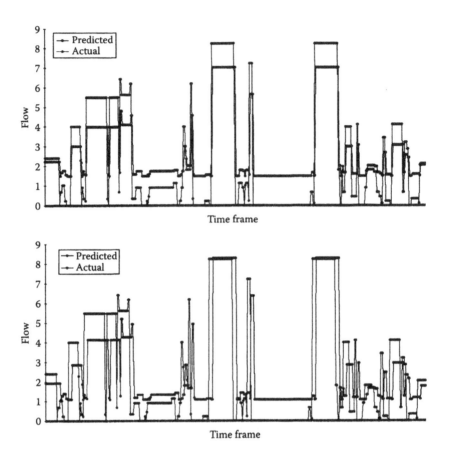

FIGURE 12.1

Time series for flow showing actual values and predicted values from Equation 12.7 (upper) at lag 1, $R^2 = .75$, and (lower) at lag 4, $R^2 = .42$.

The time series was Equations 12.7 was calculated at lags of 1, 2, 3, and 4 time intervals and compared with the linear autoregression results for those lags (Figure 12.2). The linear model was trivially better than the nonlinear linear at a lag of 1, but the gap between them widened at lag 4. The schedules of most people in the original study seemed to be organized in 1-h time blocks, and it appeared that the nonlinearity was visible at 1h intervals. Thus, the lag of 4 (1h) was pursued here also. R^2 coefficients for the linear and nonlinear models at a 1-h lag were .31 and .42, respectively.

Final parameter values and CIs for the nonlinear model are shown in Table 12.1 at all four lags. Because there were only two parameters in this model, the correlations between the two parameters are shown in the same table. The CIs show that both parameters were significantly different from 0.00 at lags 1–3. The exponent was significant at lag 4 but the constant was not.

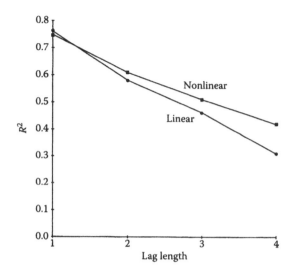

FIGURE 12.2

Comparison of R^2 coefficients for the linear and nonlinear models at lags 1–4.

TABLE 12.1

Parameter Values for Flow Data, Equation 12.7, at Lags 1–4

Parameter	Value	Standard Error	95% CI Lower	95% CI Upper	Correlation with b
Lag 4, R^2 = .42					
a	0.227	0.006	0.215	0.240	−.57
b	0.512	0.146	0.226	0.799	
Lag 3, R^2 = .51					
a	0.237	0.005	0.226	0.247	−.57
b	0.377	0.132	0.117	0.638	
Lag 2, R^2 = .61					
a	0.245	0.004	0.237	0.254	−.56
b	0.240	0.117	0.011	0.470	
Lag 1, R^2 = .75					
a	0.255	0.003	0.248	0.262	−.56
b	0.087	0.093	−0.095	0.269	

The exponent was positive, which indicated that an expanding function was occurring, which was theoretically meaningful in the original context. The small constant b was included in the model that generated predicted values of flow in Figure 12.1. The predicted and actual values tracked closely at lag 1, with apparently a systematic inaccuracy located at the low end of the flow scale.

One of the assets of NDS, and perhaps one of its challenges, is that an entire time series can be generated for one participant in a study where a conventional research strategy might have produced a single number. Experimental contexts, such as the original flow study, could involve calculating NDS metrics on a time series for each participant and then analyzing differences in those metrics across experimental conditions, or associating them with interesting individual differences. In situations where many times series are analyzed and compared, it is important to keep the analytic methods consistent for the individual time series so as not to confound any interpretations about how they would differ.

Exponential Model Structures

The following series of experimental models was first introduced in Guastello (1995) to address some analytic problems that were salient at the time. The technique was subsequently refined, based on experience with using the models (Guastello, 2002, 2005). The problems were as given below:

a. Measurements of the fractal dimension were in demand for the assessment of the complexity of time series and as a test for chaos.

b. The nonstatistical nature of the correlation dimension, which was and probably still is the most often used calculation for a fractal dimension, made it unreliable because of its susceptibility to noise, sampling problems, and so on as discussed in Chapter 2.

c. Chaotic attractors have fractal basins, but the presence of a fractal is only suggestive of chaos; the Lyapunov exponent is better suited for the determination of chaos.

d. The calculations of the Lyapunov exponent were also nonstatistical and thus susceptible to the same distortions as the fractal dimension.

e. Statistical analysis in the social sciences, particularly the applied topics, needed a means of analyzing chaotic data that allowed point estimation as well as an effect size for the level of prediction, neither of which were possible with the nonstatistical NDS metrics.

The solution took the form of a statistical representation of the Lyapunov exponent that could be extracted statistically from a time series. Functions of e were attractive as a basis for the model structure for this purpose and because they could be converted into a fractal dimension. The estimation of exponents as statistical parameters is possible through NLR of course. Furthermore, there should be a series of models ranging from simple to complex structure in which each distinct model contained a new dynamical feature as it progressed up the hierarchy.

Model Series

The derivation of the exponential structural equations for NDS started with the largest model in the series. Simpler models were obtained by progressively dropping out features. The starting point was the response surface for a wave crest catastrophe (Equation 12.9), which contains two order parameters, a bifurcation variable that affects both order parameters, and a known dimension associated with the order parameters. In other words, it had the raw ingredients that were important for developing models with interesting NDS features. A Laplace transform was applied to convert the wave crest function to a function of e and t:

$$u_2 + v_2 = \exp(cu_1 t) + \exp(cv_1 t), \tag{12.12}$$

where
  u and v are the order parameters measured at successive points in time
  c is the bifurcation parameter
  t is time[7]

Equation 12.12 becomes a statistical expression by inserting NLR weights such that

$$u_2 + v_2 = \theta_1 \exp(\theta_2 cu_1 t) + \theta_3 \exp(\theta_4 cu_1 t) + \theta_5, \tag{12.13}$$

where θ_i are the NLR weights and the raw observations of u and v have been transformed by location and scale before analysis.

Equation 12.13 is fully saturated with weights. Ideally all weights should attain statistical significance at $p < .05$. If the criterion is not met, the next procedure would be to drop the least necessary weights, which are in this case θ_1, θ_3, and θ_5. There was no particular need to carry over the asymmetry parameters from the wave crest because it is simply a linear component and not part of the manifold itself, which determines dimension. If an asymmetry variable was needed, however, it would be a simple matter to replace the constant θ_5 with $\theta_5 a$ and $\theta_5 b$. The constant could be just as readily replaced with a complex function.

Some variations on Equation 12.13 are possible. For many nonlinear time series, one might not have a hypothesis for a bifurcation effect, so c can be dropped out of the model (treated as a constant, 1.0), leaving all other elements in place. If all time units are equal for all pairs of observations in a series or an ensemble, $t = 1$, and can be dropped out of the analysis also. u_2, and v_2 can be treated as a single variable, z_2 (raw observations of an order parameter y, corrected for location and scale) on the left hand side. One would then substitute z_1 for u_1 on the right hand side. v_1 could be changed to a lag function of z, or it could be a transfer function from another variable.

As an example, Guastello et al. (2006) used the transfer function with $t = 1$ and $c = 1$ to model electrodermal responses of two people in conversation. Their adaptation of Equation 12.13 was

$$z_2 = \theta_1 \exp(\theta_2 z_1) + \theta_3 \exp(\theta_4 P_1) + \theta_5, \tag{12.14}$$

where
   z was the electrodermal response for one person at successive points in time
   P_1 was the electrodermal response for the other person in the conversation at the previous point in time

Guastello et al. (2009) used Equation 12.14 to model hand movements toward a target where z is the position on one Cartesian axis and P is the position along a second axis.

There are two simpler models in the series. Reverting to Equation 12.13 and dropping the function of the second-order parameter, we obtain a model for a time series of one variable with one order parameter and a bifurcation term

$$z_2 = \theta_1 \exp(\theta_2 c z_1) + \theta_3. \tag{12.15}$$

An alternative form of Equation 12.15 is the Laplace transform of the logistic map that May and Oster (1976) introduced for population dynamics:

$$z_2 = \theta_1 c z_1 \exp(\theta_2 z_1) + \theta_3. \tag{12.16}$$

If the bifurcation effect is suspected but the variable itself is unknown, c is replaced with a constant and absorbed into the regression weight, and all other elements of the model remain the same as shown in Equation 12.16.

If one were to drop the bifurcation hypothesis altogether, Equations 12.15 through 12.16 reduce to

$$z_2 = \theta_1 \exp(\theta_2 z_1) + \theta_3. \tag{12.17}$$

Equation 12.17 is a simple Lyapunov model. Ignore θ_1 and θ_3, and compare the result with Equation 12.6, where the Lyapunov exponent, λ, is now estimated as a regression weight θ_2. For all intents and purposes, θ_2 can be interpreted as the largest Lyapunov in the nonstatistical spectrum: positive values denote chaos, and negative values denote a dampened oscillator and a fixed point in the extreme. It is calibrated relative to moments of the order parameter instead of time, which is a common transposition in time series modeling.

Unlike the nonstatistical approach to calculating λ, θ_2 is a single mean value. In NLR, there is a parameter estimate associated with each observation, but a significant effect indicates that 95% of parameter estimates are all positive or all negative. Thus, if the goal is to test for chaos, there should be signs of both exponential expansion and contraction. The combination of results in Equation 12.17 would be a positive value of θ_2 and a negative value of θ_3.

The null hypothesis for θ_2 indicates that the exponent would be 0, which corresponds to a perfect limit cycle and a dimensionality of 1.00 (see below). Small but significant positive values of θ_2 are slightly asymmetric oscillators, and small but significant negative values tend toward dampened oscillators.

The fractal dimension, D_F, is asymptotically a function of λ (Ott, Sauer, & Yorke, 1994). Thus,

$$D_F = e^{\lambda} = e^{\lambda 2}. \tag{12.18}$$

For models that contain a bifurcation term, the D_F would be the amount of Equation 12.15 plus 1.0 for the bifurcation variable. For models that contain two order parameters without a bifurcation variable, D_F is the codimension, which is sum of the D_Fs associated with each order parameter.

Notice here that each Lyapunov exponent can accommodate time series in any fractal dimension. In the robotic model developed in Guastello et al. (2009), however, the movement trajectories need to be specified along as three different *Cartesian* dimensions in order to communicate to a machine that was looking for coordinates in 3-D space over time, each of which could hold trajectories in a fractional number of dimensions. The results showed for each Cartesian axis, two Cartesian axes expressed as fractals specified the movements across three dimensions much more accurately than other computational models that were prominent in the literature for therapy robots. A comparison of the models' fit with actual observations is shown in Figure 12.3.

Bottom-Up Modeling

When encountering a new time series, the usual objective is to account for as much variance as possible with as few parameters and logical assumptions as possible. The best place to start with the foregoing system of models is Equation 12.17—simple Lyapunov with all the ancillary constants. If statistical significance of all three parameters is not attained, try dropping θ_1; the result is Equation 12.7. If a lack of significance of θ_2 is not attained, try dropping θ_3; an inspection of the correlations among parameter estimates will suggest ahead of time whether dropping a parameter could be the likely solution.

If the results are murky with regard to R^2 or the tests on parameters, move up to Equation 12.16. If the results are still murky, consider dropping (θ_3); θ_1 tends to be more useful in this model structure. Beyond that point, the

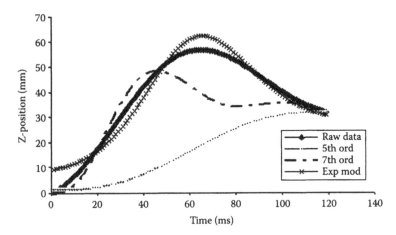

FIGURE 12.3
Graph of x–z position on a randomly selected participant and trial. (Reprinted from Guastello, S.J. et al., *Nonlinear Dynam. Psychol. Life Sci.*, 13, 114, 2009. With permission of the Society for Chaos Theory in Psychology & Life Sciences.)

research design would need to have collected measurements on a plausible bifurcation variable or a second-order parameter in order to produce a better model. On the other hand, the unknown model that produces the data might contain other types of complications. Compound models are one type of complication that is considered next.

The results of the analysis of flow data in Table 12.1 were not murky, but the R^2 for lag 4 showed room for improvement. A test of model 12.16, bifurcation with the bifurcation variable treated as a constant appears in Table 12.2. All three weights were statistically significant, but there was no improvement in R^2. One might conclude in this case that the simpler model was better, all other things being equal. The results in the table concerning Equation 12.20 are discussed later in the chapter.

TABLE 12.2

Parameter Values for Flow Data, Equations 12.16 and 12.20, at Lag 4

Parameter	Value	Standard Error	95% CI Lower	95% CI Upper	Correlation With b	With c
Equation 12.16, $R^2 = .42$						
a	0.191	0.075	0.044	0.338	−.99	−.67
b	0.152	0.047	0.059	0.244		.60
c	1.549	0.189	1.178	1.178		
Equation 12.20, $R^2 = .40$						
a	0.813	0.029	0.755	0.870	.10	−.07
b	10.226	2.039	6.215	14.237		−.70
c	−3.195	0.666	−4.504	−1.885		

Compound Models

Three cases are known where two dynamical functions were operating concurrently in a system. One example involved a periodic driver and a chaotic slave (Guastello, 1998):

$$z_2 = g(z_1) + \beta_1 a + \beta_2 b + \beta_3 c, \qquad (12.19)$$

where
    $g(z)$ was an exponential function from the model series above
    a, b, and c were control parameters
    β_i was regression weight that held the model together

The function $g(z)$ displayed a positive Lyapunov exponent. Control variables a and b were categorical variables that originated within the experimental design. Variable c, however, was itself a different exponential function over time with a negative exponent.

The designation of "driver" and "slave" had to be made in the context of the problem, but see the discussion of Granger causality below. A procedural annoyance, however, was that the linear comparison model contained four additive elements, of which one was nonlinear over time. Thus, the linear model "cheated" a bit because of the nonlinear function inherent in c. As it turned out, the R^2 for the linear comparison model was relatively high, but not greater than the R^2 for the nonlinear model.

The second case was identified in the group coordination learning experiment (Guastello, Bock, Caldwell, & Bond, 2005; Guastello & Bond, 2007; Guastello & Guastello, 1998). The group readily learned the coordination task for two of the experimental conditions, one experimental condition in the coordination study was very difficult for the participants. The first function that was extracted was a chaotic function, which indicated in this context that the self-organizing response did not occur for some of the groups. The analysis of residuals, however, showed a second function containing a negative exponent. Thus, an incomplete self-organized response may contain two kinds of functions, and an analysis of residuals would be required to find them both.

The third case involved a combination of a cusp catastrophe model and chaotic attractors (Rosser, Rosser, Guastello, & Bond, 2001). The attractors in a cusp are fixed point, although the presence of limit cycles would not usually impair the analysis. The theory that Rosser et al. were working with indicated a cusp manifold for hysteresis, but the attractors were chaotic (Figure 12.4). There were no control variables available, so they simply analyzed the dynamics of their two economic indicators using the exponential series. One variable displayed a bifurcation effect and a negative exponent. The other displayed no bifurcation and a positive exponent. Although the dynamics that were obtained did not show a clear illustration of what was hypothesized in simpler terms, the dynamics did display portions of the intended dynamics.

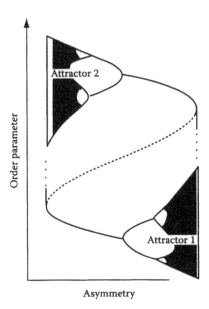

FIGURE 12.4
Hybrid function of a cusp catastrophe with chaotic attractors. (Reprinted from Rosser, J.B. et al., *Nonlinear Dynam. Psychol. Life Sci.*, 5, 350, 2001. With permission of the Society for Chaos Theory in Psychology & Life Sciences.)

Dimensionality Estimates

Johnson and Dooley (1996) explored Equation 12.17 with the Lorenz and Rossler data to assess the impact of sample size, lag, and noise for their impact on the calculation of D_F. The known D_F for the Lorenz attractor is 2.07; they obtained 1.94 and 1.84 using the NLR model with x-axis and y-axis data, respectively. The known D_F for the Rossler attractor is 2.00+e, very slightly greater than a perfect 2.00. They obtained 1.68 and 1.92 using the NLR model with x-axis and y-axis data, respectively (p. 86).

For the lag analysis, they used data from the z-axis of the Lorenz attractor, and varied lag length. D_F started at 2.20 at lag 1, dropped gradually to 2.06 at lag 8, and increased thereafter. For the y-axis on the Rossler attractor, the statistical D_F came closest to the real value at lag 10, fluctuated greatly as lag increased further, then dropped slowly toward the actual value again at around lag 35.

The sampling analysis started with a data set of 350 points from the z-axis of the Lorenz attractor and the lag that produced $D_F = 2.06$. They reduced the sample in four steps, dropping out an expansion loop of the attractor each time. D_F ranged from 2.10 to 1.76.

The analysis for noise started with the same Lorenz data. Johnson and Dooley added noise in six increments ranging from 10% to 150%. D_F dropped from 2.04 to 1.84 across that range.

Some reasonable conclusions from the foregoing analysis can be drawn as follows. Equation 12.17 can come very close to the target depending on the lag length and axis of the attractor which was being assessed. A lag length of 8 appears large relative to what is usually encountered in real data, but it is important to remember that the Lorenz and Rossler attractors are formalized but artificial systems. Data are generated in very small increments of virtual time so as to appear continuous. The importance of the attractor axes depends on the actual equations generating an attractor; for a system of x, y, and z order parameters, it is possible for one axis to be more representative of the impact of all three variables than another.

The sampling results, as Johnson and Dooley (1996) noted, were closely tied to the number of evolutions of the attractor; for example, whether the full topological range was present in the data. Full representation cannot be automatically assumed in real data. Equation 12.17 produces D_F values that are fairly resilient in the presence of noise, however.

Other Nonlinear Structures

Other nonlinear autoregression models that were advanced in the 1980s were summarized in Tong (1990, p. 96–120). Several options were variations of a linear model with complex lag. In some respects they were not especially different from the NARMAX model (Equations 2.25 through 2.26) introduced by Aguirre and Billings (1995), although an important distinction is that NARMAX is saturated with polynomials. What made most of them "nonlinear" was the recognition of the possibility of multiplicative error. Other than that feature they did not capture any NDS processes for attractors, bifurcations, chaos, and so on. There was one notable exception, which is considered next.

Exponential-Quadratic Models

The exponential-quadratic model (Equation 12.20), is designed to capture fluctuations of y as it deviates around a central axis, as a distorted oscillator might do (Tong, 1990). For that reason, the variables are normalized with 0.00 in the center.

$$y_t = \sum_{j=1}^{k} \left[\theta_1 + \theta_1 \exp\left(-\theta_3 y_{t-1}^2\right) \right] y_{t-j} + \varepsilon_t, \qquad (12.20)$$

where
  y is the time series variable
  θ_i are regression weights
  θ_3 is constrained positive

The quadratic variable is y at a lag of 1. The argument in square brackets is distributed over several lags of y, and the result is summed over lags. ε_t looks like an error terms, but it is intended as a set of IID variables; all the non-IID variance is pushed into the function of y.

Model Comparison

Johnson and Dooley (1996) compared Equations 12.20 and 12.17 for their respective abilities to model time series generated by the Lorenz, Rossler, and Henon horseshoe attractors. Equation 12.20 did the best job of modeling the Lorenz attractor ($r^2 = .996$), followed by the linear model ($r^2 = .885$), and then followed by Equation 12.17 ($r^2 = .766$). It also did the best job of the Rossler attractor ($r^2 = .969$), although Equation 12.20 was very close behind ($r^2 = .925$). The results for the Henon horseshoe were very different, with $r^2 < .10$ for Equations 12.17, 12.20, and the linear autoregressive model. They surmised that the function itself was quadratic and would probably be represented best by a quadratic function, which was the case ($r^2 = .919$). Although both models showed high predictive accuracy for the Rossler model, Equation 12.20 does not produce a fractal dimension.

Table 12.2 shows the results for 12.20 when applied to the flow data at lag 4. For this test, only that one lag length was used in order to maintain parity with 12.16, which only uses one lag length at a time. It is also noteworthy that 12.20 contains a strict linear term. The parameter estimation procedure was unconstrained NLR (the usual); because θ_3 is supposed to be negative, a starting value of -0.5 was used, whereas $+0.5$ was used for the other two parameters. All three parameters were statistically significant for 12.16, although the R^2 was slightly lower.

If one did wish to pursue the accumulated lag length up to lag 4, the value of R^2 to beat would be .75, which was the value obtained for Equation 12.7 at lag 1. A 12-parameter model would also be involved. The usual warning concerning generalizability of models would apply.

Granger Causality

It is axiomatic in conventional statistics that correlation is not causation. Causality can be inferred in an experiment, however, where values of an independent variable are manipulated prior to the collection of the dependent variable. One might design studies around the use of correlations rather than ANOVA, nonetheless, so long as the temporal and manipulation features are maintained. *Granger causality* is the equivalent issue in time series analysis where it is assumed that only events occurring at a previous point in time can possibly be the cause of events in a future point in time.

Thus, it is assumed and reasonably so in the social sciences that the future does not influence the past (Granger, 1969).[8]

Time series analysis does not usually contain the manipulation of variables found in standard experiments. Instead one is more likely to have a dependent measure in time series with a discrete event introduced within the time series, such as the onset of a governmental policy; from there, the analyst is trying to determine whether the policy changed the course of the dependent measure. Experimental designs that are formed around time series, rather than single dependent observations, are possible; however. Delignières, Torre, and Lemoine expand on this point in Chapter 26.

In linear time series analysis, one must parse the roles of autocorrelation and transfer effects. An exogenous variable X_1 can only be said to influence the target value Y_2, if it accounts for more variance in Y_2 than Y_1, the autocorrelation. If X_1 passes the test, it is said to have a Granger-causal relationship to Y (Ostrom, 1978).

In nonlinear time series with NLR, adding variables to a model does not necessarily contribute to an increase in R^2. Thus, there is a greater reliance on the significance of the regression weights. If R^2 increases and the new variable has a statistically significant weight, the results are clear, however. The nature of causality is different in NDS and the models that usually represent those processes. The conventional concept of causality is replaced by *control*; the initial conditions work together with external influences to produce a result. This theme is very salient in catastrophe models as shown in the previous chapter.

Some recent work on Granger causality attempts to go beyond the two-variable cases that we have considered thus far. For instance, EEGs can produce a minimum of eight possible time series, depending on the equipment that is used, and the goal of interpretation is to determine how they go together. The standard approach is to take time series two at a time, and assess the autocorrelations and transfer functions using linear analysis.

The new approaches are designed with network-type relationships in mind, similar to what is implied by the path analysis diagram in Figure 2.2. Reciprocal causation would be allowed (Eichler, 2006; Kamiński, Ding, Trucado, & Bressler, 2001). Directed transfer function (DTF) analysis begins with the spectral frequencies (see Chapter 16) from EEG data, and a matrix of correlations $|H_{ij}(f)|^2$ between all possible pairs of power spectra. The correlation matrix would be asymmetric. The DTF is the ratio of a particular function to the sum of all possible influences across channels (Kamiński et al., 2001, p. 147); $0 < DTF < 1$.

The limitations of the concept, however, are that the EEG functions over time are nonlinear, and thus the probability distribution of DTF is unknown. Thus, significance tests are not possible, making DTF neither a necessary nor sufficient statistic for detecting Granger causality (Eichler, 2006). Eichler was able to develop a point-wise probability calculation for

DTF that actually varied for each $\mathbf{H}_{ij}$ in the set; any alpha level associated with a DTF is dependent on alpha levels associated with the contributing correlations.

There are other approaches to the assessment of multivariate Granger causality due to Geweke (1982), Saito and Harashima (1981), and Baek and Brock (1992). They all appear to have the same limitations related to normalization, compounded alpha levels, and the use of linear versus nonlinear relationships among time series. Gourévitch, LeBouquin-Jeannès, and Faucon (2006) found that, at best, these techniques were not adversely influenced by introducing nonlinear autoregressive functions, but they could only detect linear linkages.

A possible solution to the dilemma could be related to the choice of nonlinear model structure. Guastello et al. (2006) found that the transfer function in Equation 12.13 could detect a much greater proportion of linkages than a linear model. On the other hand, no extension of the technique to multivariate Granger causality has been made yet.

Notes

1. The acronym that is adopted here is the same as the primary syntax command for SPSS.
2. If the range of y was not restricted to 0 and 1, a different real number would be inserted here.
3. In mathematical notation, it is not always easy to distinguish whether the author intends a glyph such as $-\theta x$ to mean that a subtraction is always made regardless of the sign of θ, or whether θ is constrained to be negative. The approach taken here of replacing a negative sign with a positive addition operator allows the possibility of either positive or negative signs of the weights to become the final result. The implication for point estimation is thus straightforward: add a positive or negative value as indicated by the final regression weight. There is an option of *constrained* NLR where the user can set restrictions on the range of θ. Unconstrained NLR is assumed throughout this chapter and the next one, however.
4. The horizontal form of the equation lends itself more readily to SPSS syntax than the stacked form.
5. Experience with numerous applications of the exponential models considered later in this chapter indicates that, if the data are well described by the hypothetical function and do not contain transients that are essentially competing values of the parameter estimates, differences in one's choices of initial values do not make any difference in the final values of the parameter estimates. If there is a deterministic function in the data other than the one specified, however, the algorithm could have a difficult task of finding the intended function instead of the decoy. Different sets of initial estimates could sway the program toward one solution or another.

6. $N = 74$ models, one for each participant in the study. Of further interest, they also found a variable that correlated with the R^2 for each model, accounting for 7% of the variance among coefficients.
7. Actually, the transformation rendered the exponent 2 as a variable exponent c; c regained its role as a bifurcation variable when it was split between an NLR weight and a variable in the next step.
8. Postmodern reconstructions of sociopolitical history are perhaps a notable exception, although this type of thinking rarely translates into a time series analysis of anything. Special cases within physics might be said to produce other types of exceptions.

References

Albert, A. E. & Gardner, L. A. Jr. (1967). *Stochastic approximation and nonlinear regression.* Cambridge, MA: MIT Press.

Baek, E. & Brock, W. (1992). *A general test for nonlinear Granger causality* (Technical report). Iowa City, IU: University of Iowa.

Csikszentmihalyi, M. (1990). *Flow: The psychology of optimal experience.* New York: Harper-Collins.

Eichler, M. (2006). On the evaluation of information flow in multivariate systems by the directed transfer function. *Biological Cybernetics, 94,* 469–482.

Geweke, J. (1982). Measurement of linear dependence and feedback between multiple time series. *Journal of the American Statistical Association, 77,* 304–313.

Gourévitch, B., LeBouquin-Jeannès, R., & Faucon, G. (2006). Linear and nonlinear causality between signals: Methods examples and neurophysiological applications. *Biological Cybernetics, 95,* 349–369.

Granger, C. W. J. (1969). Investigating causal relations by econometric models and cross-spectral methods. *Econometrica, 37,* 424–438.

Guastello, S. J. (1995). *Chaos, catastrophe, and human affairs: Applications of nonlinear dynamics to work, organizations, and social evolution.* Mahwah, NJ: Lawrence Erlbaum.

Guastello, S. J. (1998). Creative problems solving groups at the edge of chaos. *Journal of Creative Behavior, 32,* 38–57.

Guastello, S. J. (2002). *Managing emergent phenomena: Nonlinear dynamics in work organizations.* Mahwah, NJ: Lawrence Erlbaum Associates.

Guastello, S. J. (2005). Nonlinear methods for the social sciences. In S. Wheelan (Ed.), *The handbook of group research and practice* (pp. 251–268). Thousand Oaks, CA: Sage.

Guastello, S. J., Bock, B., Caldwell, P., & Bond, R. W., Jr. (2005). Origins of group coordination: Nonlinear dynamics and the role of verbalization. *Nonlinear Dynamics, Psychology, and Life Sciences, 9,* 175–208.

Guastello, S. J. & Bond, R. W., Jr. (2007). The emergence of leadership in coordination-intensive groups. *Nonlinear Dynamics, Psychology, and Life Sciences, 11,* 91–117.

Guastello, S. J. & Guastello, D. D. (1998). Origins of coordination and team effectiveness: A perspective from game theory and nonlinear dynamics. *Journal of Applied Psychology, 83,* 423–437.

Guastello, S. J., Johnson, E. A., & Rieke, M. L. (1999). Nonlinear dynamics of motivational flow. *Nonlinear Dynamics, Psychology, and Life Sciences, 3,* 259–274.

Guastello, S. J., Nathan, D. E., & Johnson, M. J. (2009). Attractor and Lyapunov models for reach and grasp movements with application to robot-assisted therapy. *Nonlinear Dynamics, Psychology, and Life Sciences, 13,* 99–121.

Guastello, S. J., Pincus, D., & Gunderson, P. R. (2006). Electrodermal arousal between participants in a conversation: Nonlinear dynamics for linkage effects. *Nonlinear Dynamics, Psychology, and Life Sciences, 10,* 365–399.

Hartley, A. O. (1961). The modified Gauss-Newton method for the fitting of nonlinear regression functions by least squares. *Technometrics, 3,* 269–280.

Huet, S., Bouvier, A., Gruet, M. A., & Jolivet, E. (1996). *Statistical tools for nonlinear regression.* New York: Springer-Verlag.

Johnson, T. L. & Dooley, K. J. (1996). Looking for chaos in time series data. In W. Sulis & A. Combs (Eds.), *Nonlinear dynamics in human behavior* (pp. 44–76). Singapore: World Scientific.

Kamiński, M., Ding, M., Crucolo, W. A., & Bressler, S. L. (2001). Evaluating causal relations in neural systems: Granger causality, directed transfer function and statistical assessment of significance. *Biological Cybernetics, 85,* 145–157.

Marquardt, D. W. (1963). An algorithm for least-squares estimation of nonlinear parameters. *SIAM Journal of Applied Mathematics, 11,* 431–441.

May, R. M. & Oster, G. F. (1976). Bifurcations and dynamics complexity in simple ecological models. *American Naturalist, 110,* 573–599.

Nicholls, D. F. & Quinn, B. G. (1982). *Random coefficient autoregressive models: An introduction.* New York: Springer-Verlag.

Ott, E., Sauer, T., & Yorke, J. A. (Eds.). (1994). *Coping with chaos.* New York: Wiley.

Ostrom, C. W. Jr. (1978). *Time series analysis: Regression techniques.* Beverly Hills, CA: Sage.

Ratkowski, D. A. (1983). *Nonlinear regression modeling.* New York: Marcel Dekker.

Rosser, J. B., Jr., Rosser, M. V., Guastello, S. J., & Bond, R. W. Jr. (2001). Chaotic hysteresis and systemic economic transformation: Soviet investment patterns. *Nonlinear Dynamics, Psychology, and Life Sciences, 5,* 345–368.

Saito, Y. & Harashima, H. (1981). Tracking of information within multichannel EEG record. In N. Yamaguchi & K. Fujisawa (Eds.), *Recent advances in EEG and EMG data processing* (pp. 133–146). Amsterdam: Elsevier.

Stortelder, W. J. H. (1998). *Parameter estimation in nonlinear dynamical systems.* Amsterdam: CWI Tract.

Seber, G. A. F. & Wild, C. J. (2003). *Nonlinear regression* (2nd ed.). New York: Wiley.

Tong, H. (1990). *Non-linear time series: A dynamical system approach.* New York: Oxford.

Tong, H. (1994). Comments on prediction by nonlinear least-squares methods. In F. P. Kelley (Ed.), *Probability, statistics, and optimization: A tribute to Peter Whittle* (pp. 219–224). New York: Wiley.

13

Catastrophe Models with Nonlinear Regression

Stephen J. Guastello

CONTENTS

The method of dynamic difference equations for catastrophe models that was described in Chapter 11 is particularly good for situations where many people or objects are measured at two points in time, before and after a critical event has taken place. The method of dynamic difference equations can be used for time series where many observations are taken on one entity or system, although the advisements concerning lag length and local linearity would be applicable there as well. The issue of local linearity may be more distinct with a catastrophe time series model because of the regions of local stability and hysteresis that are inherent in the models.

This chapter considers static models for catastrophe functions that can finesse the problems of lag versus local linearity and also work well in situations where the dependent variable is measured at one point in time even though the full range of attractor regions, bifurcation manifold, and the repellor are present in the data set. Static models rely on the use of the unique probability density function (pdf) associated with the catastrophe model. If the data fit the model well, then it is possible to conclude that all the local dynamics associated with the model hold true to the extent that R^2 is high and the control variables are identified. The models are tested through nonlinear regression.

Direct Method for the Cusp Model

The nonlinear regression procedures for the cusp fall into two categories where the control variables in the model are tested directly or indirectly. Although the indirect method is historically older (Cobb, 1981a,b; Cobb, Koppstein, & Chen 1983; Cobb & Zacks, 1985) than the direct method (Guastello, 2002), the direct method is presented first here for a more tractable exposition and because it is simply a more efficient method of testing the model.

Recall from Chapter 3 that virtually any differentiable function of *y* can be transformed into a pdf using Equation 3.8. As an example, the function for the cusp catastrophe response surface (Equation 3.9) was transformed into a pdf for the cusp.

The probability distribution for the cusp catastrophe appears in Figure 13.1. It is unimodal in the region around the cusp point where the bifurcation variable is low, and it becomes increasingly bimodal as bifurcation strength increases.

$$\text{pdf}(z) = \xi \exp\left[\frac{-z^4}{4} + \frac{bz^2}{2} + az\right] \tag{3.10}$$

where
  z is the dependent measure that was corrected for location and scale according to Equation 3.11
  pdf(z) signifies the cumulative probability associated with the standardized variate
  ξ is a constant that ensures unit density

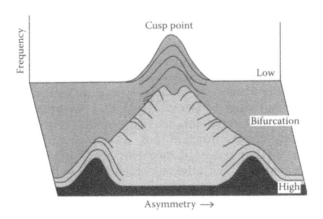

FIGURE 13.1
Probability density function for the cusp catastrophe.

Equation 3.10 can be rendered into usable form by introducing nonlinear regression weights and a cubic term:

$$\text{pdf}(z) = \xi \exp[\theta_1 z^4 + \theta_2 z^3 + \theta_3 bz^2 + \theta_4 az] \tag{13.1}$$

The nonlinear regression weights θ_1 are all treated as positive for purposes of specifying initial values of the parameters for the nonlinear regression program. The constant ξ is also treated as a nonlinear regression weight.

The cubic term plays the same role here as the quadratic term in the dynamical difference equation for the cusp (Equation 12.17): If the data for one side of the cusp is better developed in the data than the other side, the cubic term will account for variance associated with the lopsided representation of the surface. The rationale for keeping or deleting terms in the model is the same as with the dynamic difference equations: The quartic term and the bifurcation term $(\theta_3 bz^2)$ are the most important. The cubic and ξ terms are the least important. The R^2 for the cusp model is, finally, compared with the R^2 for a suitable linear model.

SPSS Commands

The first step requires a frequency distribution on the raw scores of the dependent measure, y that will produce the cumulative probability (or percentile) of y. In the second step, ua RECODE command would be used to substitute the cumulative probabilities of y for y and give the result a new variable name, PDFZ (S13.1).

```
RECODE Y  (0 = .500)  (1 = .617)  (2 = .683)  (3 = .725)
(4 = .750)  (5 = .775)  (6 = .800)  (7 = .825)  (8 = .842)
(9 = .842)  (10 = .883)  (11 = .933)  (12 = .950)  (13 = .967)
(14 = .999)  (15 = .999)  (16 = .999)  into PDFZ.         (S13.1)
```

This example starts with a range of numbers $0 < y < 16$, and all the numbers to the right of the equal signs were cumulative frequencies.[1] Note that a pdf(y) and a pdf(z) are the same sets of numerical values.

The third step converts y to z. SPSS sometimes encounters a computational overflow when running nonlinear regression if, on one of the iterations, the numerical argument to the exponent exceeds 88. The problem is solved by multiplying S (scale) by 10, as shown in (S13.2).

```
COMPUTE z = (Y - L)/(S * 10)         (S13.2)
```

If a computational overflow still occurs, change 10–100. Make the same transformation on the bifurcation (BIFUR) and asymmetry (ASYM) variables. L is the location parameter.

The next three statements execute the nonlinear regression model. Statement 13.3 defines the nonlinear regression weights, where x corresponds to ξ in Equation 13.1, and a, b, c, and d correspond to $\theta_1, \ldots, \theta_4$, respectively. Once again, the analysis is conducted with the least squares error definition.

```
MODEL PROGRAM x =0.5, a =0.5 b =0.5 c =0.5 d =0.5          (S13.3)
COMPUTE PRED = x*exp(a*(z**4) + b*(z**3) + c*BIFUR*(z**2) + d*ASYM*z)
                                                          (S13.4)
NLR PDFZ with z BIFUR ASYM                                (S13.5)
```

In the event that statistical significance is not obtained for each of the regression weights, the least essential element in the model can be dropped. In this case we would drop "$b = 0.5$" from (S13.3), and "$+b^*(z^{**}3)$" from (S13.4); $x(\xi)$ could also be dropped if needed. Statement 13.5 would not change.

Example: Binge Drinking
In this example, (Smerz & Guastello, 2008) a cusp catastrophe model for binge drinking (of alcohol) among college students was developed using attitude toward alcohol and peer pressure as the control variables ($N = 1197$). On the one hand, the theory indicated that a cusp would be a good way to describe the hysteresis associated with binge drinkers and separate binge drinkers from others. Peer pressure would act as a bifurcation variable; some people would succumb to pressure, while others would exhibit a reactance effect and behave in the opposite manner. The hypothesized relationship among the variables is shown in Figure 13.2. There was already an empirical precedent for a similar model (Clair, 1998). The survey only measured attitudes and behavior at one point in time, however. Thus, the static approach to analysis was used.

The variables that constituted peer pressure and attitude toward alcohol were composed of survey items that were aggregated into scales through factor analysis. Binge drinking was defined as the number of occasions in the previous 2 weeks the person had five or more alcoholic beverages; this definition followed from the definition adopted in other large-scale studies. The cusp catastrophe analysis for the binge drinking criterion appears in Table 13.1.

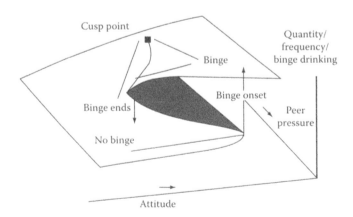

FIGURE 13.2
Cusp catastrophe model for binge drinking among college students. (Reprinted from Smerz, K.E. and Guastello, S.J., *Nonlinear Dynam. Psychol. Life Sci.*, 12, 216, 2008. With permission.)

TABLE 13.1

Results for Cusp Catastrophe Analysis of Binge Drinking

	R^2	β or θ	Model F
Binge drinking			
Basic linear	0.34		293.90**
Peer pressure		0.396	
Attitude		0.279	
Linear interaction	0.34		
Peer pressure		0.392*	200.11**
Attitude		0.292*	
Interaction		0.071*	
Cusp model	0.90		
$\theta_1 z^4$		−0.035*	22296.19**
$\theta_2 z^3$		0.129*	
$\theta_3 b z^2$		−0.003*	
$\theta_4 a z$		0.024*	
Frequency, past 30 days			
Basic linear	0.39		374.76**
Peer pressure		0.429*	
Attitude		0.299*	
Linear interaction	0.39		
Peer pressure		0.428*	249.94**
Attitude		0.303*	
Interaction		0.017	
Cusp model[a]	0.64		
$\theta_1 z^4$		−0.003*	4083.84**
$\theta_3 b z^2$		0.012*	
$\theta_4 a z$		0.099*	
Average drinks per week			
Basic linear	0.37		345.37**
Peer pressure		0.399*	
Attitude		0.307*	
Linear interaction	0.38		
Peer pressure		0.405*	241.85**
Attitude		0.320*	
Interaction		0.110*	
Cusp model	0.59		
$\theta_1 z^4$		−0.025*	3897.20**
$\theta_2 z^3$		0.099*	
$\theta_3 b z^2$		−0.001	
$\theta_4 a z$		0.035*	

Source: Smerz, K.E. and Guastello, S.J., *Nonlinear Dynam. Psychol. Life Sci.*, 12, 216, 2008. Reprinted with permission of the Society for Chaos Theory in Psychology & Life Sciences.

$*p < .05; **p < .001.$

Scale was multiplied by 10 for the conversion in (S13.2). The parameter ξ was dropped from the analysis; statistically significant weights were obtained for all four parts of the cusp model. The degree of fit was high overall ($R^2 = 0.90$), and compared favorably with the linear model that predicted binge drinking from attitude and peer pressure only ($R^2 = 0.34$) or the linear model with the interaction term ($R^2 = 0.34$).

The results supported the conclusion that the distribution of binge drinking behavior was cusp catastrophic. The binge drinking phenomenon is captured by the hysteresis region of the surface. At one stable state the binges merge into a constant level of excess. At the other stable state binge drinking does not occur. Moderate and relatively constant drinking levels occur at the low-bifurcation side of the surface. Drinking level would be deflected toward one of the attractors if the attitude or peer pressure variable were strong enough.

Peer pressure was negatively phrased such that high scores indicated that peers were attempting to persuade the individual *not* to drink. Highly critical peers, when combined with positive expectations about alcohol, produce a reactance effect.

Two other dependent variables were included in the study: frequency of drinking occasions in the past 30 days and the average number of drinks per week.[2] A well-formed cusp was obtained for frequency of use once the cubic term was dropped from the analysis, although the final model was not as strong as the one obtained for binge drinking. The bifurcation terms were missing for average number of drinks per week, and it could not be recovered by dropping the cubic term; the results were thus interpreted as not a good example of a cusp.[3]

Indirect Methods for the Cusp Model

The indirect method, which was introduced by Cobb (1978, 1981a), tests a cusp catastrophe model as a two-step process. The first step tests the shape of the cusp distribution. The second step evaluates the role of the hypothesized control variables. The analysis can be performed using least squares or maximum likelihood. The least squares method is described first.

Least Squares

The test of the cusp pdf begins with a version of Equation 13.1 in which the control variables are tested as constants of 1.0 and thus become part of the regression parameter estimates. The cubic term is always included.

$$\text{pdf}(y) = \xi \exp\left[\theta_1(a_1 - a_0)\frac{(y - y_0)}{\sigma_s} + \theta_2(b_1 - b_0)\left[\frac{(y - y_0)}{\sigma_s}\right]^2\right.$$
$$\left. + \theta_3\left[\frac{(y - y_0)}{\sigma_s}\right]^3 + \theta_4\left[\frac{(y - y_0)}{\sigma_s}\right]^4\right] \tag{13.2}$$

In Equation 13.2, the corrections for location (y_0) and scale (σ_s) of y are incorporated into the main model and are estimated as additional nonlinear regression parameters. a_1 and b_1 are the surrogate control parameters and a_0 and b_0 are their lower limits, respectively. Note that if the research variables a and b are scaled so that their lower limits are both 0.00, then Equation 13.2 simplifies to

$$\text{pdf}(y) = \xi \exp\left[\theta_1\frac{(y - y_0)}{\sigma_s} + \theta_2\left[\frac{(y - y_0)}{\sigma_s}\right]^2 + \theta_3\left[\frac{(y - y_0)}{\sigma_s}\right]^3 + \theta_4\left[\frac{(y - y_0)}{\sigma_s}\right]^4\right]$$
$$\tag{13.3}$$

The goodness of fit for Equation 13.2 with the actual data was summarized by an R^2 statistic, which was named pseudo-R^2 because the model did not permit hypothesis testing of the entire model in the usual sense; the hypothesized control variables are not yet represented in the model. Once the pdf model has been calculated, one can generate parameter estimates for each observation. The hypothesized variables, $v_1, \ldots, v_n$, in the study are all correlated with the parameter estimates for a and b. The results resemble those of a factor analysis where the original research variables are correlated with underlying factors. The interpretation of parameter "loadings" is similar to that of a factor analysis output. All the weights are used here as they are in factor analysis to produce factor scores, such that the final model for the response surface takes the form

$$f(z) = \theta_1 z^3 + \theta_1 z^2 + \theta_3(\theta_4 v_1 + \theta_5 v_2)z + \theta_6(\theta_7 v_1 + \theta_7 v_2) \tag{13.4}$$

Equation 13.4 is written for the simple case where there are two research variables v_1 and v_2. If there is a larger number of research variables, Equation 13.4 expands accordingly; if we let v^* be the number of research variables, the total number of parameters in the final model is equal to $2v^* + 4$, plus parameters associated with the scale of y and the lower limits of all the variables.

The procedure does not make any particular interpretation of statistical significance for the parameter estimates.

A linear model is also calculated as a multiple regression between v_1 and v_2 with z or y. The R^2 coefficients for the linear model and the pdf can be compared visually. An illustrative example (Cobb, 1981a) was based on original data and a cusp model published by Zeeman (1976) that had originated with a study published 20 years earlier. Human participants completed a measure of the introversion–extroversion personality trait and then performed a simulated driving task under varying degrees of alcohol intoxication. The hypothesis was that changes in driving speed were cusp distributed such that the introversion–extroversion trait would correspond to the bifurcation parameter, which it did: Extroverts showed little change in driving speed whereas introverts would drive faster or slower under intoxication. The pseudo-R^2 for the cusp and R^2 for the linear model were 0.60 and 0.00, respectively, which was a very decisive difference.

Maximum Likelihood Method

The maximum likelihood method (Cobb et al., 1983; Cobb & Zacks, 1985) incorporates a multiplicative error function such as those that might be encountered in a time series. The relative merits of maximum likelihood versus least squares were discussed in Chapter 2. For the cusp model, the argument to the exponential operator in Equation 13.2 is

$$f(z) = \left(\frac{z^4}{4} - \frac{bz^2}{2} - az \right) + \sigma_s w_t \qquad (13.5)$$

where w_t is a random error term that interacts with the scale estimate. Equation 13.5 is simplified for clarity by converting y to z and not showing the regression weights. The first three steps of the analysis are otherwise the same as those used in the least squares analysis: Compute the pdf and compute the R^2, correlate research variables with parameter estimates, and compute the linear regression for the research variables with y. The final cusp model again takes the form of Equation 13.4. There is a fourth step which is a χ^2 test that compares the degree of fit provided by the cusp and linear models to the actual distribution of y.

Computer programs to run the indirect methods were first introduced in the early 1980s when programs for nonlinear regression were relatively rare. The programs for the cusp produced a cusp model only; they did not have the full generalizability of nonlinear regression programs that appeared in comprehensive statistical packages such as SPSS or SAS. The original programs went into disuse as the Fortran code became obsolete and the comprehensive programs became more widely available. Two examples, nonetheless, appeared in Fischer and Jammernegg (1986) and Hanges, Braverman, and Rentsch (1991) that involved applications in economics and psychology, respectively.

Programs dedicated to the indirect method of testing the cusp using maximum likelihood reemerged in Van der Maas, Kolstein, and van der Pligt (2003). Their version computes the parameter estimates and their correlations between research variables and parameter estimates. The algorithm then proceeds to test for all possible cusp models with the variables associated with the parameter estimates in place, as shown in Equation 13.4. The models are compared with AIC and BIC statistics rather than R^2. Thus, there is no notion of effect size. Barunik and Vosvrda (2009) used this version of the procedure to illustrate that the S&P500 index did display catastrophic behavior in the stock market crash of 1987. Their application did not include any tests of control parameters that were implicated in previous studies of market crashes, although they did reinstate the use of R^2 for the cusp distribution itself.

Direct Method for the Swallowtail Catastrophe

The direct method for testing a catastrophe model through nonlinear regression extends to the swallowtail model. There are no published counterparts to the indirect maximum likelihood method for the swallowtail, however. Applications of the swallowtail model in the behavioral sciences are sparse overall, and the applications using the direct method through nonlinear regression currently revolve around one set of problems, which involve the emergence of leaders from a leaderless work group. This section of the chapter describes how the hypothesis was first developed around the swallowtail catastrophe's probability distribution, and then expanded to test hypotheses concerning control parameters through the direct method.

The research program started with the observation that group members' ratings of each other's leadership behavior displayed the characteristic probability distribution of the swallowtail catastrophe (Guastello, 1998) which was first recorded by Cobb (1981b). The swallowtail distribution for leadership emergence is shown in Figure 13.3. There were three distinct groupings

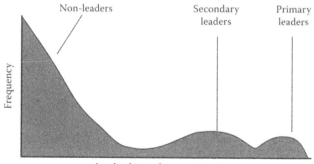

FIGURE 13.3
Swallowtail distribution of leadership endorsement ratings after a leaderless group has self-organized. (Reprinted from Guastello, S.J., *Am. Psychol.*, 62, 606, 2007b. Reprinted with permission of the American Psychological Association.)

of participants in the studies: primary leaders, secondary leaders, and non-leaders. For further information about the substance of the leadership theory, see Guastello (2007a,b).

If one were to measure the group members' perceptions of each other's behavior as a leader at the beginning of the experimental session (assuming they have no past history with each other), the initial score for everyone on the leader perception measurement would be zero. Thus, the polynomial regression method would not be appropriate because of its reliance on non-zero variation in the time-1 measurement of the order parameter. The static model that is testable through nonlinear regression is very appropriate, however, because of its reliance on the pdf.

The regression equation for the swallowtail catastrophe model is produced by applying the transformation in Equation 2.8 to the deterministic model for the swallowtail response surface (Equation 12.7) and inserting regression weights:

$$pdf(z) = \xi \exp[\theta_1 z^5 + \theta_2 z^4 + \theta_3 c z^3 + \theta_4 b z^2 + \theta_5 a z] \tag{13.6}$$

In this model, the term $\theta_2 z^4$ is the expendable component. The control parameters a, b, and c can be set to constants equal to 1.0 if the objective is only to test the swallowtail distribution.

SPSS Commands

The first two steps in (S13.1) and (S13.2) remain the same here. Statements 13.3 through 13.5 are modified as follows:

```
MODEL PROGRAM x = 0.5, a = 0.5 b = 0.5 c = 0.5 d = 0.5   e = 0.5
                                                          (S13.6)
COMPUTE PRED = x*exp(a*(z**5) + b*(z**4) + c*(z**3) + d*(z**2) + e*z)
                                                          (S13.7)
NLR PDFZ with z                                           (S13.8)
```

Statement 13.7 is testing for the shape of the pdf only, and does not include the control variables. To test for the control variables, (S13.6) remains the same, but (S13.7) becomes more elaborate:

```
      COMPUTE PRED = x*exp(a*(z**5) + b*(z**4) + c*BIAS*(z**3) +
          d*BIFUR*(z**2) + e*ASYM*z)                      (S13.9)
```

Statement 13.8 is also amplified:

```
          NLR PDFZ with Z BIAS BIFUR ASYM                 (S13.10)
```

The last statement is automatically understood by the program, however, if the dialog box in the interactive version of the program is used.

Example: Leadership Emergence

The following is a synopsis of results that were obtained in a study that included versions of the swallowtail catastrophe model with and without hypothesized control variables (Guastello, 2010). The participants in the study were 228 undergraduates who were organized into groups of 4–12 participants; groups worked against an adversary in a board game simulation of an emergency response problem. After about 90 min of activity, the group members were asked to rate each other as to who acted most like the leader of the group and second-most like the leader of the group. Scores were summed for all participants and divided by the number of group members participating. Similar questions were posed regarding 16 other specific social behaviors. A factor analysis indicated that two factors represented the 16 items: competitive behaviors and group maintenance behaviors. It was also shown from a separate analysis that the larger groups outperformed the smaller groups against the adversary, so there was a simple relationship between those two variables.

Table 13.2 shows the result for the nonlinear regression analyses. The third column, no control variables, shows the connection between the data and the swallowtail structure, without the control parameters. The weight for z^5 is small because the z scores taken to the fifth power become rather large. Statistical significance $p < .05$ was registered because the standard error associated with that term was very small; R^2 was 0.998, which left little residual error variance.

For the analyses containing control parameters, competitive behavior was tested as the asymmetry parameter, group size as the bifurcation parameter, and there were two variables hypothesized as the bias or swallowtail parameter, group maintenance, and group performance which were tested separately. The hypotheses for the control parameters were based on relationship that had been gaining consistency in previous studies: Asymmetry would be a variable that contained a relatively large constellation of traits or trait-related behaviors. Bifurcation would be a variable specific to the type of task, or experimental manipulation thereof, that was involved and would have the

TABLE 13.2

Nonlinear Regression Results for Leadership Emergence in an Emergency Response Situation

Variable	c = Group Maintenance	c = Performance	No Control Variables
ξ	55.44*	55.96*	47.41*
z^5	0.05*	0.05*	0.04*
z^4	−0.32*	−0.35	−0.52*
cz^3	0.01	0.07*	2.58*
bz^2	0.19*	0.07	−5.82*
az	1.49*	1.52*	8.20*
R^2	0.728	0.747	0.998

Source: Reprinted from Guastello, 2010, p. 194. With permission of the Society for Chaos Theory in Psychology & Life Sciences.

* $p < .05$.

effect of bringing both primary and secondary leaders to the foreground. The bias or swallowtail factor would distinguish primary and secondary leaders. Group performance was such a candidate from a previous study showing the less-than-central role of performance in the emergence of leaders (Guastello & Bond, 2007). On the other hand, another line of reasoning supported group maintenance activities as characteristic of secondary leaders (Bales, 1999).

When group maintenance was tested as the bias factor, significant weights were obtained for the quintic (necessary) and quartic (optional) components, and for the bifurcation and asymmetry variables. The effect of group maintenance was not significant, however. When performance was tested as the bias factor instead, significant weights were obtained for the bias and asymmetry variables, but not for the bifurcation variable (group size). The interpretation was that all three variables behaved as hypothesized, but the results were less definitive because of the correlation between group size and group performance. Perhaps with a larger sample, the distinct effect of each variable would be more prominent.

It is also noteworthy that the R^2 coefficient for the model with the control variables dropped by about 20% of variance accounted for compared to the model without the control variables. This attenuation is the result of substituting a hypothetical variable for a free parameter. In the best of situations, the hypothetical variable would work about as well as the free parameter. Results such as these would inform the researcher that additional variables could be contributing to the latent control parameters, a point that should be investigated in future studies. Alternatively, if the task were more challenging or lasted a longer period of time, full ranges of the catastrophic response surface and the control variables would be in better alignment.

Comparison of Direct and Indirect Methods

Both the direct and indirect methods of testing cusps through nonlinear regression have some advantages over the method of dynamic difference equations. The nonlinear regression approaches can be used where the data set is static, or in the case of the driving simulation study, the before-and-after measurements (of driving speed) cannot be recovered. The nonlinear regression approaches also circumvent the problem of finding the optimum lag length for the measurements. The limitation, however, is that the static model is not as appealing intuitively for data that really do contain interesting changes over time; some of that information appears to be lost when the data are treated as static distribution.

Indirect testing might be an advantage if one is relatively confident of the cusp structure as a hypothesis, but not so sure about the control parameters per se. If variables are tested as potential control parameters, however, the number of regression weights in the final model is much greater than what is needed for the direct method. The generalizability of a model becomes less viable as the number of weights becomes larger. If the research problem involves many variables, such as many items from a survey, a more efficient strategy would be to factor analyze the measurements first, then test the factors as variables in the cusp model. The GEMCAT method, which is based on principles of least squares and described in Chapter 14, is another viable alternative that provides a direct model that is optimized from a large number of research variables.

Similarly, the indirect methods calculate location and scale as empirical weights, which, on the one hand, could provide more precise values than what is produced with the direct method or the dynamic difference equations. On the other hand, the method adds more parameters to the model once again.

Cobb and Zacks (1988) did advance the idea that the indirect method for cusp could generalize to chaotic data that had cubic structure. The connection is reasonable, perhaps, particularly if it is unclear whether the instability in a time series is ambiguous as to whether it is chaotic or the result of fluctuations around a cusp manifold. Given the large number of chaotic structures that are known to exist, however, the modeling approach in Chapter 13 is more versatile. In principle, the indirect methods should generalize to other catastrophe models, but programs for doing so have not surfaced, nor have any applications. One could anticipate, however, that the phase of the procedure that Van der Maas et al. (2003) used to test all possible cusp models would explode in complexity if three or four control parameters were involved. The direct method has been useful for assessing swallowtail catastrophe models.

Notes

1. The set of values shown in this particular command statement originated with a problem that was actually a swallowtail catastrophe (Zaror & Guastello, 2000). The left-hand sides of the cusp Equation 13.1 and other catastrophes are all the same.
2. See the original text for discussion of issues related to the measurement of the variables used in the study.
3. See Guastello, Aruka, Doyle, and Smerz (2008) for further developments within this approach to the binge drinking problem.

References

Bales, R. F. (1999). *Social interaction systems: Theory and measurement.* New Brunswick, NJ: Transaction Publishing.

Barunik, J. & Vosvrda, M. (2009). Can a stochastic cusp catastrophe model explain stock market crashes? *Journal of Economic Dynamics & Control, 33,* 1824–1836.

Clair, S. (1998). A cusp catastrophe model for adolescent alcohol use: An empirical test. *Nonlinear Dynamics, Psychology, and Life Sciences, 2*, 217–241.

Cobb, L. (1978). Stochastic catastrophe models and multimodal distributions. *Behavioral Science, 23*, 360–374.

Cobb, L. (1981a). Parameter estimation for the cusp catastrophe model. *Behavioral Science, 26*, 75–78.

Cobb, L. (1981b). Multimodal exponential families of statistical catastrophe theory. In C. Taillie, G. P. Patil, & B. Baldessari (Eds.), *Statistical distributions in scientific work* (Vol. 4, pp. 67–90). Hingam, MA: Reidel.

Cobb, L., Koppstein, P., & Chen, N. H. (1983). Estimation and moment recursion relationships for multimodal distributions of the exponential family. *Journal of the American Statistical Association, 78*, 124–130.

Cobb, L. & Zacks, (1985). Applications of statistical catastrophe theory for statistical modeling in the biosciences. *Journal of the American Statistical Association, 78*, 124–130.

Cobb, L. & Zacks, (1988). Nonlinear time series analysis for dynamic systems of catastrophe type. In R. R. Mohler (Ed.), *Nonlinear time series and signal processing* (pp. 97–118). North Holland: Springer-Verlag.

Fischer, E. O. & Jammernegg, W. (1986). Empirical investigation of a catastrophe theory extension of the Phillips Curve. *Review of Economics and Statistics, 68*, 9–17.

Guastello, S. J. (2002). *Managing emergent phenomena: Nonlinear dynamics in work organizations*. Mahwah, NJ: Lawrence Erlbaum Associates.

Guastello, S. J. (1998). Self-organization in leadership emergence. *Nonlinear Dynamics, Psychology, and Life Sciences, 2*, 303–316.

Guastello, S. J. (2007a). Nonlinear dynamics and leadership emergence. *Leadership Quarterly, 18*, 357–369.

Guastello, S. J. (2007b). How leaders really emerge. *American Psychologist, 62*, 606–607.

Guastello, S. J. (2010). Self-organization and leadership emergence in emergency response teams. *Nonlinear Dynamics, Psychology, and Life Sciences, 14*, 179–206.

Guastello, S. J., Aruka, Y., Doyle, M., & Smerz, K. (2008). Cross-cultural generalizability of a cusp catastrophe model for binge drinking among college students. *Nonlinear Dynamics, Psychology, and Life Sciences, 12*, 397–407.

Guastello, S. J. & Bond, R. W. Jr. (2007). A swallowtail catastrophe model of leadership in coordination-intensive groups. *Nonlinear Dynamics, Psychology, and Life Sciences, 11*, 235–351.

Hanges, P. J., Braverman, E. P., & Rentsch, J. R. (1991). Changes in raters' perception of subordinates: A catastrophe model. *Journal of Applied Psychology, 76*, 878–888.

Smerz, K. E. & Guastello, S. J. (2008). Cusp catastrophe model for binge drinking in a college population. *Nonlinear Dynamics, Psychology, and Life Sciences, 12*, 205–224.

Van der Maas, H. L. J., Kolstein, R., & van der Plight, J. (2003). Sudden transitions in attitude. *Sociological Methods and Research, 32*, 125–152.

Zaror, G. & Guastello, S. J. (2000). Self-organization and leadership emergence: A cross-cultural replication. *Nonlinear Dynamics, Psychology, and Life Sciences, 4*, 113–120.

Zeeman, E. C. (1976). A mathematical model for conflicting judgments caused by stress applied to possible misestimation of speed caused by alcohol. *British Journal of Mathematical and Statistical Psychology, 29*, 19–32.

14

Catastrophe Model for the Prospect-Utility Theory Question[*]

Terence A. Oliva and Sean R. McDade

CONTENTS

Anomalies have played a big part in the analysis of decision making under risk. Both expected utility and prospect theories were born out of anomalies exhibited by actual decision-making behavior. Since the same individual can use both expected utility and prospect approaches at different times, it seems there should be a means of uniting the two. This chapter focuses on nonlinear dynamical systems (NDS), specifically a catastrophe model, to help suggest an "out-of-the-box" line of solution toward integration. We use a cusp model to create a value surface whose control dimensions are involvement and gains versus losses. By including "involvement" as a variable, the importance of the individual's psychological state is included, and it provides a rationale for how decision makers' changes from expected utility

---

[*] The original article was edited for format and first published as Oliva, T. A. and McDade, S. (2008). Catastrophe model for the prospect-utility theory question. *Nonlinear Dynamics, Psychology, and Life Sciences*, 12, 261–280. (Reprinted with permission of the Society for Chaos Theory in Psychology & Life Sciences.)

to prospect might occur. Additionally, it provides a possible explanation for what appear to be even more irrational decisions that individuals make when highly emotionally involved. We estimate the catastrophe model using a sample of 997 gamblers who attended a casino and compare it to the linear model using regression. Hence, we have actual data from individuals making real bets, under real conditions.

Ever since Kahneman and Tversky (1979) developed prospect theory (PT) to help address anomalies that occurred when individuals faced decision making under risk that were not captured by expected utility theory (EUT; von Neumann & Morgenstern, 1947), it has stimulated what amounts to a cottage industry in papers, examining the various aspects of PT and EUT (Arrow, 1971; Fischoff, 1983; Kahneman & Tversky, 1979; Tversky & Kahaneman, 1992; Wakker & Tversky, 1993). The observation that PT was descriptively more "accurate" at characterizing decisions than EUT was a significant problem (e.g., Bleichrodt, Pinto, & Wakker, 2001; Currim & Sarin, 1986), since PT deals more with how the world works, "what is," rather than how the world should work, "what should be." From this perspective, PT has been a significant force in bringing psychology into economics and helping ground it in reality.

Yet for all the papers being published on PT, EUT, and PT versus EUT, there seems to be no real progress at trying to move toward an overarching conceptualization or theory. Some researchers have tried to deal with the issues (Bleichrodt et al., 2001; Currim & Sarin, 1989). Weights have been added to bring PT curves in closer line with EUT prescriptions called cumulative PT (CPT; Bleichrodt et al., 2001; Chateauneuf & Wakker, 1999; Tversky & Kahneman, 1992; Wakker & Tversky, 1993). More creative and detailed mathematical derivations have been done along with more cleverly designed experiments and applications in different contexts of risk-taking activity (Chateauneuf & Wakker, 1999; Gonzalez & Wu, 2003; Schmidt, Starmer, & Sugden, 2005). However, if the same individual can exhibit EUT-type decision behavior in one situation, and PT-type in another, it would seem there needs to be some explanation regarding how this happens. An integrative theory would be useful to account for how these differences in decision behavior occur.

This chapter examines the use of a cusp catastrophe model to suggest a possible means of integrating EUT and PT in a broad sense. As such it is not fully developed or complete, but is intended to generate "out-of-the-box" thinking on the relationship between PT and EUT. Also, it is not the $n + 1$ mathematical derivation of the next twist or tweak to these theories. While very carefully developed experiments and mathematical derivations have provided many improvements and insights to examining PT and EUT, they can also "lock in" one's thinking about the issue. Our approach is to present a descriptive model from the NDS literature that tries to capture a global look on what is going on. Additionally we will estimate the proposed model using real consumer in a real gambling context. This is consistent with the notion that PT "... as a descriptive model cannot be established solely by its more general mathematical structure or by demonstrating that in principle it can explain preference patterns that the utility model cannot" (Currim & Sarin, 1989, p. 23). Rather it is demonstrated by empirical research. Our use of consumers in an actual and "normal," not staged, environment is in contrast to a lot of PT and UT research where the gambles are carefully articulated and controlled to get at the specific behavior. However, actual behavior is often messy and does not always conform to what we expect. Therefore, the probability of getting the clean results found in typical utility experiments is less likely.

Expected Utility and Prospect Theory

EUT, in part, arose out of a solution to the *St. Petersburg Paradox* by Bernoulli (1954), which was about a bet with an infinite payoff. von Neumann and Morgenstern developed a more formal approach in their expected utility theorem presented in their landmark work, the *Theory of Games and Economic Behavior* (1947). In contrast to PT, in EUT individuals make rational choices for gambles based on the probabilities by estimating the expected value of the gamble. Therefore, each person has a utility function $u(x_i)$ such that utility values are assigned to all possible outcomes x_i for $i = 1, 2, ..., j$. Next, the sum of all the outcome utilities are multiplied by their respective probabilities p_i for $i = 1, 2, ..., j$, to determine the expected utility. Rational people are assumed to choose options with the highest expected utility. Also we assume that most individuals are risk-averse most of the time (Byrne, 1996). Gains and losses from this perspective are treated in the same way for EUT, as shown in Figure 14.1. For example, if an individual can play a game where the payoff for drawing the Ace of Spades from a single deck of cards is $100.00 for one pick, one time, and it costs $2.00 to play, is it a good gamble (i.e., should the person take it)? In this case it is no, because the expectation (1/52 × $100 = $1.92) is less than the cost to play $2.00. In short, it would not be considered a "fair" bet or gamble. However, if you increase the payoff to $1000.00, the evaluation process stays the same and it becomes a good gamble because the expected value is now $19.23, well above the $2.00 to play. If the stakes and payoffs in the original game were multiplied by 10,000, would it change anything? Since the evaluation process would be the same, the answer is no. In this context Figure 14.1 demonstrates a utility function for a risk-averse individual and there is a decreasing marginal utility. Curves for risk-neutral and risk-taking individuals are not shown in Figure 14.1, but would be a straight line and convex curve through the origin, respectively. The important point is that it would be the same rational evaluation by the individual for gains and losses. From this perspective we could say the curves would be well behaved for the risk-averse, risk-seeker, and risk-neutral individual.

Like EUT, PT was also born out of paradox. In particular, the Allais paradox and other studies produced results inconsistent with the EUT (Allais, 1953;

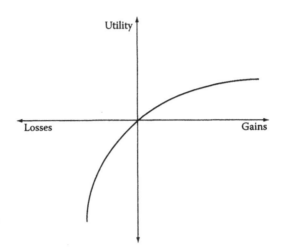

FIGURE 14.1
Expected utility theory: risk-averse individual.

Bleichrodt & Pinto, 2000; Camerer & Ho, 1994; Wu & Gonzalez, 1996). Allais argued that it is not possible to evaluate choices of a gamble independently; it depended on what the other choices were available (Allais, 1953). Kahneman and Tversky (1979) noticed that there were numerous contradictions and anomalies in how people managed risk depending on the situation, even when expected outcomes might be the same. In particular, they noticed that attitudes toward risk related to gains and losses were not the same. Consider the experiment by Thaler and Johnson (1990) where students were given the following two situations and choices:
You have just won $30.
Now choose between

a. No further gain or loss
b. A gamble in which you have a 50% chance to win $9 and a 50% chance to lose $9

or,
Choose between

a. A sure gain of $30
b. A 50% chance to win $39 and a 50% chance to win $21

In the first condition, approximately 70% of the students chose (b) (the gamble), and in the second condition only about 40% chose (b) (the gamble), even though the outcomes are the same. These sorts of results over numerous experiments have solidified the idea that gains and losses are not viewed in the same way. It is how decisions are framed that matters and generated what might be called

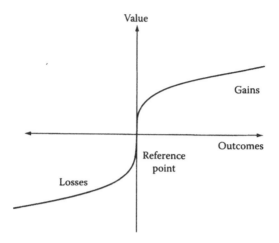

FIGURE 14.2
Prospect theory model.

non-expected utility theories. It turns out that in many situations people adjust both outcomes x_i and probabilities p_i that violate EUT axioms.

This leads to Figure 14.2, which presents the basic PT curve and illustrates the idea of how attitudes to risks concerning gains are different relative to attitudes to risks concerning losses. The figure illustrates how gains and losses are evaluated differently. The value function is S-shaped, which indicates that given the same value, there is a bigger impact of losses than of gains. The differences come from how choices are framed in this and the Thaler and Johnson (1990) experiment. Clearly, this has economic impact for things like discounts from sticker prices, and the like, which basically put the theory into practice. The key here is that the individual subjectively frames a transaction or outcome in his or her mind, which affects the utility that is expected.

The decision behavior in PT is not that of a "rational" person according to standard utility theory. On the other hand, it makes perfect sense to the individual making the choice; hence it is "rational" in a relative sense (Bernstein, 1998). But as originally conceived, PT also has problems. Under some conditions it is possible for a sure thing that was worse than other options to be chosen. A solution has been to use a weighting function to generate CPT (Gonzalez & Wu, 2003; Tversky & Kahnemann, 1992; Wakker & Tversky, 1993).

The argument is that people tend to think about various outcomes relative to a reference point rather than the final state. This framing effect creates problem that tend to interfere with what would be considered rational choice. A solution is found by introducing a weighting system. Figure 14.3, taken from Rieger and Wang (2006), shows an example of how the PT curves change under CPT weights. Rao, Zhang, and Wang (2006) use the concept of maximum loss value and mental accounting to modify the PT value function shown in Figure 14.4. The above additions and modifications as well as those

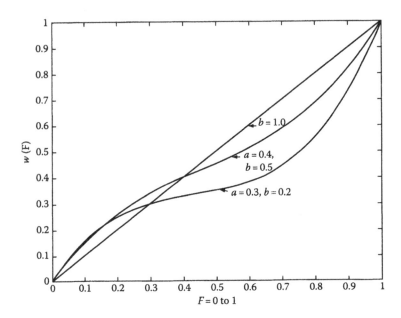

FIGURE 14.3
Cumulative prospect theory weighting. (Based on Rieger, M.O. and Wang, M., *Econ. Theory*, 28, 665, 2006.)

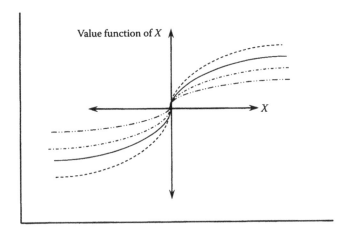

FIGURE 14.4
Value function changes with mental accounting. (Based on Rao, Y. et al., Modification of value function by incorporating mental account, in *European Financial Management Symposium*, Durham Business School, April 20–22, 2006, http://www.efmaefm.org/0EFMSYMPOSIUM/durham-2006/sympopart.shtml)

developed by others have helped keep PT a viable approach to behavioral economics.

In the end, what this comes down to is that people make decisions based in part on their personalities, attitudes toward risk, emotional state, and the total set of choices they have. And while they may be "rational" in a EUT sense, we know that there are situations where people are not consistently one way or the other. When we think about it in terms of acquaintances and others we know or have heard of, we know that people often do not exhibit a fixed risk profile. It may be true that most individuals are risk averse most of the time, yet in certain situations they become risk seekers as when a person tries any high-risk, high-cost medical procedure to save a family member. Or, similarly when rabid sports fans make what most would say are terrible bets in support of a team that virtually cannot win. The issue is to try and determine what is driving the change? Why does a careful investor in the stock market go to the race track and bet exclusively on long shots? It suggests that people operate sometimes, as predicted by EUT, and at other times according to PT. This, in turn, suggests the following questions: "Are these two theories somehow related?" Or "Can they be related in some way?" "Can we develop a more unified model, or approach, or at least suggest a new direction?"

Catastrophe Theory as a Potential Integrator of PT, CPT, and EUT

A Cusp Model

Most of the enhancements to PT have come in the form of mathematical extensions to deal with anomalies, inconsistencies, or violations discovered by researchers, even though it is considered a descriptive theory. And while the elegant analyses have improved the theories, they tend to keep the thought processes "within the box." However, moving beyond "standard" approaches is difficult in most disciplines, and perhaps more so for economics (Dore & Rosser, 2007). They have labeled what they call a Kuhnian crisis in economics and point out the need for a paradigm shift. Similarly, Rosser (2007) examines the use of catastrophe theory in economics and argues that the criticisms of the 1970s and 1980s were overdone. In particular, the Zeeman (1974) model of the stock market was heavily criticized. In contrast to the criticisms, Guastello (1993, 1995, 2002) properly tested the Zeeman model. He points out that price changes are gradual, showing little bifurcation behavior, for "value-based" trading, which is predicated on the company's profits and positions to make future profits soon. There is always some risk involved with those forecasts, but not as much as in speculative trading. He argues that speculative trading and the rationality or irrationality of other traders leads to the bifurcated region of the cusp surface. For him it is not a matter of strict rationality versus irrationality, but a continuum with regard to how much irrationality is in play at a given time. These notions tie into our approach, in that this suggests that emotions are tied to irrational decision behavior at times and affect decision-making approaches within the individual. Taken together, the foregoing papers provide support and groundwork for the logic of this kind of application.

Figure 14.4 suggests a different way to look at the weights in CPT. However, if we move beyond two dimensions to three dimensions, where the weights occupy a third dimension (say *Y*, in an *X*, *Y*, *Z* system), it generates a three-dimensional value surface as opposed to the superimposed curves representing different value curves. The weights then would vary with the decision-maker type or situation, and we could compare differences on the evolution of the value surface. Similarly, Figure 14.4 shows changes in the shape of the value function. Again, if this were translated to a value surface, it would give us a richer indication of how things changed as the decision maker changes. It is the gain from a three-dimensional characterization that provides impetus to our approach to use a cusp model to look at PT and EUT.

This creates a more complete value surface by mapping EUT, PT, and CPT onto the catastrophe structure in Figure 14.5. By doing so, we also add what we call emotional-based prospect theory (EBPT) to complete the description. This may be related to what some researchers are calling PT³ (Schmidt et al., 2005). At the back of the figure, but stopping at the singularity, the surface captures EUT, PT, and CPT in that order. The curves at the back are similar to Figure 14.1, but morph into S-shaped curves as one moves out. Changes in the S-shapes follow that of CPT shown in Figures 14.3 and 14.4 depending on the surface. As a starting point for thinking about this issue, we argue that what drives people to change behavior with regard to how they make risky decisions is due in no small part to their emotional involvement or state at the

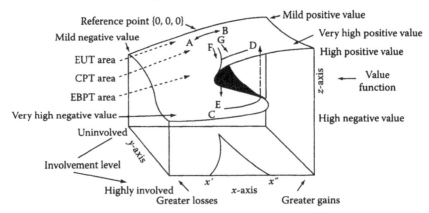

FIGURE 14.5
Catastrophe model extension of the value surface.

time. When people are relatively uninvolved, they are assumed to be more rational and careful. They will tend to think things through. On the other hand, when they are heavily (and likely emotionally) involved, they might make what an outsider would consider irrational decisions. Therefore, as the state of involvement changes from low to high, we assume the decision making moves from being driven by EUT to that by CPT. Keep in mind that part of changing involvement is related to the decision context and that is how PT experiments work. Framing usually effects involvement in some way or decision making would not change. Furthermore, given that this is an initial provisional approach we assume that the nature of the involvement can be driven by any number of things. For example, it can be the nature of the choice set, or something the individual is very interested in, versus something of no interest. In a more developed approach we would assume that emotional involvement is really a latent variable that combines a number of characteristics that would drive the behavior of the individual. For example, it might contain some form of the weighting system like that used in CPT and shown in Figure 14.4.

Past the singularity shifts in value become dichotomous and the middle and neutral values are not taken. It is much akin to jailed career criminals who suddenly find religion. Often they do not slowly change their values; they get the equivalent sensation of St. Paul being knocked to the ground and change instantly. As the surface starts to overlap along the cusp boundary, gains and losses are not evaluated "accurately." Losses may be still seen as gains and gains as losses. A sports example of this is the so-called moral victory. They should have beaten us by 40 points, but they only beat us by 5, so it's really a win. Or, people who have an "irrational" fear and evaluate a gain as a loss. Or the accident-prone person who believes that if something bad can happen, it will happen to him or her. Similarly, a sports fan believes his team can never beat another team ever regardless of the current situation where his or her team is heavily favored. Hence, Figure 14.5 integrates EUT, PT, CPT, and EBPT in a single response surface where degree of involvement is the splitting factor. In the next section, we briefly look at the model dynamics.

Model Dynamics: Low Involvement

For decisions where there is no or very low involvement, we would assume the decision maker acts more rationally, trying to figure out odds and the like. This seems to be what happens if you pose a problem in an area that an individual is not familiar with and give him or her just the basic facts. For example, you pose an outer space-related decision on choices between establishing a base on the moon or on Mars. This is in contrast to what you might get for someone asked about risk decisions related to the Iraq War, where emotions are running high on all sides.

As drawn, the very back of the surface represents an uninvolved risk-averse person. It could have been for a risk-seeker or a risk-neutral person as well, but we are assuming most start out risk-averse (Byrne, 1996). Note that gains and losses in this range are treated in the same way, and that the changes in the utility or value levels in this range are proportional to the variations (changes) in the gains and losses. This is shown by line segment AB in Figure 14.5, which shows the transition of a risk-averse individual between mild negative and mild positive values because the individual is relatively uninvolved.

For very small changes in involvement (say near or behind the AB line in the figure), we expect small changes in the shape of the value function, but they would be consistent with the risk profile of the individual and are not expected to deviate much from the basic shape. Therefore, small changes in involvement will result in small changes in how the decisions are evaluated but the decision maker will be basically using an EUT approach.

Model Dynamics: Moderate Involvement

As the level of involvement continues to increase, the decision maker moves farther out on the surface and the shape of the surface starts to change (deform) toward S-shaped curves similar to Figure 14.3 or 14.4 (e.g., at some point past the AB line segment but before the singularity). This reflects how the increasing effects of involvement affect values. Exact shapes will depend on the individual. At this point the decisions are starting to become more personal for some reason. It is possible that in the transition area the decision maker would use some sort of mix of EUT and CPT approaches, or may simply migrate to CPT. This type of change would also be true for risk neutrals and risk seekers. The point at which this happens depends on the individual decision maker. As involvement increases, the S-shaped curves become more accentuated as they are in Figures 14.3 and 14.4. In this area, changes in the value surface are not proportional to changes in gains and losses and reflect the risk-seeking and risk-aversion patterns found in PT and CPT.

Model Dynamics: High Involvement

If involvement grows to the point where the individual is highly emotionally invested in the decision, then the value surface becomes dual-valued. An example would be making a bet with a neighbor on a sporting event which involves a favorite and disliked team (a major rivalry). This occurs when the emotional level reaches and passes the singularity (cusp) and the value surface bifurcates into a high negative value or high positive values. There are no in-between levels, as there is a gap that is opening up around what might be called neutral or zero value. Depending on the involvement, this gap grows as involvement increases and it becomes a two-state world of high negative value or high positive value with no in-between level.

We call this area EBPT because the behavior here may seem irrational to an outsider and go beyond normal CPT. For example, a diehard (irrational) fan may make highly emotional bets because of being "one of the faithful" for given team, even when the experts say the team will lose and has no chance.

Often, such fans will take a straight bet or even give odds on his or her team winning. That is, the location on the gain loss axis is well into the loss area. But the individual is evaluating the bet as having a high value, when a "more rational" person would evaluate it as having a high negative value. Yet we see this behavior fairly regularly and characterize it as irrational or crazy fan behavior. On the other side, there are those who may have a similar "irrational" fear of a good bet. Hence, they evaluate a gain situation as highly negative when it should be considered highly positive. This is reflected by the overlap in the surface. For our decision makers, when emotions run high and start to take over the decision process, they tend to end up in a two-state world where there are only high and negative value outcomes. Note that initially, at the singularity, the gap in value is small. One might not even notice or pay attention to it very much. But as emotions start taking over with increased involvement, the gap widens and the overlap and apparent irrationality becomes greater.

Empirical Estimation of the Model

Whereas Figure 14.5 is intended as an initial descriptive approach, we still want to determine if this conceptual approach has any sort of empirical validity. Unlike much of the experimentation in PT which is piecemeal in nature, our approach will be to attempt to fit the entire model across EUT, PT, CPT, and EBPT.

The Data and Sample

We were generously supplied a random subsample of 1000 (997 useable) observations from consumers who attended a racing and slots gambling establishment by PeopleMetics™. As such, we had no expectations that we would get as clean a result as many of the controlled EUT and PT studies do. On the other hand, we have real consumers making real decision and losing real money.

As is often the case when data are provided by a market research business, we had no input into the design of the study or the questions that were administered electronically. The questionnaire had 45 questions and was expected to take customers no more than 10 min to complete. Of the 45 questions, 6 were relevant for our study. Most of these were 5-point Likert-type scales, except for the question regarding how much did they spend out of pocket, which was a 7-point scale. There was an open-ended question that was useful for helping to confirm both their level of involvement and intrinsic value of the trip. We believe this helps validate our evaluations, even though we have less-than-perfect data. Customers attending the facility were gamblers but were not "high-stakes" gamblers. Exact demographic numbers are not known because of the way the data were collected (in categories with open-ended ranges for the last category). However, reasonably close estimates were made by using the center points of the ranges times their frequencies. Our sample was about 67% women and 33% men, with 11% not specifying their gender; the average number of visits per subject was around nine times; 86% said the primary reason for their attendance was to play the slot, while 3% said it was to bet races; the average out of pocket spent (money from the customer not "comps," spent on drinks, entertainment, etc.) on gambling was near $237 and the average age was about 51 years. It should be noted that for the most part, the 11% who did not provide an answer for gender also did not mention the amount they spent, or their ages.

Operationalization of the Variables

Components of the X-Variable: Gains and Losses

1. Considering everything, how would you rate your overall experience today? (x_1) 4—excellent, 3—good, 2—average, 1—poor, 0—very poor.
2. My experience today at this facility... (x_2). 4—definitely exceeded my expectations, 3—exceeded my expectations, 2—met my expectations, 1—did not meet my expectations, 0—definitely did not meet my expectations.
3. How much do you typically spend "out-of-pocket" on slots and, or racing each time you visit our facility? (x_3) Less than $50, $50–$99, $100–$149, $150–$199, $200–$249, $250–$299, $300 or more.

The answers to the foregoing questions were collapsed into an index as follows. Questions 1 and 2 were averaged in a repeat measures sense. The logic of these is that experience will be a proxy for whether or not they have had gains or losses. It also helps with the problem of using 5-point scales by generating some midpoints and increasing the variation. Once the averages were obtained, they were then multiplied by the amount of out-of-pocket gambling costs, which were reverse-scaled. Losers were expected to have more out-of-pocket costs while winners were expected to have fewer. In the situation where a consumer rated the experience high and yet spent a lot out of pocket, he or she would fall in the middle of the range. This seems like a reasonable net rating for someone who lost but had fun. As we indicated above, we were able to cross-check each individual on an open-ended question. A sample of the open-ended responses that indicates we are tapping into gains and losses is presented in Table 14.1. While we have close to 950 comments, we only chose a few. These represent the extremes of some of those who had an excellent experience and some of those who did not. As evident by the comments, winning or losing seemed to matter. We did not show any of those in the middle however; most of the people with more in-between experiences or evaluations, tended to focus on things like the service, food, and the like rather than winning or losing. Hence, we feel this is a reasonable approach, given the data taps into the real monetary gains and losses, as modified by their perceptions.

TABLE 14.1

Sample of Comments from Patrons

Losers whose experience was bad

1. I used to be able to have a decent win occasionally, but even though I continue to spend the same large amount of money each month, I haven't had a night that I left with a win since August of 2006.
2. After 1500 bucks played never hit more than 100, that sucks.
3. It's very, very discouraging and I'm getting tired of giving and not getting back.
4. I could not win any money and it was my birthday.
5. No wins over $800.00 loss not one win very bad.
6. I played an average of about 25 different machines, about $400.00, and barely win with a cherry occasionally.
7. When I do come with $500.00, I go home with zero dollars. You know how I feel, (sad).

Winners whose experience was good

1. I hit the jackpot.
2. Because I finally won $900.00.
3. I had fun and the poker machines paid off.
4. Had a good time will return again soon.
5. For once I came home with extra money to put back in my bank account!
6. I won.
7. Won a little money and got the money quick with the ticket machine.

Component of the Y-Variable: Involvement

4. How many times have you visited our facilities in the past 12 months? (y_1) First visit, 1–5, 6–12, 13–23, 24, or more.

This is a direct measure of involvement. The more times the people have visited the facility, the more involved they are. People who have been to the facility 20 or more times clearly have a relationship of some sort with the facility. On the other hand, those who are on their first visit may simply be there because of a promotion or hearing about it from a friend. For them, it is an open question as to how they feel.

Components of the Z-Variable: Value

5. If one of your friends or family wanted a recommendation of a fun way to spend some time, how likely would you be to mention this establishment? (z_1) 4—very likely, 3—likely, 2—neither likely nor unlikely, 1—unlikely, 0—very unlikely.
6. How likely is it that you will return to this establishment in the next 6 months? (z_2) 4—very likely, 3—likely, 2—neither likely nor unlikely, 1—unlikely, 0—very unlikely.

Recommendations and behavior are the reasonably good indicators of how people value the facility. If customers recommend it to others and are returning within a short period, we can assume there is an underlying value to the experience. Similarly, if they do not recommend it and are not returning shortly, we can assume it has a negative value. Some of the comments in Table 14.1 would support this. As with questions 1 and 2 above, we averaged the answers to the questions to increase the variance and to get more solid accurate measure of value. Clearly, a customer picking a 5 for each answer has a different value than someone who picks a 5 and a 4. We understand there may be external reasons that affect the answers; however, given the large sample size and the number of regular customers, we think this is a reasonable measure for a preliminary approach.

We acknowledge that the above measures are not perfect. However, they have the advantage of not being from carefully controlled experiments that are difficult to make realistic. These respondents are spending their own money and time at an activity of their own choosing in a natural setting. Furthermore, with 997 respondents, the sample size is sufficient from an analysis standpoint to suggest any results that might be gotten are meaningful.

Estimation Issues

While there are a number of ways to estimate cusp catastrophe models, generally speaking there are two approaches that are more commonly found in the literature. The first is by Guastello (1982, 1995), and the second by Oliva et al. (1987). Only a brief overview of catastrophe model estimation issues is presented as these have been discussed elsewhere in the literature (e.g., Cobb, 1978, 1981; Guastello, 1982, 1992, 1995; Lange, McDade, & Oliva, 2004; Oliva, Oliver, & MacMillan, 1992; Oliva, 1991). When set to zero, Equation 14.4 is parsimonious in terms of its ability to capture complex behavior of the cusp model relative to the number of variables used. Unfortunately, estimating such surfaces is difficult because they are nonlinear, bimodal, and set of infinite tangents along the cusp. The dependent variable is inextricably intertwined with one of the independent variables, which is great for description but bad from an estimation standpoint.

By extending Cobb's (1978, 1981) work, Guastello (1982, 1992, 1995) developed a statistical approach for estimating the cusp model in Equation 14.3 by setting it equal to $dz = 0$, then inserting beta weights and setting dt equal to 1. The result is $\delta Z = Z_2 - Z_1 = \beta_0 + \beta_1(Z_1)^3 + \beta_2 Z_1 Y + \beta_3 X + \varepsilon$.

In contrast, Oliva, DeSarbo, Day, & Jedidi (1987) develop their approach by treating each variable as a latent construct made up of a set of indicator variables as follows

$$X = \sum_{i=1}^{I} \theta_{xi} x_i \tag{14.1}$$

$$Y = \sum_{j=1}^{J} \theta_{yj} y_j \tag{14.2}$$

$$Z = \sum_{k=1}^{K} \theta_{zk} z_k \tag{14.3}$$

where
  X, Y, and Z are the values of latent variables
  θ_{xi}, θ_{yj}, and θ_{zk} are the weights (impact coefficients) of the indicators x_i, y_j, and z_k of the cusp surface shown in Figure 14.5

Substitution of Equations 14.1 through 14.3 into Equation 14.4 yields the more general form

$$dZ = Z^3 - X - ZY \tag{14.4}$$

The approach used to fit Equation 14.4 is essentially the same as that found in other commonly used multivariate techniques. GEMCAT estimates a cusp model by finding those values for the weights θ_{xi}, θ_{yj}, and θ_{zk} which across all observations $t = 1, ..., T$ minimize the objective function Φ relative to Equation 14.5; that is

$$\text{Min}_{\theta_{xi}, \theta_{yj}, \theta_{zk}} = \Phi = \left\| e_t^2 \right\| = \sum_{t=1}^{T} (Z_t^3 - X_t - Z_t Y_t)^2 \tag{14.5}$$

where e_t denotes the error for case t. A major issue in solving Equation 14.9 is to find an optimization method that could deal with the peculiarities of the cusp model, while avoiding trivial solutions like $\theta_{xi} = \theta_{yj} = \theta_{zk} = 0$. Oliva et al. (1987) solved this problem through the use of a modified version of the controlled random search (CRS) proposed by Price (1979). This algorithm does not require the objective function Φ to be differentiable, nor does it require the indicator variables to be continuous. Note that the parameters are not identifiable as a whole since Φ is unchanged if weights $\theta'_{xi} = -\theta_{xi}$ and $\theta'_{yj} = -\theta_{yj}$ are used. However, setting $\theta'_{y1} = 1$ helps avoid this and other degeneracies (Oliva et al., 1987). The program was redesigned as GEMCAT II (Lange et al., 2000) to make it more user friendly, and it now includes jack-knifing and bootstrapping options to provide information on impact coefficient stability, allows more flexibility on the choice which impact coefficient to hold constant and its magnitude, and generates a pseudo R^2 for the model. Technical details on the upgrade and the algorithm are found in Lange et al. (2000), which details its development and tests on both simulated catastrophe and non-catastrophe data and a small application of software adoptions by organizations.

Results

Initially all 977 observations were standardized to provide zero means. This was followed by a simple correlation analysis of the variables X, Y, and Z shown in Table 14.2. There is virtually no multicollinearity among the variables. The little that exists occurs between X and Z, which we would expect since X is the control variable in the model and tends to drive Z.

Next, regressions were run on the data. The logic for this is simple, with 997 observations. If regression produces a good result then there is no point in looking at NDS and a catastrophe model, since a linear model works well. Hence, the regression model provides a "straw man" of sorts. After the regression was run, we used GEMCAT II to estimate the catastrophe model. Results for both are shown in Table 14.3. The regression produced an adjusted R^2 of 0.169. Additionally, the regression coefficient for Y is not stable. A low R^2 and unstable coefficient is strong indication that the linear model is not appropriate, particularly given there were 997 observations. By contrast, the catastrophe model estimation produced a pseudo R^2 of 0.8504

TABLE 14.2

Pearson Correlation Matrix
of Observations

	X	Y	Z
X	1.000		
Y	−0.016	1.000	
Z	0.412	0.004	1.000

TABLE 14.3

Catastrophe Estimation and Linear Regression Results

Catastrophe Model: GEMCAT II Results			
Indictor Measures	**X**	**Y**	**Z**
Impact coefficient	0.024*	0.028*	0.220*
Pseudo R^2	0.850		

Linear Regression Results, Dependent = Z	
Variables	**Std. Coefficient**
Constant	0.000
X	0.413*
Y	0.010
Adjusted squared multiple, R	0.169
Standard error of estimate	0.912

* $p < .0001$.

(with $df1 = 1$, $df2 = 995$, F-value 5655.063) and the impact coefficients (similar to regression coefficients) were stable, as shown in Table 14.3. The stability was determined by jackknifing. However, we note that the Y impact coefficient is reported to be a bit less stable when a number of different bootstrapping runs were used. In the end, the results are much stronger than for regression.

The impact coefficients indicate that the latent variables are approximately equal in determining the surface. Interestingly, the impact coefficient for Y (involvement) has slightly more impact than the coefficient for X on their respected latent variables (0.028–0.024). Since these variables are orthogonal, we feel that the original specification and introduction of an involvement variable is justified. Given these results, we also feel that there is more than enough evidence to suggest this approach warrants a closer examination.

Discussion

Our initial intent has been to see if NDS can provide any insight into helping resolve issues surrounding anomalies that occur when individuals make decision under risk. In particular, we wanted to look at the relationship between EUT and PT. It is our contention that if the same individual uses EUT under some conditions and different flavors of PT under others, understanding what creates the change from decision to decision would be important. Clearly, something goes on within the individual. The approach has been to try and step "out of the box" by seeing if an NDS technique can help shed light on the issue. Our rationale has been simple, in that when we look at decision making where there are gains and losses, we see that individuals are not always "rational." Over an entire spectrum of levels of rationality to irrationality, we believe that it is the level of involvement and its attendant emotions that drives people to use different evaluative approaches. Clearly, rabid fans that make totally emotional and often foolish bets on sports events are clearly outside what most would call the rational sphere. Our means of incorporating involvement into a model is to use catastrophe theory to develop a value surface. Catastrophe theory in the past has been characterized as a qualitative approach. It is in this context that we suggest it as a potential way to stimulate thinking about the decision-making under risk issue. Also, it is consistent with how PT is characterized in contrast to EUT, namely it is descriptive.

In the process, we suggest that there is another kind of decision making that we call EBPT that accounts for what appear to be the irrational decisions made when involvement runs very high. The model we propose is not left as a conjecture, but estimated on a large sample of actual consumers and a gambling establishment who have their money on the line. We present some sample statements of how they feel when they are winning or losing that suggest how involvement can drive how decisions might be made. The results are well superior to the linear model.

Limitations

At the same time, the estimation is not perfect. The good news is the use of 997 actual consumers in a real setting. The bad news is we did not have input into the design of the questionnaire. This creates a problem in two ways. We were not able to get at the information we needed directly. So at best our variables are approximations. Secondly, the use of 5-point scales is problematic. Even 7- or 9-point scales would have been better. But what we really needed was true interval data. Even some questions that could have given us ratio level data were categorized. This is clear from the statement in Table 14.1, where the amounts of money lost or one could have been discussed directly. Similarly, the number of visits could have been asked directly. In the absence of these changes it would have been nice to have more measures, so that we could have tried to develop scales from repeated measures. In our opinion, where this leaves us is with a possibly intriguing result that bears more scrutiny. Furthermore, by coming at the problem from an NDS perspective clearly sheds some light on ways in which researchers could move forward.

Future Research

Clearly, the starting place would be to conduct a study with carefully developed questions for real consumers making real decisions under risk. While these might not have to involve gains and losses of personal wealth, gambling probably provides the easiest means. On the other hand, truly high-involvement emotional risk decision might be found in other areas. Another area might be medicine where either families or patients must make various calls that are extremely high risk and involved.

Returning to the kinds of careful experiments constructed in EUT and PT, the surface suggests some other ways to construct experiments to see if the surface is in fact descriptive. This is consistent with the work in the field and is a way to marry NDS to standard research in PT. For example, one could construct experiments to see if decisions change as involvement changes. If involvement is high enough, will an individual who normally uses EUT approaches take to making irrational choices based on emotion? Will people tend to get locked into a positive or negative evaluation as involvement increases? Experiments constructed for very high involvement could also test the ideal of hysteresis or lag in switching as there are changes in the potential gains and losses.

The goal here has not to have been to present a complete solution to the decision-making under risk. Nor have we tried to tweak or otherwise extend

EUT or PT through extensions and enhancements or apply it to another area. Instead we have attempted to see if using an NDS approach can stimulate new thinking in this area. Clearly, we believe our results have shown EUT and PT can benefit from using a different perspective. We do not claim that this is "the" solution or that it is the only way to go to bring the general area decision making under risk together. Rather, we hope this will stimulate others to take different looks at the problem (get outside the box) so that a more complete theory can be developed. Clearly, for a given decision maker, EUT, CPT, and EBPT approaches are all used, depending on the level of involvement, the situation, and emotional state. The goal should be to account for these adjustments in a more global approach so that they are not simply another type of anomalous behavior.

References

Allais, M. (1953). Le comportement de l'homme rationnel devant le risque: critique des postulates et axiomes de l'ecole americaine. [The behavior of rational man under risk: Critiques of the postulates and axioms of the American school.] *Econometrica, 21,* 503–546.

Arrow, K. J. (1971). *Essays in the theory of risk bearing.* Chicago: Markham.

Bernoulli, D. (1954). Exposition of a new theory on the measurement of risk. (Translated by Dr. Lousie Sommer.) *Econometrica, 22,* 22–36.

Bernstein, P. (1998). *Against the gods: The remarkable story of risk.* New York: Wiley.

Bleichrodt, H., & Pinto, P. (2000). A parameter-free elicitation of the probability weighting function in medical decision analysis. *Management Science, 46,* 1485–1496.

Bleichrodt, H., Pinto, J., & Wakker, P. (2001). Making descriptive use of prospect theory to improve the prescriptive use of expected utility theory. *Management Science, 47,* 1498–1514.

Byrne, P. (1996). *Risk, uncertainty and decision-making in property.* London: Taylor & Francis.

Camerer, C., & Ho, T.-H. (1994). Violations of the betweenness axiom and nonlinearity in probability. *Journal of Risk and Uncertainty, 8,* 167–196.

Chateauneuf, A., & Wakker, P. (1999). An axiomatization of cumulative prospect theory for decisions under risk. *Journal of Risk and Uncertainty, 18,* 137–145.

Cobb, L. (1978). Stochastic catastrophe models and multimodal distributions. *Behavioral Science, 23,* 360–374.

Cobb, L. (1981). Parameter estimation of the cusp catastrophe model. *Behavioral Science, 26,* 75–78.

Currim, I., & Sarin, R. (1989). Prospect versus utility theory. *Management Science, 35,* 22–41.

Dore, M. H. I., & Rosser, Jr., B. (2007). Do nonlinear dynamics in economics amount to a Kuhnian paradigm shift? *Nonlinear Dynamics, Psychology, and the Life Sciences, 11,* 119–148.

Fischoff, F. (1983). Predicting frames. *Journal of Experimental Psychology: Learning, Memory, and Cognition, 9,* 103–116.

Gonzalez, R., & Wu, G. (2003). *Composition rules in original and cumulative prospect theory*. Working paper, Department of Psychology, University of Michigan, Ann Arbor, MI.

Guastello, S. J. (1982). Moderator regression and the cusp catastrophe: Application of two-stage personnel selection, training, therapy, and policy evaluation. *Behavioral Science, 27*, 259–272.

Guastello, S. J. (1992). Clash of the paradigms: A critique of an examination of the polynomial regression technique for evaluating catastrophe theory hypotheses. *Psychological Bulletin, 111*, 375–379.

Guastello, S. J. (1993). *Catastrophe and chaos theory for the NYSE stock prices: The crash of 1987 and beyond*. Paper presented to the annual conference of the Society of Chaos Theory in Psychology & Life Sciences, Orillia, Ontario.

Guastello, S. J. (1995). *Chaos, catastrophe, and human affairs: Applications of nonlinear dynamics to work, organizations, and social evolution*. Mahwah, NJ: Lawrence Erlbaum Associates.

Guastello, S. J. (2002). *Managing emergent phenomena: Nonlinear dynamics in work organizations*. Mahwah, NJ: Lawrence Erlbaum Associates.

Kahneman, D., & Tversky, A. (1979). Prospect theory: An analysis of decision under risk. *Econometrica, 47*, 263–291.

Lange, R., Oliva, T. A., & McDade, S. R. (2000). An algorithm for estimating multivariate catastrophe model: GEMCAT II. *Studies in Nonlinear Dynamics and Econometrics, 4*, 137–168.

Oliva, T. (1991). Information and profitability estimates: Modeling the firm's decision to adopt a new technology. *Management Science, 37*, 607–623.

Oliva, T., DeSarbo, W., Day, D., & Jedidi, K. (1987). GEMCAT: A general multivariate methodology for estimating catastrophe models. *Behavioral Science, 32*, 121–137.

Oliva, T., Oliver, R., & MacMillan, I. (1992). A catastrophe model for developing service satisfaction strategies. *Journal of Marketing, 56*, 83–95.

Price, W. L. (1979). A controlled random search procedure for global optimization. *The Computer Journal, 20*, 367–370.

Rao, Y., Zhang, L., & Wang, P. (2006). Modification of value function by incorporating mental account. *European Financial Management Symposium*, Durham Business School, April 20–22, http://www.efmaefm.org/0EFMSYMPOSIUM/durham-2006/sympopart.shtml

Rieger, M. O., & Wang, M. (2006). Cumulative prospect theory and the St. Peter's paradox. *Economic Theory, 28*, 665–679.

Rosser, Jr., J. B. (2007). The rise and fall of catastrophe theory applications in economics: Was the baby thrown out with the bathwater? *Journal of Economic Dynamics and Control, 31*, 3255–3280.

Schmidt, U., Starmer, C., & Sugden, R. (2005). *Explaining preference reversal with third-generation prospect theory*. Nottingham, U.K.: The Centre for Decision Research and Experimental Economics, School of Economics, University of Nottingham, #2005-19, 1–29. Retrieved February 13, 2008 from http://www.nottingham.ac.uk/economics/cedex/papers/2005-19.pdf

Thaler, R., & Johnson, E. (1990) Gambling with the house money and trying to break even: The effects of prior outcomes on risky choice. *Management Science, 36*, 643–660.

Tversky, A., & Kahneman, D. (1992). Advances in prospect theory: Cumulative representation of uncertainty. *Journal of Risk and Uncertainty, 5,* 297–323.

von Neumann, J., & Morgenstern, O. (1947). *Theory of games and economic behavior* (2nd edn.). Princeton, NJ: Princeton University Press.

Wakker, P., & Tversky, A. (1993). An axiomatization of cumulative prospect theory. *Journal of Risk and Uncertainty, 7,* 147–176.

Weintraub, E. R. (1983). Zeeman's unstable stock exchange. *Behavioral Science, 28,* 79–83.

Wu, G., & Gonzalez, R. (1996). Curvature of the probability weighting function. *Management Science, 42,* 1676–1690.

Zeeman, E. C. (1977). *Catastrophe theory: Selected papers, 1972–1977.* Reading, MA: Addison-Wesley.

Zeeman, E. C. (1974). On the unstable behavior of stock exchanges. *Journal of Mathematical Economics, 1,* 39–49.

15

Measuring the Scaling Properties of Temporal and Spatial Patterns: From the Human Eye to the Foraging Albatross

Matthew S. Fairbanks and Richard P. Taylor

CONTENTS

Over the past 50 years, the sophistication of equipment used to measure nature's processes has escalated, along with the computer power available to analyze the resulting data, allowing scientists to explore the role of order in these processes. Many of nature's processes—both in our body and in our natural and social environments—have been found to consist of an intermediate balance between perfect order and complete disorder. For example, the healthy heart is now known to have fluctuations around a periodic beat (Goldberger et al., 2002). Moreover, many phenomena feature fluctuations that occur across many frequencies, where their contributions to the overall process follow a power law scaling relationship.

This power law establishes a type of uniformity across the various time scales. In particular, the data traces that chart the behavior as a function of time exhibit so-called scale invariance and are referred to as "scale-free." Power laws do not feature a characteristic time scale. As a result, the data traces display similar characteristics over many time scales. This repetition across scales is also observed in many of nature's spatial patterns, where the analogous power law behavior is referred to as being fractal (Mandelbrot, 1982).

For both the temporal and spatial processes, the repetition of structure across many scales builds an immense complexity in the resulting pattern, and this complexity is often central to the behavior of the system under investigation. Two parameters are important for quantifying the relative amounts of coarse (slow frequency) and fine (fast frequency) structures in the trace and the associated complexity—the exponent of the power law and the range of scales over which the power law holds.

A traditional approach to determining these parameters is to perform a Fourier spectral analysis of the pattern. Consider, for example, employing this analysis to probe a temporal pattern. Fourier analysis pictures the building blocks of any temporal pattern as a set of sinusoidal waves with a range of frequencies. The analysis determines the amplitude and phase of each of the waves and this information reflects the spectral content of the pattern: a plot of amplitude against frequency reveals the extent to which the pattern follows a power law, the range of frequencies over which the power law persists and the exponent that charts the fall off of amplitude as a function of frequency.

However, as fractal studies have spread through the research disciplines (Bassingthwaighte, Liebovitch, & West, 1994; Mandelbrot, 1982), an alternative technique for assessing these parameters has gained momentum. This method is known generally as the "box-counting" technique and determines the above parameters by assessing the amount of area the pattern covers at different scales. It does this by covering the temporal or spatial pattern with a mesh of identical squares (boxes) and counting the number of squares that contain part of the pattern. This count is then repeated for increasingly small squares in the mesh in order to examine how the pattern evolves across different scales. Analogous to the Fourier analysis, the scaling behavior is then quantified from a plot of box count against box size.

In this chapter, we demonstrate the relative strengths of the two techniques (Fourier and box-counting) and show that an integration of these two compatible techniques is a versatile approach for studying scale invariant processes in natural and social environments. For the first part of the chapter, the two techniques are applied to computer-generated traces to highlight their responses to idealized data. In the second part, we move on to a physical system—movements of the human eye as a person looks at patterns projected on a screen. The human eye is chosen because it highlights the challenges of interpreting real data and also because this visual system simultaneously generates the two families of pattern discussed above. As the eye moves, the location of its gaze on the screen traces out temporal patterns: the horizontal and vertical coordinates of the gaze can each be plotted as a function of time and analyzed for invariance across time scales. At the same time, the trajectories linking the gaze locations trace out a spatial pattern that can be recorded and analyzed for invariance across size scales.

Finally, we discuss the significance of the results of the scaling analysis for the performance of the eye—in particular, how scale invariance leads to an enhanced ability to search for information in nature's fractal scenery. In this sense, the eye's motion shares appealing similarities with the motions of foraging animals such as the albatross. In both cases, the main goal is to conduct a search that covers space efficiently. The eye's search for visual information on a screen is analogous to animals searching for food. We will discuss how the mathematical properties of the power law scaling behavior achieve this through an inherent characteristic called enhanced diffusion.

Comparison of Scaling Measurement Techniques for Temporal Patterns

Fourier Spectral Analysis

We begin our investigations by considering Fourier spectral analysis, which is a traditional approach for analyzing temporal patterns. The analysis is based on the idea that any signal can be reproduced by a summation over a series of sinusoidal oscillations, each with a particular amplitude, frequency, and phase. Fourier analysis is the practice of taking an arbitrary signal and decomposing it into this series of sinusoidal components. The general technique is one of the most widely used in science and has broad technological application, thus the breadth and depth of the field surrounding it is too extensive to cover here; see Bracewell (2000) for further details.

In this chapter, we concentrate on the particular case of an experimentally determined temporal data set, that is, a signal made up of discrete data points where the length of the data set is not infinite in time. Furthermore, we are interested in the scaling behavior of the data set, which is generally charted using a power spectrum analysis, a simple variant of the more general Fourier spectral analysis.

The power spectrum analysis proceeds as follows for our case: First, a discrete Fourier transform (DFT) is applied to the data set. This breaks down the data set into its constituent oscillations. The DFT is based on the fast Fourier transform algorithm, and is implemented in a wide variety of scientific computing environments. These include commercial software like Mathematica® or MATLAB® as well as open-source C libraries like FFTW. The implementations do vary slightly (input/output data formatting, the specific form of the DFT, etc.) and will systematically affect the results, so the researcher is advised to thoroughly understand their software of choice. As a further word of caution, there are (quite literally) a multitude of techniques for applying the DFT. The "proper" technique is mostly dependent on the data available (resolution, length, etc.) and also what characteristics of the data set are being examined. For scaling analysis, it is important to preserve the relative contribution of each constituent oscillation, and thus preconditioning techniques like apodization should be avoided or only sparingly applied. See Bracewell (2000) and the wide existing body of literature on the DFT for further reference.

Second, the result of the DFT is a series of complex numbers $(X_0, X_1, X_2, ..., X_k)$ representing the amplitude and phase of the signal's constituent oscillations and a corresponding series $(f_0, f_1, f_2, ..., f_k$ with $f_k = k/N$, N equal to the length of the transformed data set, and $k = 0...N - 1$) indicating the frequencies of these oscillations. The power spectrum $(S(f))$ is computed by taking the modulus squared of each X_k i.e., $(S(f) = |X_k|^2)$ and plotting these values vs. frequency.

Finally, this plot of power or "spectral density" vs. frequency can be used to determine how the component oscillations' amplitudes fall off with frequency, that is, the "scaling behavior" of the spectrum. This can be done by simply fitting a line to the power spectrum over the region of interest. Further details on this part of the analysis can be found in the following sections, but first we will discuss the chapter's other technique for studying scaling behavior.

The Box-Counting Technique

The underlying methodology of the box-counting technique is shown schematically in Figure 15.1a,b. The data set represents a generic process in which a measured parameter (y-axis) evolves with time (x-axis), producing an aperiodic signal.

The box-counting technique performs an examination of scaling behavior similar to the power spectrum's by counting the number of squares (shaded dark in Figure 15.1), N, that contain part of the data trace. In this approach, N assesses the amount of space covered by the data set at increasingly fine time scales (since the width of each square, L, has units of time). A "scaling plot" of log N plotted as a function of log $(1/L)$ is frequently employed to chart the data.

Two magnification procedures are commonly adopted for constructing the scaling plots. For the first procedure, the square size L is reduced iteratively using the inverse expression $L = H/n$, where H is the trace length and n is the number of iterations ($n = 1, 2, 3 ...$). For the second procedure, the exponential expression $L = HC^{-n}$ (where C is a selected magnification factor that remains constant) is applied iteratively. The first procedure has the advantage of generating a larger number of data points, while the second procedure reduces computation time and produces equally spaced points across the resulting log–log scaling plot.

This basic box-count algorithm has proven to be very popular for assessing both temporal and spatial patterns, in part because of its computational simplicity. (Implementation in one's computing software of choice is straightforward, though commercial options like the Benoit fractal analysis system also exist.) However, it overestimates the space coverage of the trace, as highlighted

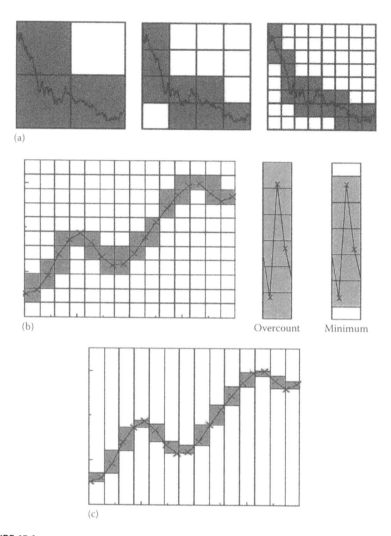

FIGURE 15.1
(a) Shows three iterations of the box-count fractal analysis technique; (b) shows a detailed single iteration for an arbitrary profile, along with a schematic of how the box count will often overestimate the box coverage of the profile; and (c) shows the same profile measured with the variation method, showing a more accurate covering estimate.

in Figure 15.1b. The left-hand image shows an example trace at a chosen box size. The two right-hand images consider a small section of the trace and demonstrate that shifting the mesh vertically results in different counts (N values of 6 and 5, respectively). Consequently, one approach to minimizing the overcount would be to repeat the count for various shifts in the mesh position and to select the minimum N value obtained. Rotating the grid is an alternative approach that effectively provides both horizontal and vertical shifts.

An even more accurate method for minimizing the overcount is to adopt the variation method (Dubuc, Quiniou, Roques-Carmes, Tricot, & Zucker, 1989), which is shown in Figure 15.1c for the same scale considered in Figure 15.1b. For this technique, columns replace the boxes. The height of the shaded area within a given column is determined by the difference of the heights at which the data trace enters and leaves the column. The total area (i.e., the sum of the shaded areas of all the columns across the whole trace) is then divided by the individual box area of Figure 15.1b to get the N value. Similar to the basic box-counting technique, this is then repeated for increasingly narrow columns, and so assesses space coverage of the trace at increasingly fine time scales.

Scaling Plots

Figure 15.2 provides a comparison of this variation method (right-hand column) and the power spectrum analysis (middle column) for four simulations of time series data (left-hand column). The first row considers data that follows a simple line as a function of time t. The power spectrum analysis plots the power, $S(f)$, against frequency, f, and reveals a peak at zero frequency, as expected. The graph for the variation method plots N against $1/L$, where L is the column width. Since L is measured in units of time, the x-axis assumes a role analogous to frequency. The data is plotted on log–log axes to highlight that the data follows a power law of the form

$$N \sim \left(\frac{1}{L}\right)^{D_t} \tag{15.1}$$

In log–log space, this power law generates a straight line with a gradient equal to the exponent D_t, where subscript "t" is used to denote that temporal rather than spatial data is being analyzed. The measurement of the gradient reveals $D_t = 1$, highlighting the mathematical significance of the exponent—it represents the dimension of the data trace.

The second row of Figure 15.2 considers data that follows a sine wave as a function of time. As expected, the power spectrum analysis reveals a peak, corresponding to the inverse of the wave's period. The variation method again generates a straight line with $D_t = 1$ for small time periods (i.e., large $1/L$ values). This is a consequence of the fact that the narrow columns are sampling the smooth, one-dimensional character of the sine wave. Only at wider columns (i.e., small $1/L$ values) does the periodicity of the sine wave become apparent. This appears in the variation graph as a deviation of the data from the power law line, beginning at a scale indicated by the arrow.

A similar situation occurs for the third row, where a data set composed of two sine waves with different frequencies and amplitudes is considered. The power spectrum analysis reveals peaks at the two relevant frequencies.

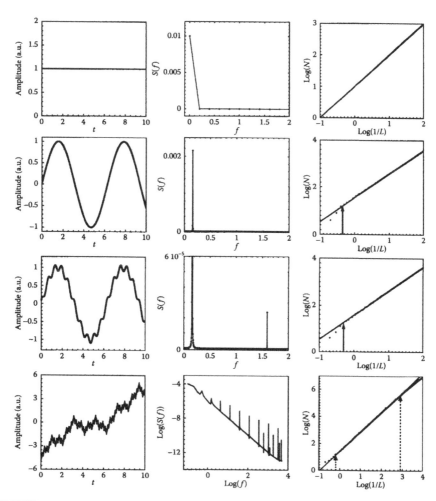

FIGURE 15.2
Each row shows a data trace (first column), its spectrum (second column), and its scaling behavior using the variation method (third column). Arrows in the third column indicate where the scaling behavior ceases to hold. The fit lines (black) are fit to the entire data set in first three rows. In the fourth row, the line fits only to the data points in the scaling region between the indicated cutoffs (dotted arrows).

The variation method again deviates from the power law line. This deviation, indicated by the arrow, occurs at approximately the same $1/L$ value as for the second row, because the additional, fast-frequency sine wave does not significantly affect the overall form of the trace.

The data considered in the fourth row consists of a superposition of many cosine waves. The function is called a Weierstrass–Mandelbrot function ($W(t)$), named after the mathematician who first studied this fractal form (Weierstrass) and Mandelbrot who generalized the function to make it analytically scale invariant (Berry & Lewis, 1980). The amplitudes of the component waves scale with frequency according to the following power law:

$$W(t) = \sum_{n=-\infty}^{\infty} r^{-n(2-D_t)}(1 - \cos(r^n t)) \tag{15.2}$$

where
   $r > 1$ defines the function's constituent frequencies (within the cosine) and amplitude scaling factor
   n is an integer

To demonstrate the sensitivity of the power spectrum analysis to this fractal power law relationship, we have plotted the results of the power spectrum analysis in log–log space. The peaks clearly occur at even intervals of $\log(f)$, with the peak heights falling off according to the straight slope expected of the power law in $W(t)$.

Whereas the power spectrum analysis follows expectations, the variation graph of the Weierstrass function reveals an intriguing result—the data once again follows a straight line, but the D_t value determined from the slope has lifted from its integer value and now assumes the value of 1.5. If we maintain the same physical interpretation as before (that the slope measures the dimension of the data), then this signifies that the Weierstrass function has a non-integer dimension. Indeed, this quality is a defining characteristic of fractals. The "covering" dimension (extracted by covering assessment techniques such as the variation method) of a fractal trace or data set must be higher than its topological dimension, which in this case is one dimensional, since the data is strung together to form a line rather than, for example, a zero-dimensional dot or a two-dimensional area. An intuitive mathematical interpretation of this result is that the repeating structure embedded in the line causes it to start to occupy area, but not to the extent of filling the two-dimensional plane. This is reflected by a covering dimension that lies between 1 and 2.

Another important result, highlighted by Figure 15.2, is that the fractal power law does not extend to infinitely small nor infinitely large frequencies. This restriction is due to the limits of measurement resolution. At the highest frequencies (i.e., large values of $1/L$), the narrowing column widths start to approach the data resolution limit, that is, the separation between data points. The scaling behavior of the data then deviates from the fractal power law and the data no longer follows the straight line, as indicated by the right-hand arrow in the variation graph. At low frequencies (i.e., small values of $1/L$), the column width starts to approach the length of the trace and the number of columns becomes severely restricted. The resulting limitation in counting statistics causes the data to deviate from the straight line. In particular, the low number of columns hinders the technique from distinguishing the fractal line from a filled two-dimensional space, with the consequence that the filled columns generate a line with a gradient approaching $D = 2$.

These "cutoffs" (indicated by the two arrows) limit the frequency range over which the fractal power law behavior is observed. It should be stressed that the frequency range does not feature inherently in the definition of a fractal. However, a limited range impedes the confidence level of determining if the line is in fact straight (e.g., many functions can appear straight over one order of magnitude when plotted on a log–log graph) and also limits the level of accuracy in determining D_t from the line's slope. As an empirical guide, column widths narrower than five data points should be excluded from fractal analysis. This is also true of columns with widths larger than one-fifth of the length of the trace itself (for experimental data, one-tenth is often used). These measurement requirements set the frequency range observed in Figure 15.2. For this mathematically generated fractal, the "upper" (i.e., high-frequency) cutoff could have been extended by reducing the data point separation. Similarly, the "lower" (i.e., low-frequency) cutoff could have been extended by generating a longer trace. For experimental data, more limiting considerations also factor into determining these cutoffs and we will return to this discussion in the next section.

It should be noted that experimental traces rarely have the exact pattern repetition observed in these simulated traces. This rather artificial form of fractal scaling is highlighted in Figure 15.3 (left column) by repeatedly magnifying one section of the trace, revealing an exact repetition of the structure at increasingly fine time scales. Because of this exact repetition, the function is labeled as an exact fractal. The right column shows another form of fractal referred to as a statistical fractal. For this fractal, the pattern does not repeat exactly. Instead, only the pattern's statistical qualities repeat at different magnifications.

This statistical form of the Weierstrass function was generated using Equation 15.2 and randomizing the phases of the constituent cosine waves. This randomization preserves the power law behavior, and hence the basic fractal quality, of the trace. Thus the power spectrum and variation graphs are very similar to those for the exact fractal shown in Figure 15.2. Many natural (temporal and spatial) processes generate statistical fractals because of nature's integration of randomness with an underlying power law scaling behavior. Because of the role of randomness, these fractals are also sometimes referred to as random or stochastic fractals.

Power Law Exponents

We are now in a position to consider how the two important parameters of power law behavior—the frequency range and the power law exponent—compare for the two measurement techniques. The observed scaling range of the fractal is determined by the cutoffs discussed above, and these occur at approximately the same frequencies for the two methods. How, though,

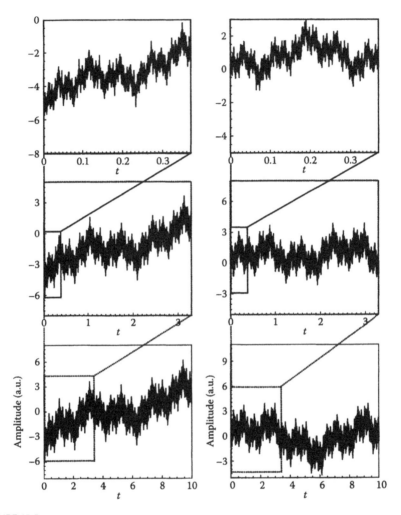

FIGURE 15.3
Two sets of Weierstrass fluctuations with exact self-similarity (left column) and statistical self-similarity (right column). The generating function is identical in each case except for the addition of a random phase to each sinusoidal component in the statistical case.

do the slopes of the fractal power laws extracted by the two analysis methods compare? This is considered in more detail in Figure 15.4, where three Weierstrass fractal traces are generated in the left figure with D_t values of 1.2, 1.5, and 1.8. The middle figure shows the associated power spectrum behavior, where the data points indicate the frequencies of the constituent cosines (determined using Equation 15.2). The amplitude values ($S(f)$) of these peaks drop off according to the following power law:

$$S(f) \propto f^{-\alpha} \qquad (15.3)$$

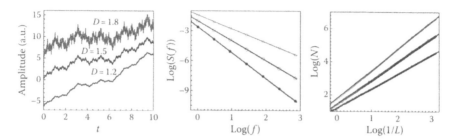

FIGURE 15.4
Three sets of Weierstrass fluctuations with differing D_t values (left panel), the associated analytical power spectra (center), and the results of the variation method (right) for each set. The analyses are only plotted over the region where the scaling behavior holds. The data points on the power spectra indicate the positions in frequency of the Weierstrass' constituent sinusoids.

Thus, the gradient is $-\alpha$ for the log–log graph shown in Figure 15.4 (middle). The right-hand graph of Figure 15.4 shows the variation method result. This graph charts the behavior $N \sim (1/L)^{D_t}$ (see Equation 15.1). Therefore, the gradient is D_t for the log–log graph shown. D_t, the fractal dimension of the temporal data set, and α, the spectral exponent, are linked by the following mathematical relation, which is derived for fractional Brownian motion (Barnsley et al., 1988):

$$D_t = \frac{(5-\alpha)}{2} \tag{15.4}$$

Note that, for a temporal trace to be fractal, D_t must have a fractional value lying in the range $1 < D_t < 2$, and so the corresponding α must satisfy $1 < \alpha < 3$.

It is informative to check that the inverse relationship between D_t and α, revealed both in Equation 15.4 and Figure 15.4, agrees with intuition. First, consider the three traces of the power spectrum graph (Figure 15.4(middle)). The α value determines the falloff in the power of the cosine wave components when moving to higher frequencies. Thus a larger α value will lead to a lower power at high frequencies. This can be seen in the Weierstrass traces (Figure 15.4(left)): high-frequency structure is less dominant in the trace with the higher α value (bottom trace). Now consider the variation graph (Figure 15.4(right)). The D_t value sets the rate at which N increases as the column size L is reduced. A higher D_t trace will therefore have a higher N value at large $1/L$ values, translating to there being more fine structure in the high D_t trace than an equivalent low D_t trace. This can be seen plainly in the three Weierstrass traces.

Whereas Equation 15.4 provides the mathematical relationship between D_t and α, the values of D_t extracted from the variation scaling plots are often

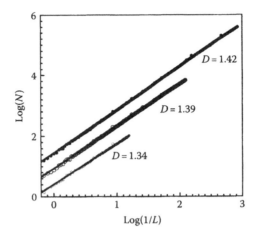

FIGURE 15.5
A comparison of variation method scaling plots for three sets of Weierstrass fluctuations. The
analyzed data sets were generated with three, four, and five orders of magnitude in data
resolution with fractal power law scaling over the plotted ranges.

found to be less than the values calculated from α using Equation 15.4. What
is the source of this discrepancy? Figure 15.5 addresses this question by
returning to the issue of cutoffs and the associated frequency range. This figure shows the result of a variation analysis performed on three Weierstrass
traces, where the analytical D_t value from Equation 15.2 is 1.5 but the frequency range over which this power law is obeyed varies.

It is clear that, as the frequency range increases, the accuracy of the D_t
value measured from the gradient rises toward the input value. For the
trace where the fractal power law holds for three orders of magnitude in
frequency, the measured value of 1.42 almost matches the input value of 1.5.
Table 15.1 shows similar results for different values of input D_t values of 1.2,
1.5 and 1.8 (left column). In each case, increasing the range of frequencies
results in a more accurately measured D_t value, as seen moving left to right
in the latter three columns of Table 15.1.

TABLE 15.1

A Summary of the Measured D_t Values Obtained
from the Variation Method for Fractal Traces
with Different Scaling Ranges

Analytical Dimension	Dimension (1.4 Orders)	Dimension (2.3 Orders)	Dimension (3.1 Orders)
1.2	1.12	1.14	1.15
1.5	1.34	1.39	1.42
1.8	1.57	1.62	1.67

Figure 15.6 considers a further approach to improving the accuracy of the box-counting method. The horizontal structured elements (HSE) method also assesses the coverage of the temporal or spatial data trace but does so in a different manner than the box-counting and variation methods. The HSE method places a mesh of squares over the trace with a width L_R that typically matches the spacing between data points, that is, the resolution of the trace. Squares in the mesh that contain part of the data trace are assigned a value of 0. Values are assigned to other squares based on their horizontal distance to the nearest square of value 0. In other words, if a square is next to a square of value 0 it is assigned a value of 1, if it is the next square along it is assigned a value of 2, and so on, as demonstrated in the schematic representation of Figure 15.6a. Once the distance values have been assigned, a count is made, first of the number of squares of value 0, $N(0)$, then of the number of squares of value 0 or 1, $N(0,1) = N(0) + N(1)$, and so on. Although beyond the scope of this chapter, it can be shown (Dubuc et al., 1989) that the following equation can be used to calculate N:

$$N = \frac{(L_R)N(0,1,2,\ldots,k)}{L^2}$$ (15.5)

where
$L = (k - 1)L_R$
k is the maximum horizontal distance value considered

Thus it is possible to plot the usual scaling plot of $\log(N)$ against $\log(1/L)$ and obtain a straight line with the gradient given by D_t.

A crucial feature of the HSE method lies in the distribution of the data points in the resulting scaling plot. Instead of having an increasing data point density in the high ($1/L$) direction (as is the case for the basic box-counting and variation methods when using the magnification equation $L = H/n$ discussed above), the data point density of the HSE method is highest at the small ($1/L$) end of the scaling plot. This difference in the point distribution between the variation and the HSE methods is exploited in Figure 15.6b, where the data from the two methods has been combined in one scaling plot. As shown more clearly in Figure 15.6c, where only every 20th data point has been shown, this provides a high data point density across all frequencies. This high density is sufficient to apply a derivative analysis to detect the frequency values of both the upper and lower cutoffs, as shown in the insets of Figure 15.6b. This more accurate location of the cutoffs (marked L_1 and L_2 in the insets), combined with the larger and more even data density between the cutoffs, improves the accuracy of the fit to the data and, thus, the D_t value extracted from the fit.

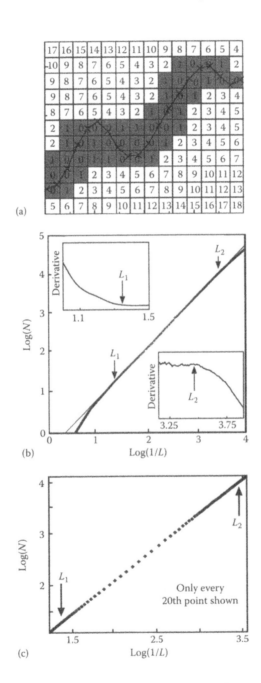

(a)

(b)

(c)

FIGURE 15.6
A demonstration of a more detailed fractal analysis on a generic data trace. (a) Shows a schematic of the horizontal structured elements (HSE) analysis method. In combination with the variation method, the analysis gives (b) a scaling plot with high data densities over all size scales, which enables derivative analysis to detect the appropriate cutoffs (insets). (c) Shows only the power law scaling region with the calculated cutoffs L_1 and L_2.

Scale Analysis of a Physical System: Human Eye Movement

We now move on to investigations of data generated by physical systems. The temporal behavior of many natural systems has been successfully quantified using power laws, spanning the physical, natural, and social sciences (Bassingthwaighte et al., 1994; Mandelbrot, 1982). A common nomenclature has emerged across these disciplines to categorize the scaling behavior of systems according to the spectral scaling exponent α, as summarized in Table 15.2.

This nomenclature is based, in part, on analogies with the color spectrum of light. For example, white power laws have equal power at all frequencies, analogous to white light featuring contributions from all visible frequencies. Pink power laws have a higher α value indicative of an increase of power at lower frequencies (e.g., see Figure 15.2), analogous to pink light. Brown (so named because time series data generated from Brownian random statistics are quantified by $\alpha = 2$) and black power laws have even higher α values. These α-value-to-color definitions correspond to those used commonly in the physical sciences, but shift somewhat in other research areas. White noise is sometimes expanded to include $0 \le \alpha < 1$, thus reducing the pink noise's range, and brown noise is often subsumed by the black noise category as a result of its close proximity in α.

The second row of Table 15.2 shows the D_t value of temporal trace obtained using Equation 15.4. While discussing this second row, it is opportune to return to our earlier definition of fractals: for time series data, the D_t must have a fractional value lying in the range $1 < D_t < 2$. This restriction arises because the D_t value must be higher than the trace's topological value (which equals 1) but cannot exceed the dimension of the graphical space that the data is embedded into—the so-called embedding dimension (which equals 2). For this reason, white power laws have $D_t = 2$, and not greater, because they are confined to a two-dimensional plane.

Note that the D_t values listed in Table 15.2 describe the dimensions of the temporal data traces. If these temporal processes correspond to an object moving in a two-dimensional plane, then the spatial pattern generated by the object's trail will have a D_s value given by the following relationship

TABLE 15.2

A Summary of α, D_t and D_s Values for the Common Categories of Power Law Scaling Behaviors

Name	White	Pink	Brown	Black
α	0	$0 < \alpha < 2$	2	$2 < \alpha$
D_t	2	$2 \ge D_t > 1.5$	1.5	$1.5 > D_t \ge 1$
D_s	2	2	2	$2 > D_s \ge 1$

(derived for the fractal process, fractional Brownian motion) to the spectral exponent α from its temporal trace (Barnsley et al., 1988; Sprott, 2003):

$$D_s = \frac{2}{(\alpha - 1)} = \frac{1}{(2 - D_t)} \qquad (15.6)$$

Note also that it is possible for the temporal trace to be fractal and the equivalent spatial pattern not to be fractal as seen in Table 15.2. This occurs, for example, with Brownian processes ($\alpha = 2$). The temporal trace is quantified by $D_t = 1.5$, while the spatial data is quantified by $D_s = 2$. So-called fractional Brownian processes ($\alpha > 2$) are good examples where both the temporal and spatial patterns are fractal. We will return to this difference between Brownian and fractional Brownian motion in the discussion section.

Let us now consider the physical significance of the α value of the temporal trace. A high α value (and corresponding low D_t value) indicates a bias toward components with long time periods. For biological, physiological or social systems, this increased correlation across long time scales implies a "memory" within the system. At the other extreme of scaling behavior, white power laws describe systems with no memory—there is no correlation between events and processes occurring at different times.

What, then, is the correlation—the "color" and dimension—of a physiological system such as the human eye? Not surprisingly, this will depend on the task being performed by the visual system. Whereas the gaze behavior when looking at figurative stimuli such as the human body or human faces has been the subject of much research (Hyona, Munoz, Heide, & Radach, 2002), the gaze patterns activated by more abstract visual stimuli are considerably more subtle and less understood. In particular, until recently, it has been unclear as to what happens when the eye goes into "search mode," seeking out valuable information embedded, or even hidden, in a clutter of highly complex background information. How does the eye search?

Figure 15.7 shows an example of the spatial (a) and temporal (b) patterns traced out by the human eye when in search mode.

The data was recorded using the eye-tracking equipment shown in Figure 15.8. This "eye-gaze system" (from LC Technologies) integrates infrared and visual camera techniques to determine the location of the eye's gaze when looking at a pattern formed on a computer screen (Hyona et al., 2002). This remote tracking is unobtrusive (requiring no attachments to participants) and is functional for a range of participants, including those wearing contact lenses and glasses. The sizes of the images on the screen were 290 mm by 290 mm, corresponding to 1024 by 1024 screen pixels (i.e., the image resolution was 35.3 pixels per cm). The eye-tracker can locate the gaze with an accuracy of 4 pixels. The distances traced out by the gaze on the screen can be converted to the corresponding change of the eye's viewing angle using the separation distance between the eye and the screen (56 cm).

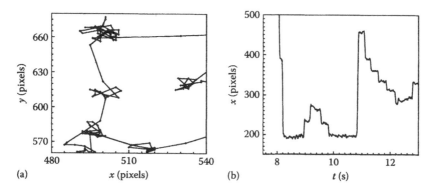

FIGURE 15.7
Subsections of a single eye-tracking data set: (a) the spatial pattern and (b) the time series *x* vs. time.

FIGURE 15.8
A photograph of one of the authors (MSF) using the eye-tracking apparatus.

Figure 15.7a shows a magnified section of one region of the spatial pattern traced out by the eye's gaze as it moves across the screen. As expected, the pattern is composed of long ballistic trajectories as the eye jumps between the locations of interest, and smaller motions called micro-saccades that occur during the dwell periods to ensure that the retina does not desensitize (Hyona et al., 2002). Figure 15.7a plots the horizontal (*x*) and vertical (*y*) locations of the gaze in units of screen pixels. Micro-saccades are expected to occur over an angular range of typically 0.5°. This angle translates to a distance of 15 pixels on the screen and, as expected, this approximately matches the typical width of the dwell regions observed in Figure 15.7a. Given that the saccades and micro-saccades are produced by different physiological mechanisms and serve different purposes within the visual system, it is expected that their scaling behaviors will be different. The scaling analysis of the spatial patterns will therefore be expected to reveal a crossover between the two processes—the saccades and micro-saccades—at a size scale of approximately 15 screen pixels.

Figure 15.7b shows the corresponding temporal pattern by plotting the x position against time t. The periods of relative motionlessness are the dwell periods at a given location, during which time the eye is undergoing micro-saccades. The typical dwell time is approximately 0.4 s. The time scale of the individual micro-saccades is expected to be approximately 10–20 ms. We note that this is on the same order as the sampling rate of the eye-tracking equipment (16 ms, 60 Hz). This measurement limitation would therefore impact on any studies of the micro-saccades. However, the focus of our investigations lies with the saccades, since these larger motions are the ones that dictate the search motion, and these operate on longer time scales than the equipment's sampling rate. The basic form of the temporal trace observed in Figure 15.7b occurs over time scales of interest (i.e., times longer than the micro-saccades and longer than the equipment resolution limit) and is formed by an interplay between dwell periods and ballistic jumps.

Recent experiments, using a similar experimental set-up to that of Figure 15.8, investigated the temporal patterns when observers were asked to search for visual icons embedded in a visually complex pattern on a computer screen. These temporal patterns followed a power law scaling behavior (Aks, Zelinsky, & Sprott, 2002). We have built on this intriguing result by investigating the temporal behavior as the eye searches through the visual complexity of a fractal pattern. Figure 15.9 (left column) shows the underlying fractal patterns (light gray) that were displayed on the computer screen. The dark trajectories are the saccades of the observer's eye during an observation period lasting 60 s. In each case, the observer was instructed to memorize the pattern in order to induce the search activity. The D_s values of the displayed monochrome fractal patterns were 1.11 (top), 1.66, and 1.89. The fourth image (bottom) is a composite of four differently colored interlocking fractal patterns, each with a D_s value of 1.6. All of the fractal images are taken from paintings created by the artist Jackson Pollock (the significance of which will be discussed in detail below).

The right-hand column features the scaling plots resulting from the spatial box-counting analysis of the saccades. However, we first concentrate on the analysis of the associated temporal patterns, as shown in Figure 15.10. The left column shows the time series traces, together with their spectral analysis (middle column) and box-counting analysis (right column). For both forms of analysis, the scaling plots examine the scaling behavior between the lower (slow frequency) and upper (fast frequency) cutoffs. In each case, as a guide we have indicated the typical dwell time (0.4 s) with an arrow. The cutoffs correspond to the time scales of 5 s and 80 ms. These are set by the same measurement principles outlined earlier for the simulated Weierstrass function: 80 ms corresponds to being five times larger than the resolution limit (set by the equipment's sampling rate of 16 ms) and 5 s corresponds to being approximately one-tenth of the length of the trace (set by the observation period of 60 s).

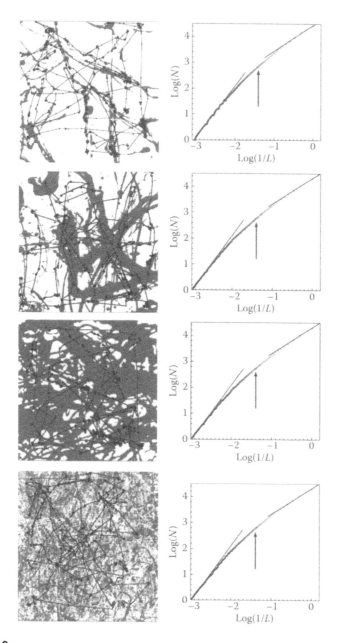

FIGURE 15.9
The complete spatial eye-tracking data sets (left column) and their associated box-count scaling plots (right). The eye-tracks are overlaid on the observed fractal patterns, which have dimensions of $D_s = 1.11$ (first row), $D_s = 1.66$ (second), and $D_s = 1.89$ (third). The final pattern is a colored composite of four $D_s = 1.6$ patterns. The scaling plots show the entire box-count data set with typical fixation size (arrows), $D_s = 1$ (right-hand line), and $D_s = 2$ (left-hand) trends.

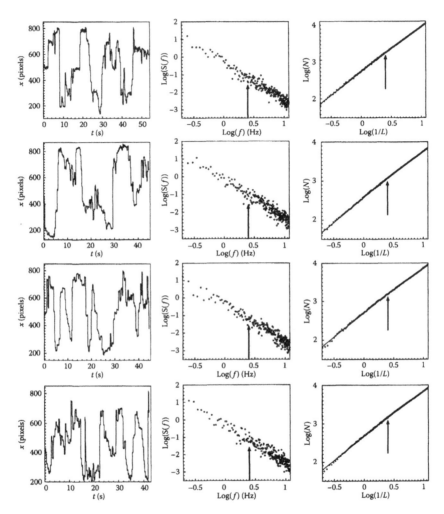

FIGURE 15.10
Time series of position *x* vs. time (left column) for each of the eye-tracks pictured in Figure 15.9 with corresponding power spectrum (center) and variation method (right) analyses. Arrows indicate the typical fixation time length (~0.4 s) on each scaling plot. Fit lines of uniform gradient show the region of strict power law scaling in each variation method plot.

However, whereas the observation range of the simulated fractal traces could be expanded by increasing the data resolution and the length of the trace, this is not the case for the experimental data of Figure 15.10. Even if the equipment sampling rate were to be increased (the latest eye-trackers can sample at 240 Hz), the experiment would be measuring a different physical process at these fine time scales, one dominated by the micro-saccades. Similarly, if we were to lengthen the data trace by increasing the observation period, a different physical process would again take over—after 60 s, observers lose concentration and their gaze tends to wander off screen.

These factors highlight a crucial characteristic of physical fractals. Unlike their mathematical counterparts, physical processes (like gaze movements) have inherent upper and lower cutoffs in addition to the measurement cutoffs. For this reason, physical fractals are often referred to as "limited-range fractals." The magnification range of nature's physical fractals is surprisingly small—the typical range is only 1.25 orders (Avnir, Biham, Lidar, & Malcai, 1998). Researchers still debate whether the term *fractal* is appropriate to describe these patterns with such limited power law scaling. It is perhaps better to concentrate on whether an observed scaling behavior offers useful information about the system in question. For instance, the gradient of scaling plots is effective for quantifying the relative contributions of different frequency oscillations in a data set, which is useful information for a variety of systems, from stock markets to eye movements. To avoid the associated semantic issues, we generally refer to the D_t and D_s values extracted from the gradients of limited-range scaling plots as being "effective dimensions."

We note that the power spectrum scaling plots of Figure 15.10 are inherently noisier than the equivalent box-counting plots. For this reason, we take the effective dimension measured from the box-counting plots ($D_t = 1.2$), and use Equation 15.4 to calculate the equivalent α value (2.6). We expect this D_t value to be an underestimate of the actual value due to the limited magnification range (see the earlier discussions on the Weierstrass function). It is interesting to compare this result to the experiment where observers were asked to search for embedded icons, where $\alpha = 2$, extracted directly from the power spectrum analysis, indicated Brownian power law behavior (Aks et al., 2002).

We now return to the scaling analysis of the spatial patterns shown in Figure 15.9. Consider the size scales plotted in the box-counting plots shown in the right column. (This box-counting technique is identical to that applied to the temporal traces with scaling behavior charted by Equation 15.1 with D_s replacing D_t.) The left-hand extreme of these plots matches the width of the screen: 1024 pixels (290 mm), which corresponds to an angular motion of the eye of 29°. The right-hand extreme corresponds to 1 pixel (0.3 mm), which corresponds to an angular motion of 0.03°. The arrow indicates an angle of 0.5° (5 mm, 18 pixels on the screen) and corresponds to the expected crossover size scale from motion dominated by saccades (small $1/L$ scales) to motion dominated by micro-saccades (large $1/L$ scales). At this size scale, we anticipate a "knee" in the data since the two processes should exhibit different scaling properties.

Figure 15.9 therefore emphasizes the challenges of measuring physical fractals. At the left-hand extreme of the graph, the data follows $D_s = 2$ (as indicated by the fit line) because the limited number of boxes cannot distinguish the fractal trace from a filled space. At the left-hand extreme, the data follows $D_s = 1$ (as shown by the second fit line) because these small box sizes are approaching the data resolution limit and the analysis picks up the one-dimensional quality of the data line. This leaves a highly limited range of size scales (spanning two orders of magnitude) over which the data switches from one scaling behavior (saccades) to another (micro-saccades)!

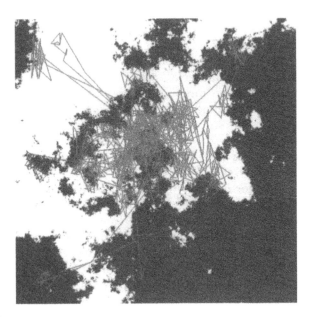

FIGURE 15.11
An example of the computer-generated fractals viewed by the subjects for the eye-tracking results shown in Table 15.3.

In terms of investigating the saccades, there is little flexibility for expanding the magnification range since the upper (fast frequency) cutoff is determined by the existence of micro-saccades (appearing at 0.5°). Only the lower cut-off (29°) could be improved by using a slightly bigger screen. However, at large angles of 65° and above, observers will start to search by using head movement rather than eye-movement and this will provide a fundamental limit to the lower cutoff. Thus the observation range could be improved by only 0.5 orders.

Despite this limited range, we can determine that the effective dimension, determined from the slopes at large scales (i.e., the saccade behavior), appears to be insensitive to the D_s value of the fractal pattern being observed: the saccade pattern is quantified by $D_s = 1.4$, even though the under-lying pattern varies over a very large range from 1.11 to 1.89. We note that this characteristic value of $D_s = 1.4$ holds for observations of multicolored fractals (bottom image).

To further test this result, we considered another form of fractal pattern for observers to search through—the computer-generated fractals shown in Figure 15.11. Table 15.3 compares the D_s values

TABLE 15.3

A Comparison of the D_s Values of the Fractal Images Being Viewed, and the D_s Values of the Patterns Traced Out by the Saccades

D_s Image	D_s Saccade
1.1	1.5
1.2	1.5
1.3	1.5
1.4	1.6
1.5	1.5
1.6	1.6
1.7	1.5
1.8	1.5
1.9	1.5

of the spatial patterns traced out by the saccades and the D_s values of the observed fractal patterns. In each case, the D_s values of the saccades are averaged over the results from six observers, each of whom observed the nine fractal images for 30 s, separated by a checkerboard pattern observed for 30 s. Although preliminary, the results confirm that the saccades trace out an inherent search pattern set at $D_s = 1.5$, regardless of the scaling properties of the pattern being observed.

Discussion

In this chapter, we have compared and contrasted two major forms of scaling analysis (power spectrum and box-counting analysis) and have applied them to the motions of the human eye—a physiological system that simultaneously generates temporal and spatial patterns. We will spend the remainder of the chapter discussing the results from this analysis and how scaling analyses applied to other natural processes offer a possible explanation for our results.

Although the effective dimension of the saccade motion can only be fitted over limited size scales, its insensitivity to such a wide range of D_s values in the observed pattern is striking. It suggests that the eye's search mechanism follows an intrinsic mid-range D_s value. Why would the eye adopt a fractal trajectory with a D_s value of 1.5? An appealing explanation lies in the mathematical properties highlighted in Figure 15.12. This figure compares three trajectories quantified by different dimensions. The left-hand image is that of a simple, straight line with $D_s = 1$. The right-hand image is a simulation of Brownian motion, in which a random process generates a trace with $D_s = 2$. The middle trace is a simulation of a fractal trace, with $D_s = 1.5$. This fractal

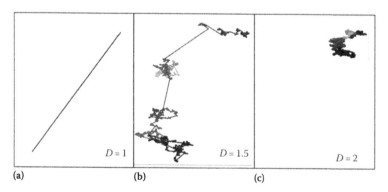

(a)	(b)	(c)
$D = 1$	$D = 1.5$	$D = 2$

FIGURE 15.12

Potential search patterns, including (a) a single line of $D_s = 1$, (b) a Lévy flight with $D_s = 1.5$, and (c) Brownian motion with $D_s = 2$.

trace is called a Lévy flight, named after the mathematician who first developed its statistical qualities (Taylor, 2005). Note that the Lévy flight shown in Figure 15.12 is embedded in a two-dimensional plane: the equivalent flight in three-dimensional space would be quantified by $D_s = 2.5$.

Lévy flights were used to model the flights of the albatross in a pioneering experiment that measured the bird's foraging behavior across the Antarctic skies (Viswanathan et al., 1996). Although the model has recently been called into question (Edwards et al., 2007), the experiment triggered a number of other investigations of foraging animals in which Lévy flights were proposed to describe the animals' behavior. Within this foraging model, the smaller trajectories allow the animal to look for food in small region and travel to neighboring regions. Longer trajectories correspond to the animal moving onto undisturbed (and possibly food-filled) regions further away, where the smaller trajectories begin the process anew.

When the results of the albatross investigations were first published, one of us (RPT) proposed that humans might subconsciously follow Lévy flights when forced to cover vast spaces. The proposal was made within the context of artists, in particular for explaining how the abstract painter Jackson Pollock adopted his infamous pouring techniques to cover his large (sometimes 30 ft wide) canvases with complex fractal patterns (Taylor, Micolich, & Jonas, 1999). Recently, researchers have started to apply scaling analysis to cell phone data to examine if people also trace out fractal "foraging" behavior during their daily lives (Gonzalez, Hidalgo, & Barabasi, 2008).

Significantly, the fractal flight has an "enhanced diffusion" compared to the equivalent random motion of a Brownian flight, and this might explain why it is adopted for the search strategies in our experiment. The Brownian flight lacks the longer flights of the fractal trace and thus its trajectories are confined within a small location, gradually filling space. The amount of space covered by the fractal trace in the same amount of time increases for smaller D_s values. However, there is a limit to how small the D_s value should be to create this enhanced diffusion. Although a trajectory of $D_s = 1$ (Figure 15.10a) covers ground quickly, it does not explore much of the two-dimensional plane. It appears then that a mid-range D_s value might be optimal for covering the terrain efficiently.

This provides an appealing explanation for why the human eye follows a fractal trajectory with an inherent D_s value set at 1.5—the mathematics of its geometry dictate that it covers terrain efficiently when searching for information. This model raises an obvious question—what happens when the eye is made to view a fractal pattern of $D_s = 1.5$? Will this trigger a "resonance" when the eye sees scaling behavior that matches its own basic characteristics? Our previous research examining people's perceptual and physiological responses to observing fractal patterns indicates a marked response when people observe mid-D_s fractals. This includes the reduction in the physiological stress-levels of the observer (Hagerhall, Laike, Taylor, Kueller, & Martin, 2008, Taylor, 2006). These results emphasize the important role of understanding the scaling properties of the human visual system and serve as an excellent demonstration of the box-counting and spectral scaling analyses and their efficacy.

Acknowledgments

We thank P. Van Donkelaar, C. Boydston, and N. Kuwada at the University of Oregon for their help with the experimental investigations of the motion of the human eye. We also thank A. P. Micolich, T. P. Martin, and R. Montgomery for their helpful discussions on the analysis of fractal patterns. R. P. T. was a Cottrell Scholar of the Research Corporation during this project.

References

Aks, D. J., Zelinsky, G. J., & Sprott, J. C. (2002). Memory across eye-movements: 1/f dynamic in visual search. *Nonlinear Dynamics, Psychology, and Life Sciences, 6*, 1–25.

Avnir, D., Biham, O., Lidar, D., & Malcai, O. (1998). Is the geometry of nature fractal? *Science, 279*, 39–40.

Barnsley, M. F., Devaney, R. L., Mandebrot, B. B., Peitgen, H.-O., Saupe, D., & Voss, R. F. (1988). *The science of fractal images.* New York: Springer-Verlag.

Bassingthwaighte, J. B., Liebovitch, L. S., & West, B. J. (1994). *Fractal physiology.* New York: Oxford University Press.

Berry, M. V., & Lewis, Z. F. (1980). On the Weierstrass-Mandelbrot fractal function. *Proceedings of the Royal Society of London. Series A, Mathematical and Physical Sciences, 370*, 459–484.

Bracewell, R. N. (2000). *The Fourier transform and its applications* (3rd ed.). Boston: McGraw Hill.

Dubuc, B., Quiniou, J. F., Roques-Carmes, C., Tricot, C., & Zucker, S. W. (1989). Evaluating the fractal dimension of profiles. *Physical Review A, 39*, 1500–1512.

Edwards, A. M., Phillips, R. A., Watkins, N. W., Freeman, M. P., Murphy, E. J., Afanasyev, V., Buldyrev, S. V., da Luz, M. G. E., Raposo, E. P., Stanley, H. E., & Viswanathan, G. M. (2007). Revisiting Lévy flight search patterns of wandering albatrosses, bumblebees, and deer. *Nature, 449*, 10441–1048.

Goldberger, A. L., Amaral, L. A. N., Hausdorff, J. M., Ivanov, P. C., Peng, C.-K., & Stanley, H. E. (2002). Fractal dynamics in physiology: Alterations with disease and aging. *Proceedings of the National Academy of Sciences, 99*, 2466–2472.

Gonzalez, M. C., Hidalgo, C. A., & Barabasi, A.-L. (2008). Understanding individual human mobility patterns. *Nature, 453*, 779–782.

Hagerhall, C. M., Laike, T., Taylor, R. P., Kueller, R., & Martin, T. P. (2008). Investigations of human EEG response to viewing fractal patterns. *Perception, 37*, 1488–1494.

Hyona, J., Munoz, D. P., Heide, W., & Radach, R. (2002). *The brain's eye: Neurobiological and clinical aspects of oculomotor research* (Vol. 140). Amsterdam: Elsevier.

Mandelbrot, B. B. (1982). *The fractal geometry of nature.* San Francisco: W. H. Freeman.

Sprott, J. C. (2003). *Chaos and time-series analysis.* New York: Oxford University Press.

Taylor, R. P. (Ed.) (2005). *Encyclopedia of nonlinear science* (Vol. 1). New York: Routledge/Taylor & Francis Group.

Taylor, R. P. (2006). Reduction of physiological stress using fractal art and architecture. *Leonardo, 39,* 245–251.

Taylor, R. P., Micolich, A. P., & Jonas, D. (1999). Fractal analysis of Pollock's drip paintings. *Nature, 399,* 422.

Viswanathan, G. M., Afanasyev, V., Buldyrev, S. V., Murphy, E. J., Prince, P. A., & Stanley, H. E. (1996). Lévy flight search patterns of wandering albatrosses. *Nature, 381,* 413–415.

16

Oscillators with Differential Equations

Jonathan Butner and T. Nathan Story

CONTENTS

In this chapter, we explore a relatively novel way of modeling oscillatory phenomena. Our approach involves using equations from idealized physics models through regression, structural equation modeling (SEM; in this context also known as differential structural equation modeling, or latent differential equation modeling), and multilevel modeling. The approach incorporates a smoothing function, either separately and prior to the analysis (as is done in regression and multilevel modeling), or built into the estimation procedure itself (as in SEM). Further, this approach can model a single time series or several time series that are believed to share the same equation, or many time series that are believed to share the same form of equation but differ in the actual terms involved. The end result is a description of oscillations in traditional terms of frequency, damping, coupling, and common nonlinear divergences, as was first recognized by Raliegh, Van der Pol, and Duffing (Abraham & Shaw, 1992). In the case where there are multiple time series and differentiating terms in the equation, the models can be expanded to examine additional variables that may account for the unique differences observed in the oscillatory patterns (e.g., what control parameters influence the shape of the phase space, or the trajectory of the system).

From a dynamical systems perspective, our approach has two appealing features. The first is that it is a form of phase space (time delay) reconstruction, which allows us to map out the shape and characteristics of a system in terms of limit cycles, or oscillatory dynamics (oscillations are an integral feature of systems theory and understanding how a system changes in time). Further, limit cycles are one of the four most common topological features in understanding phase-space plots. Second, while comparable to cycle decomposition methods based on Fourier transformations (Chow, Ram, Boker, Fujita, & Clore, 2005), our method diverges and attempts to describe additional patterns in time using escapement terms, or terms that represent injections of energy into the system (while still identifying the primary frequency of a pattern in time). Escapements counter the constant transference of energy in open systems to other unstudied parts of the system. Further, these terms have a theoretical consistency with systems thinking (Kugler & Turvey, 1987).

This chapter is divided into several sections. The section "What does it mean to be oscillatory?" links the notion of a limit cycle to expressions of force. This provides a basis for the equations modeled. The section "Nonlinear forms" expands the force equations to include nonlinear terms, providing ways to interpret their meaning. The section "Coupling" then accounts for coupled processes, where two or more variables are oscillating. The section "Using the local linear approximation to estimate derivatives" examines the smoothing function, also known as the *local linear approximation*, used to generate the terms for the equations laid out in the previous sections. The section "Ordinary least squares regression" shows how to use this method in regression for a single time series. The section "Multilevel modeling for expanding to multiple people simultaneously" expands this into the multiple person circumstance requiring the same equation form across time series, but including the possibility for variability across the time series. The section "Structural equation modeling approaches" shows the approach using SEM, which incorporates the smoothing function into the estimation procedure. The section "Additional considerations" discusses additional considerations needed to apply this technique.

Throughout this chapter, we use data taken from a longitudinal study on adolescents (and their families) coping with type 1 diabetes and on their families. Specifically, we utilize a daily diary measure of mothers' daily affect as it related to their child's type 1 diabetes. Reports were collected over 14 consecutive days, submitted every evening. While a single person's time series of data is relatively short for modeling oscillations, the ability to simultaneously model multiple time series (in this case, 252 individuals each with their own time series) can account for the lack of power.

What Does It Mean to Be Oscillatory?

In phase-space plots of time-varying systems, *limit cycles* depict the circular motion of change around an attractor (set point) and though the ideal limit cycle makes a perfect circle (or spiral), they can be far from perfect in form. In time series terms, this usually translates into *periodic* (where a small portion of the time series repeats again and again, depicting a base pattern for the entire time series) or *quasi-periodic* patterns (where a small portion of the time series seemingly repeats but each repetition is slightly different). Oscillations from a time series perspective are commonly described as having several qualities: the *frequency* is the number of oscillations that are completed in a single unit of time (*period* is also common, defined as 1/frequency, or the number of time units it takes to complete one cycle), the *amplitude* is the maximal value of an outcome within a given oscillation, and *damping* is the rate of exponential increase or decrease in the amplitude over time, across oscillations.

There are several ways of expressing oscillations mathematically (many of the more common approaches use trigonometry; see Warner, 1998). The method herein parallels expressing oscillations in terms of derivatives generated from the logical expansion of forces for a *damped oscillator* where three forces combine together to depict the force function (and thus motion) of an object on a flat surface attached to a sidewall by a spring:

$$m\frac{d^2x(t)}{dt^2} + \beta\frac{dx(t)}{dt} + \kappa x(t) = 0 \tag{16.1}$$

The first term is the mass of the object (m) times acceleration ($d^2x(t)/dt^2$). The second term is friction (β) times velocity ($dx(t)/dt$), and the third term is the spring constant (the strength of a linear spring, κ) times position ($x(t)$). Note that the right side of the equation is fixed to zero, which is consistent with *stationarity*—that the system oscillates around a fixed value of the outcome (this can be changed, but it is the starting point for the logic). In many circumstances, the data is first detrended to parallel this logic (see Chapter 2 for discussions of detrending and advanced alternatives).

Also note that Equation 16.1 is a linear expression of change in that the equation excludes any interactions or polynomials in the force function.

This linear form implies that each force is independent of the others. When we move to include nonlinear terms, this will no longer be true.

This equation is then expressed to where acceleration (the second derivative, $(d^2x(t)/dt^2)$ is on one side of the equation, with the other derivatives and their constituent constants on the other:

$$\frac{d^2x(t)}{dt^2} = -\zeta\frac{dx(t)}{dt} - \eta x(t) \tag{16.2}$$

Zeta (ζ), the ratio of friction to mass, becomes damping—the exponential rate of amplitude loss or gain as a function of time. Eta (η), the ratio of the spring constant to mass, becomes the frequency squared, in radians. Only amplitude is left out of the expression. Equation 16.2 is in a form usable in regression analyses and other model fitting methods.

Notice in this new form that both damping and the squared frequency have a negative sign before them. This will result in damping having the reverse sign than commonly discussed in physics books (in this case, positive values will indicate an increase in amplitude over time) and that the sign for frequency should just be ignored (it is almost always going to be negative, meaning a positive frequency). Throughout our presentation, we will always interpret results consistent with Equation 16.2.

If this equation were to be utilized in a regression analysis, for example, note that it is lacking an intercept. If the data has been detrended beforehand, then the intercept will be zero and is thus unnecessary. However, if data has not been detrended or if a technique was used, which continues to leave a drift in the set point the oscillations are moving around (nonstationary—the forces are not summing to zero and the oscillations are around some point that is also changing in time), then an intercept should be included as a tuning term. The choice of including or excluding this term is at the discretion of the user, but can greatly impact the number of parameters being estimated in some of the later models (i.e., once we account for multiple time series, including an intercept can add several covariance estimates).

Abraham and Shaw (1992) provided a graphical way to understand these force relationships, consistent with Equation 16.2. In each case, they plotted force on the *y*-axis with velocity or position on the *x*-axis, showing the slope between them. We replace force with acceleration, but the graphical interpretation is the same (it is off by a scaling constant). Figure 16.1 expresses these linear relationships where the coefficients would be the slopes of the lines. Note the linear relationship implies that force (or change) linearly increases (as one diverges from a position of zero, acceleration increases but at higher velocities, acceleration decreases; together these create an oscillator spiraling in toward a set point attractor). This is the assumption made by the linear dynamic model. It is also plausible, however, to have force increase nonlinearly as oscillations move farther away as a function of position (this is what soft and hard springs do in physics models). Also important to note

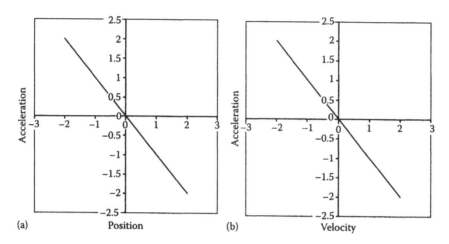

FIGURE 16.1

(a) Plot of acceleration as a function of position. The linear relationship indicates a constant spring force at all positions. (b) Plot of acceleration as a function of velocity. The linear relationship indicates a constant loss of amplitude (energy), as indicated by the negative slope.

is that the velocity and position relationships characterize the differences between *conservative* terms (which depict changes to the phase space that are unrelated to the total energy of the system) and *nonconservative* terms (which depict changes to the phase space that indicate a gain or loss of energy in the system). The relationship between acceleration and velocity is nonconservative in that it implies a loss of energy with greater velocity—this is the attractive or repulsive nature of the set point indicative of a limit cycle spiraling inward or outward. The relationship between acceleration and position is conservative; that is, across oscillations, no volume of motion is lost (but rather is more related to the speed at which the system moves through oscillations).

Nonlinear Forms

Several expansions of Equation 16.2 have been used to model nonlinear dynamic oscillations. In each case, polynomials and polynomial interactions are added, which change the phase space from a perfect circle (or spiral) to something less circular, or idealized. The changing relationship between acceleration and position in hard and soft spring models illustrates this point. Figure 16.2a exemplifies a nonlinear relationship where force increases faster once it surpasses a threshold in position; this is a *hard spring*. Figure 16.2b exemplifies a nonlinear relationship where the spring no longer continues to increase in force once surpassing a threshold in position; this is a *soft spring*.

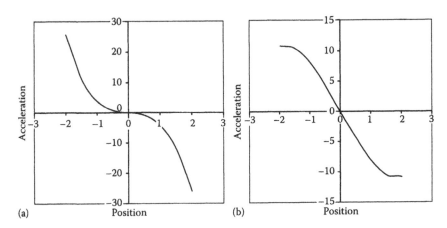

FIGURE 16.2
(a) Plot of acceleration as a function of position for a hard spring. (b) Plot of acceleration as a function of position for a soft spring.

Both of these nonlinear circumstances can be modeled by adding a cubic form of position to Equation 16.2:

$$\frac{d^2x(t)}{dt^2} = -\zeta \frac{dx(t)}{dt} - \eta x(t) - \delta(x(t))^3 \tag{16.3}$$

The only new term in Equation 16.3 is delta (δ), the cubic nature of the spring force as a function of position ($x(t)$). The sign of delta indicates the determination of the soft vs. hard spring on the cubic term. Notice that Equation 16.3 lacks the quadratic term of position, which is normally required when including a higher order term in regression models. Traditional expansions of oscillations only consist of cubic terms or terms where powers sum to cubic terms (Butner, Amazeen, & Mulvey, 2005). It has been argued that further powers are possible; the next plausible set would be to the sixth power, and so forth (Beek & Beek, 1988). This builds on the logic of the spring; for example, having symmetrical influence at a high position vs. a low position value. Because in the social sciences, it is possible to have a relationship that is not symmetrical, one should consider including quadratic terms depending on the phenomenon in question. It is unlikely that it would be necessary to go beyond cubic relationships, however, given that they would be difficult to distinguish from the lower order cubic term.

Four cubic terms have been modeled in the study of motor coordination: $dx(t)^3/dt$, $(dx(t)/dt)x(t)^2$, $x(t)^3$, and $(dx(t)^2/dt)x(t)$. The first of the terms is a function of velocity (velocity-cubed, $dx(t)^3/dt$) indicating a limit cycle's attraction where movement away in position or velocity is drawn back to the limit cycle pattern. The second term (velocity by position-squared, $(dx(t)/dt)x(t)^2$) also influences the area of the phase space and is very similar to the cubic velocity term in regard to the shape of the phase space (limit cycle); however, this term rotates the shape of the phase space such that it is slightly more angular (a different set of quadrants than the velocity-cubed term). Positive values for these terms indicate an increase in energy and negative values indicate a loss of energy in comparison to the starting point in the time series. Turning to conservative terms, the first is a function of position only (position-cubed, $x(t)^3$) and the second is again a function of velocity and position (in this case, velocity-squared*position, $(dx(t)^2/dt)x(t)$). The first term squeezes the phase space such that it becomes oval or a rounded rectangular shape. This indicates longer or shorter periods of sustained velocity prior to change as a function of position. The second term again squeezes the phase space influencing the frequency of the oscillations and indicates that change in velocity is abrupt, both in terms of increases and decreases, and as a result yields a diamond-shaped phase space or one more elliptical.

There can be some difficulty in distinguishing the impact of each nonlinear term on the oscillatory phenomenon, particularly when multiple terms contribute to a description of the oscillations. We therefore suggest making plots similar to those in Figure 16.1 where one can see the relationships between position and acceleration, and velocity and acceleration. These can be in the form of prediction plots as well, which generate the idealized relationships, or scatterplots of the data itself. Figures 16.3 through 16.6 show exemplars of each term in a series of prediction plots with equivalent phase-space plots (the series of *b* plots) generated using a Runge–Kutta fourth-order algorithm (a method of estimating data from a series of ordinary differential equations).

Coupling

In many circumstances, researchers are interested in the relationship between two or more state variables, rather than a single state variable. We limit our discussion to just two state variables, however, expanding the equations beyond two state variables is relatively straightforward (though data demanding; we visit the issue of sample size and time series length in the final section). A simplifying approach to *coupling* (linked variables through synchronizing forces) is to create a *phase variable* (a variable that describes the current location in the phase space) that is then modeled either as an oscillatory process or as some other function (merely utilizing the equations already discussed). Examples of this are common in the motor coordination literature where the relative positions of two pendulums are converted into radians (Kelso, 1994; Haken, Kelso, & Bunz, 1985). This has also been used recently to understand pain control as a function of expectant to actual pain, using a simple difference score between Likert-style scales (Finan et al., 2010). The advantage of a phase variable is that it can simplify the dynamic

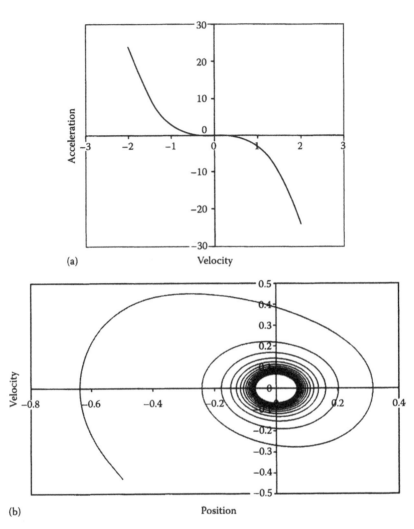

FIGURE 16.3
(a) Plot of acceleration as a function of velocity where velocity has a cubic relationship. Note
that a hypothetical line from the upper left to the lower right represents the attractive portion
of the cycle and a hypothetical line from the lower left to the upper right represents the repulsive portion of the cycle. (b) Phase space generated by a Runge–Kutta fourth-order algorithm
assuming a fixed frequency term of –1 and a velocity³ term of –3. All other terms are zero.
Extreme values of position or velocity are pulled back into the limit cycle pattern. Positive
terms depict an increase in energy and negative terms depict a loss of energy.

one is trying to represent, essentially collapsing dimensions that constitute
the phase space—a trick known to simplify even chaotic systems into easier
to manage representations (Abraham & Shaw, 1992). The disadvantage of a
phase variable is a loss of both the original metric of measures and individual divergences from phase. Haken (2006), in his theory of synergetics,

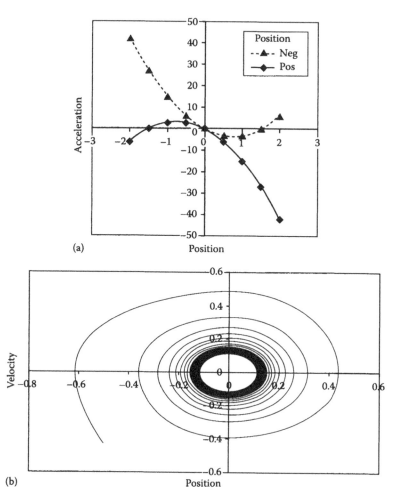

FIGURE 16.4

(a) Plot of acceleration as a function of position with separate lines for positive (2) and negative (−2) velocity to show the impact of a position² × velocity term. A hypothetical line from the upper left to the lower right represents the frequency through the limit cycle, but the locations where acceleration approaches zero are indicative of attractiveness where the cycles dwell for longer periods. (b) Phase space for position² × velocity term showing a limit cycle attractor (using the Runge–Kutta fourth-order algorithm); extreme position or velocity is pulled back into the limit cycle pattern. The term rotates the shape of the phase space such that it is slightly more angular.

argues for the importance of dynamics hierarchically, which a phase variable partly ignores.

Alternatively, one can simultaneously model both state variables, including a coupling relationship between the two. The disadvantage of this approach is that it increases the number of terms, becoming more data demanding. It also introduces some interpretation ambiguities we address below.

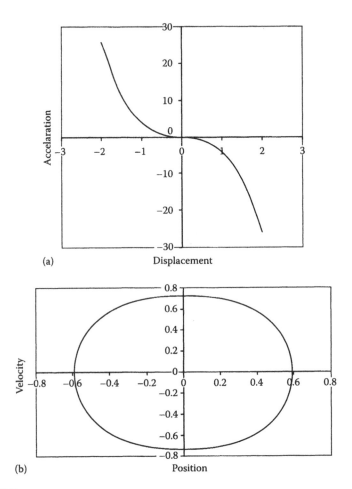

FIGURE 16.5

(a) Plot of acceleration by position showing the impact of the position³ term. This is a hard spring. (b) Phase space for the position³ term where the trajectory is squeezed, conserving the area of the space and yielding an oval or rounded rectangular shape; indicates longer or shorter periods of sustained velocity prior to change as a function of position.

When modeling two simultaneous equations, coupling has been represented in two different ways. The first is a direct descendent of Von Holst (1939, 1973) coupling logic (Butner et al., 2005). Specifically, Von Holst described coupling akin to a third party causal variable that drives two variables to become more similar. This decomposed motion into two constituent parts, a cooperative component known as the *magnet effect* and a competitive component known as the *maintenance tendency*. The second is a function of maintaining a Von Holst decomposition into cooperative and competitive components while capitalizing on covariation implicit in estimation (Boker & Graham, 1998). The differences between these two approaches parallel the

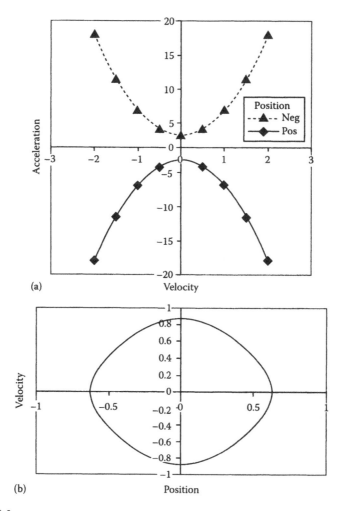

FIGURE 16.6
(a) Prediction plot of acceleration as a function of velocity depicting the position × velocity² term. The hypothetical diagonal axis from the upper left to lower right and lower left to upper right indicates a maintenance of energy over the course of the cycle though portions are loosing energy where others are gaining. (b) Phase space for position × velocity² term where the trajectory is again reshaped while conserving the area of the space yielding a diamond or elliptical shape. The term influences the frequency of oscillations such that change in velocity is abrupt.

arguments from the 1970s, 1980s, and 1990s of using raw difference scores vs. residualized difference scores (a raw difference is anchored to the individual score and thus is easier to interpret while a residualized difference score is a function of covarying which has more statistical power via reliability but must always be interpreted relative to the average pattern of change and tends to fit patterns of data beyond those it is designed to characterize; see

Rogosa, 1988, for a discussion of raw vs. residualized difference scores within the context of longitudinal data analysis). That is, each has its strengths and weaknesses. However, these two approaches, as with difference score methods, will have greater impact when they are on the outcome side of a model. Thus, we introduce both.

We first illustrate the Von Holst coupling approach where a difference in positions between the two oscillatory processes (x and y) is added to each equation

$$\frac{d^2x(t)}{dt^2} = -\zeta_1 \frac{dx(t)}{dt} - \eta_1 x(t) - \kappa_1 \left[y(t) - x(t) \right] \tag{16.4}$$

$$\frac{d^2y(t)}{dt^2} = -\zeta_2 \frac{dy(t)}{dt} - \eta_2 y(t) - \kappa_2 \left[x(t) - y(t) \right] \tag{16.5}$$

The first part of Equations 16.4 and 16.5 are identical to Equation 16.2, but for two processes (with added subscripts to distinguish the equations for x and y). Only the kappas (κ) are new: they are the coupling estimates. Notice that the calculation of the difference scores reverses between the two equations. Reversing one of the difference scores makes it so that the estimates are interpreted in the same direction, in terms of their signs (this is important to note since one is often unable to include two variables that are correlated at −1 due to multicolinearity, so the user must recognize the reversal of sign necessary for comparing the two relationships—this often leads to a preference for the second approach). The second approach merely consists of adding in the position of the other oscillator as a variable:

$$\frac{d^2x(t)}{dt^2} = -\zeta_1 \frac{dx(t)}{dt} - \eta_1 x(t) - \kappa_1 y(t) \tag{16.6}$$

$$\frac{d^2y(t)}{dt^2} = -\zeta_2 \frac{dy(t)}{dt} - \eta_2 y(t) - \kappa_2 x(t) \tag{16.7}$$

The kappas (κ) have the same definition as from Equations 16.4 and 16.5. However, this capitalizes on the covariation between the positions of the two oscillators to generate a residual-based form of difference. Residual forms of differences are known to be statistically powerful, but also tend to encounter more interpretation problems (Rogosa, 1988).

In both cases, it is important to note that nonzero estimates in kappa (κ) indicate coupling. Both positive and negative coupling terms are possible as the signs are related to a parameter that is implicit in coupled systems but not actually estimated here and is known as the *phase* (or timing) between the two oscillators. Thus, the difference between a positive and negative sign for a coupling term becomes a gross indication for the phase relationship and be best thought of as similar to a comparison of a sine function and a cosine function in adding together two wavelets (describing a shift in phase; see Morris & Carroll, 2006). We therefore recommend only interpreting the magnitude of a coupling term.

Symmetry

A comparison of the magnitude of coupling terms can be used to examine the *symmetry of influence* between the two oscillators—the extent to which one oscillator may be more dominant in driving the motion of the two. This logic assumes that the coupled influences are bidirectional in nature and that the timing of measurement and delays used to generate derivatives (see the local linear approximation later in this chapter) properly reconstitute system relationships (similar to phase space reconstruction methods). Assuming this to be true, the two coupling estimates can be compared, where equal influence results in equal coupling terms (since the coupling terms are in a squared frequency metric, the coupling terms are comparable even if the two oscillators are functioning at different scales as long as the timing of measurement is the same) and asymmetrical influence results in one coupling term being larger than the other (Butner, Diamond, & Hicks, 2007). In all cases, the larger coupling term will coincide with the oscillator being more influenced.

One could also test to see if the coupling terms differed from zero. Such circumstances imply the ability to test for unidirectional relationships or a complete lack of coupling. We warn readers to be wary of this interpretation since bidirectional relationships can detect as unidirectional or unrelated as a function of measurement timing, especially in oscillatory phenomena. Additionally, there is interchangeability between coupling and escapements (the earlier nonlinear terms) that can be thought of as a function of Takens' theorem (as discussed in Chapter 2) where expansions of how derivatives of a variable (or delays) relate to itself can account for other external known and unknown influences.

Synchrony

In a coupled system, the two frequency terms (now *eigen-frequencies* or *natural frequencies* representing the competitive portion of Von Holst's coupled systems; these are the frequencies we would expect to see if the two oscillators were not coupled) and the two coupling terms (the cooperative portion of Von Holst's coupled systems) combine together to holistically depict the phasic movement of the oscillators. It is possible to generate a *generalized Reynolds number* (a dimensionless number that indicates key properties of a system; Iberall, 1987) that signifies the *degree of synchronization* between the two oscillators. In all cases, the signs of the original estimates are ignored prior to entry into the equation. Further, we log the solution for simpler interpretation resulting in the equation

$$\text{Synchrony} = \ln \frac{|\eta_1 - \eta_2|}{|\kappa_1| + |\kappa_2|} \tag{16.8}$$

The Etas (η) and Kappas (κ) are the eigen-frequencies and coupling estimates taken from Equations 16.4 and 16.5 or 16.6 and 16.7. We include absolute values in the equation to remove the ambiguity in signs mentioned earlier. In this logged form, $-\infty$ indicates perfect *phase locking* (the two oscillators are moving at a perfect ratio of motion) since the difference in eigen-frequencies will be zero. It is important to note that this perfect circumstance will also likely cause any estimation procedure that includes a coupling term to fail to estimate due to multicolinearity in the positions. Additionally, even if it were to provide a solution, the coupling terms will appear as zero (having nothing to estimate them from).

Values of ∞ indicate complete *deregulation* where the two oscillators are unrelated to one another. Again, the coupling terms would estimate as zero, but the two eigen-frequencies would be distinct. This shows an ambiguity in the estimation in that complete deregulation and phase locking can have the identical decomposition.

The true power of the coupling procedure (and the synchrony ratio) is under loosely coupled relationships. Values slightly above zero are indicative of *drift* in a loosely coupled system where more of the patterns in time are unrelated than related. Values less than zero are indicative of *entrainment* in a loosely coupled system where more of the pattern in time holds some degree of synchrony with periods of slippage. This continuum can thus be used as a way to indicate the synchrony of the two variables.

Using the Local Linear Approximation to Estimate Derivatives

As noted in Equation 16.2, the analytic procedure uses first- and second-order derivatives to estimate the parameters. In structural equation models (SEM; a statistical approach for estimating the parameters of a theoretical model as a function of observed data; see Ullman & Bentler, 2003), these derivatives are estimated as part of the overall approach. For regression and multilevel modeling procedures (multilevel modeling is a statistical approach designed to account for data dependencies, in this case, the time series within each individual; it is analogous to conducting a regression analysis within each case and then a series of regression analyses across cases to predict differences in the within-case equations), they must be estimated prior to the procedure. Boker and Nesselroade (2002) tested two different methods of estimating derivatives locally so that a time series could generate multiple values in time. They found that the local linear approximation provided acceptably unbiased results. The equations for acceleration and velocity both follow:

$$\frac{dx_{(t_i+\tau)}}{dt} \approx \frac{x_{(t_i+2\tau)} - x_{(t_i)}}{2\tau} \tag{16.9}$$

$$\frac{d^2_{(t_i+\tau)}}{dt^2} \approx \frac{x_{(t_i+2\tau)} - 2x_{(t_i+\tau)} + x_{(t_i)}}{\tau^2} \tag{16.10}$$

Equation 16.9 generates estimates for velocity. Tau (τ) is a delay in measurement (equivalent to L in Chapter 4, but tau is commonly used in this body of literature). Thus, Equation 16.9 is merely the difference of the two values of x surrounding a specific point in time divided by 2 and accounting for the delay. Equation 16.10 follows the same logic, but for acceleration. The success of this technique is greatly influenced by the choice of tau in that it should be the same as the ideal value for phase space reconstruction (see Chapter 3 for a review of methods). Tau also results in a reduction of data points where one will lose two times tau measurements after the transformation. When applied to relatively short time series (as will be the example described here), it is not uncommon to use a tau of 1 to maintain the maximal amount of data. This choice can bias the results and thus should be conducted with caution and awareness of the trade-off between power and accuracy.

It is also noteworthy that any analyses conducted on these transformed variables are idealized in that no error is maintained in the acceleration and velocity estimates. This is one of the key advantages to the SEM approaches in that Equations 16.9 and 16.10 are implicit in the model and thus maintain errors in estimation of velocity and acceleration, respectively. Maintaining these errors in estimation allows for unbiased tests that take all of the data into account. However, one can also think of this transformation as a smoothing function where the resulting derivatives are more representative of the underlying changes that make up the system. Techniques that ignore error in estimation of acceleration and velocity can be thought of as a bleaching process where noise can be removed to some degree.

Example in SPSS

Any general statistical program can be used to create the local linear approximation estimates. We utilize SPSS, a program relatively common in the social sciences. The data first need to be structured such that each individual's data take up multiple lines, sorted by individual (if there is more than one individual) and sorted by time. We then create a lag and lead as a function of the chosen tau to generate the values around the outcome at each point in time. It is important for this step to only be conducted within each time series, if there is more than one. Finally, acceleration and velocity are created. Our syntax to generate these estimates for analyses of mothers' negative affect are shown below (the data was detrended prior).

```
*The Sort and split file commands conduct analyses within each
  dyadid only.
SORT CASES BY dyadid.
SPLIT FILE LAYERED BY dyadid.
*Use the lag and lead functions to create delays as a function
  of tau=1.
create /mnegr_1 = lead(mnegr 1).
create /mnegr_2 = lag(mnegr 1).
execute.
*Compute velocity and acceleration as a function of the
  original x and the delays.
*Again we use a tau of 1.
compute mnegrvel = (mnegr_1-mnegr_2)/2.
compute mnegracc = (mnegr_1 - 2*mnegr+mnegr_2)/1.
execute.
```

Ordinary Least Squares Regression

The regression approach allows for examining a single time series with the advantage being that it can include a variety of nonlinear escapements. To demonstrate this, we chose a mother's time series of daily ratings of negative affect, as it related to her child's type 1 diabetes, and estimated velocity and acceleration for her individual time series using the local linear approximation (after removing the linear trend from the data). We then used velocity and acceleration to calculate the four nonlinear escapement terms described earlier: $(dx(t)^3/dt$, $(dx(t)/dt)x(t)^2$, $x(t)^3$, and $(dx(t)^2/dt)x(t))$. Given that we had only 14 measures in time, we chose a tau of 1 to maximize our remaining data, while also recognizing that it can bias our results. We then conducted an ordinary least squares regression using the equation below:

$$\text{Acc} = b_0 + b_1\text{pos} + b_2\text{vel} + b_3\text{pos}^3 + b_4\text{pos} * \text{vel}^2 + b_5\text{vel}^3 + b_6\text{pos}^2 * \text{vel} + e \quad (16.11)$$

where
  "Acc" is acceleration in negative affect
  "vel" is velocity in negative affect
  "pos" is raw negative affect after detrending

We chose to include an intercept (b_0) as a tuning term (i.e., acceleration when all predictors equal 0, which is normally zero and thus unnecessary), b_1 is squared frequency, b_2 is damping, whereas b_3 through b_6 are the four nonlinear escapements mentioned earlier. e is the error in prediction. Table 16.1 shows the estimates from this equation. We provide detailed interpretation in the interpretation section below.

Example in SPSS

To conduct this analysis in SPSS, utilize the ordinary least squares regression module. Our syntax for this analysis is provided below (*Note*: to exclude an intercept, change "/noorigin" to "/origin."

TABLE 16.1

Mother's Negative Affect for Single Participant (All Terms Included)

Variable	Estimates
Intercept	−0.043*
Position	−2.209**
Velocity	0.21
Position³	−1.108
Velocity³	−6.795*
Position² × velocity	14.141
Position × velocity²	−31.149**

* $p < .05$.
** $p < .001$.

```
*Create the four possible nonlinear terms.
compute mnegcu = mnegr**3.
compute mnegVcu = mnegrvel**3.
compute mnegxVsq = mnegr*mnegrvel**2.
compute mnegsqxV = mnegr**2*mnegrvel.
execute.
*Conduct a regression with Accelerations as the DV.
REGRESSION
  /MISSING LISTWISE
  /STATISTICS COEFF OUTS R ANOVA
  /NOORIGIN
  /DEPENDENT mnegracc
  /METHOD=ENTER mnegr mnegrvel mnegcu mnegVcu mnegxVsq
    mnegsqxV.
```

Interpretation of Results

The squared frequency (coefficient for position, b_1) was −2.209. Taking the square root of this and dividing by 2π converts the term into frequency as a function of measurement timing (number of cycles per measurement). Specifically, the estimated frequency of this mother's negative affect was 0.236 cycles per day or a period of 4.23 days to complete one cycle (1/frequency). The linear damping term (b_2) was nonsignificant but still large in size. Both regression and the multilevel modeling approach tend to overestimate the size of this coefficient (Beek, et al., 1995; Butner et al. 2005). The overestimation can be resolved in the SEM approach. In the case of negative affect, linear damping is not a very realistic term as it estimates a constant rate of growth or loss in amplitude that would be uncommon in regulatory processes like daily negative or positive affect. As a result, the term is better left excluded from model.

Among the nonlinear terms, we focus only on those that reached conventional significance ($\alpha = 0.05$, two-tailed) for interpretation. The velocity-cubed term (b_5) indicates the attractive nature of the oscillations in that they are settling into a limit cycle (beginning the cycle spiraling inward and then stabilizing). At low values (near 0), the term distorts the idealized cycle and an oval-like pattern is observed. At larger values (in our case negative), there is a loss of energy until the cycle stabilizes in a limit cycle. The term also creates an inward- and outward-directed spiral such that there is both an outside attractor and inside repeller space. In this case, the energy of the cycle is changing over time and moving toward dissipation before settling into

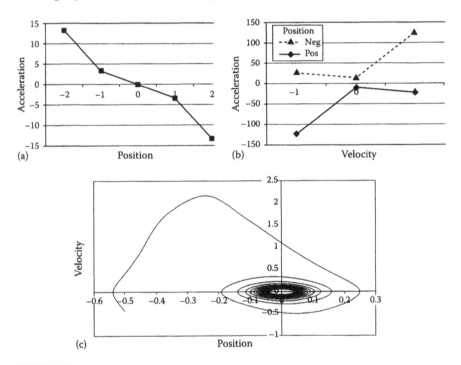

FIGURE 16.7
(a) Prediction plot for acceleration as a function of position (holding velocity at 0) for negative affect of a single mother over 14 days. Both the linear position and cubic position terms reached conventional significance. The pattern is consistent with a hard spring. (b) Prediction plot for acceleration as a function of velocity (at high and low position) for negative affect of a single mother over 14 days. The hypothetical diagonal from the upper left to the lower right indicates a fairly weak attractive portion. The hypothetical diagonal from the lower left to the upper right indicates a comparatively strong repulsive portion. (c) Phase space for the individual case using coefficients for position, velocity^3, and position × velocity^2 terms. Notice both the attractor space resulting from the velocity^3 term and the altered frequency (diamond shape) resulting from the position × velocity^2 term.

a limit cycle. The position x velocity-squared term (b_4) is conservative and squeezes the phase space such that it becomes diamond shaped rather than circular as the coefficient increases, indicating unequal rates of change as a function of where one is in the cycle. In terms of position and velocity, the position × velocity-squared term relates to change in frequency across the cycle, and can be interpreted to mean that high velocity in negative affect is followed by a sudden shift or deceleration. This pattern is much like the metaphor of a "lead foot" driver who is either accelerating full-throttle or decelerating with both feet on the brake, but never mitigating the rate of change. To illustrate these findings, Figure 16.7 shows prediction plots and idealized phase space depictions using the extracted equations and the Runge–Kutta fourth-order algorithm.

Multilevel Modeling for Expanding to Multiple People Simultaneously

We then expanded the regression approach to multiple individuals in a multilevel (mixed- or random-coefficient) model using the Mixed procedure in SPSS 17. Multilevel modeling is equivalent to conducting a regression analysis within each individual using an equation like that of Equation 16.11, saving out the coefficients for each individual and then conducting a series of regressions predicting each coefficient as a function of individual difference variables. Unlike going through this arduous series of analyses, multilevel modeling utilizes a maximum likelihood-Bayesian hybrid to simultaneously estimate the parameters we would observe if we had done this procedure. We believe multilevel modeling is quite promising for the study of systems since it is capable of estimating both the average relationship in time and across individuals (fixed effects), using both linear and nonlinear dynamic terms. It is also able to estimate variability in each of the terms across the individuals (random effects). The fixed effects have the advantage of drawing power across all measurements, addressing the dearth of measurement within a single time series by compensating through measurements across time series. The random effects have the advantage of maintaining error structure within the model so that tests of what accounts for the variability across individuals are not inflated beyond those already inflated by the local linear approximation.

To illustrate this, we expanded the prior regression model to include all 252 mothers in a multilevel model. We began with all nonlinear terms in the model (except linear damping due to it not being theoretically expected, and the intercept in order to reduce the number of estimated covariance parameters). Thus, the regression analysis conducted on the single individual becomes one case within this model, but new trends may be detectable that cannot be observed in that short time series. For time i, person j,

$$\text{Acc}_{ij} = \gamma_{10}\text{pos}_{ij} + \gamma_{20}\text{pos}_{ij}^3 + \gamma_{30}\text{pos} * \text{vel}_{ij}^2 + \gamma_{40}\text{vel}_{ij}^3 + \gamma_{50}\text{pos}^2 * \text{vel}_{ij}$$

$$+ \omega_{1j}\text{pos}_{ij} + \omega_{2j}\text{pos}_{ij}^3 + \omega_{3j}\text{pos} * \text{vel}_{ij}^2 + \omega_{4j}\text{vel}_{ij}^3 + \omega_{5j}\text{pos}^2 * \text{vel}_{ij} + e_{ij}$$

$$(16.12)$$

Fixed effects (represented by the gammas, γ) indicate the average regression equation across individuals and random effects (represented by the variance in omegas, ω) indicate how the equation varies across individuals. Notice that these directly parallel Equation 16.11 with the exception of the intercept and linear damping (b_2; there are just two terms for each effect we had before). Significant fixed and random effects were found for position and the position x velocity-squared terms (γ_{30} and ω_{3j}) and only fixed effects were observed for the cubic position term (γ_{20}). Table 16.2 contains the fixed effects. Table 16.3 contains the variance in random effects, which we discuss in the interpretation section.

TABLE 16.2

Fixed Effects Mother's Negative Affect

Variable	Average Estimate
Position	−1.954**
Position³	−0.090**
Velocity³	0.034
Position² × velocity	0.011
Position × velocity²	−0.747*

$*p < .05.$
$**p < .001.$

TABLE 16.3

Random Effects Mother's Negative Affect

Variable	Variance in Estimate
Position	0.178**
Position³	0.001
Velocity³	0.004
Position² × velocity	0.001
Position × velocity²	1.442*

$*p < .05.$
$**p < .001.$

Example in SPSS

The following syntax conducts a multilevel model in SPSS.

```
MIXED
mnegracc WITH mnegr mnegcu mnegxVsq mnegVcu mnegsqxV
  /FIXED = mnegr mnegcu mnegxVsq mnegVcu mnegsqxV | noint
  SSTYPE(3)
  /METHOD = REML
  /PRINT = SOLUTION TESTCOV
  /RANDOM mnegr mnegcu mnegxVsq mnegVcu mnegsqxV |
    SUBJECT(dyadid).
  /repeated | SUBJECT(dyadid).
```

Note that individual difference variables can be added to account for the variability in random effects. In this case, the individual difference variables are added to the model statement as interactions with the varying terms. For example, if we were interested in testing if the position x velocity-squared term differed as a function of self-esteem (depicting self-esteem as a control parameter), then the syntax would be expanded to include a self-esteem by position x velocity-squared interaction. All the issues of centering and scaling for interactions apply.

Interpretation

The fixed effect of position indicated that individuals on average were oscillating at a frequency of 0.22 cycles per day (complete one cycle every 4.5 days). However, the random effect of position indicated that this frequency was not the same for all individuals. To identify average high and low frequencies, convert the random effect from a variance to a standard deviation (square root of the frequency random effect), then add and subtract this to the fixed effect. Finally, convert these into frequency metric. In this case, average variations of frequency were a 0.21 (4.8 days to complete a cycle) and 0.24 (4.1 days to complete a cycle), respectively.

The significant position-cubed effect suggests a pattern consistent with a soft or hard spring but the lack of random effect implies this to be universal. The significant position by velocity-squared fixed effect indicates the same "lead foot" as seen in the regression example, but the significant random effect on this term suggests that this pattern is only true for some of the individuals. To graph these cases, we created prediction plots of one standard deviation above and below in Figure 16.8a and b and a plot of just the position effect in Figure 16.9.

The analyses showed that across mothers, there is a conservation of energy in the cycles of negative affect relating to their child's type 1 diabetes. Most interesting is the way in which the position × velocity-squared term describes the way that negative affect is cycling in a fashion that resembles pumping

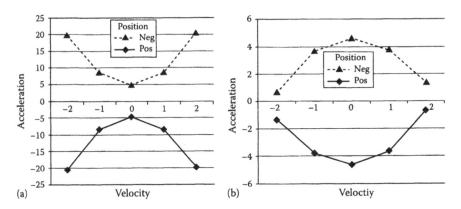

FIGURE 16.8
(a) Prediction plot of acceleration as a function of velocity from the mixed model where the position × velocity² term is at 1 standard deviation below the mean. Separate lines are shown for high vs. low position. Note the symmetrical influence of energy loss and gain over the cycle. (b) Prediction plot of acceleration as a function of velocity from the mixed model where the position × velocity² term is at 1 standard deviation above the mean. Separate lines are shown for high vs. low position. Notice that the diagonals from upper left to lower right and lower left to upper right (representative of attraction and repulsion, respectively) are on the reverse positions than in previous graphs.

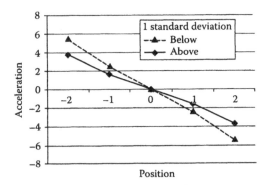

FIGURE 16.9
Prediction plot of acceleration as a function of position with separate lines for individuals 1 standard deviation above and 1 standard deviation below in the position term (cubic term was included as a fixed effect only). Individuals are slightly hard spring in nature (the fixed cubic effect) but differ in their rate of cycling.

a gas pedal rather than gradually increasing and decreasing speed while moving about the cycle. The velocity of negative affect is punctuated by sudden decreases rather than a continuous and smooth trajectory. This may be capturing something about diabetes-related critical events. An advantage to this approach is that additional psychosocial variables can be included in the model to explain the variability observed across individuals in the various nonlinear terms; however, we do not present those data here.

Coupling

The multilevel model approach has the added bonus of being able to properly model coupled systems. This involves creating a new dependent variable (DV) that switches off being the DV for the first coupled system and the DV for the second coupled system (thus each person's data at a point in time takes two rows in the data file). One then adds a pair of dummy codes that are redundant with one another. The first is coded one when the DV is the first of the coupled system and the second is coded one when the DV is the second coupled system. The intercept is suppressed to allow for redundant dummy codes (the all-zero group does not exist) and the main effects for the dummy codes take on the value of the intercepts for each process (also not necessary, and dropped here). The estimated equation is then a series of interaction terms with these dummy codes, making two parallel equations that can have different equations for each. To keep this simple, we only retain linear effects for individual i, time j, and variable d:

$$\text{Acc}_{ijd} = \gamma_{101}\text{pos}_{ij1} * \text{Dummy}_1 + \gamma_{201}\text{pos}_{ij2} * \text{Dummy}_1$$

$$+ \gamma_{302}\text{pos}_{ij2} * \text{Dummy}_2 + \gamma_{402}\text{pos}_{ij1} * \text{Dummy}_2$$

$$+ \omega_{1j1}\text{pos}_{ij1} * \text{Dummy}_1 + \omega_{2j1}\text{pos}_{ij2} * \text{Dummy}_1$$

$$+ \omega_{3j2}\text{pos}_{ij2} * \text{Dummy}_2 + \omega_{4j2}\text{pos}_{ij1} * \text{Dummy}_2 + e_{ijd} \quad (16.13)$$

Dummy_1 and Dummy_2 are dummy codes indicating the variable ($\text{Dummy}_2 = 1 - \text{Dummy}_1$). Note that when Dummy_2 equals zero (Dummy_1 would equal 1), the equation above becomes the same as Equation 16.6 (except we left out the velocity effect). When Dummy_1 equals zero ($\text{Dummy}_2 = 1$), the remaining terms generate the reverse equation, Equation 16.7. One can utilize either form of coupling using Equations 16.4 and 16.5 instead (discussed earlier). It is also important to model the error structure across the coupled system. In this case, one should estimate a separate error variance of the DV at each level of one of the dummy codes.

As with the earlier multilevel model, we can examine the estimates of the frequencies and variability in the frequencies. Now they are eigen-frequencies—the frequency we would observe if there is no coupling. Unlike previous models, one can test to see if the frequencies are the same across the two variables. This is a standard equality contrast using a weight vector with values of 1 and –1 for the two parameters being compared. A similar comparison can be made for differences in any nonlinear terms included in the model. Applying this same equality contrast to the coupling terms (γ_{201} and γ_{402}

for the equality of the fixed effects, and ω_{2j1} and ω_{4j2} for the equality of the random effects). Note that Equation 16.13 uses the coupling method where the frequency of the other variable predicts acceleration (the signs should be in the same direction); this can be applied to the other coupling term as well (be certain the signs are in the proper direction).

Structural Equation Modeling Approaches

SEM programs allow for the estimation of complex relationships that characterize a theoretical model through a series of simultaneous regression analyses. Given its potency, it can be used to calculate the local linear approximations for acceleration and velocity in conjunction with the relationships between them that constitute frequency, damping, and coupling. Like the multilevel modeling procedures described earlier, it also tends to utilize maximum likelihood estimation requiring that there be an analytic solution for each parameter but does not necessitate knowledge of the particular analytic solution. This has two advantages over the previously mentioned approaches. First, the SEM approach maintains error structures at all levels of estimation (error in estimating acceleration and velocity as well as the estimates of frequency, damping, and coupling), improving the accuracy of tests. Second, it can capitalize on multiple ways to estimate the same parameter, thereby generating estimates that are less likely to be biased. In exchange, the SEM approach generally has much more trouble with nonlinear oscillatory models in that it requires complex latent variable interactions beyond the current scope of latent variable interaction methods applied by Marsh and colleagues (Marsh, Wen, & Hau, 2006). Furthermore, modeling differences across time series is possible, but very difficult to estimate. It is more common, therefore, to ignore the inherent dependency and simultaneously model both within time series and across time series. The end result is an unbiased average effect across individuals, but standard errors that are too small (alpha inflation).

Parallels to Growth Curve Modeling

Estimating position, velocity, and acceleration directly parallels a quadratic latent variable growth model with a minimal number of measures (Boker, 2001). The common structure for growth curve modeling requires that the multiple measurements in time each be treated as separate observed variables in the model. To use the minimal number of measures, we express the model with only three measures. The observed intercepts are fixed to zero. The position latent variable is represented by a factor where all the observed variables have loadings fixed to 1. The velocity latent variable is represented by a factor where the observed variables have loadings fixed to incrementally increase one unit per measure (zero is scaled to be the middle measure Y at T). Thus, with only three measures in time, the loadings are –1, 0, and 1, respectively. The acceleration latent variable is represented by a factor where the observed variables have loadings fixed to the squared values of those used for the velocity latent variable (1, 0, and 1). Under quadratic growth models, residual variances are normally estimated for the observed measures. However, with only three measures, the position, velocity, and acceleration latent variables are complete decompositions of the three measures. The residual variances are therefore fixed to zero for identification. With four measures, they can be equated; beyond three, however, it becomes a growth model rather than an oscillatory model. In the case of four measures, they must be clearly represented in the same part of the oscillation for all individuals. The equation for the damped harmonic oscillator is generated by regressing the acceleration latent variable on the velocity and position latent variables, which constitute damping and squared frequency, respectively.

The structure described so far requires three measures for a time series. For more than three measures in time, the data file is restructured so that each line of data contains y at time t, y at time $t - \tau$, and y at time $t + \tau$. If one wishes to estimate the residual variances of the outcome, four measures are needed at a point in time instead of three (utilize two measures prior to time t and two measures after time t, excluding the measure at time t, itself). However, one must be careful in that the overall approach assumes that only a portion of the oscillation is observed within the measurement window. Add too many measures and the process will smooth out the oscillations across time, losing the very properties you are attempting to capture! We therefore choose to remain at a single measure prior to time t and a single measure after time t. This data structure will generate a great deal of missing data. Most SEM programs utilize Full Information Maximum Likelihood to account for missing data (which we recommend using to incorporate all the available data points).

Once the data have been restructured in this format, use the same model as when there are only three points in time. The model then applies for a single time series. The entire model is illustrated in Figure 16.10. When there are multiple time series, one can either ignore between vs. within differences (generating alpha inflation, but a correct estimate of the average oscillatory properties) or move to a hybrid multilevel/SEM model (a structural equation model estimated at two separate levels of the data essentially combining multilevel modeling and SEM). Several programs can conduct these hybrid models, but only Mplus can currently allow for *random slopes*—the case where variability in damping and frequency across individuals (or the set of time series) can be predicted by control parameters or individual differences of interest. Unfortunately, this model is quite data demanding and unstable, and as such, it has not resulted in an example in the literature of which we are aware (though it may possibly exist).

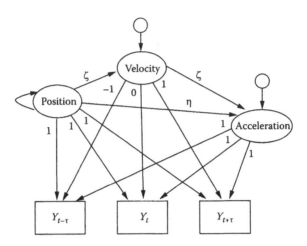

FIGURE 16.10
Structural equation modeling path diagram of a coupled damped oscillator.

Chow et al. (2005) improved on the estimation of this base model by noting that the relationships between derivatives are constant. Thus, one can get an estimate of damping not only from acceleration predicted from velocity, but also velocity predicted from position. Thus, one can improve estimation of each relationship by adding additional regression weights and equating them with the other equivalent terms. Using four measures in time (y_{t-2}, y_{t-1}, y_{t+1}, and y_{t+2}, the loadings convert to −1.5, −0.5, 0.5, and 1.5 for velocity, and these values squared for acceleration) and creating a further derivative (*jerk*; the loadings are the velocity loadings cubed), one can generate three estimates of damping and two estimates of frequency. The only assumption is that the four measures occur in the same cycle.

Under coupling, one can use the methods stated earlier (though one cannot enter the variable twice and thus would need to conduct the classic Von Holst logic with a single variable where the signs would reverse) or could alternatively take advantage of the latent variable capabilities of SEM to create forms of a phase variable within the model. For example, one can model each variable simultaneously and capture the coupled relationship by allowing the position of one variable to predict the acceleration of another. This directly parallels the third option mentioned earlier. Or, one can create a latent construct of the coupled variables and model the position, velocity, and acceleration of this new latent variable (position, velocity and acceleration become second-order latent variables where their observations are the first-order latent variables at each point in time). Presumably, the latent construct then represents some sort of phase relationship amongst the coupled processes and logically functions if the phase of the phenomena can be captured as the shared variance of the outcomes in time (see Chow et al., 2005 for an example). Lastly, one can construct latent difference

TABLE 16.4

Estimates from the Structural Equation
Model of Coupled Positive and Negative
Affect

Variable	Estimate
Positive	
Positive position (η_p)	−1.075*
Positive velocity (ζ_p)	−0.009
Negative position (κ_p)	0.010
Negative	
Negative position (η_n)	−1.135*
Negative velocity (ζ_n)	−0.009
Positive position (κ_n)	−0.004

* $p < .000$.

scores between coupled processes at each point in time and model the position, velocity, and acceleration of these latent difference scores. This form of modeling would suggest that the phase relationship is not a function of the shared variance, but more a function of the multiplicative relationship between the coupled systems, consistent with emergence. "While the third option is a logical expansion, it is complicated by combining additional family of SEM models (latent difference score models—see McArdle, 2001)." All of this taken together, venturing down this road is not for the faint of heart.

As a coupled example, we modeled the simultaneous oscillations of positive and negative affect for mothers. To capture coupling, we allowed the position of positive affect to predict the acceleration of negative affect and the position of negative affect to predict the acceleration of positive affect. Note that we chose to only use three measures to represent each time point rather than four. This decision came from our earlier analyses where the average period was close to 4.5 with some people having a shorter period. Using four measures would potentially smooth out the function weakening estimation. To be consistent with the previous analyses, we also maintained a tau of 1 (using the adjacent lag and lead for each time point). Table 16.4 contains the estimates.

Example in Mplus

To run the model collapsing across individuals, we followed the steps of generating negative affect reports from both the adolescent and mother at time t, $t-1$, and $t+2$. Thus, in this example, tau was fixed to 1.

```
TITLE: Example Coupling Model Collapsing Within and Between
  Levels
DATA:
  FILE IS "affectlaggeddet.dat";
  FORMAT IS f8,f11/7f7.3/7f7.3/7f7.3/7f7.3;
VARIABLE:
  NAMES ARE dyadid time
      tpos_tn3 tpos_tn2 tpos_tn1 tpos_t tpos_t1 tpos_t2 tpos_t3
      tneg_tn3 tneg_tn2 tneg_tn1 tneg_t tneg_t1 tneg_t2 tneg_t3
      mpos_tn3 mpos_tn2 mpos_tn1 mpos_t mpos_t1 mpos_t2 mpos_t3
      mneg_tn3 mneg_tn2 mneg_tn1 mneg_t mneg_t1 mneg_t2 mneg_t3
      ;
  USEVARIABLES ARE
!Each measures is at t-tau, t, and t+tau where tau is one
  (adjacent measures);
      tneg_tn1 tneg_t tneg_t1
      mneg_tn1 mneg_t mneg_t1
;
  MISSING IS BLANK; !Mplus 5 defaults to Full Information
    Maximum Likelihood;
ANALYSIS:
  ESTIMATOR IS ML;
  ITERATIONS = 1500;
MODEL:
!Teen's negative affect first;
tneg_tn1@0 tneg_t@0 tneg_t1@0; !Fix residual variances to zero;
Disp_tn by tneg_tn1@1 tneg_t@1 tneg_t1@1; !Create position
  factor;
Vel_tn by tneg_tn1@-1 tneg_t@0 tneg_t1@1; !Create velocity
  factor;
Acc_tn by tneg_tn1@1 tneg_t@0 tneg_t1@1; !Create acceleration
  factor;
[tneg_tn1@0 tneg_t@0 tneg_t1@0]; !Fix intercepts to zero;
[Disp_tn Vel_tn Acc_tn]; !Estimate intercepts for factors;

Acc_tn on Vel_tn (1) !Regress Acc on Vel for damping;
Disp_tn; !Regress Acc on Disp for frequency;
Vel_tn on Disp_tn (1); !Regress Vel on Disp for second damping
  estimate;
!The (1) equates the two coefficients;

!Now Mom's negative affect;
mneg_tn1@0 mneg_t@0 mneg_t1@0; !Fix residual variances to zero;
Disp_mn by mneg_tn1@1 mneg_t@1 mneg_t1@1; !Create position factor;
Vel_mn by mneg_tn1@-1 mneg_t@0 mneg_t1@1; !Create velocity
  factor;
Acc_mn by mneg_tn1@1 mneg_t@0 mneg_t1@1; !Create acceleration
  factor;
[mneg_tn1@0 mneg_t@0 mneg_t1@0]; !Fix intercepts to zero;
[Disp_mn Vel_mn Acc_mn]; !Estimate intercepts for factors;
```

```
Acc_mn on Vel_mn (2) !Regress Acc on Vel for damping,;
Disp_mn; !Regress Acc on Disp for frequency;
Vel_mn on Disp_mn (2); !Regress Vel on Disp for second damping
  estimate;
!The (2) equates the two coefficients;

!Now add coupling across two oscillators;
Acc_tn on Disp_mn; !Kappa for Teen;
Acc_mn on Disp_tn; !Kappa for Mom;
OUTPUT: CINTERVAL;
```

We also provide the syntax for using four measures in time (in this case, mom's positive and negative affect), though we do not discuss the results.

```
TITLE: Example Coupling Model with Four Time Points
DATA:
  FILE IS "affectlaggeddet.dat";
  FORMAT IS f8,f11/7f7.3/7f7.3/7f7.3/7f7.3;
VARIABLE:
  NAMES ARE dyadid time
      tpos_tn3 tpos_tn2 tpos_tn1 tpos_t tpos_t1 tpos_t2 tpos_t3
      tneg_tn3 tneg_tn2 tneg_tn1 tneg_t tneg_t1 tneg_t2 tneg_t3
      mpos_tn3 mpos_tn2 mpos_tn1 mpos_t mpos_t1 mpos_t2 mpos_t3
      mneg_tn3 mneg_tn2 mneg_tn1 mneg_t mneg_t1 mneg_t2 mneg_t3
      ;
  USEVARIABLES ARE
      mpos_tn2 mpos_tn1 mpos_t1 mpos_t2
      mneg_tn2 mneg_tn1 mneg_t1 mneg_t2
;
  MISSING IS BLANK;
ANALYSIS:
  ESTIMATOR IS ML;
  ITERATIONS = 1500;
  CONVERGENCE = 0.00005;
  COVERAGE = 0.10;
MODEL:
mpos_tn2 mpos_tn1 mpos_t1 mpos_t2 (22);
Disp_mp by mpos_tn2@1 mpos_tn1@1 mpos_t1@1 mpos_t2@1;
Vel_mp by mpos_tn2@-1.5 mpos_tn1@-.5 mpos_t1@.5 mpos_t2@1.5;
Acc_mp by mpos_tn2@2.25 mpos_tn1@.25 mpos_t1@.25 mpos_t2@2.25;
jerk_mp by mpos_tn2@-3.375 mpos_tn1@-.125 mpos_t1@.125 mpos_
  t2@3.375;
[mpos_tn2@0 mpos_tn1@0 mpos_t1@0 mpos_t2@0];
[Disp_mp Vel_mp Acc_mp jerk_mp];

Jerk_mp on Acc_mp (1)
            Vel_mp (2);
Acc_mp on Vel_mp (1)
            Disp_mp (2);
Vel_mp on Disp_mp (1);
```

```
mneg_tn2 mneg_tn1 mneg_t1 mneg_t2 (23);
Disp_mn by mneg_tn2@1 mneg_tn1@1 mneg_t1@1 mneg_t2@1;
Vel_mn by mneg_tn2@-1.5 mneg_tn1@-.5 mneg_t1@.5 mneg_t2@1.5;
Acc_mn by mneg_tn2@2.25 mneg_tn1@.25 mneg_t1@.25 mneg_t2@2.25;
jerk_mn by mneg_tn2@-3.375 mneg_tn1@-.125 mneg_t1@.125 mneg_
   t2@3.375;

[mneg_tn2@0 mneg_tn1@0 mneg_t1@0 mneg_t2@0];
[Disp_mn Vel_mn Acc_mn];

Jerk_mn on Acc_mn  (11)
            Vel_mn (12);
Acc_mn on Vel_mn (11)
            Disp_mn (12);
Vel_mn on Disp_mn (11);

Acc_mp on Disp_mn (3);
Jerk_mp on Vel_mn (3);
Acc_mn on Disp_mp (13);
Jerk_mn on Vel_mp (13);
OUTPUT:    CINTERVAL;
```

Interpretation

Notice from Table 16.4 that the coupling terms are not significant and quite small. The ability to estimate this model is likely due to the inherent alpha inflation of collapsing both within and across time series results creating a dependency in the data. Notably, the coupled multilevel model described earlier failed to converge (which is why we did not provide results). This dependency largely stabilized the SEM estimation procedure. Furthermore, the multilevel model approach distinguishes fixed and random effects. The choice of collapsing and ignoring the dependency inherently treats the coupling as just having a fixed effects portion where all random effects are included as error in prediction. To add random effects, one would have to treat the model as the multilevel SEM hybrid. Our attempts at running this model failed to reach a viable solution (just like the multilevel model approach).

To understand the overall pattern of phase locking, we calculated the synchrony ratio depicted in Equation 16.8, $\ln((1.135 - 1.075)/(0.01 + 0.004)) = 1.46$. Being above zero indicated that the relationship between positive and negative affect for mothers were more in drift and could only be described as weakly coupled. This may be the result of the way affect was assessed for mothers; rather than rating their own affect each day, they were asked to rate their affect specifically in relation to how their adolescent was successfully or unsuccessfully regulating their diabetes that day. For this reason, positive and negative affect may have a relatively discordant relationship as they are dealing with regular hardships, but also recognizing the positive moments. Both are oscillating, but somewhat isolated. This decoupling may be a coping mechanism for dealing with the chronic illness.

Additional Considerations

There are a number of hidden considerations in the procedures outlined in this chapter. Foremost has to do with the treatment of missing data. In the local linear approximation, the calculations generate Tau*2 missing data points. Any missing data in the time series can lead to an additional loss of up to three missing cases per missing data point. This loss in data is relatively ignorable under long time series but can account for a tremendous loss in short time series. We therefore recommend serious considerations of modern missing data techniques.

Most SEM programs incorporate full information maximum likelihood (FIML), which accounts for the missing data simultaneously with the parameter estimates themselves. However, applying FIML after the local linear approximation (how missing data is naturally accounted for in multilevel models) will result in a less powerful approach than the SEM alternative (more missing data is generated through the local linear approximation). We therefore recommend multiple imputation (MI) methods applied prior to estimating velocity and acceleration for the regression and multilevel procedures. Under MI, common methods to account for the within-subjects nature of the data are to model the procedures via PAN (Schafer, 2001), where a growth model or equivalent is used to account for the missingness. Butner et al. (2007) used PAN in conjunction with oscillatory daily diary data where they included day of the week, day of the week^2, and day of the week^3 to maintain the primary oscillatory components. Alternatively, one could utilize more easily accessible MI programs in SAS, SPSS, Amelia, and Norm where multiple delays are used to account for the missingness. To parallel the calculations of acceleration and velocity, one should include a lag and lead of tau along with any time-varying covariates that might assist in proper imputation.

One of the strengths of the multilevel and SEM modeling procedures is the ability to draw analysis power across time series' as well as within time series. Normally, delays tend to stabilize at a minimum of 50 measures in time and time delay reconstruction methods are notoriously demanding of sample size. However, in a series of simulations of the coupled oscillator equation in multilevel modeling, we found that the estimates for frequency and coupling tended to stabilize with as few as 10 measures in time and 50 individuals (Butner & Hicks, 2004). The dangers in estimation clearly occurred on the linear damping term, which tended to be biased. Only the equating procedure in SEM is capable of fixing this. Furthermore, there was an additional limitation of the frequency estimates in that enough of the 10 measures had to be within the same cycle to get a proper frequency estimate. The choice of tau will greatly impact these estimates. Unfortunately, the nonlinear escapement terms have yet to be explored.

Choosing to detrend and how to detrend can also have a substantial impact on results. For example, it is relatively easy to unintentionally remove some of the key oscillatory information, especially when time series are relatively short. It is important to inspect the data closely after detrending to determine if some aspect of interest has been removed. On the other hand, not detrending at all can result in covariances between acceleration, velocity, and position that have little to do with the cyclical nature, and give results that are unrepresentative of the oscillations (though still fairly representative of the system itself). The danger of not detrending is that the interpretation of the coefficients is unclear: they are no longer purely frequency, damping, etc.

Simultaneously detrending data and estimating the oscillatory relationships is possible. Such an approach would involve altering Equation 16.2 by making position the dependent variable. The equation is then expanded to simultaneously include acceleration, velocity, and their cubic relationships along with functions of time. We present the expansion of a simple damped harmonic oscillator with simultaneous linear detrending as a regression equation

$$\text{Pos} = b_0 + b_1 \text{Time} + b_2 \text{Acc} + b_3 \text{Vel} + e \qquad (16.14)$$

The first part of the equation ($b_0 + b_1 \text{Time}$) accounts for a linear detrending as a function of time akin to a growth model (where b_1 will be indicative of average velocity over the time series, removing the trend). The second part of the equation then accounts for the oscillations. However, the interpretations of b_2 and b_3 have changed from traditional time series terms.

The coefficient for acceleration (b_2) is now the squared period in radians, rather than the squared frequency. The coefficient for velocity (b_3) is now a generalized Reynolds number of the ratio of the friction constant to the spring constant. This equation can still be easily expanded to account for nonlinear relationships where acceleration takes the place of position in the previous equations and while the metrics are notably different, the underlying relationships are the same. The same is true for coupling. It is noteworthy that, while this approach is possible, it has yet to be utilized or examined as to how well it properly estimates relationships in comparison to the earlier methods already mentioned.

Acknowledgments

This study was supported by grant R01 DK063044-01A1 from the National Institute of Diabetes and Digestive and Kidney Diseases awarded to Deborah Wiebe (PI) and Cynthia Berg (co-PI). We thank the families who participated, the physicians (Mary Murray, David Donaldson, Rob Lindsay, Carol Foster,

Michael Johnson, Marie Simard), staff of the Utah Diabetes Center, Mike T. Swinyard, MC, PC, and additional members of the ADAPT team (Donna Gelfand, Jenni McCabe, Marejka Shaevitz, and Michelle Skinner).

References

Abraham, R. H., & Shaw, C. D. (1992). *Dynamics: The geometry of behavior*. Redwood City, CA: Addison-Wesley.

Beek, P. J., & Beek, W. J. (1988). Tools for constructing dynamical models of rhythmic movement. *Human Movement Science, 7*, 301–342.

Beek, P. J., Schmidt, R. C., Morris, A. W., Sim, M. Y., & Turvey, M. T. (1995). Linear and nonlinear stiffness and friction in biological rhythmic movements. *Biological Cybernetics, 73*, 499–507.

Boker, S. M. (2001). Differential structural equation modeling of intraindividual variability. In L. Collins & A. Sayer (Eds.), *New methods for the analysis of change* (pp. 5–27). Washington, DC: American Psychological Association.

Boker, S. M., & Graham, J. (1998). A dynamical systems analysis of adolescent substance abuse. *Multivariate Behavioral Research, 33*, 479-507.

Boker, S. M., & Nesselroade, J. R. (2002). A method for modeling the intrinsic dynamics of intraindividual variability: Recovering the parameters of simulated oscillators in multiwave panel data. *Multivariate Behavioral Research, 37*, 127–160.

Butner, J., Amazeen, P. G., & Mulvey, G. M. (2005). Multilevel modeling of two cyclical processes: Extending differential equation modeling to nonlinear coupled systems. *Psychological Methods, 10*, 159–177.

Butner, J., Diamond, L., & Hicks, A. (2007). Attachment style and two forms of emotion co-regulation between romantic partners. *Personal Relationships, 14*, 431–455.

Butner, J., & Hicks, A. (2004, January). *Strengths and limitations of the coupled oscillator model on daily diary data: A simulation study*. Poster session presented at Society for Personality and Social Psychology, Los Angeles, CA.

Chow, S., Ram, N., Boker, S. M., Fujita, F., & Clore, G. (2005). Emotion as a thermostat: Representing emotion regulation using a damped oscillator model. *Emotion, 5*, 208–225.

Finan, P. H., Hessler, E. E., Amazeen, P. G., Butner, J., Zautra, A. J., & Tennen, H. (2010). Oscillations in daily pain prediction accuracy. *Nonlinear Dynamics, Psychology, and Life Sciences, 14*, 27–46.

Haken, H. (2006). *Information and self-organization: A macroscopic approach to complex systems*. New York: Springer.

Haken, H., Kelso, J. A. S., & Bunz, H. (1985). A theoretical model of phase transitions in human hand movements. *Biological Cybernetics, 51*, 347–356.

Iberall, A. (1987). A physics for the studies of civilization. In F. E. Yates (Ed.), *Self-organizing systems: The emergence of order* (pp. 521–540). New York: Plenum Press.

Kelso, J. A. S. (1994). Elementary coordination dynamics. In S. P. Swinnen, J. Massion, H. Heuer, & P. Casaer (Eds.), *Interlimb coordination: Neural, dynamical, and cognitive constraints* (pp. 301–318). San Diego: Academic Press.

Kugler, P. N., & Turvey, M. T. (1987). *Information, natural law, and the self-assembly of rhythmic movement.* Hillsdale, NJ: Lawrence Erlbaum Associates.

Marsh, H.W., Wen, Z., & Hau, K. (2006). Structural equation models of latent variable interactions and quadratic effects. In G. R. Hancock & R. O. Mueller (Eds.), *Structural equation modeling: A second course* (pp. 225–265). Greenwich, CT: Information Age Publishing.

McArdle, J. J. (2001). A latent difference score approach to longitudinal dynamic structural analysis. In R. Cudeck, S. du Toit, & D. Sorbom (Eds.), *Structural equation modeling: Present and future* (pp. 1–40). Lincolnwood, IL: Scientific Software International, Inc.

Morris, J. S., & Carroll, R. J. (2006). Wavelet-based functional mixed models. *Journal of the Royal Statistical Society, Series, B, 68,* 179–199.

Rogosa, D. (1988). Myths about longitudinal research. In K. W. Schaie, R. T. Campbell, W. M. Meredith, & S. C. Rawlings (Eds.), *Methodological issues in aging research* (pp. 171–209). New York: Springer.

Schafer, J. L. (2001). Multiple imputation with PAN. In L. M. Collins & A. G. Sayer (Eds.), *New methods for the analysis of change* (pp. 357–383). Washington, DC: American Psychological Association.

Ullman, J. B., & Bentler, P. M. (2003). Structural equation modeling. In J. A. Schinka & W. F. Velicer (Eds.), *Handbook of Psychology, Volume 2: Research Methods in Psychology* (pp. 607–634). New York: Wiley.

Von Holst, E. (1939, 1973). On the nature of order in the central nervous system. In: R. Martin (Ed. and trans.), *The collected papers of Erich von Holst, vol. 1. The behavioral physiology of animal and man.* Coral Gables, FL: University of Miami Press.

Warner, R. M. (1998). *Spectral analysis of time-series data.* New York: The Guilford Press.

17

Markov Chains for Identifying Nonlinear Dynamics

Stephen J. Merrill

CONTENTS

Markov chains are in some sense well understood, yet the dynamic behavior of their sample paths are more varied than one-dimensional dynamics in that one-dimensional dynamics can be embedded in a Markov chain, but cannot approximate all chains. Chains provide a powerful way to describe the nature of the dynamics that generated the chain whether the dynamics were observed in data or simulated from a model. Markov chains can be constructed from first principles, that is, time series data of numerical or categorical data, or simulations of first-order linear or nonlinear autoregressive models. They also provide a way to include necessary random effects in a deterministic model while providing an analytic and simulation framework for these models. This approach to describing dynamics is also applicable to situations in which the observed series is not from a Markov chain, but there is one in the background through hidden Markov models (HMM).

In this chapter, the construction of Markov chains from data (empirical Markov chains), simulation of these chains, and analysis of their dynamics are emphasized. Examples in several fields are described to illustrate the range of applications of the approach. The use of the HMMs is also described. MATLAB® code is provided for the analysis.

Introduction

Markov chains were introduced by Andrei Andreyevich Markov (Markov, 1906) as part of work to extend the law of large numbers to dependent events. The idea there is that when one observes a sequence, the next value may depend on previous values—unlike the case when one generates a sequence of (independent) random numbers. The first stage in investigating the consequences of this dependence is to study the case that it is only the present value that affects the next (and knowledge of older history does not change the probabilities associated with the next value). This simplification is the Markov assumption—central to the study of Markov processes in general and Markov chains in particular.

The first instance of using a chain in applications was by Markov (1913) when he applied his findings to the first 20,000 letters of Pushkin's *Eugene Onegin*. That was the start of the science of using word patterns in a work to establish authorship. There are two basic applications for chains whose transition matrices are created from data, *empirical Markov chains*. The first is to create a dynamic model for the data that can be used to characterize it and as a foundation for generating simulated data with the same dynamic properties. The second is to establish a baseline on which other data can be tested as to whether it was likely to have been generated by the original process. An example of this latter application would be the testing of authorship of an unknown text given established patterns. A second example is to detect when a system that generated the dynamics may have changed— as in a recent study by Lancashire and Hirst (2009) to investigate whether the novelist Agatha Christie was developing dementia later in life. The first application is very useful when one desires new data sets with the same characteristics as the original, to test strategies or capture other information, such as intrinsic variability of the sample paths, in the system that generated the original series.

Markov chains have recently become a standard tool in many applications, especially through Markov chain Monte Carlo methods and HMM, both of which have been applied in a variety of settings. Markov chain Monte Carlo method has a different emphasis than ours here, and the property of convergence to a limit distribution is the feature of a chain that is used—not the dynamic behavior (except that you may wish fast convergence). In HMM, assuming that the data that is observed is one step away from a Markov chain, the data is used to determine properties of this "hidden" chain (number of states and its transition matrix). The goal and applications of HMM can be similar to those mentioned above for Markov chains, and applied when a chain is found not to be an appropriate model as when the original time series and those generated from the chain do not have "similar dynamics."

In this chapter, examples are given illustrating the construction of Markov chains, both from first principles and from data. The description of the dynamic behavior of chains based on the idea of "similar dynamics" is discussed. This idea is used to test the data needs for describing a chain and to detect changes in the dynamics in a time series—leading to the idea of bifurcation in chains. Several examples of application of these methods are given. Finally, the process of using HMM is described along with examples of their use in application.

The goal of this chapter is to give a sense of the applications and questions that a Markov chain or HMM is appropriate to address, and to introduce typical computational tools to fit, analyze, and effectively use the resulting model. We will not survey the large number of applications that have employed Markov chains and especially HMM in applications, except to note that most of these were not interested in characterizing the dynamic nature of their data.

Markov Chains

Markov chains can be described in continuous time (with exponentially distributed times between transitions) or in discrete time. Continuous chains can be easily embedded in discrete chains, so our attention here will be on the dynamics of discrete chains. With this restriction, we first begin by describing a chain.

A (discrete time) Markov chain consists of a countable set of states S (finite or infinite) and corresponding matrices $P(n)$ of transition probabilities at stage (time) n that describes the evolution of a sample path from state to state over time. The states can be described in words (e.g., for, against, no opinion) or as ranges of values of measurements or responses. The determination of the states that are most useful in a particular application has an aspect of art in it and several assignments often need to be tried. The choice may also depend on the amount of data available. If there are N states, the matrices $P(n)$ will be $N \times N$, a different matrix for each stage n. The i, jth entry in the transition matrix, $p_{ij}(n)$, is the conditional probability that the next transition will be to j if one is in state i at stage n. The Markov assumption here is that it is only knowledge of the current state that is needed to determine the probability distribution for the next state. If it turns out that a few more past states are needed to determine these probabilities, an *nth order chain* can be constructed. An *n*th order chain changes the definition of "state" to include several past values at the cost of a huge increase in the size of the data set required to determine the transition probabilities as the number of states explodes (and since the transition matrix has the dimension of the square of the number of states, there are a lot more probabilities to estimate). Here we will assume that the Markov assumption holds.

If the $p_{ij}(n)$ are independent of stage n (the same for all n), the chain is *homogeneous* and only one transition matrix **P** is required to describe the chain. It is rare to have enough data to describe nonhomogeneous transition matrices, although if there are few states, it is possible to describe how the probabilities are changing in time (e.g., by modeling the changing mean of a Poisson distribution). If nonhomogeneity is expected, there may be transient phases that would need to be handled separately in order for one to feel comfortable with the assumption of homogeneity. Here we will restrict our attention to homogeneous chains. A test of this assumption can be done following our approach to see if the chain's transition matrix may have changed over time.

With states defined and transition probabilities in hand, given a starting state, a sample path can be generated. As the process is stochastic, each sample path generated will be different (with probability 1) and thus the path depends both on the sequence of random numbers used to simulate the process and the initial state. To illustrate this, we present Example 17.1. This first example is used in several ways. First the transition matrix is specified and several sample paths are generated. Then *one* of these sample paths is used to estimate the transition matrix, which generated it.

Example 17.1

Consider the Markov chain described in Figure 17.1. There are three states, 1, 2, and 3, and the nonzero one-step transition probabilities are given. The parameter α will be fixed at 1/3 for this example.

From Figure 17.1, the transition matrix is constructed row-by-row. Starting with row 1 (state 1), note that if at that state at stage n, with probability 1, one moves to state 2. Each row must sum to 1 as that is the sum of the probabilities of all possible next states and all entries are between 0 and 1. Such matrices are called *stochastic matrices*. Continuing, we find the transition matrix corresponding to Figure 17.1 is given by

$$P = \begin{pmatrix} 0 & 1 & 0 \\ 0.5 & 0 & 0.5 \\ \alpha & 0 & 1-\alpha \end{pmatrix}.$$

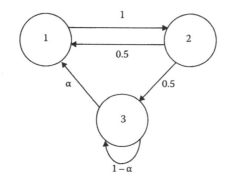

FIGURE 17.1
Three states with one-step transition probabilities specified. α is a parameter that satisfies $0 \le \alpha \le 1$.

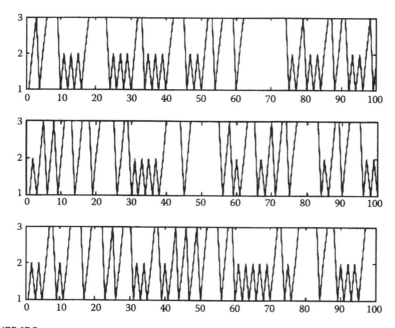

FIGURE 17.2
Three sample paths of length 100 generated from the transition matrix **P** with α = 1/3. The initial state was 1 in all cases and lines were drawn between the transitions to make them easier to see.

Simulating this chain to obtain the sample paths involves choosing an initial state and using the matrix with a random number generator to obtain the next state. The MATLAB function generate.m (code given in Appendix 17.A) uses this method by first creating a matrix that accumulates the probability in each row. It then generates a uniformly distributed number on [0,1] and generates the next state. With sample paths in hand, several can be plotted as in Figure 17.2.

```
>> P=[0,1,0;.5,0,.5;1/3,0,2/3]
P =
         0    1.0000         0
    0.5000         0    0.5000
    0.3333         0    0.6667
>> x=generate(1,P,10)
x =

    1    2    3    3    3    1    2    1    2    3
```

Another aspect of the use of an empirical Markov chain is the ability to build a chain from a time series. If the time series has values on the real line, one must define the states (usually intervals of real numbers) and estimate the transition probabilities associated with that state definition. This first aspect is sometimes called "coarse graining" or "lumping" (see Nicolis & Nicolis, 1990). We can estimate the transition probabilities through constructing a matrix of the relative frequency of observed transitions. These estimates have been shown (Zucchini & MacDonald, 2009) to be conditional maximum likelihood estimators of the true probabilities.

Example 17.2

In this example, one of the time series of length 100 generated in the previous example will be used to estimate the transition matrix (and the value of α).

```
>> % count will be a matrix which keeps track of transitions
   observed in the series
>> count=zeros(3,3);
>> for i=1:99
count(x(i),x(i+1))=count(x(i),x(i+1))+1;
end
>> count
count =
         0    30     0
        13     0    17
        16     0    23
>>% Now to compute the row sum for each row
>> for i=1:3
s(i)=sum(count(i,:));
end

>> s
s =
    30    30    39
>> % Last, divide each row by the row sum to insure to rows
   each sum to 1
>> T=zeros(3,3);
>> for i=1:3
T(i,:)=count(i,:)/s(i);
end
```

```
This is the estimate of the transition matrix that generated
  the sequence.
>> T
T =
            0     1.0000          0
       0.4333          0     0.5667
       0.4103          0     0.5897
```

A version of this procedure is the m-file transi.m in Appendix 17.A. The above process works as long as no row sum is zero (otherwise you divide by zero). This situation would mean that a possible state was not observed in the data, suggesting that a longer data sequence is needed or possibly a redefinition of the states. Note that α is approximated to be 0.4103.

The relationship between the matrix **P** (the actual transition matrix) and **T** (the estimated matrix computed from a sequence) is an interesting one and, for us, important. One approach would be to look at $\|P - T\|$ using some matrix norm (look at the term-by-term differences in the matrices and square them and add them up, for instance). Another approach might be to look at the difference in the powers of the matrices that correspond to the n-step transition probabilities as in Hoffmann and Salamon (2009). But our interest here is different. Ours question is this: "How well do the dynamics of the processes generated by **T** mimic the process generated by **P**?" In fact, do we know that having the two matrices close together in norm will mean that they have similar dynamics? To answer the question, the phrase "similar dynamics" needs to be defined. There are several properties of Markov chains that we need in order to understand the approach to be described here. A good discussion and proofs of most of these properties can be found in Heyman and Sobel (2004).

Selected Properties of Finite State Markov Chains

1. The transition matrix **P**, if $N \times N$, has N eigenvalues, all on or inside the unit circle in the complex plane.
2. The value 1 is always an eigenvalue of **P**.
3. If 1 is the only eigenvalue of modulus 1 (on the unit circle), then there is a unique limit distribution—a left eigenvector of the eigenvalue 1. This limit distribution describes the fraction of time spent in each state over the long run by any sample path. If there are a collection of paths, the limit distribution also describes the fraction of all paths in each state.
4. If a unique limit distribution has all components nonzero, then all states are ergodic, and every sample path (with probability = 1) will describe the same transition matrix **P**.
5. The eigenvalues other than 1 of **P** describe the rate and mode of approach of sample paths to the limit distribution if it exists.

Example 17.3

One way to examine the similarities and differences that will be useful for us is to examine the eigenvalues and (left) eigenvectors. To get the left eigenvectors, we use the transpose of **P** (**P'** in MATLAB). The column corresponding to the appearance of the eigenvalue 1 is a basis for the eigenspace. To find that vector, sum that column and divide the entries column by that constant.

```
>>% v is a matrix of eigenvectors of P' - seen in the columns
   and d is a diagonal
>>% matrix with eigenvalues of P' on the diagonals
>> [v,d]=eig(P')
v =
    -0.4923    -0.4851    -0.1670
     0.8103    -0.4851    -0.6087
    -0.3179    -0.7276     0.7756
d =
    -0.6076         0          0
         0     1.0000          0
         0          0     0.2743
```

```
>>% Note that the second column of d above contains the
   eigenvalue 1,
>>% so the second column of v is a basis for the associated
   eigenspace.
```

```
>> [v,d]=eig(T')
v =
     0.5059    -0.3755    -0.0345
     0.5059     0.8156    -0.6892
     0.6987    -0.4401     0.7237
d =
     1.0000         0          0
         0    -0.4604          0
         0          0     0.0501
>>% Here it is the first column of v that is the basis for
   the eigenspace.
```

From the above, the limit distribution for **P** does exist as is [0.2857 0.2857 0.4286]' while the limit distribution for **T** also exists and is [0.2957 0.2957 0.4085]'. Note that the limit distributions are very similar and the eigenvalue types, one negative and one positive, in addition to the one eigenvalue, are the same.

Example 17.4

In addition to the above, we will also use the *acf* of sample paths generated by the two series to see if their autocorrelation structures are similar. We have found using lags up to 5 or 6 is sufficient. As an example, we generated time series of length 500 from both **P** and **T** and computed the results shown in Figure 17.3.

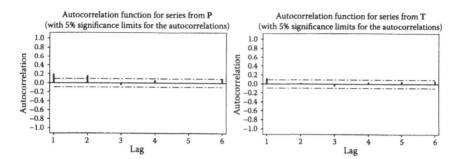

FIGURE 17.3
acf's generated through time series of length 500 from transition matrices **P** and **T**.

Note that the pattern of the autocorrelations of **P**'s time series is preserved in **T**'s time series. Remember that even two series generated from the same transition matrix will show some variation in the acf.

Our working definitions for "similar dynamics" are as follows: (a) Eigenvalues of the transition matrices away from zero show a similar blend of positive, negative, and complex values. (b) The limit distributions should have the same number of nonzero values in corresponding states and should be close as points in $\mathbb{R}^N$. (c) acf's should show similar patterns of correlation. We can add a fourth criterion that the spectral radius of the difference of the two matrices is small. In MATLAB, this is simply "norm" and for a matrix, **A** is the largest singular value, defined as

$$\text{norm}(\mathbf{A}) = (\max \text{eigenvalue}(\mathbf{A}'\mathbf{A}))^{1/2}$$

To see how this quantity varies with the time series length, 100 series of length 100 and 100 of length 500 were generated from **P**, and approximate **T** matrices were produced. Histograms of the norm of the difference were plotted for each of the time series lengths as shown in Figure 17.4.

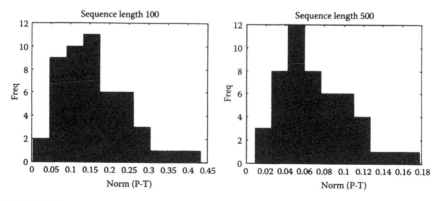

FIGURE 17.4
Distributions of norms of differences with different time series lengths to determine the approximate **T** transition matrix.

It is very clear from Figure 17.4 that more data gives a better approximation (smaller norm), a length of 500 to determine the nine probabilities in **T** is clearly better that 100—yet such a **T** was found to have similar dynamics in Example 17.2. Our rule of thumb is that the length of the series length should be at least five times the number of probabilities to be approximated (N^2) to expect similar dynamics.

Now the hard part, in Examples 17.1 and 17.2 we knew the true **P** and could compare our **T**s with it. In applications, we just have the time series and the transition matrix generated from it. We can still compare acf and other information such as lag one maps to test the applicability of the Markov chain model.

One-Dimensional Dynamics

An example of the flexibility of this approach can be seen in studying one-dimensional dynamics from a difference equation with or without noise.

Example 17.5

The discrete logistic equation $x_{n+1} = \mu x_n(1 - x_n)$ is used to generate a series for $\mu = 3.8$. This is in the chaotic region for this example. A series of length 500 is plotted in Figure 17.5a with the associated lag 1 map shown in Figure 17.5b. Using transi.m, an empirical 10×10 transition matrix is created. Position of nonzero entries in this matrix is displayed in Figure 17.6. One can see the parabolic relationship between x_{n+1} and x_n displayed there (rotated through 90°).

A time series y is generated from that matrix using generate.m. The results are shown in Figure 17.7. To compare the dynamics of the x and y series, we present the acf's in Figure 17.8. Note the near identical patterns. This suggests that the dynamic properties of the two series are similar.

Nonlinear Time Series with Error

An advantage to the Markov chain approach can now be seen as we can repeat this process with noise added to the dynamics. This will be done in two ways. First, we add noise to each x (as in measurement error) and observe only the series with noise. This looks like

$$\begin{aligned} x_{n+1} &= \mu x_n(1 - x_n) \\ z_n &= x_n + \varepsilon_n \end{aligned}$$ (17.1)

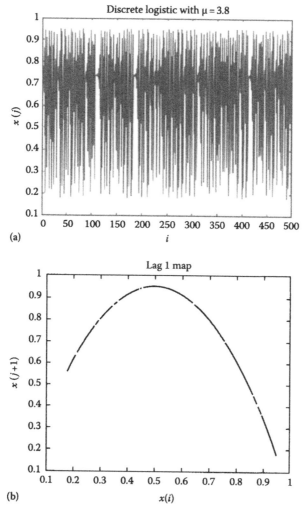

FIGURE 17.5

(a) Iterates of the logistic map. (b) Lag 1 return map of data shown in (a).

Note that the x_n series is not changed by the presence of noise. The second is to include the noise in the dynamics. This has the form

$$x_{n+1} = \mu x_n (1 - x_n) + \varepsilon_n \tag{17.2}$$

Figure 17.9 displays measurement error with x series generated as per Equation 17.2, which is a series of errors ε drawn from a normal distribution with mean 0 and standard deviation 0.1; this is a plot of z from Equation 17.1.

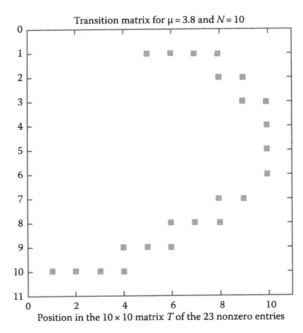

FIGURE 17.6
Position of nonzero entries in the transition matrix for the logistic map data shown in
Figure 17.5.

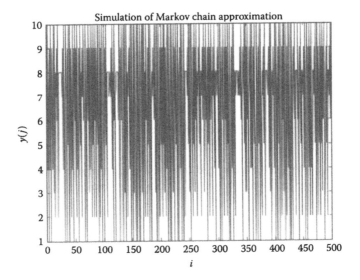

FIGURE 17.7
Iterates of y generated from the transition matrix in Figure 17.6 using generate.m.

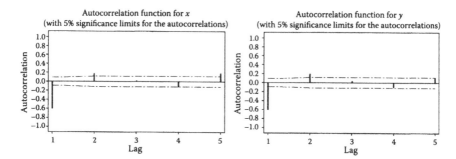

FIGURE 17.8

Autocorrelation function for the chaotic series x and the approximate series y generated from the empirical transition matrix.

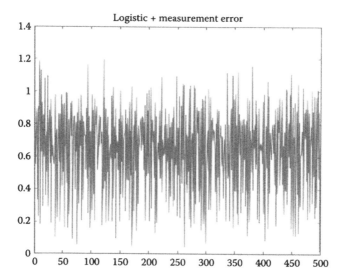

FIGURE 17.9

Measurement error added to a discrete logistic series.

Figure 17.10 is a lag 1 map of the series z from Equation 17.1. Even with noise, the parabolic relationship is seen. Figure 17.11 is a comparison of the eigenvalues from the time series x generated from the approximating Markov chain without error, with the z series with error. Note that the patterns are similar in that on the left half plane both have complex conjugate pairs, pure imaginary pairs, and a single positive real eigenvalue (other that 1). The differences are caused by the noise.

Figure 17.12 is a time series generated from Equation 17.2 containing dynamic noise with standard deviation .02. This is a case of dynamic error as the error affects the x values. A lag one map of the series in Figure 17.12 is shown in Figure 17.13. This should be compared to that in Figure 17.10. Although Figure 17.13 would suggest that the dynamics of Equation 17.2

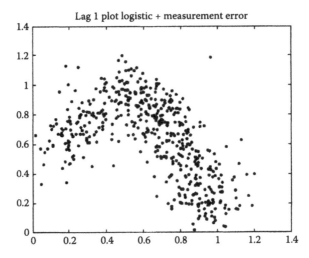

FIGURE 17.10
Lag 1 map of the series z from Equation 17.1.

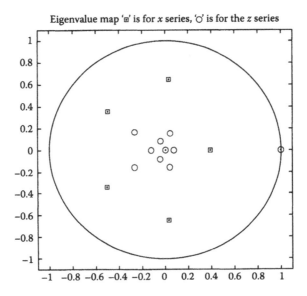

FIGURE 17.11
Comparison of the eigenvalues from the time series x generated from the approximating
Markov chain without error, with the z series with error.

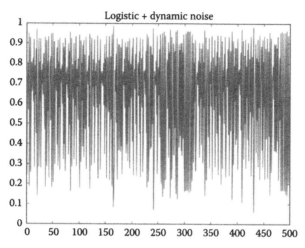

FIGURE 17.12
Time series generated from Equation 17.2.

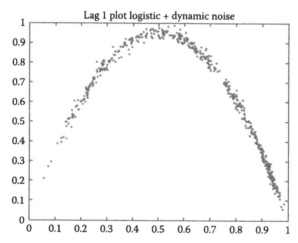

FIGURE 17.13
Lag 1 map for the series in Equation 17.2, plotted in Figure 17.12.

should be similar to that without the small amount of noise, Figure 17.14 shows that in fact they are not dynamically similar.

Bifurcation in Markov Chains

This last example suggests a definition for bifurcation in a Markov chain that reflects a change in dynamics of the underlying system. This is briefly illustrated though the bifurcation in the discrete logistic equation in this case, the

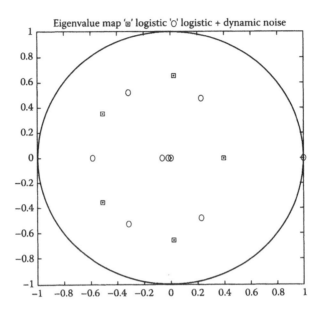

FIGURE 17.14
Eigenvalue map for the logistic without noise and that with dynamic noise Equation 17.2.

bifurcation at μ = 3 to a period 2 point. This is done through the eigenvalues of the transition matrix. Loss of the unique limit distribution and establishment of a new one is the indication of a bifurcation in the system. As a parameter changes, look for *eigenvalue(s) approaching the unit circle*. In this example, transition matrices computed with 10 states at μ = 2.9, 3.0, and 3.05. Eigen values were found at 2.9 to be {−0.81, 0.81, 1, lots of 0s}, at 3.0, {−0.92, 0.917, 1, lots of 0s}, and at 3.05, {−1, 1, lots of 0s}. From these results, it appears that a bifurcation took place just before 3.05. Moreover, because of the −1, it is a period 2 point. A more exact estimate of the bifurcation point can be had by choosing more states (and using a longer data set).

Hidden Markov Models

A HMM is one in which you observe a sequence called "emissions," but you do not observe the states the model went through to generate the emissions. At each state, it is assumed there is a probability of a particular emission. In this case, your goal is to recover the state information from the observed data as well as the emission probabilities for each state. The MATLAB® Statistics Toolbox includes five functions for analyzing HMMs: hmmdecode—calculates the posterior state probabilities of a sequence, hmmgenerate—generates a sequence for a HMM, Hmmestimate—estimates the parameters for a Markov model, hmmtrain—calculates the maximum likelihood estimate of HMM parameters, and hmmviterbi—calculates the most likely state path for a HMM sequence. To specify an HMM, one needs to specify the transition matrix and the matrix of emission probabilities. Once those are specified, a sample path can be generated.

Example 17.6

In this example, TRANS = transition matrix, here 2 × 2; EMIS = emission probabilities, here 2 × 6; EMIS takes a state and computes what will be observed.

```
>> TRANS=[.9 .1;.05 .95]
TRANS =
    0.9000    0.1000
    0.0500    0.9500
>> EMIS=[1/6, 1/6, 1/6, 1/6, 1/6, 1/6; 7/12, 1/12, 1/12,
    1/12, 1/12, 1/12];
>> [seq, states]=hmmgenerate(12,TRANS,EMIS);
>> seq
seq =
    1   1   4   1   1   1   1   4   2   2   1   5
>> states
states =
        2   2   2   2   2   2   2   1   1   1   1   1
```

Note that if one specifies the emission matrix to be the identity, then the states are exactly reported and the sequence is just the sequence from the Markov chain (our generate.m). If the emission matrix is not the identity—and here depending on the state of the chain, there is a chance of a 1 through 6 being produced, the most likely state is reported.

Now suppose the matrices are known as well as the sequence. How can one determine the likely state sequence? The results for likelystate should be compared to the actual "states" sequence above.

```
>> likelystate=hmmviterbi(seq, TRANS, EMIS);
>> likelystate
likelystate =
        2   2   2   2   2   2   2   2   2   2   2   2
```

Estimating the Transition and Emission Matrices

Suppose you do not know the transition and emission matrices in the model, and you observe a sequence of emissions, seq. There are two functions that you can use to estimate the matrices hmmestimate and hmmtrain. To use hmmestimate, you also need to know the corresponding sequence of states that the model went through to generate seq. The following command takes the emission and state sequences, seq and states, and returns estimates of the transition and emission matrices, TRANS_EST and EMIS_EST.

```
[TRANS_EST, EMIS_EST] = hmmestimate(seq, states)
TRANS_EST =
0.8989    0.1011
0.0585    0.9415
EMIS_EST =
0.1721    0.1721    0.1749    0.1612    0.1803    0.1393
0.5836    0.0741    0.0804    0.0789    0.0726    0.1104
```

Because the sequence of states is known, TRANS_EST is similar to our transi.m in that it computes the transition probabilities from the relative frequency of the observations. One can compare these outputs with the original transition and emission matrices, TRANS and EMIS, to see how well hmmestimate estimates them.

If you do not know the sequence of states, but you have an initial guess as to the values of TRANS and EMIS, you can estimate the transition and emission matrices using the function hmmtrain. In providing the guess, you are specifying the number of states in the hidden chain and the range of emissions possible to observe. For example, suppose you have the following initial guesses for TRANS and EMIS. hmmtrain uses an iterative algorithm that alters the matrices TRANS_GUESS and EMIS_GUESS, so that at each step the adjusted matrices are more likely to generate the observed sequence, seq.

```
>> TRANS_GUESS = [.85 .15; .1 .9];
>> EMIS_GUESS = [.17 .16 .17 .16 .17 .17;.6 .08 .08 .08 .08 08];
>> % You can estimate TRANS and EMIS with the following command.
[TRANS_EST2, EMIS_EST2] = hmmtrain(seq, TRANS_GUESS, EMIS_GUESS)
TRANS_EST2 =
0.2286    0.7714
0.0032    0.9968
EMIS_EST2 =
0.1436    0.2348    0.1837    0.1963    0.2350    0.0066
0.4355    0.1089    0.1144    0.1082    0.1109    0.1220
```

There are a number of good discussions of the fitting of HMM. The classic one is Rabiner (1989) where the history of these models (along with references) is given. Specific to the applications here are Visser et al. (2002) and Zucchini and MacDonald (2009). In these references, goodness of fit and appropriateness of the models constructed is discussed in detail. The analyses of the dynamics of the HMM are similar to those of the Markov chain, as in each case the transition matrix is the object of study.

Some Applications

Bipolar Disorder

Bipolar disorder is characterized by the presence of several "states," normal, mania, and depression are experienced. There have been many attempts to model aspects of this condition through nonlinear dynamics methods, for instance, Gottschalk et al. (1995), Johnson and Nowak (2002), and Ehlers (1995). The data in Gottschalk is especially interesting for us. Subjects were instructed to rate mood on a 24 point scale twice daily. Figure 17.9 in that paper shows lag one maps suggesting that a Markov chain model may be appropriate. Also, a three-state chain may be the first place to start as in Figure 17.15.

Using the time interval of 0.5 day as in Gottschalk, for each subject a transition matrix can be estimated. The three parameters α, β, and γ are the probability that one stays in the current state for the next period. Once the parameters have been estimated, it can be determined if drugs change the dynamics of the process. An example transition matrix is given below and simulated using our methods (Figure 17.16).

$$\mathbf{P} = \begin{bmatrix} 0.9 & 0.05 & 0.05 \\ 0.05 & 0.9 & 0.05 \\ 0.05 & 0.05 & 0.9 \end{bmatrix}$$

Heart Rate Variability

The time between heart beats (specifically the R-R interval) has been studied for many years as a prototypical chaotic system. In Merrill and Cochran (1997), Markov chain methods were applied when other methods had failed to describe all possible patterns found in a group of pediatric patients. Figure 17.17 shows some of the patterns observed. In particular, Patient 29 shows an "X" pattern, which cannot have come from one-dimensional dynamics as it is not the rotated graph of a function (the other two are).

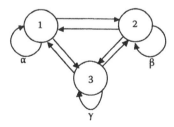

FIGURE 17.15
Proposed Markov chain model for bipolar disorder. State 1—normal, state 2—mania, and state 3—depression.

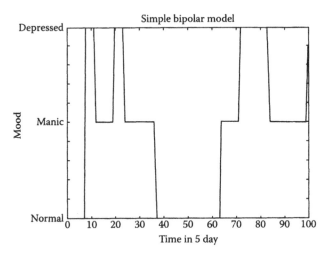

FIGURE 17.16
Transition between moods in the model given in Figure 17.15.

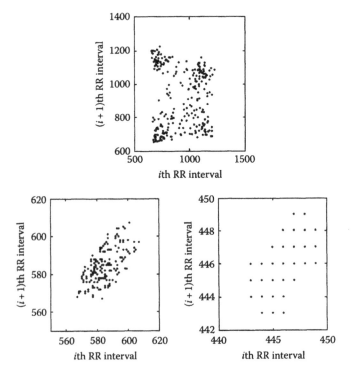

FIGURE 17.17
Figure 3 from Merrill and Cochran (1997) showing lag 1 maps for three patients. The top graph displays a pattern that cannot have come from one-dimensional dynamics. (Reprinted with permission from the American Mathematical Society.)

Summary and Comments

Data are often collected in the form of time series, and there are several approaches to analyzing and modeling them. A now classical way is through ARIMA models, where there is an assumption of a linear relationship between past values along with a moving average involving the noise component. Nonlinear analogs of that method are also available where piecewise models, or those with assumed nonlinear forms, are assumed. Deterministic methods have included assuming the presence of attractors and focused on determining fractal dimensions and other quantities of interest. These methods have difficulties in the presence of noise, since the underlying assumption is that the series is noise-free or nearly so. Using Markov chains or HMM do not have an a priori assumption of a functional form of the dependence of the next on previous values, and behave well in the presence of additive or dynamic noise. In any modeling application, the best model is usually defined by the application. A model is a good one if it can provide information or insight that illuminates the situation or provides an answer to the compelling question. If dynamic similarities between the data and the model are required, the methods presented here should help to make these approaches ones that should be considered.

Appendix 17.A

A. The m-file `generate.m`

```
%This function takes an initial state and transition matrix P
   and produces
%a sample path of length n.
function x=generate(initial,P,n)
x(1)=initial;
m=length(P);
for i=1:m
for j=1:m
cum(i,j)=sum(P(i,1:j));
end
end
for i=1:n-1
    u=rand;
    vec=sign(cum(x(i),:)-u);
    s=sum(vec);
    if prod(vec)==0
        x(i+1)=.5*(m-s)+.5;
    else
        x(i+1)=.5*(m-s)+1;
    end
end
```

B. The m-file `transi.m`

```
% This m-file takes a time series d of floating point numbers
  and creates
% an empirical mxm Markov chain transition matrix where m is
  the number of
% states (should be a positive whole number).
function Dt=transi(d,m)
k=length(d);
mi=min(d);
M=max(d);
width=(M-mi)/m;
b=zeros(m,1); % bin number computed for each data point.
for i=1:k
        b(i)=floor((d(i)-mi)/width)+1;
        if b(i)==m+1
            b(i)=m;
        end
end
% First create an mxm matrix of zeros
D=zeros(m);
% Stepping though the data set, add 1 to position i,j every
  time we see a
% transition from i to j
for i=1:k-1
    D(b(i),b(i+1))=D(b(i),b(i+1))+1;
end
Dt=zeros(m);
for j=1:m
    if sum(D(j,:))>0
        Dt(j,:)=D(j,:)/sum(D(j,:));
    else
        Dt(j,:)=0;
    end
end
```

References

Ehlers, C. L. (1995). Chaos and complexity: Can it help us to understand mood and behavior? *Archives of General Psychiatry, 52*, 960–964.

Gottschalk, A., Bauer, M. S., & Whybrow, P. C. (1995). Evidence of chaotic mood variation in bipolar disorder, *Archives of General Psychiatry, 52*, 947–959.

Heyman, D. P., & Sobel, M. J. (2004). *Stochastic models in operations research, vol.1.* Mineola, NY: Dover Publications.

Hoffmann, K. H., & Salamon, P. (2009). Bounding the lumping error in Markov chain dynamics. *Applied Mathematics Letters, 22,* 1471–1475.

Johnson, S. L., & Nowak, A. (2002). Dynamical patterns in bipolar depression. *Personality and Social Psychology Review, 6,* 380–387.

Lancashire, I., & Hirst, G. (2009, March). Vocabulary changes in Agatha Christie's mysteries as an indication of dementia: A case study. In *19th Annual Rotman Research Institute Conference, Cognitive Aging: Research and Practice.* Toronto, Canada. Retrieved April 5, 2010 from http://ftp.cs.toronto.edu/pub/gh/Lancashire+Hirst-extabs-2009.pdf

Markov, A. A. (1906). *Extension of the Law of Large Numbers to the dependent case.* In Notices of the Physio-Mathematical Society of the University of Kazan, 2nd Series, *15,* 135–156. (In Russian).

Markov, A. A. (1913). *An example of statistical investigation of the text Eugene Onegin concerning the connection of samples in chains.* In Lectures to the Royal Academy of Sciences, St. Petersburg, VI series, *7,* pp. 153–162. (In Russian).

Merrill, S. J., & Cochran, J. R. (1997). Markov chain methods in the analysis of heart rate variability. *Fields Institute Communication, 11,* 241–252.

Nicolis, G., & Nicolis, C. (1990). Chaotic dynamics, Markovian coarse-graining and information. *Physica A, 163,* 215–231.

Rabiner, L. R. (1989). A tutorial on hidden Markov models and selected applications in speech recognition. *Proceedings of the IEEE, 77,* 257–286.

Visser, I., Raijmakers, M. E. J., & Molenaar, P. C. M. (2002). Fitting hidden Markov models to psychological data. *Scientific Programming, 10,* 185–199.

Zucchini, W., & MacDonald, I. L. (2009). *Hidden Markov models for time series: An introduction using R.* Boca Raton, FL: CRC Press.

18

Markov Chain Example: Transitions between Two Pictorial Attractors*

Robert A.M. Gregson

CONTENTS

Two experiments exploring similarity judgments on pairs and triplets of stimuli drawn from pictorial series are described. The stimuli are the man–woman (M–W) and gypsy–girl (Gy–Gr) pictures that slowly change from one prototype to the other as one progresses along the series. These have been used previously to demonstrate hysteresis of category judgments on ambiguous figures; the M–W series has been both modeled as a problem in neural network theory and mapped onto part of a cusp catastrophe surface. It is

---

* The original article was edited for format and originally published as: Gregson, R. A. M. (2004). Transition between two pictorial attractors. *Nonlinear Dynamics, Psychology, and Life Sciences, 8*, 41–64. Reprinted by permission of the Society for Chaos Theory in Psychology & Life Sciences.

shown that the transition process is complicated with a zone of uncertainty and prevalence of bimodality in many of the pairwise similarity judgments. The dynamics are interpreted in terms of transitions between two saddle-node attractors that are themselves not a discrete pair but have some overlap in their composition.

There are now three ways of approaching problems in psychophysical scaling: traditional linear models with Gaussian residual noise, neural network models, and nonlinear dynamics with attractor basins. There are so far no comparative studies that treat one specific problem from all three approaches in parallel as a contrast to see what succeeds and what fails to reveal new insights. It must be noted immediately that neural networks and nonlinear dynamics are not mutually exclusive; some of the relevant deep theory involves both at once (Martinez, 1996). The models are not mutually exclusive in their predictions, and some statistical methods for comparing theory with data may be common to two or more of them. This chapter attempts to go a little way in exploration.

The problem addressed in this chapter is not new, and has already an extensive theoretical and experimental literature. Some of those literatures do make explicit reference to attractor dynamics, and to wider ideas about cognition. The intention here is to treat the problem slightly differently, by examining in more detail what dynamics actually can be observed in the transition between attending first to one and then eventually to another pictorial attractor. In short, the actual form of one or more hysteresis loops, which could resemble transition through a cusp catastrophe or to passage between two basins of attraction, is explored.

In modern dynamical theory, there is a qualitative parallel with the situation observed here. Nishiura (2002, p. 241–243) in a section on elementary transient dynamics observes that (p. 241) "If several states (not necessarily equilibria) are connected by saddle-to-saddle orbits, there exists a nearby (heteroclinic[1]) orbit that experiences successive transitions from one state to another. It is evident that saddle-to-saddle connections in the phase space are a driving factor of the transient dynamics." The prevalence of bimodal response distributions in an experiment would suggest that there is a subcritical bifurcation present, which (p. 242) "can realize the situation where several stable solutions coexist, and a jumping transition from one stable solution to another is likely to occur." A subcritical bifurcation necessarily creates a saddle-node bifurcation. This suggests that the prototypes in a perceptual task are connected by saddle-to-saddle orbits, and if they are then the observed transitions (with hysteresis) in the previously reported experiments using binary categorization responses and not similarity judgments might be thereby explained.

It is very important to make clear that this nonlinear theory is not a derivation from catastrophe theory, but more abstract and more general, though it may not be incompatible with the idea of a cusp catastrophe arising locally within the system's dynamics. This question is addressed further in the final discussion section of the chapter.

The problem of seeking a simple index representation of the dynamics in the situation explored here has received attention (McCord & Mischaikow, 1992); however, as yet no unambiguous and readily tractable solution appears to be in sight. In technical terms, the problem is one of the nonuniqueness of *the* connection matrix; there seems to be some promise in constructing transition probability matrices between a few states of the total system, but global rather than local representation of the topology is required.

Two pictorial stimulus series, the M–W (Fisher, 1967a; Figure 18.1) and the Gy–Gr (Fisher, 1967b; Figure 18.2) have been used by various workers to look at ambiguous stimulus perception, where gradual change from one image to another is effected by altering some elements of a complicated picture, deleting and replacing them progressively until the image becomes the other attractor. The process of change is reversible, and may go through a number

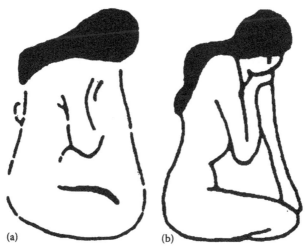

(a)   (b)

FIGURE 18.1
The end stimuli of the man–woman series.

FIGURE 18.2
The end stimuli of the gypsy–girl series.

of steps; 8 or 15 steps have been used. Obviously, not all elements in a picture are equipotential in the meaning they carry for the gestalt as a whole.

In previous studies, the situation was treated as one of seeing how categorization judgments changed when two multistable figures were used as psychophysical anchors, and a gradual transition between them was induced by presenting sequentially a series of ambiguous figures. This approach is still of considerable interest because now there are attempts to link the actual neurophysiological activity in the brain, which is induced by multistability and categorization, to observed perceptual responses (Miller, Freedman, & Wallis, 2002; Parker, Krug, & Gumming, 2002). The relevant brain dynamics are now thought to be nonlinear and unstable.

The innovation here is that the task is treated not as a discrimination or identification task, which actually raises problems if replications of judgments are elicited once the whole series has been seen and placed in some perceived order, but instead as one of similarity estimations on pairs or triples from the set of images. It is possible using this approach to explore some nonlinear relations across the ordered set of images, and also to tie the analysis into models of similarity (Gregson, 1975, 1994).

The Prototype$_1$—Ambiguity—Prototype$_2$ Paradigm

Suppose that there are at least three ordered objects, h_1, h_2, h_3, where A_p and B_p are end terms given the status of prototypes, and we construct similarities on pairs from this set. Taking S as a similarity rating, bounded $0 < S \leq 1$, where $S = 1$ implies identity (this is the notation used in Gregson, 1975), and $aSb = bSa$, then the similarities to be constructed are h_1Sh_2 and h_3Sh_2. The problem is to allocate h_2 either to A or to B, with $0 < v < 1$, where these are taken as cognitive prototypes, if they exist, A_p, or B_p, that satisfy

$$A_pSh_1 \geq v, \tag{18.1}$$

$$B_pSh_3 \geq v, \tag{18.2}$$

$$A_pSh_3 \leq (1 - v), \tag{18.3}$$

$$B_pSh_1 \leq (1 - v), \tag{18.4}$$

where v is a stringency parameter, the higher v is set the more rigidly two categories are defined as clustered about their prototypes. Defining the prototype conditions in this fashion is reminiscent of fuzzy subset logic, the setting of v also reflects the extent to which the categories A and B are not forced to be mutually exclusive. The relative similarities $h_2Rh_1|h_3$ and $h_2Rh_3|h_1$ are then the ratios

$$h_2Rh_1|h_3 = \frac{h_2Sh_1}{h_2Sh_3} \tag{18.5}$$

and

$$h_2Rh_3|h_1 = \frac{h_2Sh_3}{h_2Sh_1}. \tag{18.6}$$

These satisfy $0 \le R. \le 4$ and

$$h_2Rh_1|h_3 = (h_2Rh_3|h_1)^{-1}, \tag{18.7}$$

but if instead A_pSh_2 and B_pSh_2 are used the effect of putting $v < 1$ is to break the condition in Equation 18.7. The ancient argument about the difference between "those that are not with me are against me" and "those that are not against me are with me" hangs on the role of v in category formation in this approach.

Each h is made up of a set of physical attributes $\phi\{\mu_h\}$. If there are known measures of μ, then it is possible given two h_a, h_b to predict aSb values; call these ϕS, $R.$, and to compare the ϕS, $R.$ with observed responses ψS, $R.$. The special problem of identification that rises when

$$h_2Sh_3 = h_2Sh_1 \tag{18.8}$$

is a boundary region problem. If only Equation 18.8 is known, without any knowledge of $\phi\{\mu_h\}$, then there exists an infinite set of the possible attribute compositions of h_1, h_2, h_3. This set lies on a hyperplane normal to the continuum of hSh values; it is where some things are both equally different and equally similar, the case that Zerubavel (1991) described as extremely distasteful to the rigid mind.

Other definitions have been advanced; Estes (1994), for example, used

$$h_2Rh_1|h_3 = \frac{h_2Sh_1}{[h_2Sh_1 + h_2Sh_3]}$$

to satisfy $0 \le R \le 1$ so that R can then be treated as a predictor of choice probability for the allocation of h_2 to the prototype class A or B. Restle (1963) used a set-theoretic representation of a metathetic scale for this case; the ordered set of intersections of adjacent stimuli then have a pattern, which is analogous to the eigenvectors of a circumplex matrix.

The Man-to-Woman Experiment

Among the many ambiguous figures studied by Gestalt psychologists, a series of eight images that slowly changes from a man's face to a woman sitting sideways to the viewer has been a popular means of illustrating hysteresis in perceptual interpretations. The transition in responses from "man to woman" (or the reverse) shows hysteretic persistence according to the direction that the series is traversed. If we interpret A_p and B_p as attractors with intrinsic stability on the response surface of the dynamics, then in modern terms, the phenomenon might resemble transition across a cusp catastrophe, a possibility that is explored further in the detailed data analysis. The series of eight pictures here is taken from the version used by Stewart and Peregoy (1983) and by Haken and Stadler (1990) and discussed by Mainzer (1994, pp. 220–221).

This series of eight images was previously studied extensively by Stewart and Peregoy (1983) in a review that proposed that the dynamics of transition between the two prototypes were fully consistent with a cusp catastrophe. An experiment that predefined for the subjects the roles of the end stimuli drew on and also presented a derived range of other stimuli with the same outlines as the basic series but with progressively deleted internal detail. Their treatment of the hysteresis depended on the existence of what they called the *delay convention* (p. 348), which is deliberately designed out of the present two similarity studies.

The reasons why Haken was interested in the M–W series are made clear in a later text (Haken, 1991) where neural network theory was applied to what Haken saw as problems in cognition. Haken was able to show that a synergetic or dissipative network could perform some pattern recognition tasks, and at the same time exhibited hysteresis. Synergetic networks are characterized by macroscopic order parameters (which loosely correspond to stimulus input patterns in psychophysical theory) and with subsystems slaved to the order parameters. The features of a pattern recognized by the network correspond to the enslaved subsystems of the synergetic network during pattern formation. Features (hence, all subsystems) in the Haken simulations are actually the contents of pixels in a 60×60 lattice imposed on a prototype image. Probably, this rigid partitioning is a system feature that critically differentiates the artificial intelligence approach from human cognition.

Between the man and woman images, to which meaning can readily be given, are six more or less fuzzy images; these are perceptually ambiguous, and in the neural network modeling correspond to the phase transition of states in the synergetic computer. Each prototype image has a stable basin; these are called *global states of order* and are interpreted as attractors of the phase transitions. When an incomplete pattern is input, a competition between states, each of which corresponds to a specific prototype, starts. When a situation such as Equation 18.8 arises, recognition is a kind of symmetry breaking (Mainzer, 1994, p. 221).

It should be noted that the equations of Estes (1994) bear negligible resemblance to the network approach, and cannot, as unaugmented, predict the dynamics of shifting perception that are demonstrable and of focal interest. If a gestalt to which meaning may be given can function as an attractor in the dynamics of identification tasks, so that a set of images falls into its basin, and they are then perceived as equivalent in meaning, what is expected to happen to the stability of comparisons between images in the basin, and ones at its edge or just outside? This is an edge-of-chaos problem.

Ambiguous figures are physically fixed but perceptually unstable (e.g., Necker cubes); the perceptual instability may not be induced for a given observer until at least two images with distinct interpretations have been experienced. The similarity between a stable and an unstable image may then *exhibit* higher variance over time than a similarity rating between two gestalten, neither of which falls into a stable attractor basin, but also shows higher mean similarity. The idea here is that the meaning, once given, locks the structure of the attractor, but at the same time attractor dynamics have a fractal boundary. This situation is sometimes called *final state sensitivity*; the formal mathematics is discussed by Ott (1993, Section 5.2).

The point of invoking the concept of basins of attraction is that the dynamics can be simulated in networks, and also correlated with identifiable processes in neurodynamics in the brain. One might say that the classic psychometric mistake is to misidentify a basin of attraction as being a multivariate Gaussian probability density function in n-space.

Consider comparisons of the form $0 \leq S \leq 1$, PSh, where P is a prototype (closed, stable Gestalt, class exemplar, reference figure) and h is a variable for which, expressed as attribute sets, $\{P\} \cap \{h\} \neq \phi$. It is possible to map $\{\{P\} \cap \{h\}\}$ monotonically onto a set of intervals $\{d(P,x)\}$ without assumptions about interval properties of the ordering on d; both sides of the mapping are operationally pairwise comparisons. Attractor boundaries come into the picture if some h' lies at or near the bounds of F, but not otherwise. Then the monotonicity of M of

$$\{\{P\} \cap \{h\}\} \xrightarrow{\quad M \quad} \{d(P,x)\} \tag{18.9}$$

will be locally undetermined for $\{P\} \cap \{x'\}$.

The Haken theory is not merely concerned with matching (i.e., M or W) judgments, but also attempted within one framework derived from general synergetic equations (Haken, 1991, Chapter 5) to predict the oscillations between identification responses ($h \rightarrow M, h \rightarrow W$) and the distributions of reversal times in fluctuating perception, to prototypes allowing for response biasses. In psychophysical data, the oscillations are not strictly periodic, and their mean rate decreases to an asymptote over repeated stimulus inspection. Haken was obliged to parameterize his model with extra terms; the main interest for us here is that he had to write fluctuations in his internal attention parameters and not onto the parameters corresponding to perceived stimulus properties. He used (his Equation 5.14 et seq.) attention parameters that weight a matrix in a nonlinear fashion, so that a pattern is recognized only if the attention parameter is greater than zero; he wrote terms for recognition and discrimination, imposed a limit on exponential growth, and allowed for fluctuating inputs. The derivative (in time) of the attention parameters had to be coupled to the inputs.

Similarity Rating Experiment 1

Stimuli

The eight images used by Stewart and Peregoy and later by Haken were originally created by Fisher (1967a) and are coded as

$$M \text{ (man) } 1, 2, 3, 4, 5, 6, 7, W \text{ (woman) } = 8,$$

giving 28 possible paired comparisons and 6 triads of the form $M, W, h., h \neq M$, and W.

Subjects

Ten volunteers from the School of Psychology, Australian National University, all naïve with respect to the theories involved and unfamiliar with the stimulus materials.

Instructions and Procedure (Slightly Abbreviated)

Part one. You have 28 cards, each card shows two images side by side. Shuffle the cards, place the stack face down, and then turn up the cards one at a time. On your record sheet make a mark on the first empty line, which runs from "completely different" to "identical," to show the degree of similarity of the pair of images on the card. Turn the card face down when you have done that, and take the next card.

Part two. This part uses a set of six cards, which are different in their arrangement from the previous set. Each card shows three images; the lower two at the ends of a line are reference points, the middle image, placed higher up, is the thing to be compared with the lower two images. If the image is equally similar to (or equally different from) both of the end images, put a mark in the middle of the line. Otherwise take into account relative and unequal similarities. Consider details carefully. Repeat the procedure.

Data Analysis

It is expedient to begin by considering the correspondence, if any, with a similarity model that assumes the eight images are equally spaced on a continuum, and that consequently the pairwise similarities are monotonic, and nearly affine, on those separations on the continuum.

TABLE 18.1

Correlation with Linear Model on 28 S Pairs

Subject	Correlation	Bayes Λ^{-1}	Intercept	Slope
1	0.508	10.0	0.076	0.369
2	0.624	114.2	−0.129	0.926
3	0.691	803.8	0.050	0.703
4	0.721	2267.6	0.176	0.639
5	0.559	25.9	0.39	0.514
6	0.546	20.0	0.047	0.634
7	0.562	27.6	0.164	0.598
8	0.741	5039.5	0.145	0.480
9	0.500	8.7	−0.023	0.482
10	0.760	11605.5	−0.113	1.085

TABLE 18.2

Subdiagonal Matrix of 28 Mean Responses for Pairwise S on the M–W Series of Eight Images

	M	2	3	4	5	6	7
2	0.746						
3	0.666	0.706					
4	0.594	0.682	0.624				
5	0.638	0.586	0.702	0.708			
6	0.146	0.394	0.438	0.510	0.558		
7	0.200	0.244	0.228	0.324	0.270	0.486	
W	0.112	0.080	0.146	0.266	0.210	0.356	0.526

For each subject, separately, the 28 responses on the S task were regressed on the interval separations of the stimuli; the results are shown in Tables 18.1 and 18.2. The posterior odds on the correlations are expressed in Bayes factors as in Equation 18.10, which is Jeffreys' (1961) formula for a posterior likelihood ratio on a product–moment correlation coefficient r for n pairs of observations

$$A\left(\frac{\text{uncorrelated}}{\text{correlated}}\Big| r^2\right) = \left(\frac{2n-1}{\pi}\right)^{1/2}(1-r^2)^{(n-3)/2}. \quad (18.10)$$

The factor in favor of correlation is then Λ^{-1}.

Table 18.1 shows the heterogeneity of the subjects with respect to fitting a linear model; the model assumes both that the stimuli M, h., W are on an equal interval scale, and that $MSW > 0$. This heterogeneity is masked if the average regression of data ψS on a linear model ϕS, $d(a, b)$ approximated by

$$\phi S_{a,b} = 1 - \frac{|h_a - h_b|}{7.1} \qquad (18.11)$$

is computed, as

$$\psi S = 043 + .677 \phi S, \qquad (18.12)$$

with Bayes factor 33950.1, $r = 0.78$, %var = 61.29. Table 18.2 shows that the images are not seen as equally spaced; some steps carry more information than others, and the A/f end of the continuum is more closely spaced. It is Table 18.2 that should be a regular simplex (Guttman, 1955) in its structure, if the S is monotonic on the interval separations of the eight images. It is not strictly simplex.

Tables 18.3 and 18.4 are of interest because they offer evidence of perceptual order inversions in what would be a strict ordering if the definition of resemblance used in Haken's theory, or of similarity in a linear distance theory, held. It is predicted that no order inversions in the six triadic R. comparisons should arise; in Table 18.3, there are seven subjects with one or more inversions. For Table 18.4, the result is more extreme, with 9 of 10.

In Tables 18.3 and 18.4, the theoretical relative similarities computed from Equations 18.5 and 18.11 would be monotonically decreasing as $MRh_2|W = 5.54$, $MRh.3|W = 2.42$, and $MR.h_4|W = 1.32$. The values for $R.h_7 = R.h_2^{-1}$ and so on for h_3, h_6, h_4, $h._5$ as the series is *under that model* symmetric about its midpoint between h_4 and h_5. The theoretical R. values in Table 18.3 are hSM/hSW for each h. on a linear model. The only subject that comes near to those predicted values is Subject #10 in Table 18.4 (compare the Bayes factor

TABLE 18.3

Relative Similarities from S Values

h	Subject 1	Subject 2	Subject 3	Subject 4	Subject 5
2	28.00	24.00	44.00	3.00	6.14
3	13.00	43.00	18.00	3.00	1.00
4	3.66	21.00	1.69	2.06	3.00
5	10.50	5.28	16.00	2.00	3.00
6	2.12	0.54	0.92	1.11	1.12
7	0.50	0.14	0.36	0.32	0.64
h	Subject 6	Subject 7	Subject 8	Subject 9	Subject 10
2	13.00	16.00	3.11	33.00	21.00
3	5.33	6.14	1.91	37.00	18.00
4	10.33	3.16	2.20	18.00	0.73
5	9.33	6.00	1.87	9.00	1.00
6	2.44	3.14	0.77	3.00	0.07
7	0.17	1.00	0.71	0.36	0.05

TABLE 18.4

Relative Similarities from R Triads

h	Subject 1	Subject 2	Subject 3	Subject 4	Subject 5
2	2.57	6.14	24.00	1.94	2.57
3	1.63	5.25	1.27	3.16	3.16
4	2.57	1.94	2.57	1.00	6.14
5	2.12	1.63	1.63	0.56	1.94
6	1.63	0.92	1.94	0.47	0.78
7	1.08	0.09	0.28	0.28	0.16
h	Subject 6	Subject 7	Subject 8	Subject 9	Subject 10
2	2.12	6.14	2.12	5.25	5.25
3	2.57	5.25	1.94	2.84	3.16
4	1.38	3.16	1.27	3.54	0.78
5	1.94	2.84	1.38	4.55	1.94
6	0.47	3.16	0.38	0.61	0.56
7	0.28	3.16	0.35	0.47	0.16

R values are obtained directly from responses.

in Table 18.1). Violations of monotonicity are very obvious in Table 18.3, and less so but still arise in Table 18.4.

Discussion

This exploratory study with the M–W series shows that the series is not perceptually equally spaced, so the hysteresis obtained by Haken should not strictly have been symmetrical as he illustrates it. The regions adjacent to the prototypes, around h_2, h_3, h_6, and h_7, do show marginally higher uncertainty, which can be the outcome of local bifurcation, or what Haken called symmetry breaking.

The relationships between dyadic (S) and triadic (R) similarities are weak and difficult to interpret simply; there are large individual differences. The values in Table 18.3 are based on two degrees of freedom, whereas the values in Table 18.4 (direct R responses) are so constructed by the instructions to subjects as to have only one degree of freedom. The equal spacing from M to W is wrong phenomenologically for the subjects involved; the series is closer spaced at the M end, and Tables 18.3 and 18.4 indicate some reversals or instabilities as soon as we move from S to R judgments. Haken's model predicted one transition point between attractors, but one which moves with the direction of strict complete sequential change $M \rightarrow h \rightarrow W$, or $W \rightarrow h \rightarrow M$ in the stimulus input. It is this directionality that has been destroyed in the similarity paradigm created here. Both the data and the idea of fractal attractor boundaries suggest two boundaries and an unstable region between them. Then asking for R judgments is asking for locations mostly in a very unstable region, not putting a response simply to one or the other side of a boundary.

The Gypsy-to-Girl Experiment

Fisher's (1967b) Gy–Gr series of 13 ambiguous figures between two prototype images, 15 in all, is potentially much more interesting and more deli-
cately constructed than the series used by Stewart and Peregoy (1983) and
later by Haken. In fairness, it should be observed that with modern computer graphics it should be possible to morph images gradually with much
more control and precision than that available to the unaided artist's hand.
An example of the subtle and smooth transitions that now may be created
is shown by Wang et al. (2002). Obviously Fisher's series is a better basis for
exploring the properties of attractor boundary dynamics, though it would
require a full matrix of $15 \times 14/2 = 105$ paired comparisons if we make no
prior assumptions about where the dynamics might be most informative.
With the Fisher series, the intrasubjective consistency of local S-transitivities
on adjacent $(h_k, k, k + 1)$ and nonadjacent $(h_k, k, k + 2)$ image pairs in the full
series gives 14 and 13 pairs out of a total of 27 critical trials in an experiment
of a tractable size. The hypothesis is that if a fractal attractor boundary of θ
steps wide exists $1 \leq \theta \leq 2$ then S-intransitivity can appear at two locations
in the Gy–Gr series.

    If transitivity holds, it is predicted that: *If $h_1 Sh_2 = w_1$, $h_2 Sh_3 = w_2$, h_2 lies
between h_1 and h_3, then $h_1 Sh_3 = w_3$, and $w_3 \leq w_1, w_2$.* A violation of this can arise if
the effect of the local fractal boundary is to displace h_2 so that the betweenness
condition is not met.

Similarity Rating Experiment 2

Stimuli

The 15 images created by Fisher were presented in paired comparisons,
balanced for stimulus laterality, and the subset of pairs used is shown in
Table 18.5.

Subjects

None of the 29 volunteers from the School of Psychology had participated in
the first experiment. The subjects were first-year undergraduates who served
as part of their acquisition of course credits.

Instructions (Slightly Abbreviated)

You have 48 cards, which the experimenter has shuffled for you. Place
the shuffled stack face downwards. Then, one at a time, turn up the card.
(The next part of the procedure was identical to the first experiment).

TABLE 18.5

Cells Used in Experimental Design for Gypsy-to-Girl Experiment[a]

	1	2	3	4	5	6	7	8	9	10	11	12	13	14	15
1	–														
2	x	–													
3	x	x	–												
4	x	x	x	–											
5		x	x	x	–										
6		x	x	x	–										
7			x	x	x	–									
8				x	x	x	–								
9					x	x	x	–							
10	x					x	x	x	–						
11		x					x	x	x	–					
12			x					x	x	x	–				
13	x			x					x	x	x	–			
14		x			x					x	x	x	–		
15			x			x					x	x	x	–	

[a] The leading diagonal cells, marked as "–," which means identical pairs, are omitted by definition. Each x is a similarity between its respective row r and column c elements, $x \equiv h, Sh_c$ The five diagonals (marked with x) in the design are coded D1 (the longest) down to D5 (the shortest).

Then the instructions were emphasized as follows: *Consider details because THERE ARE NO PAIRS in which the two images are EXACTLY THE SAME, though SOME may be VERY SIMILAR to each other.*

Procedure

For each subject, the 48 cards were individually shuffled, so that no two subjects did the task in the same order. The randomization was checked by autocorrelation analysis of every stimulus and response series, and the series of residuals between predicted and observed responses under the linear model (Equation 18.11). None of these were of interest, though some autocorrelation at short lags in some response sequences was noted. The average over all subjects was negligible, some showed weak assimilation and some weak contrast (in dynamics terminology persistence or anti-persistence as quantified in tile Hurst index). The data matrix of mean similarities appears in Table 18.6.

In Table 18.7, "B" indicates marked bimodality on visual inspection, and "b" indicates weak bimodality due to outliers in a long tail distribution. Apparent bimodality is, however, so common in small samples of 29 observations that it is better for caution to compute indices of platykurtosis for each cell (Kendall & Stuart, 1969, p. 86); these are shown in Table 18.9.

TABLE 18.6

Data Matrix of Mean Similarity Responses Pooled Over 29 Subjects
in the Gypsy-to-Girl Experiment

	1	2	3	4	5	6	7	8
1	—							
2	0.671	—						
3	0.594	0.758	—					
4	0.517	0.574	0.664	—				
5		0.436	0.518	0.702	—			
6			0.473	0.638	0.709	—		
7				0.559	0.630	0.746	—	
8					0.602	0.741	0.721	—
9						0.749	0.737	0.760
10	0.322						0.638	0.632
11		0.392						0.473
12			0.258					
13	0.279			0.298				
14		0.199			0.290			
15				0.226		0.198		

	9	10	11	12	13	14	15
9	—						
10	0.763	—					
11	0.531	0.662	—				
12	0.542	0.668	0.674	—			
13		0.586	0.510	0.772	—		
14			0.379	0.634	0.694	—	
15				0.494	0.577	0.794	—

Discussion

The evidence of interest lies in any possible violation of monotonicity in the
series of diagonalizations D1 through D5. If the matrix were simplex then
the mean similarities along the diagonals, as separation from the leading
diagonal increases, should fall off monotonicallv. They do; they fall off gradually from D1 to D3, from $(k, k + 1)$ to $(k, k + 3)$, but the last two, D4 for $(k, k +$
9) and D5 for $(k, k + 12)$ show higher values than might be expected, almost
0.25. As these values have stable variances, they can be compared by a b-test,
for example, the difference between $(k, k \pm 1)$ and $(k, k + 12)$ means has $t \sim 8$,
$df = 20$. However, what is important for our purposes is the shape of the distributions of these diagonalization measures across subjects.

Table 18.8 shows the distribution of pooled S responses, that is $29 \times 48 =$
1392 responses over all subjects, which is skewed to the larger end (nearer to
identity than to complete dissimilarity). That is what one would expect given

TABLE 18.7

Cells Showing Bimodal S Distribution in the Gypsy-to-Girl Experiment[a]

	1	2	3	4	5	6	7	8	9	10	11	12	13	14	15
1	–														
2	B	–													
3	B	x	–												
4	B	B	B	–											
5		b	B	B	–										
6			B	x	b	–									
7				B	b	B	–								
8					x	X	B	–							
9						B	B	B	–						
10	B						x	b	b	–					
11		x						B	x	B	–				
12			x						B	b	B	–			
13	B				b					B	B	B	–		
14		x				x					B	B	b	–	
15			x			B						B	B	B	–

[a] Probabilities of B in diagonals, D1 through D5: 0.64, 0.54, 0.75, 0.33, 0.33. Probabilities of b in diagonals correspondingly: 0.29, 0.23, 0.08, 0.17, 0.00.

TABLE 18.8

Frequency Distributions of S on Diagonals

S	Pooled	D1	D2	D3	D4	D5
0–0.1	0.063	0	0	0	0.080	0.280
0.1–0.2	0.079	0	0	0.040	0.240	0.160
0.2–0.3	0.072	0	0.040	0.0	0.360	0.240
0.3–0.4	0.087	0	0	0.200	0.160	0.200
0.4–0.5	0.084	0.040	0.200	0.200	0.200	0.080
0.5–0.6	0.102	0.120	0.240	0.240	0	0.120
0.6–0.7	0.138	0.320	0.240	0.160	0.080	0
0.7–0.8	0.139	0.280	0.160	0.120	0	0
0.8–0.9	0.153	0.200	0.080	0.040	0	0
0.9–1.0	0.079	0.040	0.040	0	0	0
Mean of S:		0.72	0.61	0.53	0.28	0.23
s.d.		0.12	0.15	0.15	0.15	0.16

the way the pictures are drawn. Scanning the columns D1 through D5 from bottom-left to top-right displays the regions of greatest response frequencies. Tile means of the S arrays consistently increase with stimulus proximity and the distributions are homoscedastic; presenting data in this way masks the bimodalities in Table 18.7. In short, the first two moments are not sufficient statistics if the situation is nonlinear. From Table 18.8, it is seen that the

TABLE 18.9

Frequency Distributions of Higher Moments of *S*

Histogram Intervals	Probability $\lvert\sqrt{\beta_1}\rvert$	Probability μ^4
0–0.5	0.125	0.062
0.05–0.10	0.250	0.187
0.10–0.15	0.146	0.250
0.15–0.20	0.208	0.250
0.20–0.25	0.187	0.146
0.25–0.30	0.062	0.083
0.30–0.35	0.021	0
0.35–0.40	0	0
0.40–0.45	0	0.022

Distributions are collected over all 48 cells of the design.

skewness is always negative in all 48 cells, and has itself a bimodal frequency distribution as shown in Table 18.9. The fourth moment has a long tail.

It has been known for some time that similarity ratings over the interval 0.1 have a banana-shaped frequency distribution if responses ψS are plotted as ordinate against theory ϕS as abscissa; why this is so has not been investigated in detail (Gregson, 1994). The Gy–Gr series would have to be extended more beyond its Gr end toward an implicit Gr prototype for this phenomenon to be replicated here.

The fact that the similarities here are not monotonic on stimulus separation and not consistent with a linear model implies that their isosimilarity contours are not always convex about a reference point, so they are not mapped into a metric space (Gregson, 1993). As there is no available representation of the pictures as vectors of attributes it is not possible without making a lot of ancillary assumptions (including ones about stimulus dimensionality) to predict what shape the isosimilarity contours would actually take, though possible examples are available (Gregson, 1995, Chapter 7). The pictures are obviously two-dimensional on the page, but dimensionality in terms of complexity is something else. One way of approaching this question is to compute the eigenvalues of the *S* matrix in Table 18.2; that is possible because Table 18.2 is the subdiagonal triangular part of a skew symmetric matrix, where the diagonal cells would be all equal to *aSa* = 1 by definition. The first four eigenvectors only are shown in Table 18.10, the eight eigenvalues are 211, −209, 3.415, 1.163, 0.588, 0.366, 0.269, and 0.192, so we have two dominant dimensions in a bipolar pair, and two weaker ones, and of the five less than unity, the last four may be noise. These last four have moduli less than 1, so the condition for a saddlepoint is met; what is interpreted as noise in a linear model appears in this situation to be the necessary part for a saddle in dynamics.

Martelli (1992, p. 127) notes that when the modulus of some of the eigenvalues is greater than 1, and that of others less than 1, then we have a saddle

TABLE 18.10

First Four Eigenvectors for Table 18.2

M	0.0025	0.0018	−0.4254	−0.4070
2	0.0023	0.0017	−0.4737	−0.2252
3	0.0024	0.0018	−0.4574	−0.1711
4	0.0033	0.0015	−0.4392	0.0049
5	0.7070	0.7071	0.0026	−0.0022
6	0.0031	0.0006	−0.3322	0.5764
7	0.0027	−0.0009	−0.2529	0.6496
W	0.7071	0.7071	0.0070	0.0000

point. The dynamical implication of this mix is that there are copresent both stable and unstable subspaces. The values represent an interesting pattern because they suggest that the two largest eigenvalues are associated with linear separation, with some hiatus in the middle of the series, and the next two smaller eigenvectors could support the nonlinearity and bimodality found in the second experiment for a similar but more gradual transition series between prototypes. As usual, the nonlinear parts are second order compared with the dominant linear approximations that were induced by the constraints of the experimental design. However, it is the nonlinear components that characterize the subtleties of human perceptual processes as compared with the artificial intelligence neural network approach. The nonlinear components are separable from residual noise because of their regularity and persistence as shown in Tables 18.3, 18.4, and 18.7 through 18.10, but only if the data analysis does not rest exclusively on linear filtering and the first two moments of distributions, as is typical of approaches such as the popular Analysis of Variance.

Some recapitulation of the theoretical issues raised here may be helpful. A nonlinear system may be thought as, by definition, one that is hypersensitive to small variations in the conditions imposed on response choices, and on the detailed construction and diversity of the stimuli used. In modern cognitive psychological terms, the tasks used at least for the M–W series have varied in the nature and extent of *priming* of responses. One extreme form of priming is to present the stimuli only in strict serial order, and not randomly; the early gestalt interest in hysteresis is contingent upon that priming coupled with the serial response dependence based on short-term memory that facilitates priming.

The ambiguous stimulus series constructed by Fisher are composed of an invariant boundary contour and variable internal details. This invariance is relaxed a bit in the Gy–Gr series but is strongly held in the M–W series and was thus exploited by Stewart and Peregoy (1983, p. 344) to construct additional sets of almost identical stimuli made up of the boundary contour and minute internal details. These in the limit thus became identical and were interpreted as fixed level locations on the cusp catastrophe surface away from the cusp; the extreme of the bifurcation factor. The response task was like the triplet similarity comparisons in Table 18.4, but with additional information to the subject identifying the special role of two implicit prototypes. None of these instructions or additional degenerated stimuli were used in the similarity experiments, so that sort of priming was minimized here. It is thus suggested that the cusp catastrophe modeling was facilitated and even induced by the priming, which is not part of the basic similarity task.

There are in fact four theoretical approaches to modeling the perceptual phenomena, and some focus on the hysteresis and others on the region of response ambiguity. These alternatives rest on different sorts of comparative judgments, the basic yes–no binary identification response ($=A_p$ or $=B_p$), the paired comparison similarity S, and the triadic relative similarity R. Obviously, the three entail different amounts of uncertainties and thus of information.

1. Linear model: Dissimilarity $= f$ (Minkowski distance in n-dimensional metric space), where f is monotonic. This is single-valued, irrespective of the dimensionality of the Minkowski metric space, is without hysteresis, and is invalidated by Table 18.7.

2. Serial autocorrelation, $r(j, j - 1)$ of single-valued binary identification choices C, not made in the S experiments: $C = w[\pi A_p + (1 - \pi)B_p)]_j + (1 - w) [\pi A_p + (1 - \pi)B_p)]_{j-1}$ where π is the bias to one prototype and $w, (1 - w)$ is the serial effect. This can induce hysteresis if both orders $j, j - 1$ are used, without there being a cusp present. The equation is inadequate without augmentation for modeling the sequential effects between two paired comparison judgments.

3. Neural networks: This admits of discrimination learning, and involves nonlinear processes; statistically it is a form of nonlinear multivariate discriminant analysis with tuned connections between features. Each prototype and its degradations are a vector of features. In Haken's example, the number of hidden variables is much greater than in the competing nonlinear models.

4. Nonlinear dynamics and trajectories on a manifold: The manifold is the set of all possible trajectories that have the same stability. In this situation, we may neglect the attractive and repellent manifolds. Two cases, at least, need to be distinguished are as follows: (a) The cusp catastrophe is a local feature of one manifold; it does not necessarily mean that there are two attractors present in the total dynamics. One may have closed trajectories from one attractor that traverse the cusp region laterally as the control parameters vary. (b) The generic case with two saddle-node attractors. The evidence of Table 18.10 supports this interpretation. Saddle node attractors arise in far-from-equilibrium dynamics, they are not intrinsic to catastrophe theory.

There is still one more outstanding data analysis problem in reconciling the evidence for the two nonlinear models (a) and (b). Table 18.7 shows bimodalities in all adjacent pairs, including the ones with the extreme stimuli, all the stimuli are unstable to some degree. Two real stable prototypes, A_p and B_p, are not in fact physically present in the experiment. In Figure 18.17 of Stewart and Peregoy (1983, p. 355), the fitting of Cobb's program locates some of their stimuli inside and some just outside the unstable cusp fold region, and the majority completely outside the unstable region. The Gy–Gr experiment here with minimal priming and the M–W experiment of Stewart and Peregoy with explicit priming by instruction and stimulus choice are not dynamically two horses from the same stable, but their comparison reveals to some extent how much perceptual nonlinear dynamics are dependent not just on the ambiguity of the stimuli but also on the associated induced perceptual sets.

Acknowledgments

I am grateful to two anonymous referees who asked that I should also consider in detail the contrast between this study and that of Stewart and Peregoy (1983). The eigenvalue computations were done using Mathematica. Other calculations were done using Linux Fort77.

Note

1. For formal definitions of heteroclinic and homoclinic orbits, see Wiggins (1988, p. 181).

References

Estes, W. K. (1994). *Classification and cognition*. New York: Oxford University Press.
Fisher, O. H. (1967a). Preparation of ambiguous stimulus materials. *Perception and Psychophysics, 2,* 421–422.
Fisher, O. H. (1967b). Measuring ambiguity. *American Journal of Psychology, 80,* 541–547.
Gregson, R. A. M. (1975). *Psychometrics of similarity*. New York: Academic Press.
Gregson, R. A. M. (1993). The form of isosimilarity contours in nonlinear psychophysics. *Proceedings of the International Society for Psychophysics: 9th. Annual meeting*. Madrid, Spain, UNED. pp. 101–106.
Gregson, R. A. M. (1994). Similarities derived from 3-d nonlinear psychophysics: Variance distributions. *Psychometrika, 59,* 97–110.

Gregson, R. A. M. (1995). *Cascades and fields in perceptual psychophysics*. Singapore: World Scientific.

Guttman, L. (1955). A generalized simplex for factor analysis, *Psychometrika, 20*, 173–192.

Haken, H. (1991). *Synergetic computers and cognition: A top-down approach to neural nets*. Springer Series in Synergetics (Vol. 50). Berlin: Springer-Verlag.

Haken, H., & Stadler, M. (Eds.) (1990). *Synergetics of cognition*. Berlin: Springer-Verlag.

Jeffreys, H. (1961). *Theory of probability*. Oxford: Oxford University Press.

Kendall, M. G., & Stuart, A. (1969). *The advanced theory of statistics* (Vol. 1, 3rd ed.). London: Griffin.

Mainzer, K. (1994). *Thinking in complexity: The complex dynamics of matter, mind, and ordered Systems. ICoilections Travattx en. Cours* (Vol. 53, pp. 113–133). Paris: Herman.

McCord, C., & Mischaikow, K. (1992). Connected simple systems, transition matrices and heteroclinic bifurcation. *Transactions of the American Mathematical Society, 333*, 397–421.

Miller, E. K., Freedman, D. J., & Wallis, J. D. (2002). The prefrontal cortex: Categories, concepts and cognition. *Philosophical Transactions of the Royal Society of London B, 357*, 1123–1136.

Nishiura, Y. (2002). *Far-from-equilibrium dynamics*. [*Translations of mathematical monographs*, Vol. 299]. Providence, RI: American Mathematical Society.

Ott, E. (1993). *Chaos in dynamical systems*. New York: Cambridge University Press.

Parker, A. J., Krug, K., & Gumming, B. G. (2002). Neuronal activity and its links with the perception of multi-stable figures. *Philosophical Transactions of the Royal Society of London, B, 357*, 1053–1062.

Restle, F. (1963). *The psychology of judgment and choice: A theoretical essay*. New York: Wiley.

Stewart, I. N., & Peregoy, P. L. (1983). *Catastrophe* theory modeling in psychology. *Psychological Bulletin, 94*, 336–362.

Wang, K., Hoosian, R., Li, X.-S., Zhou, J.-N., Wang, C.-Q., Fu, X.-M., & Yue, X.-M. (2002). Impaired recognition of fear in a Chinese man with bilateral cingulate and unilateral amygdala damage. *Cognitive Neuropsychology, 19*, 641–652.

Wiggins, S. (1988). *Global bifurcations and chaos: Analytical methods*. New York: Springer-Verlag.

Zerubavel, F. (1991). *The fine line*. New York: The Free Press.

19

Identifying Ill-Behaved Nonlinear Processes without Metrics: Use of Symbolic Dynamics*

Robert A.M. Gregson

CONTENTS

Given ill-behaved psychological data that are unlikely to satisfy metric axioms, the use of encoding in symbolic dynamics, and hence leading into Markov analyses, is explored. Various measures of entropy are calculated. The tractability of entropic measures for categorizing the trajectories of nonlinear dynamics that may be present and chaotic is considered, with a focus on the case where there are two attractors and at least one heteroclinic orbit between them. Fast/slow dynamics are treated as a special case. The problem of identification is in other contexts the problem of diagnosis in time-varying pathologies. Some real data, selected for their psychological relevance in clinical, forensic, and psychophysical processes, that are apparently edge-of-chaos and nonstationary, are for comparison analyzed both as metric and discrete and in symbolic encoding.

---

* The original article was edited for format and originally published as Gregson, R. A. M. (2005). Identifying ill-behaved nonlinear processes without metrics: Use of symbolic dynamics. *Nonlinear Dynamics, Psychology, and Life Sciences, 9,* 479–503. Reprinted by permission of the Society for Chaos Theory in Psychology & Life Sciences.

...function imposes a specific regulatory and information structure
... certain design features are necessary for a stable phenotype. These
design constraints allude to a theoretical biology distinct from physics
and chemistry, more akin to engineering than the new physical laws that
Schrödinger originally introduced.

Rao, Wolf, and Arkin (2002, p. 236)

It is appropriate to put this survey and review of applications into some
historical context, to point out what has and what has not been achieved. Ball
(2003) in an editorial preface comments:

Over the last three or four decades interest in nonlinear phenomena has
grown nonlinearly..., the number of times the word 'nonlinear' occurs in
abstracts of articles in *Physical Review Letters*... between 1958 and 2001"
has grown (from zero to about 170 per year). "The data do not look very
linear (a) in fact a linear fit is *not* really very good (b), and a nonlinear fit
is obviously much better (c). Of course a least-squares fit does not model
fluctuations in the data.

One purpose of this chapter is to see if this trend has been matched in
experimental or theoretical psychology, and if not, then why not.

The problems of distinguishing between the trajectories of deterministic
processes and the sequential outputs of stochastic processes, and consequently the related problem of identifying the component dynamics of mix-
tures of the two types of evolution, have produced a very extensive literature
of theory and methods. One method that frequently features is so-called box
counting or cell mapping, where a closed trajectory is trapped in a series
of small contiguous regions as a precursor to computing measures of the
dynamics, particularly the fractal dimensionality (e.g., which has parallels
in the analysis of cubic maps in nonlinear psychophysics; see Udwadia and
Guttalu, 1989).

Serious difficulties are met in identifying underlying dynamical processes
when real data series are relatively short and the stochastic part is treated
as noise (Aguirre & Billings, 1995); it is not necessarily the case that treating
noise as additive and linearly superimposed is generically valid (Berthet,
Petrossian, Residori, Roman, & Fauve, 2003). Though diverse methods are
successfully in use in analyzing the typical data of some disciplines, as in
engineering, there are still apparently irresolvable intractabilities in exploring the biological sciences (particularly including psychology), and a prolif-
eration of tentative modifications and computational devices has thus been
proposed in the current literature.

The theoretical literature is dominated by examples from physics, such
as considerations of quantum chaos, which are not demonstrably relevant
for our purposes here. Special models are also created in economics, but
macroeconomics is theoretically far removed from most viable models in

psychophysiology. Models of individual choice, and the microeconomics of investor decisions, may have some interest for cognitive science, but the latter appears to be more fashionably grounded, at present, in neural networks, though again the problems of simultaneous small sample sizes, nonlinearity, nonstationarity, and high noise have been recognized and addressed (Lawrence, Tsoi, & Giles, 1996).

Much of the computational literature focuses on fractal dimensionality, Lyapunov exponents, or entropy (Mayer-Kress, 1986); however, there are paths between symbolic dynamics and entropy measures, particularly where the location of periodic saddle orbits is involved (Lathrop & Kostelich, 1989a,b). The use of Lyapunov exponents emphasizes that the predictability and controllability of processes are fundamental concerns, but even here, systems can float between uncertainty and certainty about their future evolution (Ziehmann, Smith, & Kurths, 2000). One exception, which could circumvent the difficulties in analyzing psychological data that are encoded in numbers that do not satisfy metric axioms, is to use symbolic dynamics. It is important to emphasize that the methods for examining nonlinear and nonstationary time series, developed by Casdagli (1989) and, later, by related workers (Casdagli & Eubank, 1991), do, like ARMA modeling, rely on the metric properties of the variables being observed. This is not always admissibly the case here.

The symbolic dynamical approach was advocated relatively early by Badii, Nicolis, and their collaborators (see their papers in Abraham, Albano, Passamante, & Rapp, 1989), though its historical origins go back to Hadamard in the 1890s (Wiggins, 1988). In treatments of complexity theory, as distinguished from chaos, symbolic dynamics are considered by some to be an "especially illuminating approach" (Nicolis & Prigogine, 1989, Section 4.7).

There are alternative ways of conceptualizing the mathematics that derive from the relatively simple basic assumption of symbolic dynamics, that a data string can be represented as a finite closed set of mutually exclusive nonnumerical symbols, rather than by a series of numbers. If regarded as part of coding theory (Lind & Marcus, 1995), symbolic dynamics leads into entropy measures, cyclic structures, and, particularly, shift operations; if treated as the qualitative theory of dynamics (Luo & Teng, 1993), it leads more readily into an emphasis on manifolds and topology, structural stability, and chaos. Markov chain theory can be seen as an extension from both of these two approaches (Blanchard, Maass, & Nogueira, 2000) or as a self-contained topic in its own right (Gillespie, 1992). The Markov chain approach has become centrally important in modern statistical theory and practice (Green, 1995; Robert & Casella, 1999), and it is that conceptual pathway that is being emphasized here, rather than the string-coding approach that has also been used, for example, by Guastello (2000) and by Geake and Gregson (1999).

The problem of constructing psychological measures with axiomatic bases that define some sort of metric was a continuing challenge in the later twentieth century (Krantz, Luce, Suppes, & Tversky, 1971), and the invalidity of assumptions by fiat, which had been achieved, has been described by Michell (2002). Possible foundations of psychophysical scaling were established on the basis of the long-established functional calculus (Aczél, 1966, gives a historical survey) to define what forms of scales can satisfy metric axioms (Luce, Bush, & Galanter, 1963). The assumptions therein imply functional deterministic stability, and are not of use for modeling dynamics processes without augmentation; the modern approach using Markov Chain Monte Carlo (MCMC) statistics can in some restricted cases be treated as a hybrid of Euler functions and Markov chain transitions (Winkler, 2003, p. 314), which does postulate an evolving mixture of deterministic and stochastic processes running through time. The symbolic dynamics explored here are closely related to some of the assumptions of MCMC practice, but we do not use the full apparatus of statistical estimation; rather the focus is on the ubiquitous nonstationarity of psychological time series. In effect, by using symbolic dynamics, the processes studied are taken to be in the class of discrete dynamical systems, and tests for stability that are developed are in that domain (Gumowski & Mira, 1980).

There have apparently been examples from social psychology where using symbolic dynamics instead of making metric assumptions (with ANOVA-type statistics) has produced more sensitive insights into the dynamics (Heath, 2000, p. 311). Guastello, Hyde, and Odak (1998) and Guastello (2000) on information exchanges during creative problem solving in a social group, Guastello, Nielson, and Ross (2002) in the analysis of brain activity in fMRI pattern sampling, and Pincus (2001) in family interaction dynamics, have also made valuable use of symbolic dynamics, employing some necessary variations in technical details.

If one wishes to explore nonlinear dynamics directly within the traditional framework of difference-differential equations, then the use of variables with metric properties, such as can more readily be achieved in psychophysiology than in psychophysics, is mandatory. However, a confrontation between the advocates of qualitative and quantitative methodologies has recently persisted at least in British psychology (Salmon, 2003), and neither stance makes much sense from the perspective of symbolic dynamics. A dispute between the advocates of discourse analysis, on the one hand, and the users of the general linear model in statistics, on the other, merely confirms the profound mathematical inadequacies of both factions.

The position taken here is that the identification of nonlinear dynamics, and consequently of the trajectories of attractors in psychological data, is not dependent upon the observed time series that are interpreted as realizations of samples from trajectories, being constructed from variables with intrinsically metric properties. It is necessary and sufficient that a symbolic dynamics encoding is possible, and this in turn implies that a variant of Markov analyses with associated information measures can then be constructed. It is still necessary for data sets to be sufficiently long for a stable estimation of parameters to be made, particularly if the symbolic analysis reveals the presence of nonstationarities.

Identification and Control

Before considering some examples, a brief excursion into control theory (Eykhoff, 1974) is pertinent to indicate why the examples have some didactic value. Given some dynamic process extending through time, there are three interrelated aspects that can be modeled: (a) *Identification:* There exists a model that encodes at least some aspects of the inputs and outputs of the process at any point in time, discrete or continuous. (b) *Prediction:* Some properties, local or sequential, can be predicted from a partial record of what has already evolved. (c) *Control:* It is possible to nominate a bounded window within which the process can be constrained in its outputs for any given highly probable input sequence.

This approach to systems theory is common in areas such as chemical engineering plant control, but is rare in justifying theory construction in psychology. Its relevance becomes more obvious when the identification is made formally in terms of dynamics. Misidentification invalidates prediction; paradoxically, it may not invalidate local control, but is expected to invalidate long-term control. It is fundamental that controllability does not necessarily imply predictability, and predictability does not necessarily imply controllability. Earthquakes and manic outbursts in psychopathology are cases in point. Stochastic prediction, where the probability of some event within the identified repertoire of the process can be given a value, but the precise time of its next occurrence cannot, can be controllable. Strictly deterministic prediction can define the time of the next occurrence or sequence of occurrences; chaotic systems break down deterministic predictability, but in some cases are controllable. Control of a dynamical system implies embedding it in one or more feedback loops, which may themselves contain other dynamical systems that can match, in an inverse sense, the dynamics of the systems to be controlled. The identification of the target system is a necessary precursor to constructing an inverse system to effect the control.

In the following examples, we are solely concerned with identification; the construction of feedback and control is not carried through here. It would be a different, though valuable, exercise for each example. There is no single inverse system that will control all the examples, or any others that are invented, except the equifinality of death.

Entropic and Parry Measures

One of the powerful consequences of using symbolic dynamics, on maps on the unit interval, is that Markov transition probability matrices may be created, and from those an information theory treatment is supported, leading back into entropy calculations. The deep mathematical relations between symbolic dynamics, Markov chains, and entropy measures have now an extensive literature, which has been surveyed by Blanchard et al. (2000).

This rests on some theorems of Parry (1964, 1966), showing that if a series behaves locally like a Markov process, then, from the perspective of information theory, it is Markov. This approach has been extended by Buljan and Paar (2002).

Symbolic dynamics have been linked with nonlinear psychophysics (Geake & Gregson, 1999), for encoding the existence of embedded recurrent episodes within trajectories. They are used in the generation of the entropic analog of the Schwarzian derivative, ESf, in scaling quasiperiodic psychological series (Gregson, 2002; Gregson & Leahan, 2003).

The idea of employing Markovian representations of psychological processes evolving through time is certainly not an innovation, and was used in a fundamentally different way in learning theory (Bower & Theios, 1964). There, the dependent variable was the probability of making a particular response (usually a correct one) on a trial in a learning curve that eventually entered an error-free absorbing state; a major finding was that three theoretical states in discrete time generated closer fits to observed data. Mathematical learning theory was not conceptualized as an instance of nonlinear dynamics, but rather as simpler stochastic processes with associated statistical tests.

Parry usefully distinguishes between two matrices that play a part on theory: the state transition matrix (STM), which Parry calls the structure matrix, where each cell is 0, or 1 only if a transition exists; and the transition probability matrix (TPM), where each cell is a probability. As the process is Markov, the transition is taken to be $t \rightarrow t + 1$; in real time, the increment is t to $t + \theta$, where θ is the time interval between two successive observations of the process. In simpler treatments, θ is taken as a constant, but it can be a random variable. Given a set S of states $s\ 0\ \{1,..,i,j,..,n\}$ that is exhaustive, but not necessarily ordered in terms of some measure $\mu(s)$, the elements of the TPM, $t(i, j)$, are given by

$$t(i, j) = \begin{cases} > 0 & \text{if } s(i, j) = 1; \\ = 0 & \text{if } s(i, j) = 0. \end{cases}$$

The TPM are usually taken in terms of succession, so that each $t(i, j) \in F$ is the probability of state j following state i, $t \rightarrow t + 1$. This is appropriate in a dissipative and irreversible process, such as a real psychological time series. But the reverse matrix can be computed, in which each $t(i, j) \in R$ means the probability that i is preceded by j, $t \rightarrow t - 1$. This usually has different eigenvectors. A simple example, drawn from self-report data[1] by a depressed patient keeping a daily diary of the level of experienced transient mood over some months, is as follows. The partitioning into five levels of depression is arbitrary; the raw data could admit of more levels. Probabilities have been rounded off to three decimal places. The patient was chronically depressed;

when an episode of depression occurred, it then persisted for some time; so, by definition, the process is not strictly Markov, but examination of detailed records suggested a weak periodicity of about 15 days. State 1 is "no depression experienced." Following a usual convention, the top-left cell is $t(1, 1)$.

$$F = \begin{pmatrix} 0.469 & 0.146 & 0.198 & 0.146 & 0.042 \\ 0.463 & 0.220 & 0.122 & 0.171 & 0.024 \\ 0.581 & 0.290 & 0.065 & 0.065 & 0.000 \\ 0.462 & 0.308 & 0.154 & 0.077 & 0.000 \\ 0.600 & 0.000 & 0.200 & 0.200 & 0.000 \end{pmatrix}, \quad R = \begin{pmatrix} 0.464 & 0.196 & 0.186 & 0.124 & 0.031 \\ 0.350 & 0.225 & 0.225 & 0.200 & 0.000 \\ 0.613 & 0.161 & 0.065 & 0.129 & 0.032 \\ 0.538 & 0.269 & 0.077 & 0.077 & 0.038 \\ 0.800 & 0.200 & 0.000 & 0.000 & 0.000 \end{pmatrix}$$

Both matrices are ergodic, and the stationary state vectors, V, are

$$V|F^{\infty} : 0.488, 0.201, 0.156, 0.131, 0.025$$

$$V|R^{\infty} : 0.482, 0.206, 0.156, 0.131, 0.025$$

The entropy in bits ($-\Sigma p \log_2 p$) of the F^4 vector is 1.907. The eigenvalues are to three decimal places the same for the two matrices, F and R:

$$1.000, \quad -0.109, \quad -0.072 \pm 0.057i, \quad -0.086,$$

the complex conjugate pair arises because of the skew asymmetry of the matrices. A minimal entropy case arises if there exists one and only one i-vector $\forall j$; $s(i, j) = 1$, that is, it has one absorbing state i, in dynamical terms, a point attractor onto the one symbol z. An opposite extreme is where $\forall i$; $t(i, j) < 1$ and the matrix $t(i, j)$ is ergodic. From the eigenvalues of $t(i, j)$, Parry derived a measure of the entropy ε of the system that lies between 0 and $\ln \beta$, where β is the eigenvector of the largest F eigenvalue of $t(i, j)$; $8 = \ln \beta$. For the depression example given above, the elements of the largest eigenvector are all[2] $\cong +0.447$, and the ε value is 0.804 (or in $\log_2$ units = 1.16), on the assumption that the process is homogeneous, that is, stationary, and has no local embedded episodes with different dynamics. ε may be thought of as performing a role analogous to the largest Lyapunov exponent with metric data.

If there exist two sets, S_a and S_b, which may be represented by successive subseries, length w t-units, within one trajectory, if $n_a = n_b$, then the Hamming distance between the two is

$$\Delta_{i=1}^{w}(a,b) = n^{-2} \sum_{n \times n} |S_1(i, j) - S_2(i, j)|$$

where $0 \leq \Delta \leq 1$, and if $\Delta = 0$, the two are said to be S-*equivalent* in their symbolic structure, over the subseries[3] length w. At the same time, they may not be equivalent in the sense of $\beta_1 \neq \beta_2$.

It is necessary to give a word of warning; maximal entropy solutions to Markov chains are not unique, and the number of alternative admissible solutions is up to the cardinality of the set of the symbols used (Petersen, Quas, & Shin, 2003).

Two Attractor Symbolic Dynamics

Markovian theory has been extended to encompass various constructs; implicit, intrinsic, and hidden Markov models can all be found described and distinguished, with their formal properties spelled out. Most of the published symbolic dynamical treatments explore a single attractor, involving a map on the unit interval. Here, the interest is on small local systems, probably dissipative, that have two attractors and a pathway to admit bidirectional transitions from one to the other. This case can arise in cognitive or perceptual processing, linguistics, even interpersonal psychodynamics, and has been of considerable interest (Fisher, 1967; Haken & Stadler, 1990; Gregson, 2004; Mitchener & Nowak, 2004).

Consider a dynamic system in which there are two and only two saddle-node attractors, each with homoclinic orbits, and a heteroclinic orbit connecting them. We would require a coarse scaling model of this situation, in which movement along any orbit, or local coherent bundle of orbits, is encoded as a transition through a state. It is known (Nishiura, 2002) that such a dynamic system can exist, and that if it does, then the existence of the heteroclinic orbit is assured. It is required to construct a transition matrix for a Markov chain, with the possibility of hidden Markov properties, that will serve as a model for identifying the dynamics and, in particular, the terminal stable state vectors of the system (Elliot, Aggoum, & Moore, 1995; McCord & Mischiakow, 1992).

Let us label the two attractors A and B, and the orbit bundles as a^+, a^-, h, b^+, and b^-. Then the seven states of the Markov transition matrix are one-to-one to these bundles and A and B. Using the notation $p(x, y) = prob(x \rightarrow y)$, the resulting matrix M, written as symmetric (reflected) about the heteroclinic orbit, is

$$M = \begin{pmatrix} p(a^+,a^+) & p(a^+,A) & p(a^+,a^-) & p(a^+,h) & p(a^+,b^+) & p(a^+,B) & p(a^+,b^-) \\ p(A,a^+) & p(A,A) & p(A,a^-) & p(A,h) & p(A,b^+) & & \\ p(a^-,a^+) & & & & & & \\ p(h,a^+) & & & \text{etc.} & & & \\ p(b^-,a^+) & & & & & & \\ p(B,a^+) & & & & & & \\ p(b^+,a^+) & & & & & & \end{pmatrix}$$

and it immediately follows from the dynamics that some of these transition probabilities are 0. Denoting this null matrix by M_0, and assuming some stability in any state x so that $1 > p(x, x) \geq k > 0$, and $q = 1 - k$, we have as a limiting case a square skew symmetric matrix, M_0:

$$M_0 = \begin{pmatrix} k & q & 0 & 0 & 0 & 0 & 0 \\ q/3 & k & q/3 & q/3 & 0 & 0 & 0 \\ 0 & q & k & 0 & 0 & 0 & 0 \\ 0 & q/2 & 0 & k & 0 & q/2 & 0 \\ 0 & 0 & 0 & 0 & k & q & 0 \\ 0 & 0 & 0 & q/3 & q/3 & k & q/3 \\ 0 & 0 & 0 & 0 & 0 & q & k \end{pmatrix}$$

Then the probability of passing through the path from $a+$ to $b+$ (or from $b+$ to $a+$) without reversals is $(q)(q/3)(q/2)(q/3) = q^4/27$. Obviously, if k is large, this vanishes. We next examine the eigenvalues of this matrix, and its stationary state vector most readily obtained by repeated multiplication to get M_0^∞, and hence V^∞.

The eigenvalues of M_0 are

$$k, k, k, (k-q), (k+q), k-(2/3)^{1/2}q, k+(2/3)^{1/2}q$$

The stationary state vector, V, for the range $0.2 < k < 0.8$ is

$$0.083, 0.250, 0.083, 0.168, 0.083, 0.250, 0.083$$

and we note that the convergence onto V^∞ is rather slow; 80 iterations were found to be necessary.

The matrix M_0 may be regarded as a sort of maximum entropy representation because no special status is given to the attractors A and B. If instead, as makes reasonable psychological sense, special status is given to the stability of the attractors, M_0 is rewritten with $0 < q < r < k \leq 1$ as

$$M_{ab} = \begin{pmatrix} r & q & 0 & 0 & 0 & 0 & 0 \\ q/3 & r & q/3 & q/3 & 0 & 0 & 0 \\ 0 & 1-r & r & 0 & 0 & 0 & 0 \\ 0 & q/2 & 0 & r & 0 & q/2 & 0 \\ 0 & 0 & 0 & 0 & r & 1-r & 0 \\ 0 & 0 & 0 & q/3 & q/3 & r & q/3 \\ 0 & 0 & 0 & 0 & 0 & 1-r & r \end{pmatrix}$$

As a new variable, r has been introduced. To estimate V^∞, write $\theta < 1$, $r = \theta k$; then, $V^\infty = f(\theta)|k$. Note that in the limit $(k = 1)$, A and B are then absorbing states, but as there are two of them, the process is no longer Markovian as the starting probability vector, V_0, will determine which of the two attractors is a terminal state.

The eigenvalues of M_{ab} are complicated. Writing

$$\Xi = \sqrt{(3k + 3r)^2 + 12(2q - 3kr + 2qr)} \tag{19.1}$$

and

$$\Phi = \sqrt{(-k - r)^2 + 4(kr + qr - q)} \tag{19.2}$$

the eigenvalues become

$$r,\ r,\ r,\ \frac{3k + 3r - \Xi}{6},\ \frac{3k - 3e + \Xi}{6},\ \frac{k - r - \Phi}{2},\ \frac{k + r + \Phi}{2}$$

and the constraints written on the relative values of r, q, and k may ensure for some values that $\Xi > 0$ and $\Phi > \theta$. In each case, M_0 and M_{ab}, the eigenvalues consist of a repeated scalar and two conjugate pairs. These eigenvalues are all less than unity; a treatment of this question is given at a very abstract level by Hennion and Hervé (2001). The process is still convergent.

To evaluate V^∞, k is fixed heuristically at 0.8, and θ is varied over a range. The lowest value that θ can take is $1 - k$. The effect of increasing θ is, rather obviously, to move the probability of state occupancy from the two attractors to the heteroclinic orbit. The $p^\infty(A)$ and $p^\infty(B)$ values should be compared with those from M_0. Table 19.1 shows the V^∞ values of p^∞ from M_{ab} for variable θ with $k = 0.080$.

TABLE 19.1

V^∞ Values of p^∞ from M_{ab} for Variable θ with $k = 0.80$

θ:	0.25	0.50	0.75
[a+]	0.033	0.042	0.056
[A]	0.400	0.375	0.333
[a-]	0.033	0.042	0.056
[h]	0.067	0.083	0.111
[b-]	0.033	0.042	0.056
[B]	0.400	0.375	0.333
[b+]	0.033	0.042	0.056

TABLE 19.2

Rates of Diffusion for $0.80 < k < 0.95$

k:	0.80		0.95	
V^n:	V^1	V^{80}	V^1	V^{80}
$[at]$	0.067	0.034	0.017	0.013
$[A]$	0.800	0.404	0.950	0.601
$[A^-]$	0.067	0.034	0.017	0.013
$[h]$	0.067	0.067	0.017	0.021
$[b^-]$	0.000	0.033	0.000	0.007
$[B]$	0.000	0.396	0.000	0.337
$[b^+]$	0.000	0.033	0.000	0.007

Diffusion from a Prototype

Suppose that we use as initial vector V_0: 0,1,0,0,0,0,0 or 0,0,0,0,0,1,0. As the system as written is symmetric in the attraction of A and B, we need only consider one, that from A. It is seen that the rate of diffusion over n steps, by using M_{ab}^n, depends critically on k, and is very slow if $k \simeq 1$ (Table 19.2). The system eventually moves into an equilibrium with the two attractors equally potent.

Semi-Markov with Variable Residence Times

We now extend the model to allow for the residence within a transition probability cell in M_0 to be variable, the states still being discrete. An exponential simplification used by Engbert and Kliegl (2001) is followed. This is a special case of a generalized master equation developed by Gillespie (1978, 1992).

The transition probability density function can be written as the product of two other functions:

$$p(t, n \mid m, t) = \pi(n, m\varphi(\tau \mid m)) = p(m, n) \tag{19.3}$$

where $p(n, m \mid t, t + \delta t) = p(n, m(\tau))$ is the transition dependent on the residence time τ in the state m. $\pi(n, m)$ is the stepping probability from m to n, and $\varphi(\tau \mid m)$ is the probability density function for the pausing time τ within m. The cells in M_0 that are already defined are $\pi(n, m)$.

In the case of a Markov process where

$$W_m(\tau) = \Sigma W_{nm}(\tau) \tag{19.4}$$

n is the transition probability for moving from m to any other n (one row of M_0 less its leading diagonal cell)

$W_m(\tau)$ is a constant, and the pausing time within m is exponentially distributed

$$\varphi(\tau|m) = W_m e^{-w_m \tau} \qquad (19.5)$$

where the maximum of φ occurs when $\tau = 0$. The constants W_m in Equation 19.5 enable us to adjust for some special status to be given to some states; for A and B, a longer dwell time would be expected, and the transition through h would be rapid, so that the whole system might approximate, in its temporal evolution, a two-state system with an abrupt jump between states.

The purpose of writing Equations 19.3 through 19.5 is that they make possible the estimation of $\{W_m\}$ from an observed data set with overall transition probabilities and within-state mean dwell times. The expectation of φ is $1/W_m$, and the mean is approximately equal to the standard deviation.

Hidden Markov with Nonlinear Dynamics

If the Markov chain is a discrete state in discrete time, but is obscured by noise (usually assumed to be Gaussian), it is said to be a hidden Markov process. The presence of the noise obscures the points in time where jumps between states occur, and this in turn distorts the estimates of the transition probabilities. Such models can be univariate or multivariate, and have been applied in physical, medical, and biological examples (MacDonald & Zucchini, 1997). They are suitable for the analysis of some situations where short bursts of intermittent activity are observed. This approach extends to nonlinear dynamics (Elliot et al., 1995, Section 4.4). Here, we do not consider variable residence times, but rather the observations y are linked to the states x in more than one way, so there is more than one M_0.

In the state space equation form, for N states, k is the step variable:

$$x_{k+1} = M_{k+1}(x_k) + v_{k+1} \qquad (19.6)$$

$$y_k = C(x_k) + w_k \qquad (19.7)$$

where

M is a variable $N \times N$ matrix

C is a fixed $\mathbf{N} \sim$ vector

v and w are here 2D random variables distributed as Gaussian(0, σ), with zero covariance

C in Equation 19.7 corresponds to a psychophysical function that need not be linear

The estimation of such a process involves tracking it through its evolution, the Kalman filter is a special case of such tracking methods. Recently, Bickel, Ritov, and Rydén (2002) have shown that hidden Markov processes may in some of their properties be indistinguishable from 2D time series; they considered Fisher information, Kullback–Leibler distance, and entropy functions. Such quantities can be computed explicitly at points where the TPM is degenerate.

Fast/Slow Dynamics

Interest in processes where there is a functional division between slow dynamics, which may serve as a carrier, and fast dynamics, which can resemble noise or a signal with chaotic characteristics, possibly began in engineering, but is now recognized as a paradigm for some biological applications. In fact, this area of investigation is intimately linked to the two-attractor problem just discussed above. Arecchi (1987, p. 42), from a frequency domain approach, observed: "We have shown that, whenever in nonlinear dynamics more than one attractor is present, there are two distinct power spectra: (i) a high frequency one, corresponding to the decay of correlations within one attractor; (ii) a low frequency one, corresponding to noise induced jumps."

The term fast/slow is used to label such processes.[4] It is not in fact critical whether the fast part is treated as the signal or the slow part; the important consideration is which of the two parts, if they are separable, is externally controllable over some finite time interval, at some rate of intervention.

For a very simple case, we assume that the dynamics are stationary, and first construct the STM of the slow part, which we label C_{ss}. The double suffix is to remind us that the process is assumed to be slow and stationary.

$$
C_{ss} = \begin{pmatrix}
0 & 1 & 0 & 0 & \dots & \dots & 0 & 0 & L \\
1 & 0 & 1 & 0 & \dots & \dots & 0 & 0 & 0 \\
0 & 0 & 1 & 0 & \dots & \dots & 0 & 0 & 0 \\
\dots & \dots & \dots & \dots & \dots & \dots & \dots & \dots & \dots \\
\dots & \dots & \dots & \dots & \dots & \dots & \dots & \dots & \dots \\
0 & 0 & 0 & 0 & \dots & \dots & 0 & 1 & 0 \\
0 & 0 & 0 & 0 & \dots & \dots & 1 & 0 & 1 \\
L & 0 & 0 & 0 & \dots & \dots & 0 & 1 & 0
\end{pmatrix}
$$

This matrix is defined over transitions in the real-time interval $t, t + \theta$, and θ has to be chosen by trial and error if the generator of the slow dynamics (such as a sinusoid) is not known. If θ is too small, then some terms $s(i, i) \approx 1$, and if θ is too large, then other off-diagonal cells are not zero. C_{SS} may then be part of a matrix where the elements of S are strictly ordered and the cell $t(\ldots)$ values fall off monotonically as we move away from the leading diagonal.

Let any one line (with correction for the end lines) of the TSM of C_{SS} contain the terms $t(i, i-1)$, 0, $t(i, i+1)$, and the minima over S be $min(t(i, i-1))$, $min(t(i, I+1))$. This double off-diagonal matrix with the minima substituted for all i is T_{SS}.

If the trajectory is on a closed orbit, then the cells $s(i, n)$ and $s(ri, i)$, marked L in the matrix, are also nonzero. The matrix is then a circumplex (Shye, 1978). This form corresponds to an attractor on a limit cycle. Obviously, both without and with the L cells, the C_{SS} pattern depends on finding an order of the elements of S that generates the pattern. If the n states are on a closed orbit, then there are $2n$ such orderings; one can start at any state and go in either direction round the orbit.

If the fast component is the trajectory of a chaotic attractor in its basin, then it eventually visits everywhere, has no absorbing states, and its TSM is ergodic. Call this matrix D_{ns}. The observed matrix of the process is $M_{s+n,s}$, and if the two parts add linearly within any cell, then, over some subsequence in time that is stationary:

$$M_{s+n,s} = T_{SS} + D_{ns} \tag{19.8}$$

So, subtracting cell by cell,

$$\hat{D}_{ns} = M_{s+n,s} - T_{SS} \tag{19.9}$$

It is $\hat{D}_{ns}$ that we now treat by the Parry measure to find the approximate eigenvalues of the fast part of the system.

The matrices C_{SS} and F from the depression data do have some interesting resemblance to a symbolic dynamics representation of the Belousov–Zhabotinskii chemical reaction described by Lathrop and Kostelich (1989b, p. 152). The resemblance does not reduce to saying that the causality is the same between chemistry and psychophysiology; it says merely that a problem in identifiability has a common structure. This is

$$BZ = \begin{matrix} 0100000 \\ 0010000 \\ 0001000 \\ 0000100 \\ 0000010 \\ 0000011 \\ 1000010 \end{matrix}$$

This type of matrix can be evidence of cyclic stable dynamics, on a countable space chain; a deeper analysis is given by Meyn and Tweedie (1993, p. 115). The off-diagonal array and the recurrence at $s(1, 7)$ resemble half of the symmetry of $\mathbf{C_{SS}}$, and thus the weak asymmetry of $\mathbf{F}$. The BZ dynamics are associated with closed orbits and an intermittent burst, after which the dynamics return to a periodic orbit. It is possible to employ the symbolic representation to calculate the topological entropy h_t using only relatively slow orbits and, hence, of low period, as Lathrop and Kostelich (1989) also had the maps of the reconstructed attractor and the data series 8400 points long.

If N_p is the number of periodic points for the pth iterate of the return map, then

$$h_t = \lim_{p \to \infty} \frac{1}{p} \log_2 N_p \qquad (19.10)$$

but Equation 19.10 does not work for short series. Alternatively, from the $t(i, j)$ matrix BZ,

$$N_p = \text{trace } \mathbf{M^p} \qquad (19.11)$$

The trace[5] is the sum of the diagonal elements of the TPM matrix; as it is dominated by the largest eigenvalue, λ_1, of the matrix for a large p,

$$h_t = \log_2 \lambda_1 \qquad (19.12)$$

which resembles the Parry measure derivation. In the example of BZ, $h_t =$ 0.73 bits/orbit. In situations where there are two or more attractors, and fast/slow dynamics exist, a link emerges between the two attractor situations examined above, and also apparently with nonlinear psychophysics (Gregson, 1988). This is illustrated in an example by Arecchi, Badii, and Politi (1984), who show that if there are jumps between basins of independent attractors, the system as a whole exhibits a low-frequency component in its power spectra. The Lyapunov exponent is then complex, being made up of parts corresponding to attraction and repulsion. This has some qualitative parallel with the P function used in nonlinear psychophysics; both $\mathbf{F}$ and the example created by Arecchi et al. are grounded in a cubic map, without noise. The dynamics are very complicated, and cannot be reduced to a single $1/f^b$, $1/2 < b < 2$, power spectrum.

An Ill-Behaved and Nonstationary Example

Searching for a hidden Markov modeling of real data can be based on too many gratuitous assumptions, and before any such interpretation is accepted as sufficient, a diversity of other checks of the dynamics should be explored.

There is no single magic bullet to identify dynamics. An example of murder statistics collected in Cape Town, South Africa, provides an illustration of necessary cautions to be observed if data are very nonstationary and borderline to chaotic.

Dr. Leonard Lerer collected weekly frequency returns on homicides and suicides in Cape Town from January 1986 to December 1991, a total of 313 weeks. There are five parallel series; the one on homicides using guns was analyzed by MacDonald and Zucchini (1997), and is of interest because there was an upsurge in gun usage toward the end of the period studied, so that series is not stationary in its linear trend. Here, only the series of homicides not using guns is examined; the data are not claimed to be fully exhaustive, only bodies in the morgues could be counted, and as they are simple frequencies they are not metric measures, and they are not rates unless we could correct for the size of the population at risk. There is no good a *priori* reason to think that a Markov process underlies their generation, gang warfare and domestic violence in shanty towns have complicated causalities; even if we were to postulate seasonal variations, the notion of meaningful states of the system needs some sociological justification. It is however still defensible to explore dynamics with symbolic encoding to simplify the structure.

Examination of the graphs of the series suggests that it is expedient to break the series into three successive subseries, of lengths 104, 104, and 105 weeks. Call these cases #1, #2, and #3. The basic descriptive statistics are tabled as follows: LLE is the largest Lyapunov exponent, and *ESf* is the entropic analog of the Schwarzian derivative. The autocorrelation terms (ar.) are for lags 1 through 6 (Table 19.3).

TABLE 19.3

Basic Statistics of Three Subseries
of Weekly Murders

	#1	#2	#3
Min	6	7	10
Mean	15.99	20.21	23.23
Max	45	41	52
s.d.	6.02	6.48	6.83
*ar*1	0.161	0.308	0.197
*ar*2	0.114	0.147	0.151
*ar*3	0.215	0.080	0.079
*ar*4	0.074	0.140	−0.066
*ar*5	0.029	0.207	−0.009
*ar*6	0.124	0.072	−0.121
LLE	+0.0616	+0.0698	+0.0898
ESf	−0.752	−0.475	−0.719

Kendall (1973, p. 90–91) gives explicit forms for the distribution of *arl* under zero autocorrelation, approximately $E(arl) = -0.01$, $var(arl) = 0.01$. If the series are chopped into five frequency ranges, 0–10, 11–20, 21–30, 31–40, and 41+, then the TPM matrices between ranges are as follows:

$$M\,\#1 = \begin{pmatrix} 0.083 & 0.750 & 0.167 & 0.000 & 0.000 \\ 0.151 & 0.712 & 0.123 & 0.000 & 0.014 \\ 0.000 & 0.688 & 0.312 & 0.000 & 0.000 \\ 0.000 & 1.00 & 0.000 & 0.000 & 0.000 \\ 0.000 & 1.00 & 0.000 & 0.000 & 0.000 \end{pmatrix},$$

$$M\,\#2 = \begin{pmatrix} 0.500 & 0.250 & 0.250 & 0.000 & 0.000 \\ 0.019 & 0.660 & 0.302 & 0.019 & 0.0004 \\ 0.025 & 0.400 & 0.525 & 0.025 & 0.025 \\ 0.000 & 0.200 & 0.600 & 0.200 & 0.000 \\ 0.000 & 0.000 & 0.000 & 1.00 & 0.000 \end{pmatrix}$$

$$M\,\#3 = \begin{pmatrix} 0.000 & 1.00 & 0.000 & 0.000 & 0.000 \\ 0.025 & 0.450 & 0.500 & 0.025 & 0.000 \\ 0.000 & 0.340 & 0.500 & 0.120 & 0.040 \\ 0.000 & 0.400 & 0.400 & 0.200 & 0.000 \\ 0.000 & 0.000 & 0.500 & 0.500 & 0.000 \end{pmatrix}$$

All the matrices are ergodic, and the stationary state vectors V^{∞} are

$V^{\infty}|\#1:\ 0.118, 0.716, 0.157, 0.000, 0.010$
$V^{\infty}|\#2:\ 0.039, 0.518, 0.396, 0.037, 0.010$
$V^{\infty}|\#3:\ 0.010, 0.388, 0.485, 0.097, 0.019$

As the state transition matrices are not the same for the three subseries, on the assumption that symbolic dynamics in five states is sufficient for heuristics, the nonstationarity is confirmed. The positive LLE values suggest that the process is edge-of-chaos throughout, and not random.

The eigenvalues to three decimal places for the three TPMs are

#1: 1, 0.165, −0.084, 0.026, 0
#2: 1, 0.482, 0.342, 0.031 ± 0.156*i*
#3: 1, 0.115 ± .057*i*, −0.127, 0.048

and the rank of #1 could be reduced to 4. Note that these subseries TPMs are also noisily clustered around the leading diagonal, like part of the circumplex structure. The autocorrelation structure is compatible with the small positive values of the LLE.

The bispectral form of *ESf* analysis can also be performed, giving with surrogates an indication of what is not random evolution (Gregson & Leahan, 2003). This is a form of high-pass filtering in this example; $\delta y = 0.02$ in all three cases. The layout of the $b(a, b)$ matrices is

$b(1,2)$ $b(1,3)$ $b(1,4)$ $b(1,5)$ $b(1,6)$
$b(2,3)$ $b(2,4)$ $b(2,5)$ $b(2,6)$
$b(3,4)$ $b(3,5)$ $b(3,6)$
$b(4,5)$ $b(4,6)$
$b(5,6)$

where each $b(a, b)$ is formed by the *ESf*, a function of the first, second, and third differences, from the convoluted series with terms $x_j, x_{ja}, x_{jb}, j = 1, ..., n$. As the differences become zero-autocorrelated in the random surrogate series created by the permutation of the $b(a, b)$ terms, a test for randomness is found within each triangular matrix. The lags a, b are small, so the dynamics being identified are relatively fast and the expected loci of chaotic activity.

The few italicized values in Table 19.4 lie within the 95% confidence intervals for random surrogates. We may conclude that the dynamics are not mainly random, and the magnitude of the small, and hence residual, random components is variable as the process evolves. The nonstationarity reveals itself both in the changes in the TPMs and in the higher-order dynamics; a hidden Markov modeling based on the assumption of invariance in the STM would be invalid.

TABLE 19.4

Dynamics of Random Components for Three TPMs

For # 1:	−0.3917	−0.3554	−0.6502	−0.7756	−0.4735
	−0.5369	−0.8523	−0.5648	−0.6900	
	−0.3971	−0.4381	−0.4205		
	−0.6084	−0.3307			
	−0.9449				
For # 2:	−0.4344	−0.3853	−0.7555	−0.6583	−0.6473
	−0.6049	−0.5434	−0.7389	−0.7474	
	−0.4855	−1.0801	−0.6754		
	−0.5728	−0.9719			
	−0.7528				
For # 3:	−0.6052	−0.5388	−0.6528	−0.5746	−0.5673
	−0.5934	−0.6081	−0.5059	−1.0897	
	−0.6189	−0.5911	−0.8218		
	−0.7431	−0.9452			
	−1.0659				

Conclusions

A main reason for being interested in using the approach outlined here is that psychological time series are ill-behaved, in the sense that they jump about in their structure, and can exhibit short regular repeated subseries that may be called embedded episodes (Gregson, 1983, pp. 187–196), and can be treated qualitatively as dynamics at the edge-of-chaos. Such embedded episodes are called *phantoms* by Badii and Broggi (1989, p. 67); a strange attractor series is a mix of phantoms, primitives that are not periodic, and periodic subsequences. They point out that a finite number of primitives are found also in the case where no Markov partition exists. It is assumed that a sufficient property for defining the nonlinearity of a process is the presence of a trajectory that exhibits at least one recurrent embedded episode whose start points are distributed aperiodically in time. One other approach, given sufficient data, is to treat the embedded episode as one component state in a higher-order Markov model (Raftery & Tavaré, 1994). In short, a linear stationary metric representation loses or at least distorts some information, and makes prediction by extrapolation or by modeling most uncertain. We may extend this criticism, because it implies that control by the construction of inverse processes and feedback will also be likely to fail. Averages over long sequences become meaningless.

Symbolic dynamics can usefully be employed provided that some iterative search strategy is available to find a relatively simple structure in the data. If the data are not already to hand, then variable real-time sampling (θ in the definitions used here) and search for the set of states S_n with minimum sufficient complexity have to be planned. If the data are already collected, and the interest is in nonstationarity, whose very presence invalidates the assumptions in standard packages for computing Lyapunov exponents or fractal dimensionality, then a moving rectangular or "box-car" window imposed to convert pseudo-numerical data to a symbol set S_n has to be used as an exploratory device. The width in steps t of that window is itself also a variable of the analysis. There is no general algorithm for doing this, and iterative exploration is needed.

It is after some simple representation, as in Equations 19.8 and 19.9 has been found, that indices such as Parry measure, topological entropy h_t, or *ESf* can be computed on short subseries of the observed trajectories. One may remark that the n-valued recurrence plot of S_n and the $s(i, j)$ matrix contain the same information.

Acknowledgment

The position that psychological data are not strictly metric, but only ordinal, and hence metric assumptions should be avoided in many contexts, is taken by Norman Cliff and John Keats (2002). It was John Keats who supported and inspired early Australian work in psychometrics from over 40 years ago, and my indebtedness is gratefully acknowledged here.

Notes

1. These data were kindly made available to me by Professor R. Heath. The raw data series is 201 points long. No therapeutic meaning is to be inferred from this analysis; it is purely a computational demonstration.
2. This property follows from a theorem of van den Wollenberg (1978, p. 332), on a circumplex, that is given in terms of a correlation matrix.
3. The window $t = 1$, w can be moved forward in increments of k units, and thus create a series of $\Delta_{t=1}^{w}$ values that provide a graphical view of dynamical discontinuities.
4. Guckenheimer (2003) uses the term fast/slow in a more complicated treatment of the bifurcations of such systems.
5. See Horst (1963) for an introduction to trace properties and computation.

References

Abraham, N. B., Albano, A. M., Passamante, A., & Rapp, P. E. (1989). *Measures of complexity and chaos*. New York: Plenum Press.

Aczél, I. (1966). *Lectures on functional equations and their applications*. New York: Academic Press.

Aguirre, L. A., & Billings, S. A. (1995). Identification of models for chaotic systems from noisy data: Implications for performance and nonlinear filtering. *Physica D, 85,* 239–258.

Arecchi, F. T. (1987). Hyperchaos and 1/f spectra in nonlinear dynamics: The Buridanus donkey. In Caianiello, E. R. (Ed.) *Physics of cognitive processes* (pp. 35–50). Singapore: World Scientific.

Arecchi, F. T., Badii, R., & Politi, A. (1984). Low-frequency phenomena in dynamical systems with many attractors. *Physical Review A, 29,* 1006–1009.

Badii, R., & Broggi, G. (1989). Hierarchies of relations between partial dimensions and local expansion rates in strange attractors. In N. B. Abraham, A. M. Albano, A. Passamante, & P. E. Rapp. *Measures of complexity and chaos* (pp. 63–73). New York: Plenum.

Ball, R. (Ed.). (2003). *Nonlinear dynamics from lasers to butterflies. World scientific lecture notes in complex systems,* vol. 1. Singapore: World Scientific.

Berthet, R., Petrossian, A., Residori, S., Roman, B., & Fauve, S. (2003). Effect of multiplicative noise on parametric instabilities. *Physica D, 174,* 84–99.

Bickel, P. J., Ritov, Y., & Rydén, T. (2002). Hidden Markov model likelihoods and their derivatives behave like i.i.d. ones. *Annales de l'Institut Henri Poincaré, PR 38(6),* 825–846.

Blanchard, F., Maass, A., & Nogueira, A. (Eds.). (2000). *Topics in symbolic dynamics and applications. London mathematical society lecture note series, no. 279.* Cambridge, U.K.: Cambridge University Press.

Bower, G. H., & Theios, J. (1964). A learning model for discrete performance levels. In R. C. Atkinson (Ed.) *Studies in mathematical psychology* (pp. 1–31). Stanford, CA: Stanford University Press.

Buljan, H., & Paar, V. (2002). Parry measure and the topological entropy of chaotic repellers embedded within chaotic attractors. *Physica D, 172*, 111–123.

Casdagli, M. (1989). Nonlinear prediction of chaotic time series. *Physica D, 35*, 335–356.

Casdagli, M., & Eubank, S. (Eds.). (1991). *Nonlinear modelling and forecasting.* Reading, MA: Addison-Wesley.

Cliff, N., & Keats, J. (2002). *Ordinal measurement in the behavioral sciences.* Mahwah, NJ: Lawrence Erlbam Associates.

Elliot, R. J., Aggoum, L., & Moore, J. B. (1995). *Hidden Markov models.* New York: Springer-Verlag.

Engbert, R., & Kleigl, R. (2001). Mathematical models of eye movements in reading: A possible role for autonomous saceades. *Biological Cybernetics, 85*, 77–87.

Eykhoff, P. (1974). *System identification: Parameter and state estimation.* New York: John Wiley.

Fisher, C. H. (1967). Measuring ambiguity. *American Journal of Psychology, 30*, 541–547.

Fokianos, K., & Kedem, B. (2003). Regression theory for categorical time series. *Statistical Science, 18*, 357–376.

Geake, J. C., & Gregson, R. A. M. (1999). Modelling the internal generation of rhythm as an extension of nonlinear psychophysics. *Musicae Scientiae, 3*, 217–235.

Gillespie, D. T. (1978). Master equations for random walks with arbitrary pausing time distributions. *Physical Letters A, 64*, 22–24.

Gillespie, D. T. (1992). *Markov processes.* Boston, MA: Academic Press.

Green, P. J. (1995). Reversible jump Markov chain Monte Carlo computation and Bayesian model determination. *Biometrika, 82*, 711–732.

Gregson, R. A. M. (1983). *Time series in psychology.* Hillsdale, NJ: Erlbaum.

Gregson, R. A. M. (1988). *Nonlinear psychophysical dynamics.* Hillsdale, NJ: Erlbaum.

Gregson, R. A. M. (2002). Scaling quasiperiodic psychological functions. *Behaviormetrika, 29*, 41–57.

Gregson, R. A. M. (2004). Transitions between two pictorial attractors. *Nonlinear Dynamics, Psychology and Life Sciences, 8*, 41–63.

Gregson, R. A. M., & Leahan, K. (2003). Forcing function effects on nonlinear trajectories: Identifying very local brain dynamics. *Nonlinear Dynamics, Psychology and Life Sciences, 7*, 137–157.

Guastello, S. J. (2000). Symbolic dynamic patterns of written exchanges: Hierarchical structures in an electronic problem solving group. *Nonlinear Dynamics, Psychology and Life Sciences, 4*, 169–188.

Guastello, S. J., Hyde, T., & Odak, M. (1998). Symbolic dynamic patterns of verbal exchange in a creative problem solving group. *Nonlinear Dynamics, Psychology, and Life Sciences, 2*, 35–58.

Guastello, S. J., Nielson, K. A., & Ross, T. J. (2002). Temporal dynamics of brain activity in human memory processes. *Nonlinear Dynamics, Psychology, and Life Sciences, 6*, 323–334.

Guckenheimer, J. (2003). Bifurcation and degenerate decomposition in multiple time scale dynamical systems. In J. Hogan, A. Champneys, B. Krauskopf, M. di Bernardo, E. Wilson, H. Osinga, & M. Homer (Eds.), *Nonlinear dynamics and chaos: Where do we go from here?* (pp. 1–20). Bristol, U.K.: Institute of Physics.

Gumowski, I., & Mira, C. (1980). *Recurrences and discrete dynamic systems. Lecture notes in mathematics, # 809*. Berlin, Germany: Springer-Verlag.

Haken, H., & Stadler, M. (Eds.). (1990). *Synergeiics of cognition*. Berlin, Germany: Springer-Verlag.

Heath, R. (2000). *Nonlinear dynamics: Techniques and applications in psychology*. Mahwah, NJ: Erlbaum.

Hennion, H., & Hervé, L. (2001). *Limit theorems for Markov chains and stochastic properties of dynamical systems by quasi-compactness*. Lecture notes in mathematics, 1766. Berlin, Germany: Springer-Verlag.

Horst, P. (1963). *Matrix algebra for social scientists*. New York: Holt, Rinehart and Winston.

Kendall, M. C. (1973). *Time-series*. London, U.K.: Charles Griffin.

Krantz, D. H., Luce, R. D., Suppes, P., & Tversky, A. (1971). *Foundations of measurement*, vol. I. New York: Academic Press.

Lathrop, D. P., & Kostelich, E. J. (1989a). Characterization of an experimental strange attractor by periodic orbits. *Physical Review A, 40*, 4028–4031.

Lathrop, D. P., & Kostelich, E. J. (1989b). Analyzing periodic saddles in experimental strange attractors. In N. B. Abraham, A. M. Albano, A. Passamante, & P. E. Rapp (Eds.). *Measures of complexity and chaos* (pp. 147–154). New York: Plenum Press.

Lawrence, S., Tsoi, A. C., & Cues, C. L. (1996). *Noisy time series prediction using symbolic representation and recurrent neural network grammatical inference*. Technical Report UMIACS-TR-96-27 and CS-TR-3625, Institute for Advanced Computer Studies. College Park, MD: University of Maryland.

Lind, D., & Marcus, B. (1995). *An introduction to symbolic dynamics and coding*. Cambridge, U.K.: Cambridge University Press.

Luce, R. D., Bush, R. R., & Galanter, E. (1963). *Handbook of mathematical psychology*, vol. I. New York: John Wiley.

Luo, D., & Libang, T. (1993). *Qualitative theory of dynamical systems. Advanced series in dynamical systems*, vol. 12. Singapore: World Scientific.

MacDonald, I. L., & Zucchini, W. (1997). *Hidden Markov and other models for discrete-valued time series*. London, U.K.: Chapman & Hall.

Marinov, S. A. (2004). Reversed dimensional analysis in psychophysics. *Perception and Psychophysics, 66*, 23–37.

Martelli, M. (1992). *Discrete dynamical systems and chaos. Pitman monographs and surveys in pure and applied mathematics*, vol. 62.1. London, U.K.: Longman Scientific and Technical.

Mayer-Kress, G. (Ed.). (1986). *Dimensions and entropies in chaotic systems*. Berlin, Germany: Springer-Verlag.

McCord, C., & Mischaikow, K. (1992). Connected simple systems, transition matrices and heteroclinic bifurcations. *Transactions of the American Mathematical Society, 333*, 397–421.

Meyn, S. P., & Tweedie, R. L. (1993). *Markov chains and stochastic stability*. London, U.K.: Springer-Verlag.

Michell, J. (2002). Stevens's theory of scales of measurement and its place in modern psychology. *Australian Journal of Psychology, 54*, 99–104.

Mitchener, W. C., & Nowak, M. A. (2004). Chaos and language. *Proceedings of the Royal Society of London, Series B, 271*, 701–704.

Nicolis, C., & Prigogine, I. (1989). *Exploring complexity*. New York: Freeman.

Nishiura, Y. (2002). *Far-from-equilibrium dynamics*. Translations of mathematical mono-graphs, vol. 299, Providence, RI: American Mathematical Society.

Ott, E. (1993). *Chaos in dynamical systems*. New York: Cambridge University Press.

Parry, W. (1964). Intrinsic Markov chains. *Transactions of the American Mathematical Society, 112*, 55–66.

Parry, W. (1966). Symbolic dynamics and transformations on the unit interval. *Transactions of the American Mathematical Society, 122*, 368–378.

Petersen, K., Quas, A., & Shin, S. (2003). Measures of maximal relative entropy. *Ergodic Theory and Dynamical Systems, 23*, 207–223.

Pincus, D. (2001). A framework and methodology for the study of nonlinear, self-organizing family dynamics. *Nonlinear Dynamics, Psychology, and Life Sciences, 5*, 139–173.

Rao, C. V., Wolf, D. M., & Arkin, A. P. (2002). Control, exploitation and tolerance of intracellular noise. *Nature, 420*, 231–237.

Raftery, A. E., & Tavaré, S. (1994). Estimation and modelling repeated patterns in high order Markov chains with the mixture transition distribution model. *Applied Statistics, 43*, 179–199.

Robert, C. P., & Casella, G. (1999). *Monte Carlo statistical methods*. New York: Springer.

Salmon, P. (2003) How do we recognise good research? *The Psychologist, 16(1)*, 24–27.

Shye, S. (Ed.). (1978). *Theory construction and data analysis in the behavioral sciences*. San Francisco, CA: Jossey-Bass.

Udwadia, F. E., & Guttalu, R. S. (1989). Chaotic dynamics of a piecewise cubic map. *Phsical Review A., 40*, 4032–4044.

van den Wollenberg, A. L. (1978). Nonmetric representation of the radex in its factor pattern parametrization. In S. Shye (Ed.), *Theory construction and data analysis in the behavioral sciences* (pp. 326–349). San Francisco, CA: Jossey-Bass.

Wiggins, S. (1988). *Global bifurcations and chaos*. New York: Springer-Verlag.

Winkler, C. (2003). *Image analysis, random fields and Markov chain Monte Carlo methods* (2nd. ed.). Berlin, Germany: Springer.

Ziehmann, C., Smith, L. A., & Kurths, J. (2000). Localized Lyapunov exponents and the prediction of predictability. *Physics Letters A, 271*, 237–251.

20

Information Hidden in Signals and Macromolecules: Symbolic Time-Series Analysis*

Miguel A. Jiménez-Montaño, Rainer Feistel, and Oscar Diez-Martínez

CONTENTS

We describe the conceptual background and practical implementations of some techniques for the analysis of symbol sequences and symbolic time series. We emphasize their associated software realization, the WinGramm suite of programs, that includes programs for the calculation of conditional entropies, context-free grammatical complexity, and algorithmic distance and redundancy, as well as for the generation of surrogates that preserve symbol pairs and triplets. We demonstrate the usefulness of these programs by means of two illustrative examples, taken from computational neuroscience. In the first one, we obtain evidence of the Markovian character of the cortical inter-spike intervals (ISIs) of the rat before penicillin treatment, and its disappearance afterward. In the second one, we extend previous investigations about neural spike trains generated by the isolated neuron of the

---

* The original article was edited for format and originally published as Jiménez-Montaño, M. A., Feistel, R., & Diez-Martínez, O. (2004). Information hidden in signals and macromolecules I. Symbolic time-series analysis. *Nonlinear Dynamics, Psychology, and Life Sciences, 8,* 445–478. Reprinted by permission of the Society for Chaos Theory in Psychology & Life Sciences.

slowly adapting stretch receptor organ (SAO), in order to classify sequences of different lengths of known neural behaviors. We include new spike trains, digitized employing the optimal partition procedure described by Steuer, Molgedey, Ebeling, and Jiménez-Montaño (2001).

We describe the techniques, with their associated software implementation (Jiménez-Montaño & Feistel, 2003), for the analysis of symbol sequences and symbolic time series. For the latter, a central problem consists in transforming the raw time-series measurements into a corresponding sequence of discrete symbols. Our approach complements the application of nonlinear dynamics (Heath, 2000) and statistical physics techniques to find "hidden information" in time series, defined by Goldberger et al. (2000) as *information that is neither visually apparent nor extractable with conventional methods of analysis.*

Although the symbolic treatment is closely related to the mathematical discipline of symbolic dynamics (Jackson, 1995), it is not the same thing. As clearly discussed in a review by Daw, Finney, and Tracy (2003), a general feature of the modern work in symbolic dynamics is that it is theoretical in nature, and most investigations rely on the existence of generating partitions. Unfortunately, there is no general approach for constructing generating partitions a priori when one is observing the behavior of an unknown system. Besides, in the presence of experimental noise, the concept of a generating partition is no longer well defined (Crutchfield & Packard, 1983; Ebeling, Steuer, & Titchener, 2001). Furthermore, there are special difficulties with the application of nonlinear dynamics to experimental time series of real psychophysical experiments, such as insufficient data points and nonstationarity. These series are much too short to employ some of the algorithms devised to identify properties such as Lyapunov spectra or fractal dimensionality (Gregson, 2001a). Real data tend to be shorter, by orders of magnitude, when compared with the idealized time series that arise in studies in mathematics or physics (Gregson & Leahan, 2003). To face these difficulties, several procedures have been discussed in the literature, such as recurrence quantification analysis (RQA), approximate entropy (ApEn), sample entropy (SampEn), and the entropic analog of the Schwarzian derivative (Esf).

Recurrence plots (RPs), a graphical method designed to locate hidden recurring patterns, were first introduced by Eckmann, Kamphorst, and Ruelle (1987). RPs are also an excellent tool for the visualization of high-dimensional dynamics. However, the interpretation of RPs is not as straightforward as it is with other, "conventional" types of graphs, and it requires careful analysis. In view of this shortcoming, Zbilut and Webber (1992) enhanced the technique by defining five nonlinear quantitative descriptors of the RP that were found to be useful in the quantitative assessment of time-series structures in fields ranging from molecular dynamics to physiology and molecular biology. Their approach, RQA (Webber & Zbilut, 1994), has attracted a lot of attention because it is claimed to be independent of limiting constraints, such as data set size, data stationarity, and assumptions regarding statistical distributions of data.

ApEn was developed by Pincus (1991, 1995) as a measure of regularity closely related to the Kolmogorov entropy (Boffetta, Cencini, Falcioni, & Vulpiani, 2002), the rate of generation of new information, that can be applied to the typically short and noisy time series of physiological and psychological data. The method is rooted in the work of Grassberger and Procaccia (1983) as well as Eckmann and Ruelle (1985), and has been widely applied to clinical cardiovascular studies. However, it has been criticized by Richman and Moorman (2000) who claim that ApEn lacks two important expected properties. First, ApEn is heavily dependent on the record length and is uniformly lower than expected for short records. Second, it lacks relative consistency. According to Richman and Moorman, if ApEn of one data set is higher than that of another, it should, but does not, remain higher for all conditions tested. To cope with these shortcomings, the mentioned authors developed a new family of statistics, which they called sample entropy. The name refers to the applicability to time-series data sampled from a continuous process. We refer the interested reader to their paper for a description of this approach.

The Esf, introduced by Gregson (2000, 2001b), is a new method of data analysis that has been applied successfully to measure unstable dynamics over local short time series. Besides being applicable to very short time series in which stationarity might be feasible, the Esf method may differentiate between random stochastic series and series that are strongly periodic, or are *edge of chaos in their character* (Gregson & Leahan, 2003). In this respect, one of the central quantities of our approach is the algorithmic redundancy introduced by Rapp, Celluci, Korslund, Wanatabe, and Jiménez-Montaño (2001). This quantity, calculated with the help of one of the programs in the WinGramm suite (Jiménez-Montaño & Feistel, 2003), is not sensitive to message length or to sampling frequency when stationary systems are examined (see below).

About the general question of what is the minimal length of a time series that is still amenable to extract information from it? Or, to differentiate between a regular and a pseudorandom sequence, we think that these questions can only have empirical answers (see Appendix 20.A). The reason is that, for finite sequences, randomness is not an "inherent" property of the sequence, but rather is relative to the information and computing resources at the disposal of the observer (Goldreich, 1999). As is well known, it is not possible to define the concept of randomness for finite 0–1 sequences, if the definition is to reflect the idea of irregularity (Chaitin, 1987; Li & Vitányi, 1997). Thus, for very short sequences, the question of randomness is not well defined.

In agreement with the suggestion put forth in Daw et al. (2003), the application of symbol sequence analysis techniques to discretized time series will be referred to as *symbolic time-series analysis*. This phenomenological approach is intended for real-world data and makes no a priori assumptions about the mechanism generating the data. Therefore, we consider only finite symbol sequences of highly discretized data, given as a time sequence of measurements, similar to that followed in the paper of Voss and Kurths (1998).

Related applications derive from the study of automata and formal languages (Hopcroft, Motwani, & Ullman, 2001). For example, they have been employed in statistical and algebraic linguistics worked out by Harris (1968) and Chomsky (1975), respectively. The new developments are grounded in the theory of computation and specifically in the notion of a universal language. The central notion is the algorithmic (Kolmogorov) complexity (Li & Vitányi, 1997) that measures the complexity of sequences in terms of the shortest program (for a fixed universal machine) that generates the sequence. The methods we describe are similar to those employed in statistical linguistics (Charniak, 1993) for the assignment of natural language texts to predefined categories based on their content. They have also been used in bioinformatics (Baldi & Brunak, 1998; Durbin, Eddy, Krogh, & Mitchison, 1998; Pevzner, 2000), for the detection of significant patterns in DNA, RNA, and protein sequences, and the classification of proteins into families. For a discussion of the relationship between topological and metric (Kolmogorov–Sinai) entropies and other measures of chaos in symbolic dynamics with the concept of algorithmic complexity, see Crutchfield and Packard (1982) or the review by Boffetta et al. (2002).

Data compression provides an intelligent approach to the analysis of sequences. More than 20 years ago, this was discussed at length in several papers (Ebeling & Jiménez-Montaño, 1980; Lempel & Ziv, 1976; Papentin, 1980; Wolff, 1982). This idea has been reevaluated by others (e.g., Allison, Stern, Edgoose, & Dix, 2000; Lowenstern, Hirsh, Yianilos, & Noordewier, 1995; Milosavljevic & Jurka, 1993; Rivals et al., 1997), and applied to measure information transmission by action potentials (French, Höger, Sekizawa, & Torkkeli, 2003) and to characterize the responses of primary visual cortical neurons to both random and periodic stimuli (Amigó, Szczepanski, Wajnryb, & Sánchez-Vives, 2003). The basic idea is easy to understand: by removing redundancy, data compression provides cues for detecting the structure. Incompressible strings are called algorithmically random (or, more precisely, pseudorandom), and strings that can be characterized by small programs, compared to their length in bits, are called algorithmically simple (Zurek, 1989).

Natural and artificially generated processes appear unpredictable due to the apparent randomness of the data. Crutchfield and Feldman (2001) have made a comprehensive study of the relationship between the unseen regularities and the observed randomness of different kinds of processes employing information theory methods. They identify five causes of unpredictability, some of which go back to the work of Laplace and Poincaré. Among these, probably the most familiar to the readers of this journal is the one associated with the intrinsic mechanisms that amplify unknown or uncontrolled fluctuations to unpredictable macroscopic behavior. Manifestations of this sort of randomness include *deterministic chaos* and *fractal separatrix structures* bounding different basins of attraction. Another cause for apparent randomness (not mentioned by Crutchfield and Feldman) comes from the well-known result of information theory that a message transmitted with optimal efficiency over a channel of limited bandwidth is indistinguishable from random noise to a receiver who is unfamiliar with the language in which the message is written. The cause of this apparent randomness is discussed in the companion paper by Jiménez-Montaño, Feistel, and Diez-Martínez (2004), concerning the coding of molecular structures into protein sequences. In neural spike-train analysis, the cause may be inaccurate measurements or an insufficient amount of measured data, due to the difficulty in obtaining a sufficiently long string of measurements. Another issue might be the wide range in the variability of ISIs, such as that observed in the first example we analyze.

Using compression for inductive inference goes back to the fundamental papers of Solomonoff (1964a,b), Kolmogorov (1965), and Chaitin (1966). To our knowledge, Ebeling and Jiménez-Montaño (1980) were the first to apply a particular realization of algorithmic complexity (Cover & Thomas, 1991; Li & Vitányi, 1997) to the analysis of DNA and protein sequences. In that paper, we introduced a new algorithm to estimate a complexity measure, called context-free grammatical complexity, with the purpose of revealing the information compressing a sequence by introducing new variables (syntactic categories). The length of the compressed sequence is then taken as a measure of the complexity of the sequence. This quantity refers to an individual sequence in contrast to the Shannonian measures that are related to the sequence source. A very similar complexity measure was proposed by Papentin (1980), who applied it to the determination of chemical structures (Papentin, 1982). The algorithm, introduced by Ebeling and Jiménez-Montaño (1980), to estimate the grammatical complexity has been implemented in the past by means of different programs (Chavoya-Aceves, García, & Jiménez-Montaño, 1992; Ebeling & Feistel, 1982; Jiménez-Montaño, Ebeling, Pohl, & Rapp, 2002). A new implementation of such programs is included in the WinGramm suite (Jiménez-Montaño & Feistel, 2003), which is designed to analyze symbol sequences with the methods of sub-word entropies (*n*-gram entropies), context-free grammatical complexity, and surrogate statistics. All these methods have the purpose to reveal the information content, the complexity or the redundancy embodied in biomolecular sequences (Ebeling & Jiménez-Montaño, 1980; Gatlin, 1972; Milosavljevic, 1999; Rapp et al., 2001), digital information carriers, or human writings (Ebeling & Feistel, 1982; Ebeling & Neiman, 1995; Grassberger, 1989). A problem related to sequential prediction is that of deciding what are the typical patterns in a sequence of records and what is just the result of chance (Grassberger, 1986). To approach this question, several methods to generate surrogate sequences have been suggested in the literature (Altschul & Erickson, 1985;

Jiménez-Montaño et al., 2002; Schreiber & Schmitz, 2000; Theiler, Eubank, Longtin, Galdrikian, & Farmer, 1992).

The application of algorithmic complexity to characterize spatiotemporal patterns is not new. An early application that employed the Lempel–Ziv complexity (Lempel & Ziv, 1976) for its estimation was made by Kaspar and Schuster (1987). Another application, to quantitatively characterize the dynamic patterns in patient–therapist communication, coded into nominal categories, in which the algorithmic complexity was estimated by means of the context-free grammatical complexity (Ebeling & Jiménez-Montaño, 1980), is described in Rapp, Jiménez-Montaño, Langs, Thomson, and Mees (1991). The same measure was used to characterize the heart rate variability before, during, and after cardiac surgery (Storella et al., 1996) and homeostatic drinking (Karádi & Bende, 1998). However, an important shortcoming is that the algorithmic complexity of a symbol sequence is sensitive to the length of the message. Additionally, in those cases where the sequence is constructed by the symbolic reduction of an experimentally observed wave form, as in the examples to be discussed in this communication, the corresponding value of the algorithmic complexity is also sensitive to the sampling frequency. To solve these problems, several definitions of algorithmic redundancy that are sequence-sensitive generalizations of Shannon's original definition of information redundancy were suggested by Rapp et al. (2001). In contrast to algorithmic complexity, the algorithmic redundancy is not sensitive to message length or to sampling frequency when stationary systems are examined.

We illustrate some of the capabilities of symbolic time-series analysis with two examples: The first one consists in the evaluation of the conditional entropies and context-free grammatical complexity of digitized ISI series from cortical neurons, before and after the application of penicillin (Rapp et al., 1994). We obtain evidence of the Markovian character of the cortical ISIs of the rat before the treatment, and its disappearance afterward. The second one consists in the calculation of the algorithmic complexity, distance and redundancy, as well as the block entropies of digitized individual spike-train forms to discriminate different classes of neural behavior. It extends previous investigations about neural spike trains of the slowly adapting SAOs (Jiménez-Montaño, Penagos, Hernández-Torres, & Díez-Martínez, 2000). Here, we include new spike trains, of various lengths, digitized employing the optimal partition procedure described by Steuer et al. (2001). As in the former paper, we show that the method distinguishes between different stimulated conditions. Additionally, in this chapter, we create a tree to classify different behaviors using both standard random shuffle, as well as random pair shuffle surrogates that preserve the correlations between pairs of symbols (Jiménez-Montaño et al., 2000).

The remaining part of this chapter is organized as follows. First we formulate a brief summary of the definitions of n-gram and conditional entropies, algorithmic complexity, distance and redundancy, and their corresponding mathematical expressions in order to establish the notation, referring to the literature for detailed explanations. After that, we discuss different choices for the location of partitions between symbols, specially the optimal partition we shall employ in our second example. We apply the proposed approach to the analysis of neural spike trains. Finally, we draw our conclusions and discuss the relation of symbolic time-series analysis with statistical and non-linear time-series methods (Kantz & Schreiber, 1997).

Materials and Methods

For the sake of completeness, and in order to establish the *notation*, first we recall some well-known concepts from information and formal language theories.

Entropy-Like Measures of Sequence Structure

Symbol sequences are composed of symbols (letters) from an alphabet of λ letters (e.g., for $\lambda=4$, $\{A, C, G, T\}$ is the DNA alphabet; for $\lambda=2$, $\{0, 1\}$ is the binary alphabet, etc.). Sub-strings of n letters are termed n-words. If stationarity is assumed, that is, if any word i can be expected at any arbitrary site to occur with a well-defined probability, p_i, then the n-word entropies (block entropies or higher-order entropies) are given by

$$H_n = -\sum_i p_i^{(n)} \log_2 p_i^{(n)} \tag{20.1}$$

The summation has to be carried out over all words with $p_i > 0$. The maximum number of words is λ^n, so there is a dramatic increase of the number of possible words with respect to n, which makes the estimation of higher-order entropies a difficult task (Schmitt, 1995). The entropies H_n measure the average amount of information contained in a word of length n. Defining the *self-information* of a word of length n as

$$I_n = -\log_2 p_i^{(n)} \tag{20.2}$$

then

$$H_n = \langle I_n \rangle \tag{20.3}$$

is the expected value of I_n.

The conditional entropies

$$h_n = H_{n+1} - H_n \tag{20.4}$$

give the new information of the $(n+1)$th symbol given the preceding n symbols. The *entropy of the source* (also called the Shannon entropy, h_{Sh}, by Boffetta et al., 2002),

$$h = \lim_{n \to \infty} h_n = \lim_{n \to \infty} \frac{H_n}{n} \tag{20.5}$$

quantifies the information content per symbol, and the decay of the h_n measures the correlation within the sequence. H_n and h_n are good candidates to detect the structure in symbolic sequences, since they respond to any deviations from statistical independence. In a random sequence with equi-distributed probabilities, $p^{(n)} = 1/\lambda^n$ holds for the probabilities of n-words. Therefore,

$$H_n = n \cdot \log_2 \lambda \tag{20.6}$$

For binary sequences, $\lambda = 2$ and $H_n = n$ bits. H_n exhibits a linear scaling for a random, non-equidistributed process (Schmitt, 1995),

$$H_n = n \cdot H_1 \tag{20.7}$$

the coefficient being

$$H_1 = -\sum_{i=1}^{\lambda} p_i \log_2 p_i \tag{20.8}$$

that is, the ordinary entropy or uncertainty function (Ash, 1965; Cover & Thomas, 1991). For sequences of length $n = 1$, Equation 20.7 reduces to $H_{max} = \log_2 \lambda$, which is the maximum possible value of the entropy.

For a kth-order Markov process,

$$h_n = h \quad \forall n \geq k \tag{20.9}$$

holds; therefore, deviations of h_n from its limit indicate statistical dependency on the scale of $(n + 1)$ words. The expected value of the algorithmic (Kolmogorov) complexity is asymptotically equal to the entropy of the source, h (Boffetta et al., 2002).

Shannon defines informational redundancy, R_S, as

$$R_S = 1 - \frac{H_1}{H_{max}} \tag{20.10}$$

If $p_i = 1/\lambda \forall i$, then $H_1 = H_{max}$, and $R_S = 0$; that is, when the sequence is equi-probable, the redundancy of the message is zero. Alternatively, suppose $p_j = 1$ for some j and that $p_i = 0 \; \forall i \neq j$, then $H_1 = 0$ and $R_S = 1$.

Mostly, the entropies H_n are estimated from the normalized frequencies of occurrences:

$$H_N^{\text{obs}} = -\sum_i \frac{k_i}{N} \log_2 \frac{k_i}{N} \quad \text{and} \quad I^{\text{obs}} = -\log_2 \frac{k_i}{N} \qquad (20.11)$$

which are called "observed entropies" (and observed self-information, respectively). Here, N denotes the total number of words in the sequence and k_i is the number of occurrences of a certain word i. As was shown by Herzel (1988), and Herzel, Schmitt, and Ebeling (1994), among others, in general, the naive estimation of the probabilities by means of $p_i = k_i/N$ fails, produces a finite sample effect, that is, a deviation of H_n from its true value as n increases. However, these effects will not be considered in this chapter. We refer the interested reader to the mentioned works for further details about this point. In contrast to the informational quantities defined before, which are referred to an ensemble of sequences, the observed quantities refer to *individual sequences*. In what follows, all entropies will be *observed entropies* (with the superscript "obs." suppressed for convenience).

Context-Free Grammatical Complexity

This complexity measure, introduced by Ebeling and Jiménez-Montaño (1980), represents an attempt to determine the algorithmic complexity of a sequence. The essence of this concept is to compress a sequence by introducing new variables (syntactic categories). The length of the compressed sequence is then taken as a measure of the complexity of a sequence. However, there are different ways to measure the length of the compressed sequence; in the original paper (Ebeling & Jiménez-Montaño, 1980), the number of characters of the compressed sequence was used (counting logarithmically repeated characters). We recall this approach subsequently. The set of all finite strings (words) formed from the members of the alphabet X is called the free semigroup generated by X, denoted X*. A language over an alphabet X is any subset of X*. If p and q are words from X*, then their concatenated product pq is also a member of X*.

A *context-free grammar* is a quadruple G = {N, T, P, S}, where (a) N is a finite set of elements called *nonterminals* (syntactic categories), including the start symbol S; (b) T is a finite set of elements, called *terminal symbols* (letters of the alphabet); and (c) P is a finite set of ordered pairs A → q, called *production rules*, such that q ε (N ∪ T) and A is a member of N.

Let us consider a grammar G such that L(G) = w; that is, the language generated by G consists of the single sequence w. These grammars are called "programs" or "descriptions" of the word w. The *context-free grammatical complexity of w* (Ebeling & Jiménez-Montaño, 1980) is defined as follows:

The complexity of a production rule $A \rightarrow q$ is defined by an estimation of the complexity of the word on the right-hand side, $q \rightarrow a_1v_1,\ldots, a_mv_m$:

$$K(A \rightarrow q) = \sum_{i=1}\left\{\left[\log v_i\right] + 1\right\} \tag{20.12}$$

where $a_j \varepsilon$ (N $\cup$ T), for all $j = 1, \ldots, m$. Here, $[x]$ denotes the integral part of a real number.

The complexity $K(G)$ of a grammar G is obtained by adding the complexities of the individual rules. Finally, the complexity of the original sequence is

$$K(w) = K(G(w)) = \min\{K(G) \,|\, G \rightarrow w\} \tag{20.13}$$

This quantity is a particular realization of the algorithmic complexity conceived by Solomonoff (1964a,b), Kolmogorov (1965), and Chaitin (1966).

Algorithmic Redundancy

An algorithmic definition of redundancy was introduced by Rapp et al. (2001) as

$$R_0 = 1 - \frac{C_m}{\langle C_0 \rangle} \tag{20.14}$$

where C_m is the complexity of the original message. C_0 is the complexity of a randomly shuffled equiprobable symbol sequence of the same length where $p_i = 1/\lambda$. The subscript "0" is used to indicate that this is the complexity of a random-shuffled surrogate data set. Notice that the distribution of the original sequence is not necessarily $p_i = 1/\lambda$.

$\langle C_0 \rangle$ denotes the mean value of the complexity found by averaging values from several independently constructed surrogates. Algorithmic complexity gives the highest value to random sequences. $\langle C_0 \rangle$ is, therefore, an empirical estimate of C_{max}, and R_0 is a sequence-sensitive generalization of Shannon's redundancy, R_S, defined in Equation 20.10. In this chapter, we estimate complexities with the context-free grammatical complexity (K). Therefore, we rewrite Equation 20.14 as

$$R_0 = 1 - \frac{K_m}{\langle K_0 \rangle} \tag{20.15}$$

For further details about the properties of the algorithmic redundancy, see Rapp et al. (2001).

Surrogate Sequences

To validate the results, we constructed surrogate ensembles consisting of a finite number of sequences that are the same length as the original sequence and that share with it certain statistical properties, for example, letter frequencies with the original sequence. We consider here three types of surrogate sequence ensembles. In the standard random shuffle (called *algorithm zero* in Rapp et al., 1994), the original data sequence is shuffled. This conserves the frequency of the letters, but destroys all correlations. In the RPS, the original data sequence is shuffled in such a way that the letter frequencies, the pair frequencies, and, therefore, the pair correlations are conserved. In the random triple shuffle (RTS) and higher-order shuffle, the data sequence is shuffled in such a way that the triple frequencies (or higher-group frequencies) are conserved. For further details about surrogate construction, see Jiménez-Montaño et al. (2002) and Jiménez-Montaño and Feistel (2003).

Following Theiler et al. (1992), we will take as our measure of significance the *S*-measure (also called *Z*-score), defined by the difference between the original value and the mean surrogate value of a measurement, divided by the standard deviation (SD) of the surrogate values:

$$S = \frac{|M_{orig} - \langle M_{surr} \rangle|}{\sigma_{surr}} \tag{20.16}$$

Thus, the *S*-score is the number of SD by which the quantity of interest differs from the mean value in its surrogates.

Algorithmic Distance

Zurek (1989) proposed a metric for the space of binary sequences in terms of Kolmogorov complexity, called *algorithmic distance*. It measures the least amount of information involved in transforming a sequence *s* into a sequence *t*. Li et al. (2001) introduced a normalized version of this quantity and applied it to the problem of comparing two genomes. Contrasting with traditional sequence distances that require an alignment (Mount, 2001; Pevzner, 2000), the algorithmic distance works on unaligned sequences. See the companion paper by Jiménez-Montaño, Feistel, and Diez-Martínez (2004).

Given two sequences x and y, their distance (Li et al., 2001) is defined as follows:

$$d(x, y) = 1 - \frac{K(x) - K(x \mid y)}{K(xy)}, \quad 0 \leq d(x, y) \leq 1 \tag{20.17}$$

where $K(x \mid y)$ is the conditional Kolmogorov complexity of x given y, and $K(x) = K(x \mid \varepsilon)$, where ε is the empty sequence, is the unconditional Kolmogorov complexity of x (Li & Vitányi, 1997). $K(x|y)$ measures the randomness of x given y.

The numerator $K(x) - K(x|y)$ is the amount of information in y about x. Within an additive logarithmic factor, it is true that

$$K(x) - K(x|y) = K(y) - K(y|x) \qquad (20.18)$$

where $K(y) - K(y|x)$ is the amount of information in x about y (Li & Vitányi, 1997). The denominator $K(xy)$ is the amount of information in the string x concatenated with y. It serves as a normalization factor, to get a length-independent measure. The expression

$$R(x, y) = \frac{K(x) - K(x|y)}{K(xy)} \qquad (20.19)$$

measures the relatedness of x and y (Chen, Kwong, & Li, 2000).

An algorithm for the estimation of Equation 20.17, using the context-free grammatical complexity to approximate the Kolmogorov complexity (for user-defined sequences x, y) as well as several programs to evaluate the informational and algorithmic quantities introduced in this section, can be found in the suite of programs, WinGramm (Jiménez-Montaño & Feistel, 2003).

Construction of Symbolic Partitions

For a discussion of common methods for constructing symbolic partitions and symbol trees, and the similarities between time-delay embedding and symbol-sequence analysis, see the excellent review by Daw et al. (2003). There is a theoretical optimal choice for locating partitions for noise-free, deterministic processes (Crutchfield & Young, 1989). Tang, Tracy, Boozer, deBrauw, and Brown (1995) found that reliable reconstructions can be achieved given a noisy chaotic signal, provided the general class of the model of the underlying dynamics is known. They considered both observational and dynamical noise, finding that substantial noise produces a strong bias in the symbol sequence statistics. Nonetheless, such bias can be tracked and effectively eliminated by including the noise characteristics of the model. However, for most experimental observations, when noise is present, one has to face the practical problem of choosing appropriate partitions for a data set that may have been generated by an unknown dynamic process with unknown levels of noise. Since the usefulness of the symbolic time-series analysis depends crucially on the procedure used to partition the data among a finite set of symbols, a lot of effort has been devoted to this problem. If this is done inappropriately, spurious results will be obtained. The best partition can be operationally defined as the partition that most effectively reveals the randomness of the original data, that is, the best partition is the one that gives the largest algorithmic complexity estimate. Employing the context-free grammatical complexity (Rapp et al., 1994), we empirically found that partitioning about the median meets this criterion. The binary sequences reported in our first application (see "Applications" section below) were constructed by partitioning about the median. From the informational point of view, this partitioning maximizes the ordinary entropy (h_0). In a subsequent article, Steuer et al. (2001) found that the conditional entropies h_n are strongly dependent on the choice of the binary partition. By maximizing a conditional entropy of a binary sequence with respect to the partition threshold c, although we do not get the highest of the conditional entropies for all possible partitions, we get an empirical rule to choose the partition that most effectively reveals the structure of the given sequence. For the definition of the *optimal binary measurement partition*, see Steuer et al. (2001) and Ebeling et al. (2001). However, we still have to decide among the different conditional entropies, h_n ($n = 0, 1, 2, 3,...$), the one to be maximized. Thus, having identified "good" candidate partitions, we need a criterion to compare them. This can be achieved by appealing again to the algorithmic complexity criterion above, but, in order to make it independent of the sequence length, we have to reformulate it as follows: the best partition is the one that gives the minimum algorithmic redundancy estimate. This approach was first suggested by Jiménez-Montaño (1989); however, the estimation of this quantity was not as satisfactory as the one proposed by Rapp et al. (2001).

In our second application (see "Applications" section below), we first find the partitions that maximize h_n ($n = 0, ..., 5$) and, for each partition, evaluate R_0 using (20.15). By default, we choose the one that maximizes h_0 (the partition about the median) unless there is another partition that achieves a value of R_0 that is at least 10% lower.

Applications

The spiking behavior of cells has traditionally been described as a random point process, that is, a statistical process in which the dynamical information is carried by a series of event timings (Miller & Snyder, 1991; Rieke, Warland, de Ruyter van Steveninck, & Bialek, 1997). In particular, it has been described as a renewal process with independent and identically distributed ISIs (Koch, 1999). This means that the chance of finding some particular ISI is independent of whether a short or a long ISI preceded it. In its elementary form, the analysis of ISI sequences begins with an examination of the statistical properties of the distribution formed by the $\{I_j\}$, such as the mean, median, and SD. Additional properties of the distribution, such as skewness and kurtosis, can also be determined. The value of this analysis is well established. These measures are, however, insensitive to the sequence of the spike trains. Average values and SD of ISI sequences are invariant after randomly shuffling the original sequence.

The use of a discrete representation of a continuous time series may be advantageous in the analysis of noisy systems, decreasing statistical fluctuations in estimation and inference problems (Tang et al., 1995; van der Heyden, Diks, Hoekstra, & DeGoede, 1998). In a pioneering paper, to study the effect of ethanol in the midline cerebellum of rats, Sherry and Klemm (1980) encoded the real-time ISIs by means of a three-letter dynamical partitioning, that is, the change of length of the intervals, and not the length itself, between two consecutive spikes determines the symbol onto which an interval is mapped. An interval shorter than its precursor produces the symbol 1, an interval of the same length the symbol 2, and a longer interval the symbol 3. Another early application of coarse-graining to ISI sequences was made by Dayhoff and Gerstein (1983a,b), under the name of "quantized Monte Carlo method," to find favored patterns in spike trains. Van der Heyden et al. (1998) gave Markov chain representations of an ISI sequence of an electroreceptor of the weakly electric eel *Apteronotus leptorhynchus* (brown ghost knife fish). These authors found that, both, the binary as well as the heptary representations of the ISI sequence display a Markov order of at least 4. Also, Jiménez-Montaño et al. (2002) have shown that ISIs, obtained from measurements by Pei and Moss (1996) on the crayfish caudal photoreceptor, encoded with the help of a three-letter dynamical partitioning, are close to represent a second-order Markov chain.

Here, we give evidence of the Markovian character of cortical ISI sequences of the rat, employing a coarse-graining about the median and calculating several conditional entropies, with the help of WinGramm (Jiménez-Montaño, Feistel, & Diez-Martínez, 2004). Our example starts from the experimental results obtained by Zimmerman, Rapp, and Mees (1991), on the sensitivity of neural activity to convulsants. To save space, we refer the reader to the mentioned paper for the details of the experimental methods employed. It is well known that a highly concentrated drop of penicillin applied directly to the surface of the brain cortex produces electrical activity similar to focal epileptic seizures (Walker, Johnson, & Kollros, 1945). However, Zimmerman et al. (1991) found no statistically significant differences in the mean ISI data in the pre- and post-penicillin condition. By embedding the data in two- and three-dimensional spaces, they found a dramatic sensitivity to the drug. Further analyses of the results obtained by Zimmerman's group, employing the informational and algorithmic tools described in this chapter, may be found in Rapp et al. (1994); Schmitt (1995); Jiménez-Montaño, Pöschel, and Rapp (1997); Steuer et al. (2001); and Jiménez-Montaño et al. (2002). It was established in these studies that for three of the seven neurons (1, 5, and 6), the spike trains clearly differ from pseudorandom sequences before the treatment. Two of them (3 and 7) are on the borderline of randomness and the other two are indistinguishable from pseudorandom sequences of the same composition. After the penicillin treatment, all sequences are indistinguishable from pseudorandom sequences, except neuron 6. For neuron numbering and SD of conditional entropies of the seven ISIs, digitized with respect to the median, see Table 2 in Jiménez-Montaño et al. (2002).

Figure 20.1a shows the conditional entropies of ISI sequences digitized with respect to the median (maximizing h_0) with values taken from Table 20.1. Data recorded from neurons 1 and 5 are displayed. The original ISIs are compared with a standard random-shuffled surrogate of neuron 5 (C5S) and an RTS (C5T) of the same neuron, in which the data sequence is shuffled in such a way that the triple frequencies are conserved (see above). The original data are indistinguishable from C5T. Thus, it appears that the digitized ISI sequence of neuron 5 is very similar to a second-order Markov chain. An equivalent result was obtained previously for data from neuron 1,

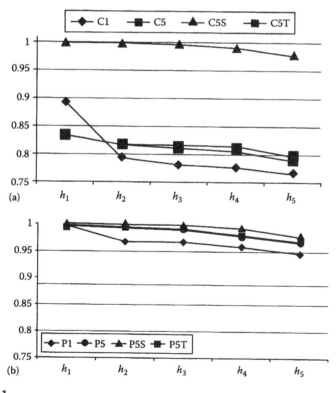

FIGURE 20.1
Graphical display of the conditional entropies in Table 20.1 (a) before the penicillin treatment and (b) after the penicillin treatment.

TABLE 20.1

Conditional Entropies, before and after Penicillin Treatment, of Digitized ISI Sequences of Two Neurons and Surrogates[a]

	h_1	h_2	h_3	h_4	h_5
C1	0.8935	0.7940	0.7808	0.7761	0.7659
C5	0.8345	0.8170	0.8105	0.8059	0.7889
C5S	0.9992	0.9982	0.9960	0.9892	0.9768
C5T	0.8345	0.8170	0.8155	0.8125	0.7974
P1	0.9966	0.9670	0.9661	0.9574	0.9450
P5	0.9957	0.9943	0.9884	0.9753	0.9649
P5S	0.9996	0.9986	0.9963	0.9911	0.9760
P5T	0.9957	0.9943	0.9909	0.9787	0.9670

[a] C1 and C5 come from neurons 1 and 5 in Jiménez-Montaño et al. (2002), together with a simple random-shuffled surrogate of neuron 5 (C5S) and an RTS (C5T) of the same neuron, before penicillin. P1, P5, P5S, and P5T are the corresponding sequences after penicillin.

digitized with the optimal binary partition that maximizes h_3 (Steuer et al., 2001). After the penicillin treatment, this behavior vanished. Therefore, these results suggest that a certain degree of structure is present in the analyzed coarse-grained ISI values during control conditions, which disappears after the treatment (Figure 20.1b).

The second application is an extension of the investigations reported in Jiménez-Montaño et al. (2000). As in the previous publication, we applied the symbolic time-series analysis to study spike trains generated by the isolated neuron of the slowly adapting SAO in order to classify different known neural behaviors (i.e., individual spike-train forms). Segundo, Diez-Martínez, and Quijano (1987) observed the influence of regularly and irregularly arriving stimuli. These were pulse-like lengthenings, that is, "tugs" applied to the muscle element of the SAO. The SAO neuron is a pacemaker cell that responds to isolated stimuli much like other cells respond to the arrival of EPSPs (excitatory postsynaptic potential) (Bryant, Ruíz Marcos, & Segundo, 1973; Diez-Martínez & Segundo, 1983). The effects of stimuli include shortening of the intervals in which they occur, that is, they excite. Diez-Martínez, Pérez, Budelli, and Segundo (1988) studied the consequences of periodically applied stimuli in pacemaker SAO neurons. Two individual spike-train forms are pervasive and straightforward: (a) locking is characterized by almost fixed phases; (b) intermittency is distinguished by discharges that shift irregularly between prolonged epochs where spike phases barely change, and brief bursts with marked variations. As stimulus frequencies change, locking alternates with intermittency. Locked domains have simple, rational spike-to-tug ratios (e.g., 1:1 and 2:1). Still, particularly with small amplitudes, other spike-train forms appear, which are less clear. These include (Segundo, Sugihara, Dixon, Stiber, & Bersier, 1998) (a) phase walk-through, that is, phases vary cyclically, increasing or decreasing,

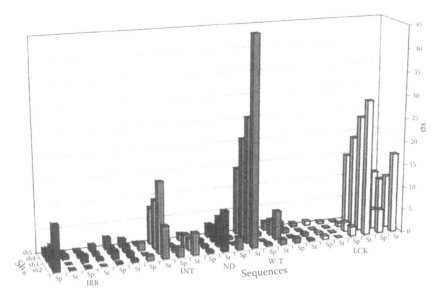

FIGURE 20.2
The S-score of the conditional entropies, $S(h_n)$, $n = 1,...,5$, for the sequences in Table 20.2, calculated with respect to 100 surrogates by the method of RPS. From left to right, we find the following spike-train forms: *irregular* (IRR), *intermittent* (INT), *not-defined* (ND), *walk-through* (W-T), and *locking* (LCK). Within each form, the results correspond to the sequences in the table, in the same order, for the stimulated (St) and spontaneous (Sp) regions. In general, the S-score has much higher values for the stimulated regions. However, for the irregular sequences, the contrary is true, indicating that the stimulus increases the sequence degree of randomness.

respectively, depending on whether stimulus frequency is larger or smaller than the natural one; and (b) "messy," that is, difficult to describe succinctly. The sequences that we call irregular (Figure 20.2) refer to spike discharges from spontaneously regular pacemaker neurons that were subject to highly irregular stimulation. Hence, discharge periodicities were weak and ISIs were noisy. Finally, sequences that we call not-defined (ND) are those which do not fit into any of the categories described previously. These probably correspond to what Segundo et al. (1998) have referred to as "messy."

We studied 32 spike-train sequences (time series and digitized versions) obtained in previous experiments (Diez-Martínez et al., 1988). Experiments were performed on abdominal segments of *Procambarus clarkii* or *bouvieri*. Dissection, recording, and data storage procedures are explained in detail elsewhere (Segundo et al., 1987). SAO spikes were recorded extracellularly from the dorsal nerve. Usually, trains of 100 or more stimuli were applied at invariant average stimulus frequencies. Many frequencies were examined in random order. In some experiments, stimuli were applied irregularly at equivalent frequencies. We analyzed twelve arbitrarily chosen individual spike-train forms; the first four were selected in our previous work (Jiménez-Montaño et al., 2000). Every spike train analyzed here, chosen from a broader set, was typical of a particular form. The selection of such trains was based on the use of a varied set of conventional neurophysiological methods (oscilloscope tracings, raster displays, ISI histograms, peristimulus interval histograms, autocorrelation, cross-correlation, etc.). Furthermore, other forms of display were also employed (i.e., ordered interval plots (Segundo et al., 1987) and Poincaré maps (Diez-Martínez et al., 1988)).

To digitize the neural spike trains, we estimated the redundancy, R_0, according to Equation 20.15 for six binary partitions that maximize the conditional entropies from h_0 to h_5, for each time series, in the stimulated (St), non-stimulated or spontaneous (Sp), and interstimulus regions. For most of the digitized series, R_0 did not diminish more than 10% of its value when h_n ($n = 1, ..., 5$) were maximized; thus, we employed the partition about the median. All the series from the spontaneous section, those of irregular behavior as well as the four series studied in our previous work (Jiménez-Montaño et al., 2000), belong to this class. However, four sequences, LCK2, LCK3, ND, and WT3 (Table 20.2), have lower values of R_0, in the stimulated and interstimulus regions, when digitized with respect to h_1, h_1 (or h_4), h_3, h_3 (or h_4), respectively. In all these cases, we employed the partition that maximizes the corresponding conditional entropy.

The resulting symbol sequences have different degrees of randomness and structure as evaluated, both, by conditional entropies and algorithmic

TABLE 20.2

Neural Spike Trains Analyzed in the Second Example

Type	Code[a]	ISIs Stimulated Section	ISIs Non-Stimulated Section	Intervals between Stimulus
Irregular	**IRR1**	182	338	474
	IRR2	191	559	403
	IRR3	111	730	447
Intermittent	**INT1**	84	185	218
	INT2	80	111	129
Undefined	**ND**	88	291	393
Walk-through	**WT1**	89	152	152
	WT2	112	106	175
	WT3	118	105	176
Locking	**LCK1**	50	806	333
	LCK2	136	264	264
	LCK3	172	293	293

[a] The codes are arbitrary labels to distinguish different time series of the same type. Time series in bold type were previously analyzed, partitioned about the median (Jiménez-Montaño et al., 2000). Return maps for all the sequences are available upon request.

complexity. The S-score of the conditional entropies, $S(h_i)$ for $i = 1, ..., 5$, has insignificant values in non-stimulated regions and for the irregular sequences in the stimulated and non-stimulated regions (Figure 20.2). The opposite is true for the series with intermittent, walk-through, and locking spike-train forms, in their stimulated regions, when compared with RPS surrogates. Similar but more pronounced results were obtained with standard random-shuffled surrogates (data not shown). This means that maintaining symbol pairs obviously produces sequences nearer to the original sequences if nearness is expressed in local autocorrelational properties. The values of the S-score of the algorithmic complexity, $S(K)$, when compared with RPS surrogates, are consistent with these results (Figure 20.3). Again, these outcomes are similar, but less prominent than those obtained with standard random-shuffled surrogates (data not shown). In general, the sequences of the stimulated regions deviate from their surrogates more than the associated non-stimulated parts. The only exceptions are the irregular sequences that have $S(K)$ values close to 3 SD in their spontaneous regions. These sequences increase their randomness after the application of the stimuli. For further details of the other behaviors, see the corresponding figure caption.

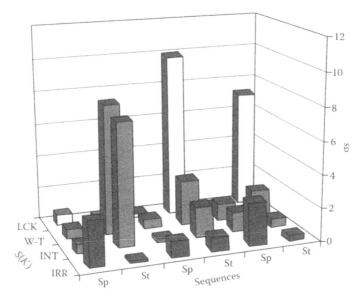

FIGURE 20.3
The S-score, $S(K)$, of the context-free grammatical complexity, K, calculated with respect to 100 surrogates by the method of RPS, for the following spike-train forms: *irregular* (IRR), *intermittent* (INT), *walk-through* (W-T), and *locking* (LCK). From left to right, the spontaneous (Sp) and stimulated (St) regions for the sequences in Table 20.2 are displayed. The results are consistent with the ones in Figure 20.2. In general, the score has much higher values for the stimulated regions. However, for the irregular sequences, the contrary is true. They separate from their surrogates more than 3 SD in the spontaneous region.

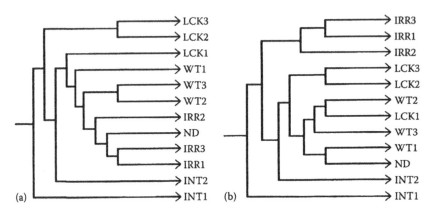

FIGURE 20.4

Classification tree of the analyzed spike-train forms, obtained with WinGramm (a), and with PHYLIP, a package of programs for inferring phylogenies by Joe Felsenstein (b). (From http://evolution.genetics.washington.edu/phylip/software.html. With permission.)

Figure 20.4 shows the classification tree of the analyzed spike-train forms obtained with the WinGramm suite (a) and with the PHYLIP (the **PHYL**ogeny **I**nference **P**ackage) by Joe Felsenstein (b), employing in both cases the algorithmic distance matrix (Table 20.3). The elements of this matrix were calculated applying Equation 20.17 to each pair of sequences displayed in Table 20.2. Although these sequences have different lengths, it was not necessary to align them to get the values. In contrast, in bioinformatics (Baldi & Brunak, 1998; Durbin et al., 1998; Pevzner, 2000), sequence comparison by alignment is employed in the vast majority of sequence comparison methods. Typical models center on the similarities and differences between the sequences, trying to maximize the number of matches between symbols, or to minimize the number of differences, or to do both in some combination. Notice that, in general, there is a similarity in the way both trees cluster sequences corresponding to the same wave form. However, there are minor differences between them (for further details, see figure caption). The observation that different clustering algorithms produce slightly different trees for the same distance matrix is well known. The results depend on the clustering procedure employed (two popular clustering software algorithms are SPSS statistical package, http://www.spss.com, and STATISTICA Mutivariate Exploratory Techniques, http://www.statsoftinc.com).

Conclusions

This chapter describes some techniques for the analysis of symbol sequences and symbolic time series. We emphasize their associated software implementation, the WinGramm suite of programs (Jiménez-Montaño & Feistel, 2003).

TABLE 20.3

Algorithmic Distance Matrix for Sequences in the Second Example

	INT1	INT2	IRR1	IRR2	IRR3	LCK1	LCK2	LCK3	ND	WT1	WT2	WT3
INT1												
INT2	0.8900											
IRR1	0.8806	0.8973										
IRR2	0.8986	0.9208	0.8472									
IRR3	0.9073	0.9083	0.8176	0.8358								
LCK1	0.8833	0.8700	0.8359	0.8301	0.8735							
LCK2	0.8788	0.9024	0.8729	0.9261	0.9644	0.8351						
LCK3	0.9340	0.9302	0.8865	0.9000	0.9136	0.8500	0.6857					
ND	0.8333	0.8154	0.8448	0.8443	0.8681	0.8121	0.8676	0.8529				
WT1	0.9103	0.8797	0.8809	0.8740	0.9119	0.8138	0.8268	0.8561	0.7829			
WT2	0.9245	0.8795	0.8500	0.8782	0.9004	0.7660	0.8659	0.8272	0.8015	0.8571		
WT3	0.9252	0.8395	0.8508	0.8955	0.9008	0.8182	0.8250	0.8148	0.8209	0.8750	0.7722	

This package complements other tools available on the Internet (e.g., the TISEAN package (Hegger, Kantz, & Schreiber, 1999) and the software developed by Sprott, for the analysis of time series and the study of chaotic dynamical systems (http://sprott.physics.wisc.edu/)). However, the approach and most of the tools described here are not to be found in other packages. In particular, these include the programs for the calculation of the context-free grammatical complexity, the algorithmic distance and redundancy, as well as the programs for the calculation of surrogates that preserve pairs and triplets. We have demonstrated the usefulness of these programs by means of two illustrative examples, taken from computational neuroscience. However, these methods have a wide range of applicability. For a review of the problem of predictability, with examples taken from meteorological data series, astrophysics, cardiology, cognitive psychology, finance, and letter sequences generated by nonlinear maps, see the work of Ebeling, Molgedey, Kurths, and Schwarz (2001b). The relationship among the informational and algorithmic quantities discussed in this chapter and the central concepts in nonlinear dynamical systems, such as Lyapunov exponents and topological and Kolmogorov–Sinai (or metric) entropy, are clearly explained in a review by Boffetta et al. (2002). We refer the reader interested in pursuing these matters to this paper for a thorough presentation.

Regarding the problem of digitizing time-series measurements into their corresponding sequences of discrete symbols, we show that, to choose the partition that most effectively reveals the structure of the given sequence, it is not sufficient to maximize each one of a set of subsequent conditional entropies, h_n ($n = 0, 1, 2, 3, \ldots$). Having identified "good" candidate partitions, we gave a criterion to compare them in pairs. The best one is that which gives the minimum algorithmic redundancy estimate. However, in general, it is not possible to obtain the simultaneous maximization of two conditional entropies. For example, let us consider h_o and h_j ($j > 1$). On the one hand, the partition that maximizes h_o (the median partition), minimizes the bias in symbol composition, but leaves symbol correlations intact. On the other hand, by maximizing h_j ($j > 1$), certain correlations (depending on j) are minimized, but a bias in symbol composition appears. Thus, the most suitable partition may differ for different data sets. Nonetheless, the differences seem to be small in the cases we have studied. For another method to quantify the regularity of complex time series, which employs a different partitioning technique and the Lempel–Ziv complexity, see Radhakrishnan, Wilson, and Loizou (2000).

Finally, we want to stress that the natural ground of application of our approach is, of course, the analysis of symbol sequences. Therefore, it may be advantageously applied to characterize strings of nominal categories, as in the study of the dynamics of creative problem solving (Guastello, Hyde, & Odak, 1998), or to characterize coded psychiatric protocols (Rapp et al., 1991). Furthermore, another field where it may be applied successfully is the analysis of DNA and protein sequences, as discussed in the companion paper by Jiménez-Montaño, Feistel, and Diez-Martínez (2004).

Appendix 20.A

In this appendix, we report the results of an empirical test to distinguish a relatively short deterministic versus random number series, in order to compare our approach with that of Zbilut, Giuliani, and Webber (2000), who employed RQA for the same purpose. Following these authors, we employed short runs of (a) a logistic map in the chaotic regime (generated with WinGramm); (b) digits of π; (c) pseudorandom numbers, PRN; and (d) physical random numbers, RAN (Figure 20.5). As is apparent from the

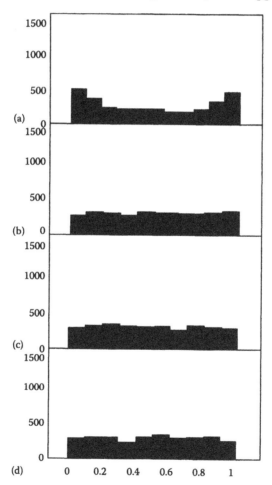

FIGURE 20.5

Distributions for (a) a logistic map in the chaotic regime; (b) digits of π; (c) pseudorandom numbers, PRN; and (d) physical random numbers, RAN, for $N = 3000$. (From Maads, H., URL www. random.org, maintained by Maads Haar, Department of Computer Science, Trinity College, Dublin, Ireland. With permission.) Distributions were similar for $N = 5000$ (not shown).

TABLE 20.4

Complexity, Conditional Entropies, and S-Values for the Sequences in Figure 20.5

Sequence[a]	Length	K	$S(K)$	h_1	h_2	h_3	$S(h_1)$	$S(h_2)$	$S(h_3)$
RAN	300	110	32,800	0.999	0.994	0.991	1,114,207	0.523	1.018
RAN	3,000	670	−0.419	1.000	1.000	0.997	2.226	0.822	1.134
PRN	300	109	−0.900	0.996	0.993	0.983	0.483	0.670	0.564
PRN	3,000	670	0.560	1.000	1.000	0.999	0.612	0.690	0.923
Pi	300	111	0.244	0.990	0.987	0.981	2.820	0.144	0.475
Pi	3,000	669	−0.120	1.000	1.000	0.997	0.583	0.934	1.845
Logistic	300	107	−1.49	0.996	0.979	0.962	1.004	5.885*	4.831*
Logistic	3,000	649	−4.274*	0.998	0.998	0.990	11.551*	15.284*	9.293*

[a] See text for the definition of the sequences. For the calculation of the S-values, we employed 10 simple random-shuffled surrogates. All sequences were digitized about the median.
* $p < .05$.

figure, except for the logistic map, all data sets exhibit a uniform distribution. Our figure is equivalent to Figure 2 in Zbilut et al. (2000).

As can be seen in Table 20.4, the S-values for the context-free grammatical complexity (K) and the first three conditional entropies clearly separate the logistic, of length 3000, from the other series of the same length. For the logistic of length 300, the value of $S(K)$ is also higher than the values in the same column, but is not conclusive. However, the values of $S(h_2)$ and $S(h_3)$, even for this short sequence, are significant.

Acknowledgments

We thank Werner Ebeling for useful suggestions to improve the WinGramm suite of programs and Paul E. Rapp for the cortical ISI sequences of the rat. We also thank M. Reigosa Pardavila, J. M. Trejo-Vargas, and F. Cordoba-Valdés for their help with the figures and tables. The development of WinGramm was partially supported by a Grant-in-Aid for Scientific Research (#32201-E) from CONACYT, Mexico. Finally, we thank the anonymous referees for helpful suggestions to improve the manuscript.

References

Allison, L., Stern, L., Edgoose, T., & Dix, T. I. (2000). Sequence complexity for biological sequence analysis. *Computers and Chemistry, 24*, 43–55.

Altschul, S. F. & Erickson, B. W. (1985). Significance of nucleotide sequence alignments: A method for random sequence permutation that preserves dinucleotide and codon usage. *Molecular Biology and Evolution, 2*, 526–538.

Amigó, J. M., Szczepanski, J., Wajnryb, E., & Sánchez-Vives, M. V. (2003). On the number of states of neuronal sources. *BioSystems, 68,* 57–66.

Ash, R. (1965). *Information theory.* New York: Interscience Publishers.

Baldi, P. & Brunak, S. (1998). Bioinformatics, the machine learning approach. *Bioinformatics* 17, 149–154.

Boffetta, G., Cencini, M., Falcioni, M., & Vulpiani, A. (2002). Predictability: A way to characterize complexity. *Physics Reports, 356,* 367–474.

Bryant, H. L. Jr., Ruíz Marcos, A., & Segundo, J. P. (1973). Correlations of the neural spike discharges produced by monosynaptic connections and by common inputs. *Journal of Neurophysiology, 36,* 205–225.

Chaitin, G. J. (1966). On the length of programs for computing finite binary sequences. *Journal of the Association of Computer Machines, 13,* 547–569.

Chaitin, G. J. (1987). *Algorithmic information theory.* New York: Cambridge University Press.

Charniak, E. (1993). *Statistical language learning.* Cambridge, MA: The MIT Press.

Chavoya-Aceves, O., García, F., & Jiménez-Montaño, M. A. (1992). Program to estimate the grammatical complexity of a sequence. In *Proceedings of IX national meeting on artificial intelligence* (pp. 243–254). México: Grupo Noriega Editores (in Spanish).

Chen, X., Kwong, S., & Li, M. (2000). A compression algorithm for DNA sequences and its applications in genome comparison. In R. Shamir, S. Miyano, S. Istrail, P. Pevzner, & M. Waterman (Eds.), *Proceedings of the fourth annual international conference on computational molecular biology, RECOMB 2000,* Tokyo (p. 107). New York: ACM Press. Available at: http://www.cs.ucsb.edu/~mli/#bpapers

Chomsky, N. (1975). *Syntactic structures.* Berlin: Mouton de Gruyter.

Cover, T. M. & Thomas, J. A. (1991). *Elements of information theory.* New York: John Wiley & Sons, Inc.

Crutchfield, J. P. & Feldman, D. P. (2001). Randomness observed: Levels of entropy convergence. Available from: http://arxiv.org/abs/cond-mat/0102181

Crutchfield, J. P. & Packard, N. H. (1982). Symbolic dynamics of one-dimensional maps: Entropies, finite precision, and noise. *International Journal of Theoretical Physics, 21,* 433–466.

Crutchfield, J. P. & Packard, N. H. (1983). Symbolic dynamics of noisy chaos. *Physica D: Nonlinear Phenomena, 7,* 201–223.

Crutchfield, J. P. & Young, K. (1989). Inferring statistical complexity. *Physical Review Letters, 63,* 105–108.

Daw, C. S., Finney, C. E. A., & Tracy, E. R. (2003). Symbolic analysis of experimental data. *Review of Scientific Instruments, 74,* 916–930. Available from: http://www.chaos.engr.utk.edu/abs/abs-rsi2002.html

Dayhoff, J. E. & Gerstein, G. L. (1983a). Favored patterns in spike trains. I. Detection. *Journal of Neurophysiology, 49,* 1334–1348.

Dayhoff, J. E. & Gerstein, G. L. (1983b). Favored patterns in spike trains. II. Application. *Journal of Neurophysiology, 49,* 1349–1363.

Diez-Martínez, O., Pérez, R., Budelli, R., & Segundo, J. P. (1988). Locking, intermittency and bifurcations in a periodically driven pacemaker neuron: Poincaré maps and biological implications. *Biological Cybernetics, 60,* 49–58.

Diez-Martínez, O. & Segundo, J. P. (1983). Behavior of a neuron in a recurrent excitatory loop. *Biological Cybernetics, 47,* 33–41.

Durbin, R., Eddy, S. R., Krogh, A., & Mitchison, G. (1998). *Biological sequence analysis.* Cambridge, MA: Cambridge University Press.

Ebeling, W. & Feistel, R. (1982). *Physics of self-organization and evolution* (in German). Berlin: Akademie-Verlag.

Ebeling, W. & Jiménez-Montaño, M. A. (1980). On grammars, complexity, and information measures of biological macromolecules. *Mathematical Biosciences, 52,* 53–71.

Ebeling, W., Molgedey, L., Kurths, J., & Schwarz, U. (2001b). Entropy, complexity, predictability and data analysis of time series and letter sequences. In A. Bunde, J. Kropp, & H. J. Schellnhuber (Eds.), *The science of disaster: Climate disruptions, heart attacks, market crashes* (pp. 1–23). Berlin: Springer-Verlag.

Ebeling, W. & Neiman, A. (1995). Long-range correlations between letters and sentences in texts. *Physica A, 215,* 233–244.

Ebeling, W., Steuer, R., & Titchener, M. R. (2001). Partition-based entropies of deterministic and stochastic maps. *Stochastics and Dynamics, 1,* 45–61.

Eckmann, J. P., Kamphorst, S. O., & Ruelle, D. (1987). Recurrence plot of dynamical systems. *Europhysics Letters, 4,* 324–327.

Eckmann, J. P. & Ruelle, D. (1985). Ergodic theory of chaos and strange attractors. *Reviews of Modern Physics, 57,* 617–654.

French, A. S., Höger, U., Sekizawa, S., & Torkkeli, P. H. (2003). A context-free data compression approach to measuring information transmission by action potentials. *BioSystems, 69,* 55–61.

Gatlin, L. L. (1972). *Information theory and the living system.* New York: Columbia University Press.

Goldberger, A. L., Amaral, L. A. N., Glass, L., Hausdorff, J. M., Ivanov, P. Ch., Mark, R. G., et al. (2000). PhysioBank, PhysioToolkit, and PhysioNet. *Circulation, 101,* e215–e220. Available at: http://circ.ahajournals.org/

Goldreich, O. (1999). Pseudorandomness. *Notices of the American Mathe-matical Society, 46,* 1209–1216.

Grassberger, P. (1986). Toward a quantitative theory of self-generated complexity. *International Journal of Theoretical Physics, 25,* 907–938.

Grassberger, P. (1989) Estimation of the information content of symbol sequences and efficient codes. *IEEE Transactions on Information Theory, 35,* 669–675.

Grassberger, P. & Procaccia, I. (1983). Estimation of the Kolmogorov entropy from a chaotic signal. *Physical Review A, 28,* 2591–2593.

Gregson, R. A. M. (2000). Elementary identification of nonlinear trajectory entropies. *Australian Journal of Psychology, 52,* 94–99.

Gregson, R. A. M. (2001a). Responses to constrained stimulus sequences in nonlinear psychophysics. *Nonlinear Dynamics, Psychology, and Life Sciences, 5,* 205–222.

Gregson, R. A. M. (2001b). Scaling quasi-Periodic psychophysical functions. *Behaviormetrika, 29,* 41–57.

Gregson, R. A. M. & Leahan, K. (2003). Forcing function effects on nonlinear trajectories: Identifying very local brain dynamics. *Nonlinear Dynamics, Psychology, and Life Sciences, 7,* 139–159.

Guastello, S. J., Hyde, T., & Odak, M. (1998). Symbolic dynamic patterns of verbal exchange in a creative problem solving group. *Nonlinear Dynamics, Psychology, and Life Sciences, 2,* 35–58.

Harris, Z. (1968). *Mathematical structures of language.* Melbourne, FL: Krieger Publishing Company.

Heath, R. A. (2000). *Nonlinear dynamics: Techniques and applications in psychology*. Mahwah, NJ: Lawrence Erlbaum Associates, Inc.

Hegger, R., Kantz, H., & Schreiber, T. (1999). Practical implementation of nonlinear time series methods: The TISEAN package. *Chaos, 9,* 413–435. Available from: http://www.mpipks-dresden.mpg.de/~tisean/

Herzel, H. (1988). Complexity of symbol sequences. *Systems Analysis Modelling Simulation, 5,* 435–444.

Herzel, H., Schmitt, A. O., & Ebeling, W. (1994). Finite sample effects in sequence analysis. *Chaos, Solitons & Fractals, 4,* 97–113.

Hopcroft, J. E., Motwani, R., & Ullman, J. D. (2001). *Introduction to Automata Theory, Languages, and Computation,* 2/E. Reading, MA: Addison-Wesley.

Jackson, E. A. (1995). *Perspective of nonlinear dynamics*. New York: Cambridge University Press.

Jiménez-Montaño, M. A. (1989). Pattern recognition, molecular sequence analysis and biological information. In W. Ebeling & H. Ulbricht (Eds.), *Irreversible processes and self-organization* (pp. 136–151). Leipzig: Teubner.Texte zur Physik.

Jiménez-Montaño, M. A., Ebeling, W., Pohl, T., & Rapp, P. E. (2002). Entropy and complexity of finite sequences as fluctuating quantities. *BioSystems, 64,* 23–32.

Jiménez-Montaño, M. A. & Feistel, R. (2003). WinGramm: *Grammatical Complexity Analysis of Sequences*. (Program and user manual). Retrieved August 1, 2010 from http://www.julio.sandria.org/archivos/archivos/WinGramm2/Wingramm 2.0-UG.pdf

Jiménez-Montaño, M. A., Penagos, H., Hernández-Torres, A., & Díez-Martínez, O. (2000). Measures of complexity in neural spike-trains of the slowly adapting stretch receptor organs. *BioSystems, 58,* 117–124.

Jiménez-Montaño, M. A., Pöschel, T., & Rapp, E. P. (1997). A measure of the information content of neural spike trains. In E. Mizraji, L. Acerenza, F. Alvarez, and A. Pomi, (Eds.), *Biological complexity, a symposium* (pp. 133–142). Montevideo: Dirac.

Kantz, H. & Schreiber, T. (1997). *Nonlinear time series analysis*. Cambridge: Cambridge University Press.

Karádi, K. & Bende, I. (1998). The indirect verification of minimal two-generator model of homeostatic drinking by complexity examination. *Journal of Theoretical Biology, 192,* 55–60.

Kaspar, F. & Schuster, H. G. (1987). Easily calculable measure for the complexity of spatiotemporal patterns. *Physical review A, 36,* 842–848.

Koch, C. (1999). *Biophysics of computation. Information processing in single neurons.* New York: Oxford University Press.

Kolmogorov, A. N. (1965). Three approaches to the definition of the concept of quantity of Information. *IEEE Transactions on Information Theory, IT14,* 662–669.

Lempel, A. & Ziv, J. (1976). On the complexity of finite sequences. *IEEE Transactions on Information Theory, 22,* 75–81.

Li, M., Badger, J. H., Chen, X., Kwong, S., Kearney, P., & Zhang, H. (2001). An information-based sequence distance and its application to whole mitochondrial genome phylogeny. *Bioinformatics, 17,* 149–154.

Li, M. & Vitányi, P. (1997). *An introduction to Kolmogorov complexity and its applications*. Berlin: Springer.

Lowenstern, D., Hirsh, H., Yianilos, P., & Noordewier, M. (1995). DNA sequence classification using compression-based induction. *DIMACS Technical Report, 95–04,* 1–12.

Maads, H. (2010). URL www.random.org, maintained by Maads Haar, Department of Computer Science, Trinity College, Dublin, Ireland.

Milosavljevic, A. (1999). Discovering patterns in DNA sequences by the algorithmic significance method. In J. T. L. Wang, B. A. Shapiro, & D. Shasha (Eds.), *Pattern discovery in biomolecular data: Tools, techniques and applications* (pp. 3–23). New York: Oxford University Press.

Milosavljevic, A. & Jurka, J. (1993). Discovery by minimal length encoding: A case study in molecular evolution. *Machine Learning, 12*, 69–87.

Miller, M. & Snyder, D. (1991). *Random point processes in time and space.* New York: Springer.

Mount, D. W. (2001). *Bioinformatics, sequence and genome analysis.* Cold Spring Harbor, New York: Cold Spring Harbor Laboratory Press.

Papentin, F. (1980). On order and complexity. I. General considerations. *Journal of Theoretical Biology, 87*, 421–456.

Papentin, F. (1982). On order and complexity. II. Application to chemical and biological structures. *Journal of Theoretical Biology, 95*, 225–245.

Pei, X. & Moss, F. (1996). Characterization of low-dimensional dynamics in the crayfish caudal photoreceptor. *Nature, 379*, 618–621.

Pevzner, P. A. (2000). *Computational molecular biology, an algorithmic approach.* Cambridge, MA: MIT Press.

Pincus, S. M. (1991). Approximate entropy as a measure of system complexity. *Proceedings of the National Academy of Sciences USA, 88*, 2297–2301.

Pincus, S. M. (1995). Approximate entropy (ApEn) as a complexity measure. *Chaos, 5*, 110–117.

Radhakrishnan, N., Wilson, J. D., & Loizou, P. C. (2000). An alternate partitioning technique to quantify the regularity of complex time series. *International Journal of Bifurcation and Chaos, 10*, 1773–1779.

Rapp, P. E., Celluci, C. J., Korslund, K. E., Watanabe, T. A. A., & Jiménez-Montaño, M. A. (2001). An effective normalization of complexity measurements for epoch length and sampling frequency. *Physical Review E, 64*, 016209.

Rapp, P. E., Jiménez-Montaño, M. A., Langs, R. J., Thomson, L., & Mees, A. I. (1991). Toward a quantitative characterization of patient–therapist communication. *Mathematical Biosciences, 105*, 207–227.

Rapp, P. E., Zimmerman, I. D., Vining, E. P., Cohen, N., Albano, A. M., & Jiménez-Montaño, M. A. (1994). The algorithmic complexity of neural spike trains increases during focal seizures. *Journal of Neuroscience, 14*, 4731–4739.

Richman, J. S. & Moorman, J. R. (2000). Physiological time-series analysis using approximate entropy and sample entropy. *American Journal of Physiology—Heart and Circulatory Physiology, 278*, H2039–H2049.

Rieke, F., Warland, D., de Ruyter van Steveninck, R., & Bialek, W. (1997). *Spikes: Exploring the neural code.* Cambridge, MA: MIT Press.

Rivals, E., Delgrange, O., Delahaye, J. P., Dauchet, M., Delorme, M. O., Hénaut, A., et al. (1997). Detection of significant patterns by compression algorithms: The case of approximate tandem repeats in DNA sequences. *Computer Applications in Biosciences, 13*, 131–136.

Schmitt, A. (1995). *Structural analysis of DNA sequences.* Berlin: Dr. Köster.

Schreiber, T. & Schmitz, A. (2000). Surrogate time series. *Physica D, 142*, 346–382.

Segundo, J. P., Diez-Martínez, O., & Quijano, H. (1987). Testing a model of excitatory interactions between oscillators. *Biological Cybernetics, 55*, 355–365.

Segundo, J. P., Sugihara, G., Dixon, P., Stiber, M., & Bersier, L. (1998). The spike trains of inhibited pacemaker neurons seen through the magnifying glass of nonlinear analysis. *Neuroscience, 87,* 741–766.

Sherry, C. J. & Klemm, W. R. (1980). Entropy correlations with ethanol-induces changes in specified patterns of nerve impulses: Evidence for "byte" processing in the nervous system. *Progress in Neuro-Psycophamacology, 4,* 261–267.

Solomonoff, R. J. (1964a). A formal theory of inductive inference, Part 1. *Information & Control, 7,* 1–22.

Solomonoff, R. J. (1964b). A formal theory of inductive inference, Part 2. *Information & Control, 7,* 224–254.

Steuer, R., Molgedey, L., Ebeling, W., & Jiménez-Montaño, M. A. (2001). Entropy and optimal partition for data analysis. *The European Physical Journal B, 19,* 265–269.

Storella, R. J., Shi, Y., Wood, H. W., Jiménez-Montaño, M. A., Albano, A. M., & Rapp, P. E. (1996). The variance and the algorithmic complexity of heart rate variability display different responses to anesthesia. *International Journal of Bifurcation and Chaos, 6,* 2169–2172.

Tang, X. Z., Tracy, E. R., Boozer, A. D., deBrauw, A., & Brown, R. (1995). Symbol sequence statistics in noisy chaotic signal reconstruction. *Physical Review E, 51,* 3871–3890.

Theiler, J., Eubank, S., Longtin, A., Galdrikian, B., & Farmer, J. D. (1992). Testing for nonlinearity in time series: The method of surrogate data. *Physica D, 58,* 77–94.

van der Heyden, M. J., Diks, C. G. C., Hoekstra, B. P. T., & DeGoede, J. (1998). Testing the order of discrete Markov chains using surrogate data. *Physica D, 117,* 299–313.

Voss, H. & Kurths, J. (1998). Test for nonlinear dynamical behavior in symbol sequences. *Physical Review E, 58,* 1155–1158.

Walker, A. E., Johnson, H. C., & Kollros, J. J. (1945). Penicillin convulsions. The convulsion effects of penicillin applied to the cortex of monkey and man. *Surgery Gynecology Obstetrics, 81,* 692–701.

Webber, C. L., Jr. & Zbilut, J. P. (1994). Dynamical assessment of physiological systems an states using recurrent plot strategies. *Journal of Applied Physiology, 76,* 965–973.

Wolff, J. G. (1982). Language acquisition, data compression and generalization. *Language & Communication, 2,* 57–89.

Zbilut, J. P., Giuliani, A., & Webber, C. L., Jr. (2000). Recurrence quantification analysis as an empirical test to distinguish relatively short deterministic versus random number series. *Physics Letters A, 267,* 174–178.

Zbilut, J. P. & Webber, C. L., Jr. (1992). Embeddings and delays as derived from quantification of recurrence plots. *Physics Letters A, 171,* 199–203.

Zimmerman, I. D., Rapp, P. E., & Mees, A. I. (1991). The geometrical characterization of neural activity displays a sensitivity to convulsants. *International Journal of Bifurcation & Chaos, 1,* 253–259.

Zurek, H. (1989). Thermodynamic cost of computation, algorithmic complexity and the information metric. *Nature, 341,* 119–124.

21

Orbital Decomposition: Identification of Dynamical Patterns in Categorical Data

Stephen J. Guastello

CONTENTS

Symbolic dynamics is a class of mathematical analyses for identifying and manipulating patterns in nominally scaled string variables of varying length and for writing functions that describe any recurring patterns. For instance, suppose that a temporal (or spatial) series contained the occurrence of five state variables A, B, C, D, and E in no known order. A search of the data series could reveal recurring patterns, such as ADD, ECB, and so on. The isolation of such patterns signifies a reduction in randomness of the series; a totally random series would not produce any common patterns. Furthermore, it may be possible to write hyperfunctions that describe patterns of elementary clusters. For instance, if we let $F = ADD$ and $G = ECB$ in the example above, one might be able to extract a recurring pattern of F and G such as FGG, FFG, or strings of greater length (Goertzel, 1995; Nicolis & Prigogine, 1989; Ott, Sauer, & Yorke, 1994).

As Gregson noted in Chapter 19, numerical time series that are ill-behaved, possibly as a result of complex but unknown dynamics, lend themselves to symbolic dynamic analysis. Patterns of numbers would be given nominal codes, and one would then search the nominal codes for patterns. The particular procedure that is described in this chapter is capable of doing exactly this. In fact, it originated with the analysis of numerical chaotic data (Lathrop & Kostelich, 1989). Its attractiveness for the behavioral sciences is that it lends itself easily to the use of nominal categories without having to assume a continuously valued underlying metric. It was first introduced for such purposes by Guastello, Hyde, and Odak (1998).

The theory behind the technique is described next and followed by some practical tips for data analysis. Some examples are considered next in which the problems escalate in complexity from one set of categories that occur only one at a time, to categories that can co-occur, multiple categorical variables applied to one set, and some exotic data sources of physiological origin.

Orbital Decomposition

The technique got its name from its fundamental assumption that chaos results from numerous coupled periodic orbits that are all tangled in the time series of observations. The objective is to untangle them. The data are not assumed to be chaotic, however, but rather the analysis can accommodate data that is as complex as chaos.

Unlike some types of symbolic dynamics analysis, it does not assume a particular symbol length (*string length*), but determines the optimal string length for the determination of elementary patterns. There are close associations among the number of recurring patterns, measures of entropy, and the Lyapunov exponent. The analysis as it was developed for the behavioral sciences goes further to include statistical analysis that determines the extent to which the isolated patterns could have occurred by chance and the extent to which the isolated patterns overlap the original data.

Entropy and Information Functions

In the well-known Shannon (1948) theory of information, a bit of information is the amount required to determine a dichotomous outcome. For a multichotomous outcome involving N equiprobable states, $N-1$ bits are needed. There are also provisions for states with unequal a priori probabilities. If a system were truly random and arbitrarily long, one would require as many bits of information to determine the series as there were original data points. Entropy was defined as the inverse of information or $E = K - I$, where K is an asymptotic maximum as I approaches 0.

An important shift in the thinking about information, entropy, and chaos was that information is now regarded as intrinsic to the data series itself, rather than something that is applied by a knowledgeable agent from the outside.

The shift is predicated on the principle of iteration that is inherent in nonlinear dynamical systems (NDS) processes; that is, the value of X at time 2 is a function of X at time 1, the output of X_2 is then fed back into the same function to produce X_3, and so on. Simple functions are known to produce chaotic time series; thus, a small amount of information, which would be the knowledge of the function, goes a long way to explain random-appearing series. Because of the affine transformation (recursive or iterating function) inherent in chaotic processes, chaos is now said to *produce* information. Entropy *is* information in this context (Crutchfield, 1994; Nicolis & Prigogine, 1989; Ott et al., 1994).

Symbolic dynamics, information, entropy, and Lyapunov dimensionality are related as follows. Consider a set of nominal categories with unequal odds of occurrence. The Shannon entropy (H_s) for the set of categories is

$$H_s = \sum_{i=1}^{t} p_i \left[\ln\left(\frac{1}{p_i} \right) \right] \qquad (21.1)$$

where p_i is the probability associated with one categorical outcome. The topological entropy, however, is based on strings or hypothetical orbits of length C:

$$H_T = \lim_{C \to \infty} \left(\frac{1}{C} \right) \log_2(N_C) \qquad (21.2)$$

where N is the number of identifiable periodic orbits (recurring strings) of length C in the series.

In theory, Shannon entropy and topological entropy are equivalent if the categories appear in equal frequency (Crutchfield, 1996). The equal frequency assumption may be untenable in most real situations, however. Also because of the inverse relationship that exists between entropy and information in the Shannon model compared to the topological definition, it would be possible to observe increases in Shannon entropy as string length C increases, and decreases in H_T as C increases. The examples in this chapter amplify this point.

In some situations, not all possible sequences of a given length are viable. Thus, the next procedural step is to construct a transition matrix $\mathbf{M}^C$ of the remaining allowing sequences. The matrix is square with the allowable strings of length C on each axis. Matrix entries are 0 or 1 to represent whether or not a string at time 1 is followed by a string at time 2.

If we were to use this matrix to calculate H_T from Equation 21.2, a determinant of a very large matrix would be required; the computational intensity would be prohibitive. A much simpler computation requires only the trace of $\mathbf{M}^C$ (Lathrop & Kostelich, 1989, p. 4030):

$$H_T = \lim_{C \to \infty} \left(\frac{1}{C} \right) \log_2(\mathrm{tr}\mathbf{M}^C) \qquad (21.3)$$

The trace of $\mathbf{M}^C$ is the sum of 0's and 1's on the diagonal of the matrix. It is the *number of strings that are immediately followed by an identical string;* for instance, if a string called ABF is immediately followed by another ABF, the value 1 would appear in the diagonal. (The strings that appear more than once, but not immediately so, are taken into consideration later in the process.)

The limit function means that the calculation of H_T is made on strings of length 1, 2, 3, and so on, until $\text{tr}\mathbf{M}^C$ reaches 0. The solution would be the value of C obtained just before the value that produced $\text{tr}\mathbf{M}^C = 0$. This solution should be regarded as tentative, however, until the statistical parts of the procedure have also been calculated; see below.

As the string length goes to infinity, H_T approaches the base-2 logarithm of the maximum Lyapunov exponent of the time series. The maximum Lyapunov exponent is, in turn, the largest eigenvalue of $\mathbf{M}^C$, which is approximated by $\text{tr}\mathbf{M}^C$. Dimensionality is, therefore,

$$D_L = e^{H_T} \tag{21.4}$$

Likelihood Chi-Square

As we have seen in previous chapters, the direct calculation of dimension is susceptible to the influence of noise. Thus, statistical techniques for interval- and ratio-scaled metrics have been built to produce an R^2, which would be the proportion of variance accounted for by the nonlinear deterministic model. The proportion of variance attributable to noise, random shock, and the usual uncontrollable uncertainties is $1 - R^2$. For this analysis with nominal metrics, statistical companions to topological entropy calculations were developed (Guastello et al., 1998), as follows.

First, after determining the baseline probability of each interaction type, it is possible to compute the baseline probability of each string of responses of string length C. A likelihood chi-square (χ^2) test can be conducted on all string lengths of 2 to C in the following manner. For each string that actually appeared *two or more times,* the expected number of observations is the simple combinatorial probability of the events. For example, for a string A–B–C, the expected probability is $P_A P_B P_C$. Given N^* strings of length C in a set, the expected frequency of string A–B–C is

$$F_{ex} = P_A P_B P_C N^* \tag{21.5}$$

The asterisk is placed next to the N to indicate that the value changes as the computations proceed for increasing values of C. The likelihood χ^2 is preferable to the Pearson variety because of the small expected and observed frequencies for many strings where C is long and the baseline probabilities of some of the elements are low:

$$\chi^2 = 2\sum \left[F_{ob} \ln\left(\frac{F_{ob}}{F_{ex}}\right) \right] \tag{21.6}$$

There are two deviations from the otherwise-standard procedure that has been defined so far in conjunction with Equations 21.5 and 21.6. (a) If $C = 1$, the baseline probabilities that compose F_{ex} are those that would reflect that the states of $C = 1$ are all equally probable, thus replacing Equation 21.6 as a definition of F_{ex}. (b) Strings with observed frequencies of 1 or 0 are aggregated into a single category for purposes of computing χ^2. The observed frequency for that group is simply the number of strings that appeared only once. The expected frequency for that group would be the difference between the total number of strings of length C and the sum of expected frequencies for all the other cells combined.

The values of χ^2 can be compared against critical values in standard statistics books for the statistical significance of χ^2. Note that the null hypothesis for $C = 1$ is that all states have equal probability. This null hypothesis is different from analyses for $C \geq 2$, where the null is that the odds of a particular string are equal to the a posteriori combinatorial probabilities of the states. Rejection of the latter null hypothesis implies that an emergent process is taking place whereby the probabilities associated with the strings are different from the probabilities associated with the combinatorial odds of their parts.

Ideally, χ^2 should reach a maximum for the value of C that corresponds to the optimal H_T, as defined through Equation 21.3. If that were always the case, there would be no need for the statistical test. The two optima do not converge in reality, however. Thus, the χ^2 provides an indication of the quality of the solution based on H_T. Once again, there is inevitably a trade-off between explaining all the data with the maximum number of axioms or rules, which could become quite large, and explaining as much data as possible with the fewest axioms or rules.

Phi-Square

Although ϕ is usually used as a surrogate for the product-moment correlation for a 2 × 2 contingency, it can be used as a measure of crude association for any $N \times M$ contingency table (Keppel, Saufley, & Tokunage, 1992). This assertion was based on the derivation of Cramer's V, which essentially calculates ϕ for an $N \times M$ contingency table where elements of N and M are rank-ordered. Keppel et al. relaxed the restriction of rank-ordering. In orbital decomposition, Guastello et al. (1998) treated the N and M as expected and observed frequencies of category membership, rather than the membership in two independently defined sets of categories. Equation 21.7 shows the otherwise common value of phi-square (ϕ^2), the proxy for proportion of variance accounted for one categorical variable by another:

$$\phi^2 = \frac{\chi^2}{N^*} \qquad\qquad (21.7)$$

where N^* is the total number of strings being parsed for a given C.

It is well known that ϕ is affected by the proportion of observed frequencies. It cannot equal to 1.00, even with perfect association, for 2 × 2 tables, if neither variable is balanced 50% in each cell. It is unclear at the outset of an orbital decomposition analysis the extent to which the data would push the limits of robustness of the ϕ^2. Any problems would not be more than nuisances here, however, because the objective of using ϕ^2 is to compare string taxonomies that have similar constraints.

One can expect two sources of unusual results, however. One is that the ϕ^2 for $C = 1$ can often exceed 1.00, which is technically not possible because a proportion of variance accounted for cannot logically exceed 1.00. Although this outcome does not happen all the time, the aberration can be traced to the definition of χ^2 not being drawn from a 2 × 2 contingency.

Another condition where $\phi^2 > 1.00$ occurs when the data contain a long series of one event out of the possible range of events. One reported example involved a string of 14 successive episodes of police brutality—which repeated itself—in contrast to $C = 3$ for the next-longest string that repeated itself (Spohn, 2008, p. 110). ϕ^2 exceeded 1.00 for $C > 9$. The trace for $3 < C < 14$ was 1 in each case, and composed of the same events of police brutality in each case. The interpretation of the events in that data was therefore based on $C = 3$, with the understanding that the police-brutality events would be treated as an independent dynamic.

Another example of very dominant strings of events that pushed $\phi^2 > 1.00$ was reported in Parchman and Katerndahl (2009). This study involved a search for patterns in medical interviews with patients who had specific medical conditions. Whereas events were recorded as discrete events in succession in previous studies (e.g., Guastello, 2000; Guastello et al., 1998; Pincus 2001; Pincus & Guastello, 2005; Spohn, 2008), this one recorded the event that occurred in short time intervals. More successive time intervals were expended on one task than any other, producing a result similar to, but not as dramatic as, the one that Spohn reported. Nonetheless, ϕ^2 exceeded 1.00 for most values of C.

Isolating Strings

The variables that are organized into sets of categories are perhaps as critical to the research as the computations themselves. The old saying goes, "Garbage in, garbage out." The choices of coding schemes typically have less to do with the dynamics and more to do with the substance of the problem, although dynamic events might indeed be the substance of the problem (e.g., Crutchfield, 1994; Lathrop & Kostelich, 1989). For problems involving the analysis of conversations among people (e.g., Guastello, 2000; Guastello et al., 1998; Pincus, 2001; Pincus & Guastello, 2005), the communications literature is replete with possibilities.

The final step in the analytic process is to isolate the specific strings that have a qualitative meaning for the problem under study. One can either settle on the set of strings associated with C just before H_T reaches its limit, or on a shorter C that is associated with better values of Shannon entropy or χ^2.

One Vector of Mutually Exclusive Categories

Consider the case where a conversation among people in a group has been coded such that each utterance can contain only one category of behavior from the set. This is a very likely situation where the norm of spoken conversation results in relatively brief utterances. Again, the set of categories depends on what is psychologically relevant to the researcher. Give the categories letter codes. The Roman alphabet runs out of letters after 26; so it would be advisable to keep the number of codes less than 26, which is usually more than enough.

At the time of this writing, a computer program is under construction to perform the orbital decomposition analysis. The good news, however, is that the computations can be performed with hand calculations and frequency distributions that can be obtained from a spreadsheet.[1]

For our first example, consider a (fictitious) problem consisting of 20 observations that have been coded A, B, C, D, and E. The normal habit would be to make a column in a spreadsheet consisting of 20 rows. Unfortunately, frequency analysis done through standard programs cannot jump rows when counting frequencies of a pair or triplets of observations; so we made an adaptation. Table 21.1 depicts the 20 observations that have been broken into subsets of 5 observations per row. Next, the second row of 5 observations is copied to the right of the first 5 observations, the third row of 5 observations is copied to the right of the second 5 observations, and so forth. The use of chunks of 5 observations is arbitrary, and it is based on the estimate that C will not exceed 5. Longer chunks can be used if this estimate is not desirable.

The observations and calculations for $C = 1$ are made as follows: There are five code categories A through E with frequencies, as shown in the second column of Table 21.2. Under the assumption of equal probability, the expected frequency of each is 4, as given in the third column. It might have been the case that the research was prepared for more than five codes, but only five were observed, and so only those five are analyzed. On the basis of 20 observations (admittedly few, but illustrative), the observed probabilities of each are given in the fourth column of Table 21.2. The fifth column shows the partial calculations for Shannon entropy, which are eventually summed. The sixth column shows the partial calculations for χ^2, which are eventually summed and multiplied by 2.

To continue, there were two codes that immediately followed themselves, B and D. Thus, $\text{tr}\mathbf{M}^C = 2$. Log_2 of $2 = 1$, divided by $C = 1$ still equals $1 = H_T$. $D_L = e^1 = 2.718$. The values of $\text{tr}\mathbf{M}^C$, H_T, and D_L are written into Table 21.3 along with Shannon entropy (H_s) and χ^2.

TABLE 21.1

Data Organization for Orbital Decomposition
Analysis

Original series

A	B	B	C	B	B	C	D	D	D
B	C	D	D	D	E	A	E	A	D
E	A	E	A	D	E	A	B	B	C
E	A	B	B	C					

Sequences of five repeated observations for easier viewing

A	B	B	C	B	B	C	D	D	D
B	C	D	D	D	E	A	E	A	D
E	A	E	A	D	E	A	B	B	C
E	A	B	B	C					

$C = 1$

A	B	B	C	B	B
B	C	D	D	D	E
E	A	E	A	D	E
E	A	B	B	C	

$C = 2$

A	B	B	C	B	B	C
B	C	D	D	D	E	A
E	A	E	A	D	E	A
E	A	B	B	C		

$C = 3$

A	B	B	C	B	B	C	D
B	C	D	D	D	E	A	E
E	A	E	A	D	E	A	B
E	A	B	B	C			

The observations and calculations for $C=2$ are made as follows: There was exactly one doublet that repeated itself. One would look for sequences such EAEA, etc., to determine that EA is a repeating doublet. For all $C=2$, there were ten different doublets that actually appeared. It is possible to give the doublets a new set of code names, but the present example is simple enough that it should be possible to follow the nomenclature as it is presently given. The expected frequencies of the doublets that appeared one, two, or three times are given individually in Table 21.2. The doublets that appeared only once were aggregated into one category for calculating the components of χ^2. The expected frequency of single doublets is 19 (N^*) minus the sum of all the other expected frequencies (5.605) = 13.395. P_{obs} and $p \ln(1/p)$ are given for each doublet. For the partial χ^2 calculations, $F_{ob} \ln(F_{ob}/F_{ex})$ is given separately for each doublet that appears three times or two times. Again, the doublets that appeared only once were aggregated into one category, which has a group observed and expected frequency.

TABLE 21.2

Interim Calculations for Orbital Decomposition Analysis

Code	Freq.	Expected Frequency	P_{obs}	Shannon $p \ln(1/p)$	χ^2 $F_{ob} \ln(F_{ob}/F_{ex})$
$C = 1$					
A	4	4	0.20	0.323	0.000
B	6	4	0.30	0.361	2.433
C	3	4	0.15	0.285	−0.863
D	4	4	0.20	0.322	0.000
E	3	4	0.15	0.285	−0.863
$C = 2$					
BB	3	1.710	0.158	0.292	1.686
BC	3	0.855	0.158	0.292	3.766
EA	3	0.570	0.158	0.292	4.982
AB	2	1.140	0.105	0.237	1.124
DD	2	0.760	0.105	0.237	1.935
DE	2	0.570	0.105	0.237	2.511
AD	1	13.395	0.053	0.156	−4.834
AE	1		0.053	0.156	
CB	1		0.053	0.156	
CD	1		0.053	0.156	
$C = 2$					
BBC	3	0.243	0.167	0.289	7.540
ABB	2	0.324	0.111	0.244	3.640
DEA	2	0.108	0.111	0.244	5.838
ADE	1	17.325	0.056	0.161	−4.997
AEA	1		0.056	0.161	
BCB	1		0.056	0.161	
BCD	1		0.056	0.161	
CBB	1		0.056	0.161	
CDD	1		0.056	0.161	
DDD	1		0.056	0.161	
DDE	1		0.056	0.161	
EAB	1		0.056	0.161	
EAD	1		0.056	0.161	
EAE	1		0.056	0.161	

To continue, there was one doublet that immediately followed itself, which was EA. Thus, $\text{trM}^C = 1$. Log_2 of $1 = 0$, divided by $C = 2$ still equals $0 = H_T$. $D_L = e^0 = 1.00$. The values of trM^C, H_T, and D_L are written into Table 21.3 along with Shannon entropy (H_s) and χ^2.

The observations and calculations for $C = 3$ are made as follows: There was exactly one triplet that repeated itself. One would look for sequences such

TABLE 21.3

Final Statistics for Orbital Decomposition Analysis

C	trM^C	H_T	D_L	χ^2	df	N*	ϕ^2	H_S
1	2	1	2.718	1.413	4	20	0.071	1.574
2	1	0	1.000	22.339	6	19	1.176	2.207
3	1	0	1.000	24.042	3	18	1.336	2.555
4	0	Undefined						

BBCBBC, etc., to determine that BBC is a repeating triplet. For all $C = 3$, there were 14 different triplets that actually appeared. The expected frequencies of the triplets that appeared one, two, or three times are given individually in Table 21.2. The expected frequency of single triplets is 18 (N*) minus the sum of all the other expected frequencies $(0.678) = (17.933)$. P_{obs} and $p \ln(1/p)$ are given for each triplet. For the partial χ^2 calculations, $F_{ob} \ln(F_{ob}/F_{ex})$ is given separately for each triplet that appears three times or two times. Again, the triplets that appeared only once are aggregated into one category, which has a group observed and expected frequency.

To continue, there was one triplet that immediately followed itself, which was BCC. Thus, trM$^C = 1$. Log$_2$ of 1 = 0, divided by $C = 3$ still equals $0 = H_T$. $D_L = e^0 = 1.00$. The values of trM^C, H_T, and D_L are written into Table 21.3 along with Shannon entropy (H_s) and χ^2. There were no repeating quadruplets in the series of observations; so, for $C = 4$, trM$^C = 0$, is undefined, and the rest of the analysis does not matter.

The solution to this problem would be the set of patterns with $C = 3$ based on Equation 21.3. Consider the statistical features, however. For $C = 1$, χ^2 is nowhere close to statistical significance. It reaches $p < 0.001$ for both $C = 2$ and $C = 3$, however. The ϕ^2 statistic was interesting; unlike the problem situations reported earlier where ϕ^2 exceeds 1.00 for $C = 1$ or for series with dominating single patterns, ϕ^2 was very low for $C = 1$, but exceeded 1.00 for $C > 1$. If one were trying to decide between $C = 2$ and $C = 3$ to determine which set of strings to interpret, ϕ^2 indicated that more deviation from expectation per string was associated with $C = 3$. The two triplets ABB and BBC, which appeared more than once, would be interpreted. The remaining triplets would be ignored. A larger data set, which would have to result for a longer series of real events, could make one or more of the other triplets interpretable.

Multiple Codes and Multiple Categorical Variables

It is not difficult to conceptualize problems in which more than one variable is involved, or in which the categories of behavior in the study could appear simultaneously. Two categories of problem are considered next. In one case, there is one vector of categories, but the categories can appear simultaneously. In the other, there are several categorical variances to consider and apply to each observation. The nuance is in the data setup; the orbital decomposition analysis is not really any different from the example just described.

One Vector of Nonexclusive Categories

Spoken conversations with short utterances could readily produce series of coded data where only one category or response was emitted in one utterance. The hypothetical example just presented could be prototypical of such situations. On the other hand, if the conversation were transpiring by e-mail communications, the writers or speakers would have more opportunity to express more ideas, and span more categories of utterance in one e-mail. Thus, it is possible to conceptualize a coding scheme that could apply to data where each observation is dense with a possible representation of codes.

The coding strategy would be to code each unit of observation for as many codes that could reasonably apply in the context of the research problem. In the prototypic example (Guastello, 2000), the written utterances could be coded for 1 to 9 categories present simultaneously. For the orbital decomposition analysis, the code combinations such as [2,7], [2,4,6], [9], and so forth would be given letter codes A, B, C, etc., and treated as intact units. The computational analysis would then proceed as in the example just presented.

An important finding from this particular research was that the solutions for the optimal C, D_L, and the concomitant descriptive strings were highly dependent on the coding strategy that was applied to the same source data. Two separate coding schemes were actually used. One scheme that involved nine nonexclusive categories resulted in $C = 2$ and $D_L = 1$, which suggests a simple oscillator. The other scheme involved six nonexclusive categories, and was taken from a different theoretical origin. The latter scheme resulted in $C = 3$ and $D_L = 1.948$. Thus, the patterns and any conclusions about the entropy or the complexity of the dynamics are predicated on the system of events that one chooses to view in the data.

Several Vectors of Mutually Exclusive Categories

Spohn (2008) introduced a more complex problem where political events were coded for five different variables and each variable was composed of several categories. The study pertained to collective violence in societies, and the goal was to identify patterns of behavior that could support or refute a theory that described violent actions as a sequence of step functions. The first variable was the originator of the action—the government, a church, an ethic or communal group, an outside party, etc., with 10 categories altogether. The second variable was the type of action—initiating violent action on another group, retaliating, destroying property, making plans for

a violent action, etc., with nine categories altogether. The third variable characterized the target of the action with categories similar to those in the first
variable. The fourth and fifth variables described stages of violentization in
the perpetrator and the recipient of the action. The source information came
from detailed archives of events that occurred in Yugoslavia, Peru, South
Africa, and India.

The analytic strategy for orbital decomposition was not substantially different from the strategy that was applied to written conversations. Each
event was given a set of five codes. Each set of specific codes that was used
was given a substitute letter code A, B, C, etc. The standard analysis took
over from this point.

Data from each country were analyzed separately. Spohn found that the
C and other indicators were different in each country's events, and thus
different nonlinear dynamics were involved. In Yugoslavia, both the $\text{tr}M^C$
and χ^2 criteria supported $C = 2$, with ensuing D_L and interpreted strings. In
Peru, $\text{tr}M^C$ supported $C = 4$, but the χ^2 criterion supported $C = 3$ more strongly.
H_s was also larger for $C = 3$ than it was for $C = 4$; thus, triplets were interpreted.

In South Africa, $\text{tr}M^C$ supported $C = 14$. The event sequence was heavily
dominated by repetitions of violent actions initiated by police. Leaving out
the contribution of this one sequence, the remaining events that contained
some variety of activity were interpreted at $C = 3$.

In Gujarat, $\text{tr}M^C$ supported $C = 5$, but the χ^2 criterion supported $C = 4$.
Strings of four codes were interpreted. The listing indicated that most of
them involved the same event occurring four times in a row before something else happened.

Spohn concluded that the nature of the resulting strings in each case did
support the theory that was driving the study. Her concluding remarks,
however, were that "We can see also how different *kinds* (emphasis added)
of information we glean from whole-pattern analysis are from those of linear statistical models. In the real world, what happens next and what happened last are incredibly important to the pattern, and so it is with the orbital
decomposition's strings" (p. 114).

Conversions of Metrics from Continuous to Discrete Form

Two relatively simple cases have been mentioned parenthetically so far. One
involves taking a continuous measurement and breaking it into discrete categories, and then doing the orbital decomposition analysis from that point.
This form of discretization is not substantially different from what is entailed
in the recurrence analysis (Chapter 11). Another variety involves taking
difference scores and making the same type of discretization; the implied
meanings of difference scores are usually different from those associated

with raw observation, but the analysis is virtually the same. This chapter continues with two of the more challenging types of continuous data that might benefit from the orbital decomposition analysis: electromyographs (EMGs) and functional magnetic resonance images (fMRIs).

EMGs

An EMG is a measure of muscle activity that occurs when a task is performed. The most common form involves the use of skin sensors; electrodes that are more deeply implanted are possible, however. EMG sensors are usually used in combination, such as monitoring what triceps are doing while biceps are activated, but for present purposes, only the data from a single sensor are considered.

The instances of muscle activation are plotted as complex waves. Analytically, they are not fundamentally different from electroencephalographs (EEGs) or electrocardiographs (EKGs). Figure 21.1[1] shows the common format of amplitudes for an EMG activation wave. As with EEGs or EKGs, this type of time series can be analyzed for patterns in amplitudes of time and the dynamical content therein (Aydin & Cecin, 2009; Rodrick & Karwowski, 2006). They can also be converted to spectral frequency data, which, in essence, depicts the difference between peaks; higher frequencies display shorter times between peaks, and lower frequencies consume more time between peaks. EKGs have been analyzed both ways with interesting results in each format.

The typical research problem involves a focus on a type of activity and the change in activation patterns over time, such as what might be associated

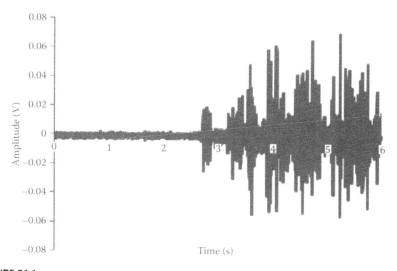

FIGURE 21.1

Raw EMG data showing the transition from a resting point to a reach-and-grasp action.

with fatigue or parts of an action that look and feel like a single integrated action. The participants in such a study would probably repeat an action many times to produce a data file that is characteristic of the action. The conventional form of analysis would involve determining the most commonly occurring frequency values and plotting that range as an "envelope" of over time. This approach is not fundamentally different from audio signal processing where a characteristic wave envelope is constructed for a particular musical sound. There is usually a particular interest in the shape of the early part of the envelope that is associated with the onset of the action, the middle portion where the action is sustained, and the last portion where the action ends. Any frequencies captured in the EMG that are not part of the envelope of central values are interpreted as noise, and the noise is not regarded as meaningful.

In light of what we now know about noise, the traditional assumption about infrequent frequencies was investigated for any inherent nonlinear dynamics (Nathan, Guastello, & Jeutter, 2009). Figure 21.2 shows the distributions of frequency by amplitude for five human subjects who repeated a reach-and-grasp task three times; only one EMG sensor was involved. A frequency distribution was made that contained the 15 replications of the task data. The 26 rarest values were isolated for further analysis; the use of 26 rare values was arbitrary, but based on the convenience of using only the 26 letters of the alphabet without resorting to additional symbols. The rare values ranged from 8 to 500; rounding reduced them to 19 values.

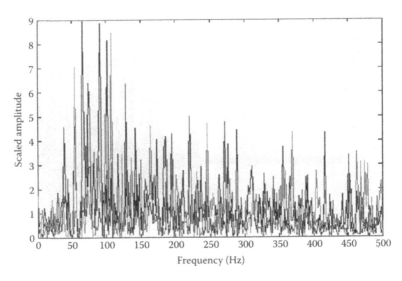

FIGURE 21.2
EMG data for five participants repeating one task three times, transformed to a plot of amplitude versus spectral frequencies.

The 19 values were compiled into bins of 40, which reduced the number of codes needed to 11.

The 11 codes A through K were observed again in the order in which they appeared over the 15 time series. Orbital decomposition proceeded from this point, with the usual calculations. The results were remarkable for both their simplicity and complexity simultaneously. No repeated codes were observed for $C = 1–3$. Repetitions were observed for $C = 4$, and none for $C = 5$ or beyond. The solution thus rested unambiguously at $C = 4$. There were actually four sequences of four values that appeared two or more times. H_S and H_T were also maximal at $C = 4$; $H_T = 0.5$ and $D_L = 1.65$.

The results for C, H_T, and D_L for the 15 replications of a task were further replicated for 8 other tasks where the participant reached and grasped objects of different shapes and orientations ($107.84 < \chi^2 < 157.54$; $0.97 < H_S < 0.99$). The specific frequencies that were associated with the surviving strings of four frequencies were not the same for all reach-and-grasp tasks, however. ϕ^2 was greater than 1.00 in each case; so no attempt was made to interpret this statistic. Under the circumstances, however, it would not have changed any conclusions about the optimal description of strings.

One important point of the study was that it illustrated how physiological events can emerge with patterns that cannot be reduced to simpler patterns (of smaller C). Another important point is that the range of spectral frequencies that traditionally have been interpreted as noise actually contains some deterministic structure. The specific meaning of the sequences of four frequencies cannot be fully deciphered on the basis of one sensor. Thus, the substantive interpretation was left for further research involving a set of sensors, and a report should be forthcoming eventually.

fMRIs

From a theoretical point of view, much of the neuroscience of the twentieth century has focused on either broad areas of the brain, at the scale of the hippocampus or temporal lobe, or the activities of individual neurons. The contemporary questions, however, are concerned with meso-level circuits that often run across the main topological structures of the brain (Freeman, 2000).

fMRIs are a noninvasive means of detecting neural activity in specific locations in the brain. The equipment allows for repeated scans while the study participant or patient is performing a task, thus making the system friendly to time series analysis. Analytic methods for isolating patterns of responses are still needed, however (Bianciardi et al., 2007; Esposito et al., 2005), and the use of orbital decomposition was suggested as a plausible alternative (Guastello, Nielson, & Ross, 2002).

In an fMRI, areas of the brain "light up" on the two- or three-dimensional images in response to oxygenation that is produced during neural activity.

Lights of different colors denote levels of activation for a specific area. The scan parses the brain's activities into a three-dimensional rectangular grid of $64 \times 64 \times 29$ *voxels*. The typical uses of fMRIs involve compiling all images from the same task, and also from multiple human participants into a single set of images that can be viewed as a three-dimensional object or as any of 29 two-dimensional slices. Thus, the capability of time series analysis is not exploited at the level of assembling images.[3]

There is a new principle that makes the organization of the data for the analysis different from the examples considered thus far: Instead of simply regarding the time series as one containing orbits tangled within other orbits over time, they are tangled together across *space* and time. Each voxel would be an observation containing four parameters: *I*, *LR*, *FB*, and *V*. *I* would be the intensity of activation. *LR* would be the coordinate of the voxel also on a range of 1–64 from left to right. *FB* would be the coordinate of the voxel on a range of 1–64 from the front of the head to the back. *V* would be the coordinate of the voxel on a range of 1–29 from the bottom of the scanned brain area to the top.

The mock time series would organize the voxels in the specific order of [*, 1, 1, 1], [*, 2, 1, 1] … [*, 64, 1, 1], [*, 1, 2, 1] … [*, 64, 2, 1] … [*, 64, 64, 1], [*, 1, 1, 2] … [*, 64, 64, 29]. In other words, the voxels would be entered from left to right, beginning with the frontal row on the lowest slice. Each row would be sequentially filled in until the lowest slice was filled, and then the next vertical slice begins. When the last vertical slice is completed, the data from the next scan starts.

The challenge that immediately arises is that there are 118,704 voxels per scan. The data set can be substantially reduced by eliminating voxels that do not activate on any of the scans in the series. The data set can be simplified further by raising the threshold for determining what is really active. The analysis that ensues could be simplified further by limiting the number of activation levels of interest. More than one activation level is probably desirable, however, because activation areas often have epicenters, and the distinction between an epicenter and a more peripheral basin could be biologically meaningful.

Unfortunately, the story ends abruptly just as it starts to become exciting. Research on the application of the orbital decomposition technique to fMRI data has been stalled for two reasons. One is the need to write a program that can fathom the computational intensity and interface it with the data file system of the standard fMRI software. The other is a practical problem for psychological and psychomotor research, which is that the fMRI machines require 2 s to make a full scan, whereas the neurological events of interest transpire in substantially less time. Thus, experimental designs that capture the mini-episodes are also needed. Nonetheless, the orbital decomposition concept can be logically extended to complex data with this type of structure.

Notes

1. The analysis was actually easier in the old-style mainframe version of SPSS. In that system, all the data showed in Table 21.1, part 2, would be entered into an ASCII file. The SPSS control program would then contain format statements that defined variables as two-characters wide, three-characters wide, etc., without the need to reformat the data itself into spreadsheet fields.
2. The author thanks Dominic Nathan for producing Figures 21.1 and 21.2.
3. Nonlinear regression is used intensively, however, to determine if an activation color should be applied to a particular voxel; however, that is a different set of questions.

References

Bianciardi, M., Sirabella, P., Hagberg, G. E., Giuliani, A., Zbilut, J. P., & Colosimo, A. (2007). Model-free analysis of brain fMRI data by recurrence quantification. *NeuroImage, 37*, 489–505.

Cecen, A. A., & Erkal, C. (2009). The long march: From monofractals to endogenous multifractality in heart rate variability analysis. *Nonlinear Dynamics, Psychology, and Life Sciences, 13*, 181–206.

Crutchfield, J. P. (1994). The calculi of emergence: Computation, dynamics, and induction. *Physica D, 75*, 11–54.

Crutchfield, J. P. (1996, June). *Is anything ever new: Discovering hidden order in chaos.* Invited lecture to the 6th annual international conference of the Society for Chaos Theory in Psychology & Life Sciences, Garden City, New York.

Esposito, F., Scarabino, T., Hyvarinen, A., Himberg, J., Formisano, E., Comani, S., et al. (2005). Independent component analysis of fMRI group studies by self-organizing clustering. *NeuroImage, 25*, 193–205.

Freeman, W. J. (2000). *Neurodynamics: An exploration of mesoscopic brain dynamics.* New York: Springer-Verlag.

Goertzel, B., (1995). A cognitive law of motion. In R. Robertson & A. Combs (Eds.), *Chaos theory in psychology and the life sciences* (pp. 135–154). Mahwah, NJ: Lawrence Erlbaum Associates.

Guastello, S. J. (2000). Symbolic dynamic patterns of written exchange: Hierarchical structures in an electronic problem solving group. *Nonlinear Dynamics, Psychology, and Life Sciences, 4*, 169–188.

Guastello, S. J., Hyde, T., & Odak, M. (1998). Symbolic dynamic patterns of verbal exchange in a creative problem solving group. *Nonlinear Dynamics, Psychology, and Life Sciences, 2*, 35–58.

Guastello, S. J., Nielson, K. A., & Ross, T. J. (2002). Temporal dynamics of brain activity in human memory processes. *Nonlinear Dynamics, Psychology, and Life Sciences, 6*, 323–334.

Keppel, G., Saufley, W. H., & Tokunage, H. (1992). *Introduction to design and analysis* (2nd edn.). New York: Freeman.

Lathrop, D. P., & Kostelich, E. J. (1989). Characterization of an experimental strange attractor by periodic orbits. *Physics Review, 40,* 4028–4031.

Nathan, D. E., Guastello, S. J., & Jeutter, D. (2009, July). *Exploring EMG pattern detection using symbolic dynamics during the performance of functional upper extremity tasks.* In Paper presented to The 19th Annual International Conference of the Society for Chaos Theory in Psychology & Life Sciences, Milwaukee, WI. Retrieved, January 15, 2009, from www.societyforchaostheory.org/conf/2009/abstracts2009.pdf

Nicolis, G., & Prigogine, I. (1989). *Exploring complexity.* New York: Freeman.

Ott, E., Sauer, T., & Yorke, J. A. (Eds.). (1994). *Coping with chaos.* New York: Wiley.

Parchman, M., & Katerndahl, D. (2009, November). *Dynamical differences in patient encounters involving uncontrolled diabetes in which treatment was and was not altered.* In Paper presented to American Primary Care Research Group, Montreal, Canada.

Pincus, D. (2001). A framework and methodology for the study of non-linear, self-organizing family dynamics. *Nonlinear Dynamics, Psychology and Life Sciences, 5,* 139–174.

Pincus, D., & Guastello, S. J. (2005). Nonlinear dynamics and interpersonal correlates of verbal turn-taking patterns in group therapy. *Small Group Research, 36,* 635–677.

Shannon, C. E. (1948). A mathematical theory of communication. *Bell System Technical Journal, 27,* 379–423.

Rodrick, D., & Karwowski, W. (2006). Nonlinear dynamical behavior of surface electromyographical signals of biceps muscle under two simulated static work postures. *Nonlinear Dynamics, Psychology, and Life Sciences, 10,* 21–36.

Spohn, M. (2008). Violent societies: An application of orbital decomposition to the problem of human violence. *Nonlinear Dynamics, Psychology, and Life Sciences, 12,* 87–115.

22

Orbital Decomposition for Multiple Time-Series Comparisons

David Pincus, David L. Ortega, and Annette M. Metten

CONTENTS

Statistical analyses that focus on categorical data tend to be underdeveloped compared to other methods. Yet, psychological phenomena are replete with processes of categorical change (Long, 1997). This disparity is reflected strongly within the context of nonlinear patterns of change in categorical time series as well. A clear example of this methodological shortage exists within family therapy research, where flows of categorical change during family conversations are of critical focus to family therapists, yet have remained off limits to most researchers. At the same time, static, independent, and readily quantifiable data tend to be of focus to researchers, yielding results that are less relevant to clinicians (Snyder & Kazak, 2005).

There have been some outstanding recent exceptions to this general rule (see Snyder & Kazak, 2005, for a recent review), with new methodologies including the use of nonlinear differential equations to model marital interactions (Gottman, Murray, Swanson, Tyson, & Swanson, 2002), the use of *state-space grids* (Lewis, Lamey, & Douglas, 1999) to understand family rigidity (Granic, Hollenstein, Dishion, & Patterson, 2003; Hollenstein, Granic, Stoolmiller, & Snyder, 2004), and the use of log-linear modeling innovations to understand multilevel dynamics in family therapy (Howe, 2004; Howe,

Dagne, & Brown, 2005). Each of these methodological advances focuses on the study of conversations, which are inherently nonlinear (involving sudden and disproportionate changes), dynamical (unfolding over time with serial dependence), and systemic (involving mutual influences among individuals). This chapter provides an introduction and empirical demonstration for the use of orbital decomposition (OD; Guastello, Hyde, Odak, 1998; Guastello, 2000), a procedure designed specifically for the purpose of identifying and quantifying nonlinear patterns within categorical time series. Building on the use of the approach to analyze a single time series, this chapter provides an extension of the approach for comparing multiple time series. The goals of this chapter are (a) to provide step-by-step guidance in the use of OD for time-series comparisons from data collection through interpretation; (b) to provide a conceptual understanding of theory underlying the approach to allow for adaptation of the approach to different research contexts; and (c) to provide a detailed illustration of how the approach may be applied to the analysis of conversation patterns, one particularly well-suited domain of inquiry for the method.

Orbital Decomposition: Empirical Demonstration and Conceptual Discussion

Beyond the task of analyzing conversation patterns, OD may be applied broadly to various situations involving changing states across time or space. For example, Guastello, Nielson, and Ross (2002) have proposed a means of using OD to study activation patterns across the brain by way of three-dimensional fMRI, and Spohn (2008) has applied the method to the examination of large-scale patterns of violent action across a range of civil conflicts. Indeed, the steps followed below may be applied to any data set involving the comparison of more than one sequence of nominal states changing over time.

The current example analysis uses the coded sequential utterances from five sessions of family therapy to derive the following outcomes: (a) the appropriate string length (C) for analysis, (b) the degree of randomness in each discussion (using information entropy, H_i), (c) the degree of deterministic chaos within each session across different string lengths (Lyapunov dimension, D_l), (d) the structural complexity (fractal dimension, D_f), (e) the structural integrity (R_t^2) of each session, (f) standardized stationarity (S_t) values for each session, and (g) the contribution of each group member to structure within sessions. Table 22.1 contains brief step-by-step instructions for the entire analysis.

Aspects of the Data: Sampling Options, Coding, and Reliability

Researchers planning to use OD to compare categorical change over time will first need to derive mutually exclusive and exhaustive categories of system

TABLE 22.1

Step-by-Step Instructions for Carrying Out an OD
Analysis on a Family Discussion

- Code conversation(s) (nominal time series)
  - that is, speaker $1 - n$
- Enter data in a column in SPSS
- Make Ns for each discussion equal study-wise (or use metric
  entropy instead of Shannon below)
- Expand string length (C) in subsequent columns $(2 - j)$
  moving one utterance at a time (i.e., 13, 132, 1323, 13234, ...)
- Run frequencies on longer strings experiment-wise to
  determine optimal length (longer is better)
  - Longest proximal recurrence
  - χ^2 analysis
  - Predetermined value for dispersion of recurrence frequencies
    (i.e., mini–max recurrence >5, at least 25% of recurrences)
- Compute frequencies (f), probabilities, Shannon entropy
  values, and number at each frequency (nf) for all patterns and
  enter as variables in data file (using compute and recode
  functions in SPSS)
- Calculate value(s) for Lyapunov dimension* (based simply on
  number of distinct proximal recurrences)
- Sum values for Shannon entropy to obtain values for each
  discussion
  - Run comparison stats as desired (i.e., ANOVA)
- Run a "curve fit," "power" regression model with nf as the DV
  and f as the IV
  - Lead beta-weight (slope of curve) in model is fractal
    dimension
  - Fit index (R^2) may be used as a measure of system integrity
- Examine stationarity within sessions using a linear regression
  of frequency over time
- Run any other analyses as desired—that is, discriminant
  analysis, state-space grids, linear regression analyses with
  recurrence as DV
- Enjoy!

states. As with any coding scheme, reliability will need to be demonstrated,
typically inter-rater reliability using Cohen's Kappa or some similar statistic.
Next, one must decide whether one is going to use a time-sampling strategy
or an event-sampling strategy. In time-sampling, change is defined based
on set units of time, where time intervals in the series are held constant. By
contrast, event-sampling examines the unfolding of sequential events. For
example, imagine that a researcher is interested in comparing movement
among dominant brain-wave frequencies (EEG, or electroencephalograph)
for individuals who have received different treatments for insomnia. One
would begin by defining four mutually exclusive (i.e., states do not overlap)

and exhaustive (i.e., all possible states of the system are included) states (i.e., alpha, theta, delta, and beta). Time-based sampling of these states would involve the selection of an optimal length of time at which to capture change among these states. If one selected a 30 min window, for example, one would be at risk of under-sampling, as several brain-wave patterns may unfold within this period of time. Important change processes may be missed if one under-samples the series. Conversely, one risks over-sampling by using too short a time window, such as 1 s or less. Here, dominant brain-wave frequencies may appear very stable across such a short window. One should have a clear empirical rationale for selecting time-based sampling and also for determining the optimal scale of change for a particular process to avoid under- or over-sampling.

Event-based sampling involves tracking actual change in series, without respect to set time intervals. The series is examined on a continuous basis and any change is recorded in a simple sequential manner, without respect to duration. There are several practical advantages to either method of sampling. For example, time-based sampling will produce series with equal Ns when series are measured for the same amount of time, while event-based sampling will not. Yet, in event-based sampling, there is less risk of over- or under-sampling, as each change is captured. However, there is a loss of potentially interesting information, namely, the duration in which the system maintains any particular state. When using event-based sampling, there is no distinction between a participant who remains in an alpha brain-wave state for 30 s or for 30 min. Therefore, if the duration is of great importance, one would likely choose to use time-based sampling. The selection of either will depend upon the phenomenon of interest and goals of the study.

The data used for the statistical demonstrations in this chapter originated from the coded turn-taking interactions within five sequential family therapy sessions (sessions 2–6 of functional family therapy; Alexander & Parsons, 1982). Session 1 was not included as it was primarily involved with paperwork. The four participants in the sessions were members of an intact Caucasian family: father, mother, and adolescent daughter, along with their family therapist. The sequential turn-taking responses were coded from a videotape using specific heuristics to determine which verbalizations were sufficient to count as an utterance (i.e., single words such as "Okay" were sufficient) and how to code order in overlapping verbalizations (see Pincus & Guastello, 2005, for the actual coding scheme). In addition to the coding scheme, visual displays of amplitudes provided assistance in identifying speaker changes. This coding scheme is relatively simple, and generates four mutually exclusive and exhaustive system states defined based on who is speaking at any given point in time. However, even in such a simple system, there may be some information lost as one derives categories. For example, in conversations, there are frequent talk-overs, where more than one person speaks at once. Indeed, researchers should be mindful of the coding process, as information will invariably be lost in comparison to the pure phenomenon unfolding over time. Of greatest practical concern in research involving comparisons among multiple time series will be the possibility that there was a systematic bias in coding effects. For example, if one conversation had many more talk-overs than another, their coding reliabilities may differ.

One graduate student in clinical psychology coded all sessions, from 2 to 6, and a second coder (also a doctoral candidate) coded an 11 min segment selected at random (beginning at minute 6 of session 5) to test for inter-rater reliability for speaker and utterance duration (the latter of which was not used in the current study). Each coder was blind to the plan to examine the discussions for patterned recurrences in turn-taking. The agreement between coders was high based on Cohen's Kappa ($K = 0.85$), similar to past research using the same coding scheme ($K = 0.82$, Pincus & Guastello, 2005).

The choice to use speaker codes alone, without regard to content, is based on prior research demonstrating validity of turn-taking as a reflection of underlying relational processes (Pincus, Fox, Perez, Turner, & McGee, 2008; Pincus & Guastello, 2005), and the practical relevance of turn-taking in most forms of family therapy (see Pincus, 2001, for an integrative model based on family response patterns). To maintain consistency in the number of utterances across sessions, only the first 218 utterances were used. This cutting off of the tail end of the series is one way to maintain equal Ns when event-based sampling is used, which would not be necessary if time-based coding had been used.

Measuring Turbulence: Topological Entropy and Lyapunov Dimensionality

One begins the analysis by *slicing* the time series into segments of increasing lengths (C). In research involving a single discussion, one may select the optimal C based on the length at which the longest proximal recurrence occurs along with a ϕ_C^2 value to examine the relative structure at longer Cs (see Chapter 23). The ϕ_C^2 statistic is analogous to R^2, representing the variance accounted for in frequency data. As such, it may be used as a measure of deterministic structure (versus random noise) present within a categorical data set. Past research using OD to examine a single discussion has shown a convergence between a local peak in ϕ_C^2 values (as C increases) and the longest proximal recurrence in the discussion (Guastello, 2000, Pincus, 2001). However, because the current data set involves comparisons among five discussions, one will expect the longest proximal recurrences to vary across sessions. Indeed, the presence of different optimal string lengths may be of particular theoretical interest. For example, Spohn (2008) found that string lengths varied across patterns of violence in countries with different governing types. Within the current analysis, C was kept constant for the purpose of comparing turbulence across sessions.

Therefore, an empirically based heuristic was used to find the optimal C to apply equally across each of the series. The heuristic should balance the ability to find structured recurrences in the discussion (limited by longer strings) with the increased clinical and empirical interest of longer patterns. For the current analysis, C is increased with the limits such that (a) the minimum–maximum (mini–max) recurrence for each session remains greater than or equal to 5 and (b) at least 25% of the patterns in any session occur only one time (Pincus et al., 2008). Conceptually, this heuristic maintains an analytic window through which the frequency distribution of patterned recurrences may be optimally examined, without losing its shape (if C is too short) or becoming flat (if C is too long).

The longest C at which the criteria above are met is C = 6, with the mini–max recurrence of 6 (pattern: Fa-Mo-Fa-Mo-Fa-Mo) occurring in session 5 and the lowest percentage of one-time patterns equal to 32% observed in session 3. Using C = 6, each series is *decomposed* into patterns moving one utterance at a time across the entire series. These patterns within the time series then become the subject of study, analogous to participants in a group-based research design.

The first quantitative measures to be derived are *topological entropy* (H_t) and its mathematically transformed metric, *Lyapunov dimensionality* (D_l). These measures describe the degree of deterministic (i.e., nonrandom) turbulence for a phenomenon that transpires over time or space (Guastello, 2000).

In order to measure topological entropy in a categorical time series, Guastello (2000) translated the concept of *periodic orbits* from Lathrop and Kostelich's (1989) physics methodology into proximal recurrences (immediately repeating patterns within the conversation) of categorical states. By equating periodic orbits and proximal recurrences, one applies a mathematical formula for topological entropy (H_t) to categorical time series:

$$H_t = \lim_{C \to \infty} \left(\frac{1}{C} \right) \log_2 \mathrm{tr}\,(M_c) \qquad (22.1)$$

where C is the string length within the time series. Guastello (2000) defines proximal recurrences as *the trace of the transition matrix of options* (M_c). Theoretically, this transition matrix is square and contains, on its perimeter, every possible categorical outcome of length C.

Lyapunov dimensionality (D_l) is then represented by

$$D_l = e^{H_t} \qquad (22.2)$$

The calculations in Equations 22.1 and 22.2 are repeated where string lengths are widened from 1 to C.

One may gain a better conceptual understanding of the meaning of periodic orbits and topological entropy by imagining the movement of a marble in basin. Imagine a marble rolling inside a large bowl, for example. If one adds energy or noise to the system (i.e., shaking the bowl) the marble will more or less circle around the edges of the bowl, rolling round and round in a single periodic orbit. This relatively simple behavioral dynamic is known as a limit cycle, and the underlying attractor is called a dampened oscillator (see Guastello, 1995, for a review of attractor dynamics). Next, imagine that the bowl has a more complex basin, specifically several lower areas within the basin as in a single large bowl with three distinct bottoms. Such a basin will result in more complex trajectories by the marble, with each of the lower areas becoming observable through the orbits of the marble around those areas. Just as the name suggests, orbital decomposition "decomposes" a time series into its component orbits. Furthermore, by examining orbits at different string lengths, one obtains a variety of orbits nested within larger orbits, analogous to the movement of a tiny marble across a basin containing bowls within bowls within bowls. In small group interactions, flows of information measured as verbal turn-taking are analogous to the movement of the marble across this deterministic surface. With increasing varieties of proximal recurrences in family turn-taking patterns, one may infer more complexity in the underlying relationships among the group members (Pincus & Guastello, 2005).

Examining the longest values for C and the D_l with C fixed, one finds that the longest proximal recurrences for sessions 1–5 were 8, 16, 16, 6, and 8, respectively (see Table 22.2), suggesting far more coordination and structure in the turn-taking process during sessions 2 and 3 compared to the other sessions. At the same time, the D_l values across the five sessions (with $C = 6$) were 1.396, 1.649, 1.396, 1.181, and 1.302, respectively, together suggesting that session 2 also involved a relatively high degree of turbulence. Along with these quantitative indices, a qualitative analysis of member involvement indicates that the longest proximal recurrences in sessions 1–4 centered around the therapist going back and forth with the mother (i.e., Tx-Mo repeating 16 times in session 1) and daughter (Tx-Dtr repeating 32 times in a row in sessions 3 and 4, and Tx-Dtr repeating 12 times in a row in session 4). In contrast, the longest proximal recurrence in session 5 was Fa-Dtr-Mo-Tx-Mo-Dtr-Mo-Dtr, a qualitatively richer and more complex pattern than long strings involving only two members. One may hypothesize that this would be a positive index of a change in the therapeutic process beginning in session 5.

Information Entropy: Interactive Randomness and Novelty

Another entropy measure that may be calculated using OD is *information entropy*, a measure originally derived from Shannon and Weaver's (1949) theory of information (see also Attneave, 1959, and Chapter 9). Information entropy was designed to measure the novelty in categorical change processes in terms of bits, with one bit defined as *the amount of information required to determine a dichotomous outcome* (Shannon & Weaver, 1949). From this definition, information may be used as a measure of the degree of uncertainty related to a categorical outcome. For example, if an outcome is certain, then

TABLE 22.2

Distinct Proximal Recurrences (R_p) at Increasing String Lengths
from $C = 4$ to $R_p = 0$, Resulting Values for Topological Entropy (H_t)
and Lyapunov Dimension (D_l)

C	Session 1 Proximal Patterns	R_p	H_t	D_l
4	1212, 2121, 3131, 1313, 1414, 4141, 3232	7	.702	2.017
6	313131, 131313, 141414, 212123,	4	.333	1.395
8	31313131, 13131313	2	.250	1.284
	Session 2 Patterns			
4	1212, 1313, 1414, 2121, 3131, 3141, 4141	7	.702	2.017
6	141414, 414141, 121212, 212121, 131212, 131313, 313131, 312121	8	.500	1.649
8	14141414, 41414141, 12121213, 21212131	4	.250	1.284
10	1414141414, 4141414141	2	.100	1.105
12	141414141414, 414141414141	2	.083	1.087
14	14141414141414, 41414141414141	2	.071	1.074
16	1414141414141414, 4141414141414141	2	.063	1.065
	Session 3 Patterns			
4	1414, 4141, 1212, 2121, 1313	5	.581	1.788
6	141414, 414141, 121212, 212121	4	.333	1.395
8	14141414, 41414141, 14141412,	3	.198	1.219
10	1414141414, 4141414141	2	.100	1.105
12	141414141414, 414141414141	2	.083	1.087
14	14141414141414, 41414141414141	2	.071	1.074
16	1414141414141414, 4141414141414141	2	.063	1.065
	Session 4 Patterns			
4	4141, 1414, 2421, 2424, 4212, 1242, 2121	7	.702	2.017
6	414141, 141414	2	.167	1.182
	Session 5 Patterns			
4	2121, 1212, 1313, 3131, 4141, 3431	6	.646	1.908
6	121212, 131313, 124212	3	.264	1.302
8	24313434	1	0	1

Nominal codes: therapist = 1, father = 2, mother = 3, and daughter = 4.

no bits of information are required to predict the outcome; while, conversely, if an outcome is truly random, then exactly one bit is required. Contemporary NDS (nonlinear dynamical system) models have used information entropy in the converse, as a measure of novelty and complexity (Guastello, 2000; Prigogine & Stengers, 1984).

As a concrete example, imagine that one is trying to identify a word in a crossword puzzle based on individual letters. The more novel a letter is, such as z or x, the more information it carries. Thus, the presence of a z or an x in a crossword is very helpful toward finding the correct word because of the fact that it is used so infrequently. A *c* or a *b* on the other hand would not provide much information to a puzzle solver because these letters are so common. So long as we know there is an actual word there to be found (a deterministic process of some sort), novel letters actually convey more information than the more common letters. As such, with increasing novelty in a categorical sequence, information entropy will grow larger. The information entropy (H_i) for a set of categories with unequal odds of occurrence is

$$H_i = \sum_{i=1}^{r} p_i \left[\ln\left(\frac{1}{p_i}\right) \right] \qquad (22.3)$$

where *p* is the probability associated with each ($i = 1$ to *r*) categorical outcome of the observation of interest (Guastello, 2000; Shannon & Weaver, 1949). Essentially, the result of Equation 22.3 is a weighted average of the degree of novelty present within a categorical time series.

With the patterns at $C = 6$ entered as a column in SPSS (Statistical Package for the Social Sciences), it is now possible to calculate information entropy (H_i) using Equation 22.3. One need only compute the frequencies (f) for each pattern (entered as a new variable in the data set), divide by the total number (i.e., 218) of utterances to obtain each pattern's probability (entered as a new variable), calculate the information entropy value for each pattern based on Equation 22.3 (entered as a new variable), and then sum these values.

Following these steps, the information entropy values for sessions 1–5 are 4.58, 4.13, 3.75, 4.36, and 5.01, respectively. As information entropy values in the current study (based on $N = 218$ and $C = 6$) have a possible range from 0 (if two speakers spoke back and forth for the entire session) to 5.38 (if there were no recurring patterns), these sessions appear to be generating a moderate (session 3) to moderately high (session 5) degree of information overall, with a progressive increase in the structure during the first three sessions of therapy, followed by a return to baseline in session 4, and then a jump to higher novelty in session 5. If one wished to more easily interpret these data based on their possible range, or compare information entropy values for this sequence of family therapy sessions with other sessions (i.e., different therapist, different family, and different Ns or Cs) or even with other types of natural systems, one could simply divide by the highest possible value (5.38) to obtain scores ranging from 0 to 1, in this case 0.851, 0.768, 0.697, 0.810, and 0.931, respectively. One may interpret these values as reflecting the percentage of volume filled by the hypothetical space defined by all the possible communication states of the family system: the family's *phrase space* (Butz, Chamberlain & McCown, 1997).

Inferential comparisons among sessions may be carried out based on the frequency data (i.e., χ^2 analyses) for each session, or based on the sample of H_i values for each session. The latter was selected for the current demonstration. A one-way analysis of variance (independent samples) of the entropy values across discussions suggested that the mean information entropy values across discussions were not equivalent ($F = 41.031$, $df = 4$, $p < 0.001$; see Figure 22.2). Nor were the variances across discussions equivalent (Levine test = 64.577; $df = 4, 1085$; $p < 0.001$). This result is not surprising as *error* is deterministic within self-organizing systems due to their interactive nature, not stochastic as in normal distributions comprised of noninteractive random observations. This *dependent error* (as opposed to random error as is assumed under the general linear model) will have a greater impact on analyses involving discussions at the extremes of information entropy, low and high. In each case, the variance in entropy values for each utterance will approach zero. By contrast, entropy values in the moderate range display the highest variance and as such a more stringent test of mean differences. In situations where extremely high or low H_i values are observed, one may choose to analyze frequencies rather than H_i values to compare the structure and complexity across sessions to bypass the problem of correlated means and variances. With H_i values ranging from moderate to high moderate in the current study, the Games–Howell statistic for post hoc comparisons among distributions with unequal variances was used; it indicated that entropy decreased across sessions 1 and 2 ($p < 0.001$), dropped further from session 2 to 3 ($p = 0.024$), rose from session 3 to 4 ($p < 0.001$), and rose again from session 4 to 5 to the highest level of all sessions ($p < 0.001$; see Figure 22.1). Researchers interested in evaluating the psychotherapy process and

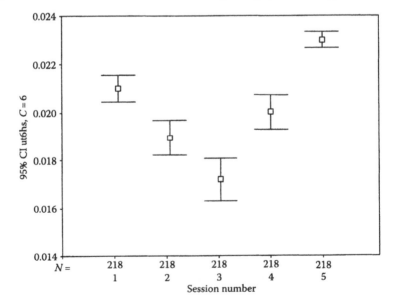

FIGURE 22.1
Error bar graph of information entropy values for patterns at $C = 6$ across five sessions of family therapy.

outcome could examine the relationships between information entropy and a number of family processes and outcomes (i.e., power, affiliation, conflict, and response to treatment).

Complexity, Structural Integrity, Inverse Power Laws, and Fractal Dimension

A final measure of system complexity that may be derived from OD is *fractal dimension* (D_f), which is defined as the slope of the curve defined by the number of different patterns (*y*-axis) at each level of recurrence (*x*-axis). Along with D_f, one may analyze the integrity of the relational system based on the fit of the resulting curve to an inverse power law (IPL) using the statistical model:

$$Y = aX^{-b} \tag{22.4}$$

where
  X is the number of recurrences for a given pattern (analogous to the *magnitude* of the recurrence phenomenon)
  Y is the frequency at which one observes each particular value of recurrence
  a is an intercept
  b is a nonlinear regression weight representing the shape of the IPL curve, which can be used as an estimate of fractal dimension (see Pincus & Guastello, 2005, for a more detailed explanation)

The discovery of IPLs in conversation patterns is interesting because IPLs are ubiquitous in self-organizing systems from across the natural sciences. Such systems generate patterns of change involving an exponentially greater number of small events compared to large events (e.g., plants, neuronal structure, neuronal connectivity, internet connectivity, bronchial tubes, rivers, and fault lines), producing branchlike structures known as fractals (Bak, 1996; Mandelbrot, 2004; West & Deering, 1995). It is expected that each of the family therapy sessions in the current analysis will produce an IPL as well consistent with prior studies of small group interaction (Pincus & Guastello, 2005; Pincus et al., 2008).

Steeper curves within an IPL (i.e., higher fractal dimension values) indicate a higher ratio of low recurrence patterns compared to high recurrence patterns, thus higher complexity, while shallow curves (i.e., lower fractal dimension values) indicate a greater structure (i.e., repetition) within the time series, thus lower complexity. Furthermore, the degree of the IPL model fit (R_i^2) may be used as a measure of integrity. This R_i^2 value reflects the degree to which the frequency distribution of the various patterns (or strings) conforms to a power law. So long as one can assume equivalence in coding reliability across sessions, more ill-fitting conversations may be interpreted as containing deformities in the fractal structure of the conversation. If one imagines, for example, visually the fractal branching structure of a tree, one example of such a deformity is having too many medium-sized branches, compared to small or large ones.

Trees with more small branches relative to large ones are more structurally complex, whereas trees with fewer small branches have a simpler and more rigid structure. An idealized fractal tree with high structural integrity will display a smooth exponential relationship between the size and frequency of branches from trunk to tips, regardless of the level of complexity. Trees lacking in structural integrity, displaying a lack of balance between large and small branches, will appear warped or distorted. Inasmuch as conversation patterns reflect important interpersonal processes (Pincus & Guastello, 2005), and their fractal structure serves an adaptive role in the development of the group (Pincus et al., 2008), one may hypothesize that the integrity of the fractal structure would have adaptive significance. For example, discussions with lower R_i^2 values may be considered to be displaying a lack of balance between high- and low-frequency-patterned recurrences in turn-taking. Theoretically, this could be expected to occur in groups containing cliques or other forms of splitting subgroups, or it could occur in groups with an abundance of complex coalitions, leading to shifts in alliances, conflicts, and leadership across a discussion. Or it could even occur in groups whose dynamics change across a conversation, for example, shutting down around a particular topic.

Plotting IPLs and calculating the fractal dimension may be done simply in SPSS, using the frequency distribution for the patterns within each session. Using the frequencies (f) variable that was used above to calculate information entropy, one next computes the frequencies of each frequency or the number of different patterns at each frequency (N_f). One should note that if the output of the *descriptives* command in SPSS is used for this calculation, it is necessary to divide each output frequency by the original pattern frequency so that the same pattern is not counted more than once. For example, if there are 20 patterns that recur in the session four times, there are in fact only five *different* recurring patterns ($N_f = 5$) at this level of recurrence ($f = 4$). To complete the analysis in SPSS, one selects a *regression* analysis using *curve estimation* with the selected model as *power* (i.e., power law), with N_f entered as the dependent variable and f as the independent variable. The output includes a fit index (R_i^2), a standardized beta-weight (the absolute value of the negative B-value) representing the shape of the IPL curve (D_f), and finally standard errors of the estimate (*SE B*), which may be used for inferential testing. However, in situations where the fit to the IPL is extremely high, as in the current analysis, these standard error values will be too small to practically allow for estimated IPLs to overlap.

The D_f and R_i^2 and values for sessions 1–5 are listed along with the frequency distributions for a patterned recurrence in Figure 22.2. These results mirror those of information entropy, suggesting a pattern of steady decrease in flexibility from session 1 to 3, followed by a sharp increase into session 5.

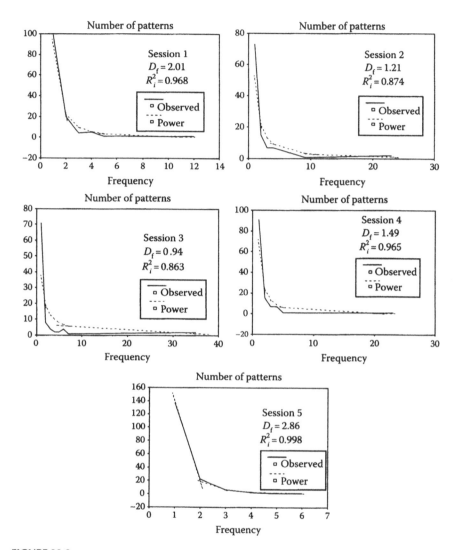

FIGURE 22.2

IPL curves depicting the number of distinct patterns (y-axis) displaying recurrence at different frequencies (x-axis). The B-weight representing shape (i.e., steepness) of the curve is used to estimate fractal dimension (D_f), while the fit index is used as a measure of integrity (R_i^2).

Furthermore, these results suggest that the integrity of the family system was broken down over the first few sessions, prior to an increase in structural flexibility in session 5 along with a reintegration process.

With respect to the current series of family therapy sessions, one may infer that these values reflect a process of breaking down dysfunctional family structures within the safe confines of the therapeutic setting in order to facilitate the emergence of more flexible patterns of interaction, as is typical in the family systems literature (Minuchin & Fishman, 1974) and the broader scientific literature pertaining to self-organizing systems (Bak, 1996; Kauffman, 1995; Prigogine & Stengers, 1984). However, additional data (i.e., pertaining directly to family structures) and analyses (i.e., regression analysis involving the prediction of changes in integrity and complexity across sessions) would be required to sufficiently test these types of interpretation.

Stationarity, Within-Session Stability, and Change

The final basic analysis one will typically wish to carry out when analyzing family interactions using OD is an analysis of the stationarity within each session. Stationarity may be defined as the degree with which a time series reflects the output of a stable underlying process, or in a more technical sense, the degree to which the transition matrix of a Markovian process is stable from beginning to end (Gottman & Roy, 1990). Stationarity is typically considered to be a methodological requirement for the analysis of Markovian processes (e.g., a categorical time-series analysis based on transition matrices; see Chapter 17).

Stationarity may be assessed through an examination of the degree to which the frequencies (f) for each pattern are consistent across each session (see Figure 22.3). If the frequencies for each pattern are relatively stable across the conversation, then one may infer that the underlying process producing those patterns is stable. If the frequencies change significantly, then the underlying dynamics producing them may be considered to have changed. Stationarity in f across the discussion may be assessed in SPSS by conducting a linear regression analysis, using f as the dependent variable and time as the independent variable.

The results of this analysis provide two pieces of important information. First, a statistically significant t-test value will suggest that the discussion is nonstationary. By contrast, a nonsignificant result indicates that the f is stationary across the discussion. Second, in cases of nonstationarity, the sign and magnitude of the t-value may be used as an index of increasing or decreasing structure across the series, or structure over time (S_t). For example, if most of the high-frequency-utterance patterns come early in the discussion, one would find a large negative S_t-value indicating nonstationarity in the direction of decreasing structure across the session. If the discussion contains a pileup of high-frequency recurrences toward the end of the discussion, one would obtain a high-positive S_t-value indicating nonstationarity in the direction of increasing structure across the session. While a low and nonsignificant S_t-value would suggest stationary dynamics, with a relatively consistent repetition of patterns (f-values) across the discussion.

The results of this analysis across these five sessions of family therapy illustrate an even distribution of higher-frequency versus lower-frequency patterns across discussion 1 ($R^2 = 0.077$, $S_t = 1.136$, $p = 0.2573$, $df = 1.216$), decreases in the structure across session 2 ($R^2 = 0.224$, $S_t = -3.376$, $p = 0.009$)

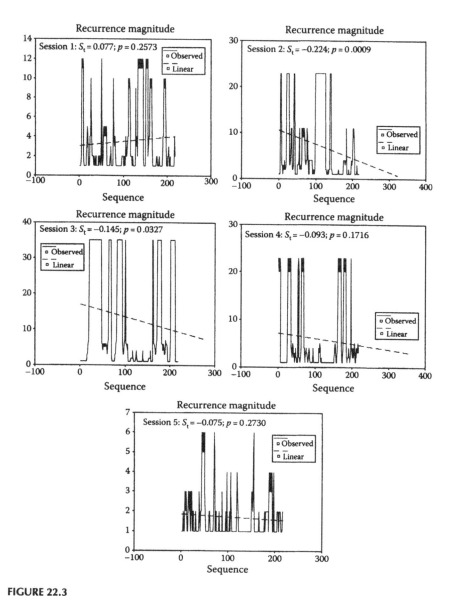

FIGURE 22.3

Recurrence magnitudes over time for sessions 1–5. Lines represent different recurrence magnitudes (*y*-axis) over time (*x*-axis). Lower values of S_t suggest greater stationarity for the underlying transition matrix of pattern generation across the session.

and 3 ($R^2=0.077$, $S_t=-2.15$, $p=0.033$), and a stable relative structure across session 4 ($R^2=0.077$, $S_t=-1.371$, $p=0.172$) and 5 ($R^2=0.075$, $S_t=-1.099$, $p=0.273$). These results suggest, for example, that the increasing relative rigidity of the session dynamics observed in sessions 2 and 3 occurred more so at the outset of the sessions than at the end, that the decreases in structural integrity observed in sessions 2 and 3 were related to processes occurring within the session involving the *breaking up* of initially rigid patterns at the outset of each session, and that by contrast the increasing relative flexibility observed in sessions 4 and 5 was a result of processes that occurred between rather than within these sessions.

Changes in Member Contributions to Structure and Complexity

Now we explore some additional hypotheses that may be tested above and beyond the identification of important patterns, pattern lengths, and measures of complexity. For example, it is possible to examine the contribution of each group member to the creation of a structure within each session using only the time series from each session. Conceptually, this is accomplished by first considering the structure of a pattern to be the number of times it repeats across the discussion. Patterns that are highly repetitive may be considered to be highly structured or rigid. Discussions filled with highly repetitive patterns will be more globally rigid. By contrast, low-frequency, non-repetitive patterns are less structured, and discussions built from a variety of low-frequency, non-repetitive patterns are more flexible and unpredictable. Next, one may wish to explore the degree to which a particular member of the discussion tends to be involved in either high- or low-frequency patterns. If a member is disproportionately involved in high-frequency patterns, one may infer that this person is contributing to rigidity within the discussion. Conversely, if a member is disproportionately involved in low-frequency patterns, one may infer a contribution toward flexibility.

Following this reasoning, a simple regression analysis may be used to test for speaker contribution to the structure. First, one calculates the expected probabilities for each pattern observed (i.e., the combinatorial probabilities of the speaker base-rates within the pattern multiplied by N) and enters them into the data file in SPSS. This will serve as a control variable within the regression analysis. Next, four additional variables are entered into the data file representing the degree of presence of each speaker within each pattern. This variable is a simple frequency count of the number of times a person speaks within a pattern. For example, in the pattern Tx-Fa-Tx-Mo-Tx-Dtr, the speaker-presence scores would be Tx = 3, because the therapist spoke three times; Fa = 1, Mo = 1, and Dtr = 1, because each of them spoke one time within the pattern. Finally, one builds a regression model with the natural log (a linear conversion) of frequency of recurrence for each pattern as the criterion variable, expected frequency entered as the first predictor variable (to control for the impact of different speaker base-rates), followed by a stepwise entry of each of the four speaker-presence scores. The resulting standardized beta-weights (B) for each speaker-presence variable may be used as indices of each speaker's contributions to deterministic patterning within each session. Higher B-values indicate more contribution to the structure and repetition across patterns, while lower B-values indicate less contribution to the structure or, by contrast, a relative contribution toward flexibility. Finally, one may examine the overall fit index for the member contribution (R^2_{mc}) to determine the degree to which unknown factors (beyond member contributions and expected values) are contributing to the structure and repetition.

Subsequent analyses following from these results include calculating an index of relative disparity in the contribution among members. For example, are their B-values equivalent, or are one or two members primary in driving the structure within a discussion? One may then go a step further and examine the relationship between this degree of member contribution disparity and the levels of entropy and integrity across sessions. In addition, one may wish to focus on particular group members who hold key hypothetical roles, for example, the degree to which therapist contribution is associated with changes in entropy across sessions. Finally, one may generate hypotheses based on the relative degree to which speaker base-rates alone predict the structure across sessions. Hypothetically, the less impact that can be attributed to individual speaker base-rates, the more one may attribute patterning to be the result of emergent relational processes among members.

The results for session 1 indicate that the regression model predicting log frequency of patterns member contribution (R^2_{mc}) within the session ($R^2_{mc} = 0.406$, $N = 212$) included the therapist ($B_{tx} = 0.454$, $t = 7.557$; $p < 0.001$) and the daughter ($B_{dtr} = -0.170$, $t = -2.389$, $p = 0.018$) above and beyond speaker base-rate probabilities ($B_{br} = 0.159$; $t = 2.040$; $p = 0.043$). These results suggest a high degree of structuring by the therapist in particular and a significant complexity influence (a negative contribution toward patterning) by the daughter. The high influence on the structure by the therapist conforms to expectations, particularly in this first treatment session, while the influence of the daughter on complexity could be interpreted in a number of different ways, depending on one's theoretical lens. The most clear-cut interpretation may be that the daughter is introducing novel information within the session, which has yet to be incorporated into the family's characteristic modes of responding.

The best model for member contribution to patterning in session 2 ($R^2_{mc} = 0.768$, $N = 212$) included the daughter ($B_{dtr} = -1.027$, $t = -8.440$; $p < 0.001$), followed by the therapist ($B_{tx} = -0.787$, $t = -7.244$, $p < 0.001$), and the father ($B_{ftr} = -0.275$, $t = -4.67$, $p < 0.001$) above and beyond speaker base-rate probabilities ($B_{br} = 1.778$; $t = 14.630$; $p < 0.001$). One may note the increase in the overall fit of the model, suggesting a greater influence of base-rates and individual speakers on the structure within session 2. In addition, the daughter has overtaken the therapist and the father has entered the model, with each now adding novelty and complexity to patterns in which they are involved as speakers (i.e., each exerting a negative influence on patterning).

The session 3 model ($R^2_{mc} = 0.912$, $N = 212$) included the daughter ($B_{dtr} = -0.225$, $t = -6.143$; $p < 0.001$), the father ($B_{ftr} = 0.158$, $t = 3.498$, $p = 0.001$), and the mother ($B_{mtr} = 0.099$, $t = 2.239$, $p = 0.026$) above and beyond speaker base-rate probabilities ($B_{br} = 1.283$; $t = 22.115$; $p < 0.001$). Here, one observes another jump in the overall fit of the model, a large influence of individual speaker base-rates, the first entrance of the mother to the model behind the father in adding structure, and the daughter continuing in her most influential role in creating novel patterns. In this most rigid session, the therapist has dropped out of the influence model. One may speculate that the parents are beginning to work together to incorporate the novel information provided by the daughter without the direct assistance of the therapist.

The session 4 model ($R_{mc}^2 = 0.599$, $N = 212$) included a return to the influence of the therapist ($B_{tx} = 0.332$, $t = 5.920$; $p < 0.001$) and the daughter ($B_{dtr} = 0.146$, $t = 2.219$, $p = 0.028$), for the first time adding positively toward the structure in the session above and beyond speaker base-rate probabilities ($B_{br} = 0.476$; $t = 6.384$; $p < 0.001$).

Finally, in session 5, where the largest jumps in entropy and integrity values were observed, the influence model displayed a drop in the overall fit ($R_{mc}^2 = 0.225$, $N = 212$), with positive structuring by the therapist ($B_{tx} = 0.222$, $t = 2.468$; $p = 0.014$), and negative structuring by the daughter ($B_{dtr} = -0.275$, $t = -3.270$, $p = 0.001$) and the mother ($B_{mtr} = -0.149$, $t = -2.102$, $p = 0.037$). Speaker base-rates were insignificant in session 5 ($B_{br} = -0.003$; $t = -0.042$; $p = 0.966$), which may be due in part to the equivalence of speaker base-rates (range = 22.5% of total utterances for the father to 29.8% of utterances for the therapist), limiting the variance of this predictor.

Exploratory analyses were conducted to examine the putative relationships among the various indices above with H_i, D_f, and R_i^2 values across each of the sessions ($N = 5$; see Table 22.3). Across the five sessions, significant correlations

TABLE 22.3

Correlations among Session Indices ($N = 5$) Including Fractal Dimension (D_f), Structural Integrity (R_i^2), Member Contribution Model Fit (R_{mc}^2), B-Weight for Base-Rate Probabilities (B_{br}), and B-Weights for Contributions of Mother (B_{mtr}), Therapist (B_{tx}), Daughter (B_{dtr}), and Father (B_{ftr})

	D_f	R_i^2	H_i	R_{mc}^2	B_{br}	B_{mtr}	B_{tx}	B_{dtr}	B_{ftr}
D_f	—	0.874*	0.974***	0.976***	-0.823*	-0.811*	0.476	-0.258	-0.007
R_i^2	—	—	0.921**	-0.939**	-0.948**	-0.635	0.725	0.592	0.078
H_i	—	—	—	0.990***	-0.817*	-0.815*	0.454	0.261	-0.154
R_{mc}^2	—	—	—	—	0.876**	0.735	-0.552	-0.365	-0.069
B_{br}	—	—	—	—	—	0.456	-0.884**	-0.759	-0.341
B_{mtr}	—	—	—	—	—	—	-0.036	0.212	0.172
B_{tx}	—	—	—	—	—	—	—	0.969***	0.632
B_{dtr}	—	—	—	—	—	—	—	—	0.632
B_{ftr}	—	—	—	—	—	—	—	—	—

* $p < 0.10$.
** $p < 0.05$.
*** $p < 0.01$.

were observed (a) between member contribution model fit values (R_{mc}^2) and each of the dynamics indices (H_i, D_f, and R_i^2), (b) between speaker base-rate contribution to patterning (B_{br}) and each of these indices, (c) between mother contribution to patterning (B_{mtr}) and H_i and D_f, (d) between therapist (B_{tx}) and daughter (B_{dtr}) contributions to patterning, and (e) between speaker base-rate (B_{br}) and therapist (B_{tx}) contributions to patterning. It should be mentioned that with only five sessions, the statistical power was quite limited. Each of the significant correlations was very large, and several other large-to-moderate correlations were not statistically significant.

The significant correlations among the entropy and integrity measures are to be expected, as rigidity distorts the overall shape of the IPL. Future research may be helpful in examining the extent to which similar distortions may occur at higher entropy values. Furthermore, the degree to which these two indices actually measure independent constructs at all is a question for future research. They may not. Also, it is worth mentioning that if one were to calculate the fractal dimension using a linear correlation between the log-transformed values of both sides of Equation 22.4, then the two indicators would be completely confounded. In the current analysis, a power-law fit was tested directly; so the two indices may vary with some independence.

The extremely strong correspondence between therapist and daughter contributions to structuring may be an interesting result. What this suggests is that in sessions where the therapist contributed strongly to patterning or flexibility, the daughter also contributed and in the same direction. One could speculate that this result is reflecting a positive therapeutic alliance. However, it could also be a reflection of conflict between the two. To make such interpretations, one would require further analysis of the data set. For example, the verbal statements or patterns could be coded for relational information (e.g., conflict or closeness), or the session content could be coded in a more global and qualitative manner.

Furthermore, the strong negative correlation between the speaker base-rate contribution to structuring and the therapist contribution suggests that when the therapist is more structuring, speaker contributions become more equal. One may speculate that this is due to the therapist working to facilitate a more open and democratic process of information exchange. Again, further analysis would be required to test such a conclusion.

It is more difficult to speculate about the mother's emergence as an important family member for predicting negative entropy (i.e., coherence) across sessions, particularly because the mother appears to contribute little influence on patterning within the sessions. One speculation would be that she is assuming the role of a relational leader (Guastello et al., 2005), participating significantly in the process of integrating novel information produced by other members (i.e., her daughter) into the family system, for example, by following other members with reflections or extending statements. When the mother does contribute to patterning, it may have a greater impact on global dynamics if she exerts this type of a strong relational influence.

The correlation between R_{mc}^2 values (i.e., member contribution model fit) and B_{br} values (i.e., speaker base-rate contribution) is likely reflecting common association between these measures and a common underlying construct related to both measures of entropy and system integrity (i.e., fit to the IPL). These relationships suggest that system integrity and complexity may increase when a relative influence is more evenly spread among members. This interpretation makes general theoretical sense inasmuch as one would expect higher flexibility along with more equal speaker base-rates in family systems with more egalitarian relationships.

Discussion and Conclusions

Although the purpose of the current set of empirical analyses was to provide a demonstration of the OD methodology, the current results may also add to the understanding of family processes. First, the values for entropy measures strongly suggest that the therapeutic process within these sessions of family therapy is a reflection of a self-organizing system of relationships among the family members and the therapist. The evidence for self-organizing dynamics is particularly strong with respect to the fits of the relative pattern recurrence frequencies within each session with IPL (fractal structure) models (R^2 ranging from 0.866 to 0.998). These results are consistent with prior family (Pincus, 2001) and group interaction studies (Pincus et al., 2008; Pincus & Guastello, 2005).

Although a complete theoretical discussion is beyond the scope of this example analysis, at a minimum the current results suggest that family systems evolve according to similar principles as other self-organizing systems in the life sciences. Specifically, the current results suggest that family dynamics emerge through flows of information exchange among members, that a family system may open (higher entropy) or close (lower entropy) in response to evolutionary demands, and that family systems will generally function best with an optimal mix of complexity and structural integrity. Future research will be necessary, however, to better test these conclusions.

On the broadest scale, methodologies such as OD may assist in bridging the science practice gap in various areas involving the use of conversation dynamics, such as the assessment of productivity in work groups, broader unit interactions within business or political structures, processes of relationship development, and the assessment of family and group therapies. Beyond conversations, OD may be applied to any number of research questions involving comparisons in the dynamics of categorical change processes.

When theoretical questions are primary, the existence of power laws and low-dimensional chaos may provide clues about self-organizing processes. When there are practical concerns about different levels of complexity in the change dynamics, each of the measures of complexity may provide useful indices. When dynamic transitions are of focus, either due to natural instabilities in the system, or due to experimental manipulations, then the examination of stationarity may be of focus. Finally, when system-to-component relationships (i.e., hierarchical self-organization processes) and component-to-component (e.g., driver–slave, coupling, and sync) dynamics are of interest, one may focus on the differential contributions of these components to global dynamics. The analytic strategies and indices demonstrated here should be considered to be a jumping off point for future exploration into the various systems within the natural world that involve qualitative changes unfolding over time.

References

Alexander, J. F., & Parsons, B. V. (1982). *Functional family therapy*. Pacific Grove, CA: Brooks/Cole.

Attneave, F. (1959). *Applications of information theory to psychology: A summary of basic concepts, methods, and results*. New York: Hult, Rinehart, and Winston.

Bak, P. (1996). *How nature works: The science of self-organized criticality*. New York: Springer-Verlag.

Butz, M. R., Chamberlain, L. L., & McCown, W. G. (1997). *Strange attractors: Chaos, complexity, and the art of family therapy*. New York: Wiley.

Gottman, J. M., Murray, J. D., Swanson, C. C., Tyson, R., & Swanson, K. R. (2002). *The mathematics of marriage: Dynamic nonlinear models*. Cambridge, MA: MIT Press.

Gottman, J. M., & Roy, A. K. (1990). *Sequential analysis: A guide for behavioral researchers*. Cambridge, U.K.: Cambridge University Press.

Granic, I., Hollenstein, T., Dishion, T. J., & Patterson, G. R. (2003). Longitudinal analysis of flexibility and reorganization in early adolescence: A dynamic systems study of family interactions. *Developmental Psychology, 39*, 606–617.

Guastello, S. J. (1995). *Chaos, catastrophe, and human affairs: Applications of nonlinear dynamics to work, organizations, and social evolution*. Mahwah, NJ: Lawrence Erlbaum.

Guastello, S. J. (2000). Symbolic dynamic patterns of written exchanges: Hierarchical structures in an electronic problem-solving group. *Nonlinear Dynamics, Psychology, and Life Sciences, 4*, 169–189.

Guastello, S. J., Hyde, T., & Odak, M. (1998). Symbolic dynamic patterns of verbal exchange in a creative problem solving group. *Nonlinear Dynamic, Psychology, and Life Sciences, 2*, 35–38.

Guastello, S. J., Craven, J., Zygowicz, K. M., & Bock, B. R. (2005). A rugged landscape model for self-organization and emergent leadership in creative problem solving and production groups. *Nonlinear Dynamics, Psychology, and Life Sciences, 9*, 297–334.

Guastello, S. J., Nielson, K. A., & Ross, T. J. (2002). Temporal dynamics of brain activity in human memory processes. *Nonlinear Dynamics, Psychology, and Life Sciences, 6*, 323–334.

Hollenstein, T., Granic, I., Stoolmiller, M., & Snyder, J. (2004). Rigidity in parent–child interactions and the development of externalizing and internalizing behavior in early childhood. *Journal of Abnormal Child Psychology, 32*, 595–607.

Howe, G. W. (2004). Studying the dynamics of problem behavior across multiple time scales: Prospects and challenges. *Journal of Abnormal Child Psychology, 32*, 673–678.

Howe, G. W., Dagne, G., & Brown, C. H. (2005). Multilevel methods for modeling observed sequences of family interaction. *Journal of Family Psychology, 19*, 72–85.

Kauffman, S. A. (1995). *At home in the universe.* New York: Oxford University Press.

Lathrop, D. P., & Kostelich, E. J. (1989). Characterization of an experimental strange attractor by periodic orbits. *Physics Review, 40*, 4028–4031.

Lewis, M. D., Lamey, A. V., & Douglas, L. (1999). A new dynamic systems method for the analysis of early socioemotional development. *Developmental Science, 2*, 457–475.

Long, J. S. (1997). *Regression models for categorical and limited dependent variables.* Thousand Oaks, CA: Sage.

Mandelbrot, B. (2004). *Fractals and chaos.* New York: Springer.

Minuchin, S., & Fishman, C. H. (1974). *Family therapy techniques.* Cambridge, MA: Harvard University Press.

Pincus, D. (2001). A framework and methodology for the study of nonlinear, self-organizing family dynamics. *Nonlinear Dynamics, Psychology and Life Sciences, 5*, 139–174.

Pincus, D., Fox, K. M., Perez, K. A., Turner, J. S., & McGee, A. R. (2008). Nonlinear dynamics of individual and interpersonal conflict in an experimental group. *Small Group Research, 39*, 150–178.

Pincus, D., & Guastello, S. J. (2005). Nonlinear dynamics and interpersonal correlates of verbal turn-taking patterns in a group therapy session. *Small Group Research, 36*, 635–677.

Prigogine, I., & Stengers, I. (1984). *Order out of chaos: Man's new dialog with nature.* New York: Bantam.

Shannon, C. E., & Weaver, W. (1949). *The mathematical theory of communication.* Urbana, IL: University of Illinois Press.

Snyder, D. K., & Kazak, A. E. (2005). Methodology in family science: Introduction to the special issue. *Journal of Family Psychology, 19*, 3–5.

Spohn, M. (2008). Violent societies: An application of orbital decomposition to the problem of human violence. *Nonlinear Dynamics, Psychology, and Life Sciences, 12*, 87–115.

West, B. J., & Deering, B. (1995). *The lure of modern science: Fractal thinking.* River Edge, NJ: World Scientific.

23

The Danger of Wishing for Chaos*

Patrick E. McSharry

CONTENTS

With the discovery of chaos came the hope of finding simple models that would be capable of explaining complex phenomena. Numerous papers claimed to find low-dimensional chaos in a number of areas ranging from the brain to the stock market. Years later, many of these claims have been disproved and the fantastic hopes pinned on chaos have been toned down as research with more realistic objectives follows. The difficulty in calculating reliable estimates of the correlation dimension and the maximal Lyapunov exponent, two of the hallmarks of chaos, are explored. Given that nonlinear dynamics is a relatively new and growing field of science, the need for statistical testing is greater than ever. Surrogate data provides one possible approach but great care is needed in generating relevant surrogates and in interpreting the results. Examples of misleading applications and challenges for the future of research in nonlinear dynamics are discussed.

    The word chaos has become synonymous with complicated systems that are beyond the realm of understanding. At worst, chaos describes a state of complete disorder and confusion. To a scientist, chaos presents a less

---

* The original article was edited for format and originally published as McSharry, P. E. (2005). The danger of wishing for chaos. *Nonlinear Dynamics, Psychology, and Life Sciences, 9,* 375–397. Reprinted with permission of the Society for Chaos Theory in Psychology & Life Sciences.

pessimistic view of life: behavior so unpredictable as to appear random because of the inherent sensitivity to small perturbations in the initial conditions. This suggests that many complex systems could possibly be described by low-dimensional deterministic mathematical models. It was the discovery of this intermediate possibility between order and disorder that brought so much hope and excitement in the 1980s, culminating with popular science books such as *Chaos* (Gleick, 1988). While most real-world systems are undoubtedly nonlinear, the quality of data available often favors traditional linear analyses. In practice, a modeling framework that incorporates a mixture of deterministic and stochastic approaches may have the greatest utility.

The tools of the nonlinear dynamicist borrow from a diverse range of disciplines including physics, mathematics, statistics, and engineering. While these traditional scientific disciplines offer a strong framework, this new field of research requires the formation of many new methods and techniques. It is interesting to note a division of research at this point: method driven and application driven. First, there are those who have focused on the construction and testing of new methods for carrying out "nonlinear time series analysis"—the name given to the investigation of systems and models that may display chaos (Kantz & Schreiber, 2003). On the other hand, there are many researchers interested in exploring whether the ideas and concepts from nonlinear dynamics can provide additional insight into their particular fields of expertise. This division has presented and continues to present an unfortunate danger for the credibility of nonlinear time series analysis. The proposition of new nonlinear methods without suitable statistical testing and the subsequent adoption of these methods by practitioners could destroy the integrity of all research, which uses nonlinear techniques. Falling into a state of disrepute is likely to cause a setback to the scientific progress of nonlinear dynamics.

In this chapter, we will argue that it is extremely difficult to show that any real system is chaotic because of the requirement of collecting a large number of high-quality observations from a period when the underlying dynamics are stationary. For this reason, all estimates of system invariants should be accompanied by an expression of the level uncertainty associated with these estimates. Nevertheless, the techniques arising from the field of nonlinear time series analysis can make a substantial contribution to science. We shall look at some of the tools available to those researchers who believe that nonlinear methods may offer a more in-depth understanding of their particular systems of interest. We will focus on the mathematical sanity checks afforded by statistical tests that can help avoid drawing misleading conclusions. Specifically, we will investigate (a) the difficulty in identifying low-dimensional chaos using empirical data; (b) the challenges in constructing, estimating, and evaluating nonlinear models; and (c) the need for statistical tests that are relevant to the specific application being proposed.

The layout of the chapter is as follows. We first review the mathematical concepts underpinning a number of nonlinear methods and the definitions of system invariants such as the correlation dimension and maximal Lyapunov exponent. Then we discuss why it is so difficult to estimate these invariants from empirical data and to conclude that any real system is actually chaotic. Next, we provide some guidelines for justifying the use of nonlinear methods and explore the use of surrogate data for statistical testing. In the penultimate section, we discuss some of the challenges facing the nonlinear dynamics research community when proposing the application of new methods and give examples from the field of biomedical research. The final section provides a discussion of approaches for promoting scientific progress in the field of nonlinear dynamics and concludes the chapter.

Mathematical Tools

The usual starting point for any nonlinear time series analysis is an attempt at reconstructing the underlying state space, which, in theory, provides a unique description of the state of the system. For univariate time series, $s_n (n = 1, \ldots, N)$, a set of m-dimensional state vectors may be defined using a delay coordinate reconstruction:

$$x_n = (s_{n-(m-1)\tau}, s_{n-(m-2)\tau}, \ldots, s_{n-\tau}, s_n),$$

where τ is known as the time delay or time lag.

The mathematical theory of embedding tells us that under certain conditions, it should be possible to use this reconstruction as a faithful representation of the underlying dynamics if m satisfies $m > 2D_F$, where D_F is the box-counting dimension (Takens, 1981; Sauer, Yorke, & Casdagli, 1991). Unfortunately, this mathematical theory tells us little about what to do when we are faced with noisy data of finite duration. For example, while the choice of τ is irrelevant in theory, it is extremely important in practice. Suggested values for τ are often based on mutual information (Fraser & Swinney, 1986) or a geometric interpretation of the reconstruction (Rosenstein, Collins, & De Luca, 1994). Similarly, the fact that we are unlikely to know the value of D_F beforehand means that we also have to estimate a reasonable value for m from the data. The method of *false nearest neighbors* provides one possibility for estimating a sufficient value for m (Kennel, Brown & Abarbanel, 1992). This is typically achieved by varying the size of m and monitoring the number of false nearest neighbors associated with areas of the reconstructed state space that have self-intersections. Unfortunately, the detection of false nearest neighbors is subject to the choice of an arbitrary constant, which will vary with position in state space. By testing for consistency between the dynamics imposed by a model and the observational uncertainty while allowing for variation in the local instabilities of the nonlinear dynamics throughout state space, it is possible to determine a robust estimate for the minimum value for m (McSharry & Smith, 2004).

Fractal Dimensions

The spate of papers aiming to measure the fractal dimensions of different systems was perhaps motivated by the hope of determining the number of active degrees of freedom. The ability to demonstrate that a particular data set was generated by a low-dimensional deterministic process implies that, in theory, there exists a model with a few degrees of freedom, which might be able to represent the dynamics. There are many examples of misleading estimates for the dimension of complicated real-world systems. For example, the weather was reported as having a dimension between three and eight; see (Lorenz, 1991) for a critique. Similarly, in the case of stock market returns, evidence of nonlinearities was found but claims of low-dimensional chaos were not well justified (Scheinkman & Le Baron, 1989).

Although there is an entire family of dimensions, known as Renyi dimensions, the correlation dimension, D_2, is the easiest to calculate from data (Kantz & Schreiber, 2003). D_2 reflects how the probability that the distance between two randomly chosen points will be less than ε, scales as a function of ε. The correlation integral, which counts the fraction of pairs (x_i, x_j) whose distance is smaller than ε, is defined by

$$C(\varepsilon, N) = \frac{2}{N(N-1)} \sum_{i=1}^{N} \sum_{j=i+1}^{N} \theta \left(\varepsilon - || x_i - x_j || \right)$$

where θ is the Heaviside step function, $\theta(x) = 0$ if $x \le 0$, and $\theta(x) = 1$ if $x > 0$ (Grassberger & Procaccia, 1983a, 1983b). C is expected to scale like a power law, $C \propto \varepsilon^{D_2}$, for an infinite amount of data where $N \to \infty$ and $\varepsilon \to 0$, and the correlation dimension D_2 is defined by

$$D_2 = \lim_{\varepsilon \to 0} \lim_{N \to \infty} \frac{d \ln C(\varepsilon, N)}{d \ln \varepsilon}$$

There are two problems when attempting to identify a scaling region: (a) the finite sample size places an upper limit on N and (b) the finite accuracy of the data and the sparseness of near neighbors when ε is small.

Unfortunately, even if it were possible to overcome the obstacles above, it is likely that temporal correlations in the data will give rise to underestimates of D_2. It has been shown that infinite-dimensional stochastic signals can lead to finite- and low-dimensional estimates (Theiler, 1986). The estimates for C can be biased by temporal correlations since small spatial separations between pairs of points can occur because they were observed closely in time. For finite data sets, this effect of points nearby in time can be limited by restricting the sums over i and j so that $|i - j| > W$ for some constant W (Theiler, 1986). Even when i is much greater than j, the distance between x_{i+1} and x_{j+1} is unlikely to be independent of the distance separating x_i and x_j leading to further biases in the estimate. While we can never be sure whether there is a sufficient amount of data, a space-time-separation diagram may be employed to ascertain with confidence that we do not have enough data (Provenzale, Smith, Vio, & Murante, 1992).

Lyapunov Exponents

The most familiar signature of chaos is displayed by the unpredictability of the future despite the system obeying deterministic equations of motion. This unpredictability appears as an increasing average forecast error with larger prediction lead times. The phrase *sensitive dependence on initial conditions* is often used to describe the inherent instability of the solutions that cause this unpredictability. Specifically, two nearby initial conditions will, on average, diverge over time. While many linear systems give rise to a slow rate of divergence, it is the exponential divergence demonstrated by some nonlinear systems that is characteristic of chaotic systems.

Consider a deterministic dynamical system described by the discrete map, $x_{n+1} = F(x_n)$. The evolution of an infinitesimal uncertainty, ε, in the initial condition, x_0, over a finite number of time steps, k, is given by

$$\varepsilon_\kappa = M(x_0, k)\, \varepsilon$$

where $M(x_0, k)$ is the linear tangent propagator formed by the product of the Jacobians along the k steps of the trajectory $M(x_0, k) = J(x_{k-1})\, J(x_{k-2}) \cdots J(x_0)$.

The linear dynamics of uncertainty growth may be analyzed using the singular value decomposition (SVD) of M, giving $M = U \Sigma V^T$, where the columns of the orthogonal matrix $U(V)$ are the left (right) singular vectors, respectively, and the entries of the diagonal matrix Σ are the singular values, $\sigma_i(x_0, k)$, usually ranked in decreasing order (Strang, 1988). From a geometrical point of view, the linear dynamics have the effect of transforming a circle into an ellipse, where the right singular vectors describe the directions of the axes of the ellipse and the singular values reflect the lengths of these axes. Finite-time Lyapunov exponents, which depend both on the initial position, x_0, and the number of time steps, k, are defined by

$$\lambda_i^{(k)}(x_0) = \frac{1}{k} \log_2 \sigma_i(x_0, k)$$

The system invariants known as the Lyapunov exponents are then defined by taking the limit as k goes to infinity,

$$\Lambda_i = \lim_{k \to \infty} \lambda_i^{(k)}(x_0)$$

A system is said to be chaotic if the leading Lyapunov exponent, Λ_1, is positive. Alternatively, a negative Λ_1 indicates the existence of a stable fixed point.

While the expression for $\lambda_i^{(k)}(x_0)$ may be rewritten as $\sigma_i(x_0, k) \propto e^{k\lambda_i^{(k)}(x_0)\ln 2}$, the effective growth rate defined by a positive $\lambda_i^{(k)}(x_0)$ does not imply exponential growth; it is only in the case of Λ_i, that a positive value implies an effective exponential growth (Ziehmann, Smith, & Kurths, 1999). There is a common misconception that the value of Λ_1 provides a measure of the unpredictability. This is untrue since Λ_1 only applies to infinitesimal uncertainties, and as such an infinitesimal uncertainty can never limit predictability. Furthermore, Λ_1 is an average defined over an infinitely long trajectory and does not reflect short-term dynamics. For some chaotic systems, it can be shown that in the short term, all uncertainties may shrink for a finite time (Smith, Ziehman, & Fraedrich, 1999).

Estimates from Data

Unfortunately, the attractiveness of finding chaos in real-world systems has the potential to cloud the better judgment of many researchers. There are three fundamental problems with seeking chaos in such systems. First, the existence of chaos can only be mathematically proven for a handful of simple toy models, such as the logistic map (May, 1976). Second, our perspective of the underlying dynamics of the system is obtained from the time series of observations or data that we collect using some measurement process. The quality of such data sets is limited by the observational uncertainty that arises from measurement errors, noise, artifacts, and missing values. Third, the underlying dynamical processes may not be stationary implying that no single mathematical model is capable of describing the dynamics throughout the entire duration of the experiment used to record the data. The combination of these problems implies that we are unlikely to obtain data of a high enough quality (resolution and duration) from a stationary epoch of the underlying dynamics to enable us to calculate the correlation dimension, D_2, or the maximal Lyapunov exponent Λ_1. In the following sections, we investigate the difficulties associated with estimating these two system invariants when faced with noisy time series of short duration. We also suggest that despite these estimation difficulties, the resulting nonlinear quantities may still be of value.

Estimating the Correlation Dimension

When computing D_2, it is important to know what type of error should be expected for a given quantity of data, N. Although it can be shown both numerically and analytically that this error generically scales as $O(1/N^{1/2})$ for $N \to \infty$, this is not always the case (Theiler, 1990). While the quantity of data required to reliably estimate D_2 for a particular system has been explored by a number of investigators, giving rise to a range of differing opinions, it suffices to say that the results are pessimistic for the majority of interesting real-world systems. Various claims for the minimum number of data points, N_{min}, include $N_{min} = D_2^{42}$ (Smith, 1988), $N_{min} = 2^{D_2}(D_2 + 1)^{D_2}$ (Nerenberg & Essex, 1990), $N_{min} = 10^{\frac{D_2}{2}}$ (Ruelle, 1990), $N_{min} = 10^{2 + \frac{2}{5}D_2}$ (Tsonis, 1992), and $N_{min} = 10^{1 + \frac{D_2}{2}}$ (Kantz & Schreiber, 2003). The variation of N_{min} with D_2 suggested by each of these estimates is illustrated in Figure 23.1. The best case scenario implies that $N = 100,000$ points are necessary for estimating $D_2 = 10$. In short, the news is bad if not completely foreboding for any researcher hoping to estimate D_2 from short noisy time series.

An alternative approach is to use Takens' estimator, which gives a maximum likelihood estimate of D_2 without estimating the slope of the correlation integral directly (Takens, 1985). An extension to this approach produces an estimate of D_2, which is consistent with measures at all smaller length scales (Guerrero & Smith, 2003). Furthermore, this approach provides

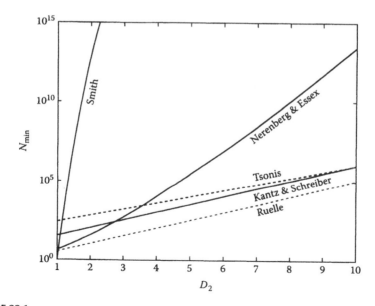

FIGURE 23.1

The suggested minimum number of data points, N_{min}, required to provide a reliable estimate of the correlation dimension, D_2.

constraints on the accuracy of the estimate. The difficulty in calculating reliable estimates of D_2 suggests that the uncertainty in such estimates should be clearly presented through the use of a distribution or error bars. Without a quantification of the uncertainty, the estimate should certainly be treated with caution.

In practice, it is likely to be easier to build a data-driven nonlinear model than to compute a reliable estimate of the dimension. If such a nonlinear model is useful for fulfilling one's objectives, whether they are prediction, classification, or the pursuit of a more in-depth understanding, and it can outperform simple linear benchmarks, then this in itself can arguably justify the use of nonlinear methods. Being unable to say for certain, whether or not a system is chaotic, is not a justification to give up and resign ourselves to traditional linear techniques, but rather a warning about what is possible and how the results of nonlinear time series analyses should be presented.

Estimating Lyapunov Exponents

In practice, when dealing with experimental data, it is usual to attempt to estimate the maximal Lyapunov exponent, Λ_1 by investigating a long but finite trajectory obtained by making k as large as possible. Some initial attempts at calculating Λ_1 gave misleading results since they blindly assumed the existence of exponential growth, and therefore can produce a finite value of Λ_1 for stochastic data where the true value is infinite (see, e.g., Wolf, Swift, Swinney, & Vastano, 1985). For high-quality experimental data, it may be possible to approximate the underlying dynamics by a model, which provides a means of estimating the entire Lyapunov spectrum (Brown, Bryant, & Abarbanel, 1991; Eckmann, Oliffson, Kamphort, Ruelle, & Ciliberto, 1986; Sano & Sawada, 1985). More recently, tests for the exponential growth have been advocated (Kantz, 1994; Rosenstein, Collins, & De Luca, 1993).

Consider the perfect scenario where the equations of motion are exactly known. For example, we will use the two-dimensional chaotic Ikeda map (Ikeda & Daido, 1980). The finite-time Lyapunov exponents, $\lambda_i^{(k)}$, may be calculated using a recursive QR decomposition (see Abarbanel, Brown, & Kennel, 1992). Figure 23.2 shows the convergence of $\lambda_i^{(k)}$ for simulations using 1000 different initial conditions. Even in this noise-free scenario, we find that a relatively large value of k is required to obtain an accurate estimate for Λ_1. The mean of the distribution converges to the Lyapunov exponent via $\langle \lambda_i^{(k)} \rangle = \Lambda_i + a_i k^{-\alpha_i}$ whereas the standard deviation scales as $b_i k^{-\beta_i}$. Scaling exponents, α_i and β_i typically range between 0.5 and 1 (Abarbanel et al., 1992). The central limit theorem does not necessarily apply to deterministic systems. Note that the distributions for $k = 2^{18}$ are nonnormal implying that uncertainty is best presented using percentiles, as shown in Figure 23.2.

For experimental data, it is usually difficult to collect sufficient high-quality data from a period during which the underlying dynamics are stationary.

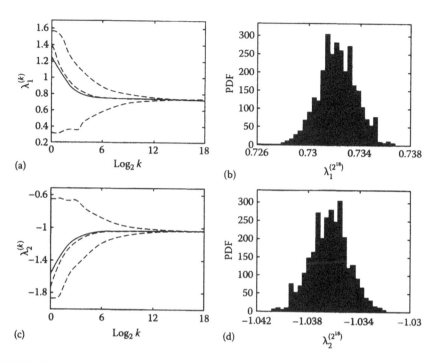

FIGURE 23.2

Finite-time Lyapunov exponents for the two-dimensional chaotic Ikeda map: (a) convergence of $\lambda_1^{(k)}$ with k, (b) convergence of $\lambda_2^{(k)}$ with k, (c) probability density function of $\lambda_1^{(k)}$ for $k = 2^{18}$, and (d) probability density function of $\lambda_2^{(k)}$ for $k = 2^{18}$. In (a) and (c), the dashed lines reflect the 5%, 50%, 95% percentiles and the solid line is the mean.

With the exception of carefully controlled electronic circuits and lasers, most real-world systems are unlikely to provide suitable data sets. In fact, it is often a challenge to obtain enough data to pursue a linear analysis. For example, recommendations for the analysis of heart rate variability suggest using a 5 min window to calculate the power spectrum, thereby providing a balance between the minimum data requirements for the fast Fourier transform and the period of time where the underlying process is likely to be stationary (Malik & Camm, 1995). In the case of the brain, the electrical activity measured by the electroencephalogram reflects extremely erratic behavior, making it exceedingly difficult to find stationary periods. Numerous studies attempting to estimate nonlinear measures have opted for windows ranging between 20 and 40 s (see McSharry, Smith, & Tarassenko 2003a and references therein). In both these examples, any attempt to detect the underlying nonlinear dynamics may be obstructed by the noise and nonstationarity. The use of intracranial electrodes as opposed to noninvasive scalp electrodes for the identification or prediction of epileptic seizures may increase the likelihood of detecting and exploiting the nonlinearity, but the applicability is obviously reduced.

The Utility of Nonlinear Measures

Despite the fact that invariants such as D_2 and Λ_1 cannot be reliably estimated for short noisy data sets, the applications of the techniques used for their computation will still provide a measure of a nonlinear quantity. It is important to recognize that this quantity no longer reflects an invariant of the system, but simply a nonlinear measure obtained from a particular segment of data. It is still possible to ask (a) if monitoring the time dependence of this quantity is useful for prediction or classification and (b) whether this quantity can outperform benchmarks given by the traditional indicators that are based on classical linear statistics.

Using the words "effective" or "approximate" as prefixes emphasizes that the quantity being calculated is based on the theory of nonlinear dynamical systems but that this quantity should not be taken as a convergent estimate of an invariant of the underlying system. Following this prescription, a number of techniques have been proposed for applications in the field of biomedical research. An effective correlation dimension was postulated as a method for anticipating epileptic seizures (Lehnertz & Elger, 1999). Approximate entropy was proposed as a measure of system complexity (Pincus, 1991) and applied to heart rate time series (Pincus & Goldberger, 1994). The difficulty in estimating the entropy of short noisy time series motivated an alternative approach with improved accuracy, known as sample entropy (Richman & Moorman, 2000). Similarly, it is reasonable to ask whether an effective Lyapunov exponent can enhance our ability to classify or detect dynamical transitions. The convergence and divergence of short-term maximum Lyapunov exponents from adaptively selected electrodes has been used to provide predictions of epileptic seizures (Iasemidis et al., 2003).

An alternative method for detecting nonlinear transitions is to measure changes in the distribution of points falling in different Voronoi partitions of the multidimensional state space. This conceptually intuitive method, multidimensional probability evolution (MDPE), is capable of detecting changes in the underlying dynamics that are invisible to linear statistics. MDPE was employed to detect epileptic seizures using electroencephalograms from noninvasive scalp electrodes. While both variance and MDPE are able to detect seizures, MDPE gave less false positives (McSharry, He, Smith, & Tarassenko, 2002). MDPE was also employed to identify partial epileptic seizures from heart rate time series and may actually be better at detecting nonlinear dynamical transitions than the effective nonlinear measures based on D_2 or Λ_1 since the latter can have identical values for different multidimensional distributions (McSharry, 2004).

The explanations as to why these nonlinear measures should outperform linear statistics range from little more than hand-waving arguments to carefully tested hypotheses. Unfortunately, it is only too easy to dream up explanations for success after the fact that often leads to a misleading approach to conducting scientific research. We will return to the question of what constitutes a useful diagnostic in the last section and propose a framework for ensuring that novel nonlinear methods are of real value. This requires both an investigation of scientific merit and an exploration of simple benchmark tests in order to justify their benefits for practical applications. First, we will investigate approaches for constructing statistical tests of nonlinear methods.

Statistical Testing

Nonlinear dynamics provide one possible explanation for the many irregularities found in time series, yet, linear stochastic processes also have the ability to generate complicated signals. Given that so much theory and expertise exists in specifying, constructing, and evaluating linear models, it is extremely imprudent to blindly assume that a nonlinear approach is best without a thorough scientific justification. In order to motivate the construction of nonlinear models, one should be confident that the data collected contains the characteristics of an underlying nonlinear process, which could not have been produced by a linear process. Both nonlinear determinism and the stochasticity arising from random shocks to the system or variations in the parameters are candidates for the irregularity and complexity apparent in empirical data. A certain degree of confidence can be obtained by performing a statistical significance test using the *method of surrogate data*. Each surrogate data set appears like the original time series but only contains some specific prescribed characteristics, for example, the observations were generated by a linear stochastic process. The idea is to test whether a given nonlinear measure when applied to the original data gives a result that is different to that obtained by applying it to a collection of these surrogates.

Given a nonlinear measure, γ, which takes on the value γ_0 when computed from the original data set, we wish to know whether this particular value suggests that the underlying dynamical process is nonlinear. Another way to think about this is to ask what distribution, $p(\gamma)$, of values would we expect to obtain from a *similar* linear stochastic process. If the value γ_0 is not consistent with $p(\gamma)$, then the data might be nonlinear. In general, we will not have any theory to determine an analytical expression for $p(\gamma)$ and so we use the surrogates to estimate this distribution using a Monte Carlo approach (Theiler, Eubank, Longtin, Galdrikian, & Farmer, 1992). We test against a chosen null hypothesis by constructing N surrogate data sets that (a) preserve certain characteristics of the original time series and (b) are also consistent with the specified null hypothesis. By specifying the probability, α that we are prepared to reject the null hypothesis although it is true, we obtain a test that is valid at the $(1 - \alpha)$ significance level. A rank-based one-sided test with significance $(1 - \alpha)$ may be employed by generating $N = 1/\alpha - 1$ surrogates in order to test whether γ_0 is smaller than expected for data obeying the null hypothesis. By computing the N values for the nonlinear measure: γ_i ($i = 1, ..., N$), we can reject the null hypothesis whenever γ_0 is smaller than all of the γ_t. For example, $N = 19$ surrogates are required for testing at the 95% significance level.

The sophistication of the surrogates employed depends on the nature of the characteristics that we want to preserve. Surrogates that preserve the empirical distribution can be obtained by randomly shuffling the original data set without repetition. These can then be used to test the null hypothesis that the data are independent random numbers sampled from some fixed but unknown distribution. Another null hypothesis is that the data comes from a stationary linear stochastic process with normally distributed inputs. Surrogates with the same power spectrum may be obtained by taking the Fourier transform of the original data, randomizing the phases and transforming back to the time domain. This technique may also be applied to multivariate data sets (Pritchard & Theiler, 1996). A more general null hypothesis is that the data was generated by a stationary linear stochastic process with normally distributed inputs, which was then subjected to a monotonic instantaneous time-independent measurement function. To test this hypothesis, we require surrogates with the same empirical distribution and power spectrum. Amplitude-adjusted surrogates are capable of preserving the empirical distribution but yield a slightly modified power spectrum (Theiler et al., 1992). A technique for polishing these surrogates so that they better replicate the power spectrum is also available (Schreiber & Schmitz, 1996). An approach that aims to preserve both the distribution and the spectrum exactly uses a normal autoregressive process and a monotonic static transform (Kugiumtzis, 2002). Constrained randomization using simulated annealing provides a means of generalizing the approach to specify a wide variety of surrogates including multivariate data (Schreiber, 1998).

The method of surrogate data may also be used for testing new techniques such as medical diagnostic tools for classification or prediction. The possibility of using nonlinear methods for facilitating medical diagnostics calls for a different kind of test. This should test the efficacy of the new method against simple linear benchmarks. To achieve this goal, we suggest the use of *clinically relevant surrogates* (McSharry, Smith, & Tarassenko, 2003b). By generating surrogates that are not relevant to the specific clinical framework, it is possible to erroneously identify promising techniques for the wrong reasons. For example, there have been a number of conflicting investigations concerning the predictability of epileptic seizures using nonlinear methods. While there are likely to be nonlinear processes active in the transition leading to an epileptic seizure, the question is whether nonlinear methods provide any additional skill, in detecting this dynamical transition, beyond that offered by traditional linear techniques. Some investigations have suggested that a nonlinear measure based on the correlation integral outperformed a linear measure using the autocorrelation function (Martinerie et al., 1998). Unfortunately, this claim was supported by a surrogate data analysis that could not distinguish between changes in the variance and those in the correlation integral. If a simple benchmark, such as moving variance, is to be outperformed before the medical community seriously adopts a new nonlinear technique, then this must be accounted for in the surrogates. One approach is to use block surrogates, which attempt to preserve the time dependence of the variance (McSharry et al., 2003a).

Challenges

The principle, known as Occam's Razor, suggests that when attempting to select from within a class of models we should aim to identify the simplest model that is compatible with the observations. While increasing the complexity of a model naturally gives more freedom to provide a better fit to the observations, a model with too many parameters will not distinguish between the generative dynamics that we wish to extract and fluctuations due to factors such as measurement errors, nonstationarity and noise. This problem, known as over-fitting, is particularly relevant when attempting to construct nonlinear models since such models have the complexity to adapt to extremely complicated data. We will now consider model complexity, parameter estimation and evaluation with regard to constructing nonlinear models.

There are two obvious routes to take when attempting to identify the optimal size or complexity of a nonlinear model. Given a set of candidate explanatory variables, one approach for constructing a model starts with no terms and then considers the addition of one term at a time (Judd & Mees, 1995). An alternative starts with a general description that includes all possible variables and then successively eliminates terms until arriving at a reduced form of the model. The objective of both approaches is to arrive at an intermediate model, which optimizes a specific cost function that measures the goodness of fit of the model to the observations. While the latter approach requires more computational effort, it has the advantage of being less likely to fall into local minima (Hendry & Krolzig, 2003). A number of different reduction paths can be searched simultaneously to prevent the algorithm from removing a relevant variable, while retaining other variables as proxies. In this way, the global minimum in the cost function can be approached more reliably. This approach has been successfully employed for constructing models of biochemical reactions (Crampin, Schnell, & McSharry, 2004).

The additional complexity and extra parameters required to specify nonlinear models also implies that while they may be better suited for reproducing time series of irregular observations, they are also more likely to fit the noise as well as the actual dynamics. This problem, known as *over-fitting*, implies that the model will provide good results on in-sample data (that is used for specifying and training the model), but is likely to fail to generalize to new data. It is extremely difficult to distinguish between fluctuations due

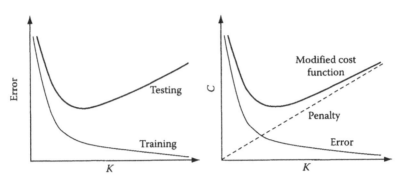

FIGURE 23.3
Analysis of model error versus model complexity as measured by the number of parameters, K, using a training and testing data set (left). Adding a penalty term to the cost function provides an alternative approach for avoiding over-fitting problems and may enable the identification of a model that generalizes well (right).

to the underlying dynamics and those due to the observational uncertainty. An obvious sign of an over-fit model is one that performs better on training (in-sample) data than testing (out-of-sample) data, as shown in Figure 23.3. Splitting the available data set in two and using one half for training and the other for testing is a good procedure for detecting such problems. An alternative method that may be employed when working with short data sets is to generate leave-one-out models and to predict the omitted point (or points), thereby providing quasi out-of-sample results.

One approach for reducing over-fitting is to add a penalty term to the cost function that penalizes non-parsimonious models. The more complex the model, measured by the number of parameters, the heavier the penalty. In this way, the addition of the penalty attempts to reproduce the effect of evaluating the model on an out-of-sample testing data set, by creating a minimum in the cost function for an intermediate model complexity, as shown in Figure 23.3. Note that such penalty terms are often referred to as regularization techniques and can also be viewed as the incorporation of priors. The simplest form of penalty function is one that is proportional to the number of parameters (Akaike, 1974). A penalty function that depends on both the number of observations and number of parameters may be determined using a Bayesian approach (Schwarz, 1978). Alternatively, this penalty function may be arrived at by assuming that the optimal model minimizes the description length of an encoding of the data (Rissanen, 1978). While the idea of having a cost function that works for all models is attractive, it is more realistic that a universal cost function does not exist and that the appropriate cost function will depend intimately on the structure of the model and its application.

When applying nonlinear models to noisy time series, one should expect the goodness of fit, measured by the discrepancy between the model and the data, to vary with position in state space. While this property of nonlinear models is generally ignored, the contribution of the local stretching factors induced by the nonlinear model can be used to identify model error by checking for the consistency between the model and the observational uncertainty (McSharry & Smith, 2004). This approach allows the user to detect why the model is inadequate and can also suggest how these imperfections can be reduced.

Parameter estimation for nonlinear models is also difficult since many of the traditional linear goodness-of-fit tests are unsuitable for applications using nonlinear models. For example, many classical techniques assume normal distributions for the measurement errors and the model error term. In contrast, even in the case of normally distributed measurement errors, the application of nonlinear models for generating predictions will convolve these measurement errors so that the effective model error term depends on the local structure of the nonlinear model. This implies that standard parameter estimation methods, such as least squares and total least squares will lead to biased estimates. For this reason, it is better to use a parameter estimation technique that directly incorporates the interplay between the nonlinear structure of the model and the observational uncertainty (McSharry & Smith, 1999). In fact, root-mean-square evaluation statistics could actually reject the very nonlinear system that generated the data in the first place.

Meteorologists are familiar with the task of generating forecasts using nonlinear weather models. A method that has recently been employed for justifying the use of nonlinear models for prediction is to generate and evaluate probabilistic forecasts (Palmer, 2000). These forecasts, known as ensemble forecasts, take account of the uncertainty in the initial conditions by sampling from the initial distribution and propagating each trajectory forward in time. In this way, the ensemble forecasts can also quantify the uncertainty in the prediction. One approach for evaluating these ensemble forecasts is based on information theory (Roulston & Smith, 2002). Indeed a true measure of the success of a nonlinear modeling framework should depend on the application of the model. For this reason, it may be necessary to incorporate the users' utility function in order to provide a model with optimal functionality.

Discussion and Conclusion

A growing interest in interdisciplinary research and the resulting interplay between traditional model-driven research and modern data-driven modeling techniques may explain why nonlinear dynamics is suddenly contributing to solving so many practical problems (Stark & Hardy, 2003). Furthermore, the increasing availability of efficient computing resources, both in terms of speed and memory, greatly facilitates the application of nonlinear time series analysis techniques. This is especially relevant in the case of large data sets, which are required if such techniques are to prove themselves useful. The ability to analyze large data sets has given rise to new fields of interdisciplinary research, such as systems biology (Kitano, 2002). New experimental techniques in biochemistry offer an increasing number of empirical data sets and nonlinear models are likely to provide a deeper understanding of the mechanisms underlying complex biochemical reactions (Crampin, McSharry, & Schnell, 2004).

At present, the utility of nonlinear methods is still being debated. A large amount of research is required before these new methods can compete with the vast body of expertise that is available for guiding the application of classical linear statistical techniques. The foremost challenge concerns the statistical testing of new nonlinear methods and the ability to be confident that similar results cannot be achieved by simpler linear approaches. Fortunately, today's researchers can easily carry out extensive Monte Carlo simulations on an average PC. Such simulations are necessary for applying statistical tests in the absence of analytical distributions, a typical challenge in the construction of relevant null hypotheses for evaluating nonlinear time series analysis techniques. The use of clinically relevant statistical tests is particularly important if these nonlinear methods are to be used for constructing new medical diagnostic tools.

In order to make scientific progress, it is necessary to have well-understood and carefully defined benchmarks. The biomedical signal-processing community is accustomed to testing new techniques on public domain databases such as Physiobank (Goldberger et al., 2000; www.physionet.org/physiobank). While the use of such databases is to be applauded, it does not fully address the possibility of over-fitting, whereby a model or technique appears to have high performance statistics but fails to generalize to new data. This failure can be circumvented, to some extent, by holding back some of the data for testing while the remaining data are used for learning. These databases contain real biomedical signals recorded from human subjects under standard clinical settings. For this reason, it is impossible to know a priori which part of the signal corresponds to the underlying dynamics and how much can be attributed to noise. This is complicated further by electrical interference, movement artifacts, muscle artifacts, and missing data.

One approach for comparing methods is to employ a database of synthetic signals. A nonlinear dynamical model has been used to generate realistic synthetic electrocardiogram signals (McSharry, Clifford, Tarassenko, & Smith, 2003c). Open-source code for a freely available algorithm provides researchers with the ability to generate electrocardiogram signals with known characteristics, both in the time and frequency domains (www. physionet.org/matlab/ecgsyn). Different realizations of the electrocardiogram signal may be generated by varying the seed of the random number generator. These realizations can then be used to compute the uncertainty inherent in signal-processing techniques and to provide error bars for estimates. The ability to easily produce such signals equips many researchers with the data necessary to compare and evaluate their new biomedical signal-processing techniques. While this is currently a useful exercise for illustrating the advantages of new techniques, it is hoped that in the future it could form the basis of a gold seal of approval for new techniques. Many journals now insist that databases used for obtaining new scientific evidence are made public domain as a prerequisite of publication. This has the advantage of allowing method-driven researchers to compare different techniques. Having access to carefully labeled public domain databases, such as those at Physionet, can reduce the time required to accumulate and investigate biomedical databases and increase the pace of scientific exploration.

By promoting good practice when applying nonlinear time series techniques, it will be possible to avoid the many dangerous pitfalls that await the unsuspecting researcher. In summary, we recommend that (a) estimates calculated from time series are supplied with confidence intervals, (b) the method of surrogates is employed before discarding traditional linear models, and (c) synthetic data is used to explore new techniques and to formulate simple benchmark tests. While the pursuit of chaos may be akin to searching for the Holy Grail, the future of nonlinear time series analysis and its application to empirical data is likely to be a rewarding one.

Acknowledgment

PEM is a Royal Academy of Engineering Research Fellow and acknowledges support from the Engineering and Physical Sciences Research Council, United Kingdom. This chapter benefited from numerous discussions with Leonard A. Smith.

References

Abarbanel, H. D. I., Brown, R., & Kennel, M. B. (1992). Local Lyapunov exponents computed from observed data. *Journal of Nonlinear Science, 2*, 343–365.
Akaike, H. (1974). A new look at the statistical identification model. *IEEE Transactions on Automatic Control, 19*, 716–723.
Brown, R., Bryant, P., & Abarbanel, H. D. I. (1991). Computing the Lyapunov spectrum of a dynamical system from observed time series. *Physical Review A, 43*, 27–87.
Crampin, E. J., McSharry, P. E., & Schnell, S. (2004). Extracting biochemical reaction kinetics from time series data. In M. Gh. Negoita, R. J. Howlett, & L. C. Jain (Eds.), *Knowledge-Based Intelligent Information & Engineering Systems* (pp. 329–336). (Lecture notes in artificial intelligence, No. 3214). Berlin: Springer-Verlag.
Crampin, E. J., Schnell, S., & McSharry, P. E. (2004). Mathematical and computational techniques to deduce complex biochemical reaction mechanisms. *Progress in Biophysics and Molecular Biology 86*, 77–112.

Eckmann, J. P., Oliffson Kamphorst, S., Ruelle, D., & Ciliberto, S. (1986). Lyapunov exponents from a time series. *Physics Review A, 34,* 4971–4979.

Fraser, A. M., & Swinney, H. L. (1986). Independent coordinates for strange attractors from mutual information. *Physics Review A, 33,* 1134–1140.

Gleick, J. (1988). *Chaos: Making of a new science.* New York: Penguin Books.

Goldberger, A. L., Amaral, L. A. N., Glass, L., Hausdorff, J. M., Ivanov, P. Ch., Mark, R. G., Mietus, J. E., Moody, G. B., Peng, C. K., & Stanley, H. E. (2000). PhysioBank, PhysioToolkit, and PhysioNet: Components of a new research resource for complex physiologic signals. *Circulation, 10,* e215–e220.

Grassberger, P., & Procaccia, I. (1983a). Measuring the strangeness of strange attractors. *Physica D, 9,* 189–208.

Grassberger, P., & Procaccia, I. (1983b). Characterisation of strange attractors. *Physical Review Letters, 50,* 346–369.

Guerrero, A., & Smith, L. A. (2003). Towards coherent estimation of correlation dimension. *Physics Letters A, 318,* 373–379.

Hendry, D. F., & Krolzig, H. M. (2003). New developments in automatic general-to-specific modelling. In Stigum, B. P. (Ed.), *Econometrics and the philosophy of economics* (pp. xxx–xxx). Princeton: Princeton University Press.

Iasemidis, L. D., Shiau, D. S., Chaovalitwongse, W., Sackellares, J. C., Pardalos, P. M., Principe, J. C. et al. (2003). Adaptive epileptic seizure prediction system. *IEEE Transactions on Biomedical Engineering, 50,* 616–627.

Ikeda, K., & Daido, H. (1980). Optical turbulence: Chaotic behaviour of transmitted light from a ring cavity. *Physical Review Letters, 45,* 709–712.

Judd, K., & Mees, A., (1995). On selecting models for nonlinear time series. *Physica D, 82,* 426–444.

Kantz, H. (1994). A robust method to estimate the maximal Lyapunov exponent of a time series. *Physics Letters A, 185,* 77–87.

Kantz, H., & Schreiber, T. (2003). *Nonlinear time series analysis* (2nd edn). Cambridge: Cambridge University Press.

Kennel, M. B., Brown, R., & Abarbanel, H. D. I. (1992). Determining embedding dimension for phase-space reconstruction using a geometrical construction. *Physical Review A, 45,* 3403–3411.

Kitano, H. (2002). Systems biology: A brief overview. *Science, 295,* 1662–1664.

Kugiumtzis, D. (2002). Statically transformed autoregressive process and surrogate data test for nonlinearity. *Physical Review E, 66,* 025201(R).

Lehnertz, K., & Elger, C. E. (1998). Can epileptic seizures be predicted? Evidence from nonlinear time series analysis of brain electrical activity. *Physical Review Letters, 80,* 5019–5022.

Lorenz, E. N. (1991). Dimension of weather and climate attractors. *Nature, 353,* 241–244.

Malik, M., & Camm, A. J. (1995). *Heart rate variability.* Armonk, NY: Futura.

Martinerie, J., Adam, C., Le van Quyen, M., Baulac, M., Renault, B., & Varela, F. J. (1998). Can epileptic crisis be anticipated? *Nature Medicine, 4,* 1173–1176.

May, R. M. (1976). Simple mathematical models with very complicated dynamics. *Nature, 261,* 459–467.

McSharry, P. E. (2004). Detection of dynamical transitions in biomedical signals using nonlinear methods. In M. Gh. Negoita et al., *Knowledge-based intelligent information & engineering systems* (pp. 483–490). (Lecture notes in artificial intelligence, No. 3215). Heidelberg/Berlin: Springer-Verlag.

McSharry, P. E., Clifford, G., Tarassenko, L., & Smith, L. A. (2003c). A dynamical model for generating synthetic electrocardiogram signals. *IEEE Transactions on Biomedical Engineering, 50,* 289–294.

McSharry, P. E., He, T., Smith, L. A., & Tarassenko, L. (2002). Linear and nonlinear methods for automatic seizure detection in scalp electroencephalogram recordings. *Medical & Biological Engineering & Computing, 40,* 447–461.

McSharry, P. E., & Smith, L. A. (1999). Better nonlinear models from noisy data: Attractors with maximum likelihood. *Physical Review Letters, 83,* 4285–4288.

McSharry, P. E., & Smith, L. A. (2004). Consistent nonlinear dynamics: Identifying model inadequacy. *Physica D, 192,* 1–22.

McSharry, P. E., Smith, L. A., & Tarassenko, L. (2003a). Comparison of predictability of epileptic seizures by a linear and a nonlinear method. *IEEE Transactions on Biomedical Engineering, 50,* 628–633.

McSharry, P. E., Smith, L. A., & Tarassenko, L. (2003b). Prediction of epileptic seizures: are nonlinear methods relevant? *Nature Medicine, 9,* 241–242.

Nerenberg, M. A. H., & Essex, C. (1990). Correlation dimension and systematic geometric effects. *Physical Review A, 42,* 7065–7074.

Palmer, T. N. (2000). Predicting uncertainty in forecasts of weather and climate. *Reports on Progress in Physics, 63,* 71–116.

Pincus, S. M., (1991). Approximate entropy as a measure of system complexity. *Proceedings of the National Academy of Sciences USA, 88,* 2297–2301.

Pincus, S. M., & Goldberger, A. L. (1994). Physiological time-series analysis: What does regularity quantify? *American Journal of Physiology: Heart Circulation Physiology, 266,* H1643–H1656.

Pritchard, D., & Theiler, J. (1994). Generating surrogate data for time series with several simultaneously measured variables. *Physical Review Letters, 73,* 951–954.

Provenzale, A., Smith, L. A., Vio, R., & Murante, G. (1992). Distinguishing between low-dimensional dynamics and randomness in measured time series. *Physica D, 58,* 31–49.

Richman, J. S., & Moorman, J. R. (2000). Physiological time-series analysis using approximate entropy and sample entropy. *American Journal of Physiology: Heart Circulation Physiology, 278,* H2039–H2049.

Rissanen, J. (1978). Modeling by shortest data description. *Automatica, 14,* 465–471.

Rosenstein, M. T., Collins, J. J., & De Luca, C. J. (1993). A practical method for calculating largest Lyapunov exponents from small data sets. *Physica D, 65,* 117–134.

Rosenstein, M. T., Collins, J. J., & De Luca, C. J. (1994). Reconstruction expansion as a geometry-based framework for choosing proper delay times. *Physica D, 73,* 82–98.

Roulston, M., & Smith, L. A. (2002). Evaluating probabilistic forecasts using information theory. *Monthly Weather Review, 130,* 1653–1660.

Ruelle, D. (1990). Deterministic chaos: the science and the fiction. *Proceedings of the Royal Society of London A, 427,* 241–248.

Sano, M., & Sawada, Y. (1985). Measurement of the Lyapunov spectrum from a chaotic time series. *Physical Review Letters, 55,* 1082–1085.

Sauer, T., Yorke, J. A., & Casdagli, M. (1991). Embedology. *Journal of Statistical Physics, 65,* 579–616.

Scheinkman, J. A., & Le Baron, B. (1989). Nonlinear dynamics and stock returns. *The Journal of Business, 62,* 311–337.

Schreiber, T., & Schmitz, A. (1996). Improved surrogate data for nonlinearity tests. *Physical Review Letters, 77,* 635–638.

Schreiber, T. (1998). Constrained randomization of time series data. *Physical Review Letters, 80*, 2105–2108.

Schwarz, G. (1978). Estimating the dimension of a model. *Annals of Statistics, 6*, 461–464.

Smith, L. (1988). Intrinsic limits on dimension calculations. *Physics Letters A, 133*, 283–288.

Smith, L. A., Ziehmann, C., & Fraedrich, K. (1999). Uncertainty dynamics and predictability in chaotic systems. *Quarterly Journal of the Royal Meteorological Society, 125*, 2855–2886.

Stark, J., & Hardy, K. (2003). Chaos: Useful at last? *Science, 301*, 1192–1193.

Strang, G. (1988). *Linear algebra and its applications*. San Diego: Harcourt College Publishers.

Takens, F. (1981). Detecting strange attractors in turbulence. In *Dynamical systems and turbulence* (pp. 366–381). (Lecture notes in mathematics, No. 898.) Berlin: Springer-Verlag.

Takens, F. (1985). On the numerical determination of the dimension of an attractor. In B. L. J. Braaksma, H. W. Broer, & F. Takens (Eds.), *Dynamical systems and bifurcations* (pp. 99–106). (Lecture notes in mathematics, No. 1125). Berlin: Springer.

Theiler, J. (1986). Spurious dimension from correlation algorithms applied to limited time-series data. *Physical Review A, 34*, 2427–2432.

Theiler, J. (1990). Statistical precision of dimension estimators. *Physical Review A, 41*, 3038–3051.

Theiler, J., Eubank, S., Longtin, A., Galdrikian, B., & Farmer, J. D. (1992). Testing for nonlinearity in time series: The method of surrogate data. *Physica D, 58*, 77–94.

Tsonis, A. A. (1992). *Chaos: From theory to applications*. New York: Plenum Press.

Wolf, A., Swift, J., Swinney, H., & Vastano, J. (1985). Determining Lyapunov exponents from a time series. *Physica D, 16*, 285–317.

Ziehmann, C., Smith, L. A., & Kurths, J. (1999). The bootstrap and Lyapunov exponents in deterministic chaos, *Physica D, 126*, 49–59.

24

Methodological Issues in the Application of Monofractal Analyses in Psychological and Behavioral Research*

Didier Delignières, Kjerstin Torre, and Loïc Lemoine

CONTENTS

* The original article was edited for format and originally published as Delignières, D., Torre, K., & Lemoine, L. (2005). Methodological issues in the application of monofractal analyses in psychological and behavioral research. *Nonlinear Dynamics, Psychology, and Life Sciences, 9,* 435–461. Reprinted with permission from the Society for Chaos Theory in Psychology & Life Sciences.

A number of recent research works tried to apply fractal methods to psychological or behavioral variables. Quite often, nevertheless, the use of fractal analyses remains rudimentary, and the goal of researchers seems limited to evidencing the presence of long-range correlation in data sets. This article presents some recent developments in monofractals theory, and some related methodological refinements. We also discuss a number of specific issues related to the application of fractal methods in psychological and behavioral research. Finally, we consider the potential use of such approach for a renewal of classical issues in psychology and behavioral science.

A number of experimental papers, in the last decade, revealed the fractal properties of psychological or behavioral variables, when considered from the point of view of their evolution over time. Fractals were evidenced, for example, in self-esteem (Delignières, Fortes, & Ninot, 2004a), in mood (Gottschalk, Bauer, & Whybrow, 1995), in serial reaction time (Gilden, 1997; Van Orden, Holden, & Turvey, 2003), in finger tapping (Gilden, Thornton, & Mallon, 1995; Delignières, Lemoine, & Torre, 2004b), in stride duration during walking (Hausdorff, Peng, Ladin, Wei, & Goldberger, 1995), in relative phase in a bimanual coordination task (Schmidt, Beek, Treffner, & Turvey, 1991; Torre, Lemoine, & Delignières, 2004), and in the displacement of the center-of-pressure during upright stance (Collins & De Luca, 1993; Delignières, Deschamps, Legros, & Caillou, 2003). Generally, these variables were previously conceived as highly stable over time, and fluctuations in successive measurements were considered as randomly distributed, and uncorrelated in time. As such, a sample of repeated measures was assumed to be normally distributed around its mean value, and noise could be discarded by averaging. This methodological standpoint was implicitly adopted in most classical psychological researches (for a deeper analysis, see Gilden, 2001; Slifkin & Newell, 1998). In other words, temporal ordering of data points was ignored and the possible correlation structure of fluctuations was clearly neglected.

Fractal analysis focuses in contrast on the time-evolutionary properties of data series and on their correlation structure. Fractal processes are characterized by a complex pattern of correlations appearing following multiple interpenetrated timescales. In such process, the value at a particular time is related not just to immediately preceding values, but to fluctuations in the remote past. Fractal series are also characterized by self-similarity, signifying that the statistical properties of segments within the series are similar, whatever the timescale of observation.

Evidencing fractal properties in empirical time series has important theoretical implications, and could lead to a deep renewal of models. Fractals are considered as the natural outcome of complex dynamical systems behaving at the frontier of chaos (Bak & Chen, 1991; Marks-Tarlow, 1999). Psychological variables should then be conceived as the macroscopic and dynamical products of a complex system composed of multiple interconnected elements. Moreover, psychological and behavioral time series often present fractal characteristics close to a very special case of fractal process, called $1/f$ or *pink* noise. "$1/f$ noise" signifies that when the power spectrum of these time series is considered, each frequency has power proportional to its period of oscillation. As such, power is distributed across the entire spectrum and not concentrated at a certain portion. Consequently, fluctuations at one timescale are only loosely correlated with those of another timescale. This relative independence of the underlying processes acting at different timescales suggests that a localized perturbation at one timescale will not necessarily alter the stability of the global system. In other words, $1/f$ noise renders the system more stable and more adaptive to internal and external perturbations (West & Shlesinger, 1989).

$1/f$ noise was evidenced in most series produced by "normal" participants, characterized as young and healthy. As such, this $1/f$ behavior could be considered as an indicator of the efficiency of the system that produced the series. In contrast, series obtained with older participants or with patients with specific pathologies exhibited specific alterations in fractality (Gottschalk et al., 1995; Hausdorff et al., 1997; Yoshinaga, Miyazima, & Mitake, 2000).

The application of fractal methods, nevertheless, often remains rudimentary: analyses are limited to the use of a unique method, the collected series are sometimes too short for a valid assessment, and more generally, the theoretical background of fractals and related methods is not fully exploited. The recent theoretical and methodological refinements of fractal analyses (see, e.g., Eke et al., 2000; Eke, Hermann, Kocsis, & Kozak, 2002) appear largely unknown in the psychological community. These methodological limitations severely restrain the potential impact of fractal approaches on psychological theories and models, and frequently the discovery of a fractal behavior in experimental series is just presented as an anecdotic result.

The aim of this paper is to present a wide overview of fractal analyses, with a special focus on the recent refinements of both theoretical background and methodological approach. We first develop the formal distinction between fractional Gaussian noise and fractional Brownian motion, which seems essential for relevant application of fractal methods. Then, we present the different methods proposed in the literature, in the frequency and the time domains. The next section will focus on the methodological problems related to the identification of long-range correlations, the classification of series, and the estimation of the fractal exponents. Finally, we evoke some more basic questions concerning the nature of the empirical series to use in a fractal approach, and we suggest some guidelines for the development of such approach in psychological and behavioral research.

FGN/FBM Model

In order to ensure better understanding of the following parts of this article, a deeper and more theoretical presentation of fractal processes is necessary. A good starting point for this presentation is Brownian motion, a well-known stochastic process that can be represented as the random movement of a single particle along a straight line. Mathematically, Brownian motion is the integration of a white Gaussian noise. As such, the most important property of Brownian motion is that its successive increments in position are uncorrelated: each displacement is independent of the former, in direction as well as in amplitude. Einstein (1905) showed that, on average, this kind of motion moves a particle from its origin by a distance that is proportional to the square root of the time.

Mandelbrot and van Ness (1968) defined a family of processes they called *fractional Brownian motions* (fBm). The main difference with ordinary Brownian motion is that in an fBm, successive increments are correlated. A positive correlation signifies that an increasing trend in the past is likely to be followed by an increasing trend in the future. The series is said to be persistent. Conversely, a negative correlation signifies that an increasing trend in the past is likely to be followed by a decreasing trend. The series is then said to be anti-persistent.

Mathematically, an fBm is characterized by the following scaling law:

$$< \Delta x > \propto \Delta t^{H}, \tag{24.1}$$

which signifies that the expected displacement $<\Delta x>$ is a power function of the time interval (Δt) over which this displacement is observed. H represents the typical scaling exponent of the series and can be any real number in the range $0 < H < 1$. The aims of fractal analysis are to check whether this scaling law holds for experimental series and to estimate the scaling exponent. Ordinary Brownian motion corresponds to the special case $H = 0.5$ and constitutes the frontier between anti-persistent ($H < 0.5$) and persistent fBms ($H > 0.5$).

Fractional Gaussian noise (fGn) represents another family of fractal processes, defined as the series of successive increments in an fBm. Note that fGn and fBm are interconvertible: when an fGn is cumulatively summed, the resultant series constitutes an fBm. Each fBm is then related to a specific fGn, and both are characterized by the same H exponent. These two processes possess fundamentally different properties: fBm is nonstationary with time-dependent variance, while fGn is a stationary process with a constant expected mean value and constant variance over time. The H exponent can be assessed from an fBm series as well as from the corresponding fGn, but because of the different properties of these processes, the methods of estimation are necessarily different.

Recently, a systematic evaluation of fractal analysis methods was undertaken by Bassingthwaighte, Eke, and collaborators (Caccia, Percival, Cannon, Raymond, & Bassingthwaighte, 1997; Cannon, Percival, Caccia, Raymond, & Bassingthwaighte, 1997; Eke et al., 2000, 2002). This methodological effort was based on the previously described dichotomy between fGn and fBm. According to these authors, the first step in a fractal analysis aims at identifying the class to which the analyzed series belongs, that is, fGn or fBm.

Then, the scaling exponent can be properly assessed, using a method relevant for the identified class. The evaluation proposed by these authors clearly showed that most methods gave acceptable estimates of H when applied to a given class (fGn or fBm), but led to inconsistent results for the other. As claimed by Eke et al. (2002), researchers were not aware before a recent past of the necessity of this dichotomic model. As such, a number of former empirical analyses and theoretical interpretations remain questionable.

Monofractal Analysis Methods

We present in this section the most commonly used methods in monofractal analysis. Some of them work in the frequency domain, and the other in the time domain. After their first introduction, a number of refinements were proposed for each of them. These refinements are explained in detail in the following paragraphs.

Power Spectral Density Analysis

This method is widely used for assessing the fractal properties of time series, and works on the basis of the periodogram obtained by the fast Fourier transform algorithm. The relation of Mandelbrot and van Ness (1968) can be expressed as follows in the frequency domain:

$$S(f) \propto \frac{1}{f^{\beta}},\tag{24.2}$$

where
  f is the frequency
  $S(f)$ is the correspondent squared amplitude
  β is estimated by calculating the negative slope ($-\beta$) of the line relating $\log(S(f))$ to $\log f$ (Figure 24.1)

Obtaining a well-defined linear fit in the double-logarithmic plot is an important indication of the presence of long-range correlation in the original series.
  According to Eke et al. (2000), power spectral density (PSD) allows distinguishing between fGn and fBm series, as fGn corresponds to β exponents ranging from -1 to $+1$, and fBm to exponents from $+1$ to $+3$. β can be converted into $\hat{H}$ according to the following equations:

$$\hat{H} = \frac{\beta + 1}{2} \quad \text{for fGn,}\tag{24.3a}$$

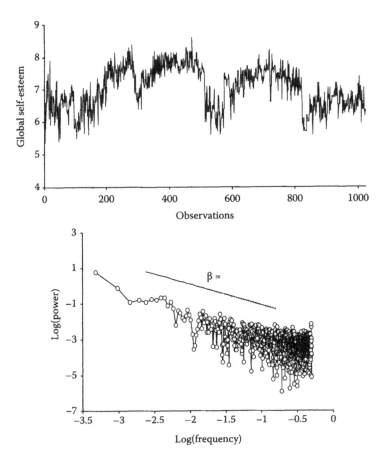

FIGURE 24.1
Upper: an example of empirical time series, obtained by a bidaily assessment of self-esteem during 512 consecutive days. Lower: bi-logarithmic power spectrum obtained by PSD from the above series. (Data from Delignières, D. et al., *Nonlinear Dynam. Psychol. Life Sci.*, 8, 479, 2004.)

or

$$\hat{H} = \frac{\beta - 1}{2} \quad \text{for fBm.} \tag{24.3b}$$

Note that in these equations and thereafter in the text, $\hat{H}$ represents the estimate provided by the analysis, and H the true exponent of the series.

Eke et al. (2000) proposed an improved version of PSD, using a combination of preprocessing operations: first the mean of the series is subtracted from each value, and then a parabolic window is applied: each value in the series is multiplied by the following function:

$$W(j) = 1 - \left(\frac{2j}{N+1} - 1 \right)^2 \quad \text{for } j = 1, 2, \ldots, N. \tag{24.4}$$

Third, a bridge detrending is performed by subtracting from the data the line connecting the first and last point of the series. Finally, the fitting of β excludes the high-frequency power estimates ($f > 1/8$ of maximal frequency). This method was proven to provide more reliable estimates of the spectral index β, and was designated as $^{low}PSD_{we}$.

Detrended Fluctuation Analysis

This method was initially proposed by Peng et al. (1993). The $x(t)$ series is integrated by computing for each t, the accumulated departure from the mean of the whole series:

$$X(k) = \sum_{i=1}^{k} \left[x(i) - \bar{x} \right]. \tag{24.5}$$

This integrated series is divided into nonoverlapping intervals of length n. In each interval, a least-squares line is fit to the data (representing the trend in the interval). The series $X(t)$ is then locally detrended by subtracting the theoretical values $X_n(t)$ given by the regression. For a given interval length n, the characteristic size of fluctuation for this integrated and detrended series is calculated by

$$F = \sqrt{\frac{1}{N} \sum_{k=1}^{N} \left[X(k) - X_n(k) \right]^2}. \tag{24.6}$$

This computation is repeated over all possible interval lengths (in practice, the shortest length is around 10, and the largest $N/2$, giving two adjacent intervals). Typically, F increases with interval length n. A power law is expected, as

$$F \propto n^{\alpha}, \tag{24.7}$$

where α is expressed as the slope of a double-logarithmic plot of F as a function of n (Figure 24.2). As PSD, detrended fluctuation analysis (DFA) allows distinguishing between fGn and fBm series, fGn corresponding to α exponents ranging from 0 to 1, and fBm to exponents from 1 to 2. α can be converted into H according to the following equations:

$$\hat{H} = \alpha \quad \text{for fGn,} \tag{24.8a}$$

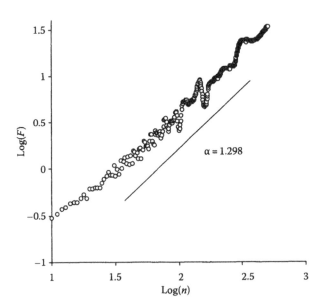

FIGURE 24.2
Diffusion plot obtained by the application of DFA to the series presented in the upper panel of Figure 24.1. The statistics $F(n)$ is plotted against the length of time intervals.

or

$$\hat{H} = \alpha - 1 \quad \text{for fBm}. \tag{24.8b}$$

Rescaled Range Analysis

This method was originally developed by Hurst (1965). The $x(t)$ series is divided into nonoverlapping intervals of length n. Within each interval, an integrated series $X(t, n)$ is computed:

$$X(t,n) = \sum_{k=1}^{t} [x(k) - \bar{x}], \tag{24.9}$$

where $\bar{x}$ is the average within each interval. In the classical version of rescaled range (R/S) analysis, the range R is computed for each interval, as the difference between the maximum and the minimum of integrated data $X(t, n)$:

$$R = \max_{1 \le t \le n} X(t,n) - \min_{1 \le t \le n} X(t,n). \tag{24.10}$$

An improved version, *R/S-detrended*, was proposed by Caccia et al. (1997): a straight line connecting the end points of each interval is subtracted from each

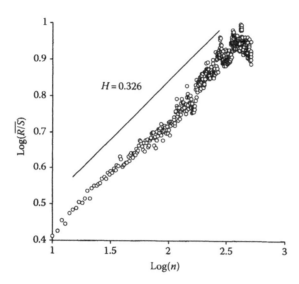

FIGURE 24.3

Diffusion plot obtained by the application of R/S analysis to the series presented in the upper panel of Figure 24.1. The statistics R/S is plotted against the length of time intervals.

point of the cumulative sums $X(t, n)$ before the calculation of the local range. In both methods, the range is then divided for normalization by the local standard deviation (S) of the original series $x(t)$. This computation is repeated over all possible interval lengths (in practice, the shortest length is around 10, and the largest $(N - 1)/2$, giving two adjacent intervals). Finally, the R/S are averaged for each interval length n. $\overline{R/S}$ is related to n by a power law

$$\overline{R/S} \propto n^H. \tag{24.11}$$

$\hat{H}$ is expressed as the slope of the double-logarithmic plot of $\overline{R/S}$ as a function of n (Figure 24.3). R/S analysis is theoretically conceived to work on fGn signals, and should provide irrelevant results for fBm signals.

Dispersional Analysis

This method was introduced by Bassingthwaighte (1988). The $x(t)$ series is divided into nonoverlapping intervals of length n. The mean of each interval is computed, and then the standard deviation (SD) of these local means, for a given length n. These computations are repeated over all possible interval lengths. SD is related to n by a power law

$$SD \propto n^{H-1}. \tag{24.12}$$

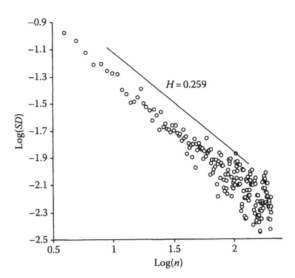

FIGURE 24.4
Diffusion plot obtained by the application of dispersional analysis to the series presented in the upper panel of Figure 24.1. The *SD* of the mean is plotted against the length of time intervals.

The quantity $(H - 1)$ is expressed as the slope of the double-logarithmic plot of *SD* as a function of *n* (Figure 24.4). Obviously, the *SD*s calculated from the highest values of *n* tend to fall below the regression line and bias the estimate. Caccia et al. (1997) suggested to ignore measures obtained from the longest intervals. As *R/S* analysis, dispersional analysis (Disp) is theoretically conceived to work on fGn signals, and should provide irrelevant results for fBm signals.

Scaled Windowed Variance Method

These methods were developed by Cannon et al. (1997). The $x(t)$ series is divided into nonoverlapping intervals of length *n*. Then the *SD* is calculated within each interval using the formula

$$SD = \sqrt{\frac{\sum_{t=1}^{n}\left[x(t) - \bar{x}\right]^2}{n-1}}, \tag{24.13}$$

where $\bar{x}$ is the average within each interval. Finally, the average standard deviation $(\overline{SD})$ of all intervals of length *n* is computed. This computation is repeated over all possible interval lengths. For a fractal series, $\overline{SD}$ is related to *n* by a power law

$$\overline{SD} \propto n^H, \tag{24.14}$$

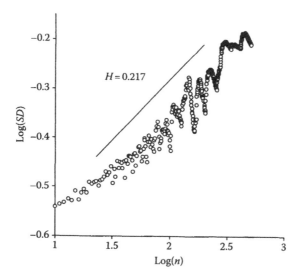

FIGURE 24.5

Diffusion plot obtained by the application of SWV analysis to the series presented in the upper panel of Figure 24.1. The average *SD* is plotted against the length of time intervals.

where $\hat{H}$ is expressed as the slope of the double-logarithmic plot of $\overline{SD}$ as a function of n (Figure 24.5). Cannon et al. (1997) showed that a detrending of the series within each interval before the calculation of the *SD* provided better estimates of H, especially with short series. Exploiting the diffusion properties of signals, scaled windowed variance (SWV) methods are conceived to work properly on fBm, but should provide irrelevant results on fGn.

SWV methods can also be used to distinguish between fGn and fBm near the $1/f$ boundary. Eke et al. (2000) proposed a method called *signal summation conversion* (SSC) method, based on the application of SWV to the cumulative sum of the original signal. If the obtained $\hat{H}$ is lower than one, the original series is a fGn (in this case, the cumulant series is the corresponding fBm). If $\hat{H}$ is higher than one, the original series is an fBm.

Identifying Fractal Processes

The first step, in a fractal analysis, is to detect the presence of long-range dependence in the series. All these methods theoretically allow evidencing such long-range dependences through the visual inspection of power spectrum in the frequency domain, or of the diffusion plot in the time domain. Usually, researchers apply a unique method (in the frequency domain or in the time domain), and base their conclusions on this visual, and qualitative, observation of a linear regression in double-logarithmic plots. This apparent simplicity is highly questionable and raises a number of methodological and theoretical problems. A simulated or experimental time series, while not possessing any long-range correlation property, can mimic in the resultant double-logarithmic plots the expected linear fit, and lead to false claims about the presence of underlying fractal processes (Thornton & Gilden, 2005).

Rangarajan and Ding (2000) highlighted the possible misinterpretations that could arise from the application of a unique method in fractal analysis. They developed a series of examples showing how spectral or time-related methods, applied in isolation, could lead to false identification of long-range dependence. They showed, for example, that a series composed by the superposition of an exponential trend over a white noise gives a perfect linear fit in the diffusion plot obtained through R/S analysis. The spectral method, conversely, provided a flat spectrum revealing the absence of serial correlation in the series. As well, an order-one autoregressive process could be interpreted as a fractal series on the basis of the application of R/S analysis: the diffusion plot presents in this case also a perfect linear fit. The absence of long-range correlation is nevertheless attested by the power spectrum, with a typical flattening at low frequencies. Rangarajan and Ding (2000) concluded with the necessity of an integrated approach, based on the consistent use of several tools, in the frequency as well as in the time domain. The identification of long-range correlation requires the obtaining of the typical graphical signature with several methods, and also the consistency of the obtained slopes (this consistency is assessable through Equations 23.3 and 23.8).

This integrated approach, nevertheless, remains limited to the qualitative analysis of spectral and diffusion plots, and does not include any test aiming at statistically evidence the presence of long-range correlation. Some authors have proposed the application of surrogate tests, in order to differentiate between long-range scaling and a random process with no long-range correlation (see, e.g., Haussdorf et al., 1995). Surrogate data sets are obtained by randomly shuffling the original time series. Each surrogate data sets has the same mean and variance as the corresponding original series, and differs only in the sequential ordering. The scaling exponents of the surrogate data sets are then statistically compared to those of the original series. Nevertheless, the interest of these tests remains limited, because considering their null hypothesis, they allow to attest for the presence of correlations in the series, but they are unable to certify their long-range nature.

This problem was addressed by several recent papers (Farrell, Wagenmakers, & Ratcliff, 2006; Thornton & Gilden, 2005; Wagenmakers, Farrell, & Ratcliff, 2004). According to these authors, the main question is to statistically distinguish between short-term and long-term dependence in the series. Short-term dependence signifies that the current value in the series is only determined by a few number of preceding values. These short-term dependence are generally modeled by the ARMA models developed by Box and Jenkins (1976), which are composed by a combination of autoregressive and moving average terms. A quite simple solution could be to compare the shape of the autocorrelation function, which is supposed to be exponential in the case of a short-term memory process, and to decay according to a power law in the case of long-term dependence. This comparison, nevertheless, remains qualitative, and autocorrelation functions do not present sufficient information to give support to unequivocal statistical tests.

Wagenmakers et al. (2004) based their approach on the so-called ARFIMA models, which are frequently used in the domain of econometry for modeling long-range dependence (see, e.g., Diebolt and Guiraud, in press). ARFIMA is the acronym of *autoregressive fractionally integrated moving average*, and these models differ from the traditional ARMA models by the inclusion of an additional term, d, corresponding to a fractional integration process. Wagenmakers et al. (2004) proposed to test the null hypothesis $d = 0$, in order to determine whether the analyzed series belongs to the ARFIMA or to the ARMA families.

Thornton and Gilden (2005) proposed to contrast, more directly, on the basis of the obtained power spectra, short-range (ARMA) processes, and long-range fractal processes. They constructed an optimal Bayesian classifier that discriminates between the two families of processes, and showed that this classifier had sufficient sensitivity to avoid false identifications.

As can be seen, the identification of true long-range correlation in a series is not so straightforward, and remains a current theoretical and methodological debate. The simple presence of a linear trend in the log–log power spectrum, or in a part of this spectrum, cannot be per se considered as a definitive proof of underlying long-range dependence.

Series Classification

The preliminary classification of series as fGn or fBm is a crucial step in fractal analysis. This procedure requires methods that can be applied to both classes of signals. Four methods, among those previously presented, satisfy this initial requirement (PSD, $^{low}PSD_{we}$, DFA, and SSC).

Delignières et al. (2006) analyzed the performances of these four methods, for distinguishing between fGn and fBm series apart from the $1/f$ boundary. They used simulated series with known true H exponents, generated according to the procedure proposed by Davies and Harte (1987). Their analyses were conducted with series of fGn with "true" H exponents ranging from 0.1 to 0.9, by steps of 0.1, and with the corresponding series of fBm, obtained by integration. 40 fGn series and 40 fBm series were generated, for each H value. The accuracy of each method was assessed through the mean of the 40 estimates of H obtained for each true H value, and the variability was estimate through the SD of the samples of estimates.

Their results showed that the four methods were able to distinguish between fGn and fBm, at least when true H exponents were sufficiently far from the $1/f$ boundary. Nevertheless, a zone of uncertainty remained: a number of series classified as fGn with exponents close to 1 were in fact fBm processes. The opposite was also observed, but to a lesser extent. This asymmetry results from the important negative bias that characterizes all methods for fBm series with low H exponents (i.e., $H = 0.1$ or $H = 0.2$). This negative bias was particularly salient for PSD: all fBm series with $H = 0.1$ were classified as fGn using this method. PSD worked better for $H = 0.2$, despite the negative bias, because of a low variability in H estimation. $^{low}PSD_{we}$ also presented a negative bias for fBm series with $H = 0.1$ or $H = 0.2$, but this bias was lesser than for PSD. DFA gave quite similar results in terms of misclassification percentages, because of a global negative bias for fBm series and a rather high variability in H estimation. Finally, SSC appeared unable to provide a better signal classification in this uncertainty range.

This difficulty to distinguish between fGn and fBm around the $1/f$ boundary is problematic, as a number of empirical series produced by psychological or behavioral systems falls into this particular range (e.g., Delignières et al., 2004a, 2004b; Gilden et al., 1995; Gilden, 2001; Hausdorff et al., 1997). Finally, the best solution when series fall into this uncertainty range could be to restrain analyses to methods insensitive to the fGn/fBm dichotomy, such as $^{low}PSD_{we}$ or DFA. In other terms, the solution could be to work directly on β or α exponents, without trying to convert them into H metrics. This could be necessary, for example, when the goal is to determine the mean fractal exponent of a sample of series, and when some series are classified as fGn, and the others as fBm (see, e.g., Delignière et al., 2004). The mean exponent can be computed in this case on the basis of the samples of β or α obtained by $^{low}PSD_{we}$ or DFA, and then possibly converted into H. DFA seems preferable in this case, as this method presents lower biases than spectral analyses. The high variability of DFA should be compensated by a sufficient number of series in the sample.

Estimating H

After the classification of the series as fGn or fBm, the next step is to accurately estimate the fractal exponent. Delignières et al. (2004c) showed that the accuracy of each method depended on the class of analyzed series (i.e., fGn or fBm), and within each class, on the localization of the underlying exponent with the H continuum (i.e., anti-persistent or persistent series).

Estimating H for fGn Series

When a series is clearly classified as fGn, a number of methods are available for a more accurate estimation of its fractal exponent. Clearly, the least-biased method for fGn series is DFA. Alternatively, one could use SWV methods on the cumulative sum of the original series. For these two methods, the bias remains limited over the whole range of H, and variability seems acceptable, at least for $H \leq 0.5$.

R/S analysis presents a positive bias for series with $H < 0.4$, but limited biases for $H \geq 0.5$, and a low variability within this range. Disp could also be proposed for the analysis of fGn series with $H \leq 0.5$. Nevertheless, the level of variability seems higher for Disp than for DFA or SWV within this H range. The results of Delignières et al. (2004c) concerning Disp are different from those of Eke et al. (2000), who selected Disp as the most relevant for the analysis of fGn series. Caccia et al. (1997) proposed improved versions of this method, reducing bias and variance in H estimation. Future tests could show if they could constitute an alternative to SWV and R/S methods for fGn series.

In conclusion, the accurate estimation of H should follow different ways according to the nature of the series. For anti-persistent noises ($H < 0.5$), the best strategy seems to calculate the cumulative sum of the series, and then to apply SWV. For persistent noises, R/S analysis provides the best results.

Estimating *H* for fBm Series

Clearly, the best method for fBm series is SWV: biases are limited over the whole range of H values, and variability remains low, especially for $H < 0.5$. In contrast, DFA presents a systematic negative bias and a high level of variability. $^{low}PSD_{we}$ could represent an interesting alternative, but is characterized by higher levels of variability than SWV, and some systematic biases, for very low and very high H values.

Means Comparisons

In some occasion, the accurate estimation of scaling exponents is not of prior interest, and the main goal of researchers is to contrast the mean exponents obtained in two or more experimental groups (see, e.g., Chen, Ding, & Kelso, 2001). The main requirement for means comparison is to obtain a low variability in H estimation. Limited biases can be accepted, if they do not interfere with the capability of the method to distinguish between exponents. Delignières et al. (2004c) proposed some guidelines for such comparisons.

Means Comparisons for fGn Series

PSD seems the best candidate for fGn series. Despite a negative bias for low values of H, and a positive bias for high values, the variability in H estimation remains low, even for short series. SWV, applied on the cumulative sums of the original series, could constitute a valuable alternative when $H < 0.5$. For persistent fGn ($H > 0.5$), R/S analysis could also be used.

Means Comparisons for fBm Series

For sub-diffusive fBm series ($H < 0.5$), SWV presents the best guarantees: variability remains limited (below 0.1) and biases are absent. The choice is more difficult concerning over-diffusive fBm series ($H > 0.5$), because all methods present high levels of variability within this range. The best choice seems to be $^{low}PSD_{we}$, but the use of time series longer than 1024 points is highly recommended.

Relevant Series for Fractal Analysis

A given system can be observed from diverse points of view, and then produce different kinds of series. Are some series more favorable than others for expressing the fractal properties of a system? Chen, Ding and Kelso (1997) analyzed the series obtained in a synchronization tapping paradigm, during which participant had to tap in synchrony with an auditory metronome. Two series were collected: the series of the successive inter-tap intervals, and the series of successive time delays between the occurrence of the signal and the corresponding tap. The analysis of the second series revealed a clear long-range process belonging to the fGn class, with a spectral exponent β, obtained by PSD, of about 0.54, and an estimate of H, obtained through R/S analysis, of about 0.79. The analysis of the inter-tap intervals series gave more inconsistent results, with a positive slope in the log–log power spectrum (corresponding to a β exponent of about −1.46), and a parabolically shaped curve in the diffusion plot of R/S analysis. The authors proposed the notion of *fundamental time series*, to qualify the dynamical variables that seem to possess relevant information for revealing the fractal properties of a given system.

Another interesting example is provided by a line of research that aimed at evidencing the fractal properties of bimanual coordination. In the bimanual coordination paradigm, participants are requested to perform simultaneous rhythmical oscillations with the two hands, according to a prescribed phase relationship between the two effectors (Kelso, Holt, Rubin, & Kugler, 1981; Kelso, 1984). Two modes of coordination were proven to be particularly stable and were extensively studied: the in-phase coordination, in which homologous muscles perform simultaneous contractions, and the anti-phase coordination, in which homologous muscles perform alternate contractions. The relevant variable for analyzing such coordination is the relative phase, that is, the difference between the instantaneous phases of each oscillator, which equals theoretically 0° for the in-phase mode, and 180° for the anti-phase mode.

Two measures of relative phase were used in the literature, and were generally considered as interchangeable. Continuous relative phase (CRP) is derived from the position (x_t) and velocity $(\dot{x}_t)$ time series of each oscillator. The phase angle is determined for each oscillator using the following equation:

$$\phi_t = \tan^{-1}\left(\frac{\dot{x}_t}{x_t}\right), \tag{24.15}$$

and the relative phase is determined as the instantaneous difference between the phase of each oscillator.

Discrete relative phase (DRP) is punctually computed as the temporal difference between similar inflexion points in the oscillation of the two oscillators, reported to the period of one oscillator. CRP was often interpreted as a higher resolution form of DRP. Nevertheless, Peters, Haddad, Heiderscheit, van Emmerik and Hamill (2003) showed that these two measures essentially differ in nature: DRP yields information regarding the relative dispersion of events in oscillatory signals, while CRP described their relationship in a higher order phase space.

The fractal properties of relative phase in bimanual coordination were first studied by Schmidt et al. (1991) on the basis of CRP series collected in trials performed in anti-phase. Spectral analyses revealed β exponents, ranging from 1.64 to 2.96, with an average value of about 2.52. Note that this result seemed particularly unrealistic, as such β value suggested a kind of over-diffusive fBm, far from the $1/f$ range typically observed in biological systems. In a more recent study, Torre et al. (2004) analyzed the fractal properties of DRP series, collected in in-phase and anti-phase trials. On the basis of $^{low}PSD_{we}$ and DFA results, DRP series were classified as persistent fGns (for comparison, the mean β estimate was 0.34 for in-phase trials, and 0.44 for anti-phase trials). The estimation of H, performed with four methods ($^{low}PSD_{we}$, DFA, R/S analysis, and SWV), gave mean values ranging from 0.67 to 0.72 for in-phase series, and from 0.72 to 0.78 for anti-phase trials. These estimates were obviously more realistic, falling in the $1/f$ range and compatible with the essential stationarity of such behavioral series. On the basis of these results, DRP could be considered as providing a fundamental time series for assessing the fractal properties in bimanual coordination.

CRP and DRP series present another important difference. CRP series are computed as genuine time series, as successive values are spaced by equal time intervals. Conversely, DRP series correspond to a cycle-to-cycle measurement, and the time interval between two successive values depends on the local period used as denominator in the calculation of the relative phase.

DRP series have to be considered as *event series*, composed of temporally ordered measures, but not as genuine time series.

In fact, most series in psychological and behavioral studies are event series, and not time series. This was the case, for example, in the experiments proposed by Gilden (2001), in which the analyzed series were composed of ordered successive performances. In the special case of continuation tapping experiments, the collected series are composed of the successive inter-tap time intervals (Delignières et al., 2004b; Gilden et al., 1995). True time series are rather uncommon in such research (see, nevertheless, Collins & De Luca, 1993; Delignières et al., 2004a; Treffner & Kelso, 1999).

This could be conceived as a formal obstacle for the application of time series analyses such as those previously presented. Most authors consider, nevertheless, that these time series analyses remain applicable, but obviously time cannot be considered here in its absolute sense. When applying spectral analyses, "frequency" should not be read in Hertz units, but rather in inverse trial number (Gilden, 2001), or in number of cycles for N trials or observations (Musha, Katsurai, & Teramachi, 1985; Yamada, 1996; Yamada & Yonera, 2001). As well, the "intervals" taken into account in all time-related methods are not time intervals, but rather lengths of samples of successive observations.

This distinction between time series and event series remains crucial in fractal analyses. Researchers aiming at undertaking a fractal approach to a given system could be naturally inclined to opt for time series, considering the nature of the statistical procedures commonly used in this domain. We believe, nevertheless, that the key variable in fractal analysis is not fluctuation in time, but rather cycle-to-cycle or trial-to-trial fluctuation. In most psychological experiments, the possible mechanisms underlying dependence in series are obviously related to the serial occurrence of trials. Supposed sequential effects such as priming, knowledge of results, suggest clearly to focus on trial-to-trial variability. In other kinds of experiments (as, e.g., the previously evoked studies on motor coordination), the choice of discrete event series is not so directly defensible. Continuous fluctuation in time could be considered per se as a variable of interest, and often the dynamics of the system under study does not present the necessary periodical key events justifying the collection of discrete series (see, e.g., Collins & De Luca, 1993, Treffner & Kelso, 1999). Frequently, nevertheless, the behavior of the system possesses a kind of periodicity that could allow collecting a discrete event series, with one measure for each successive cycle.

As explained at the beginning of this paper, the main aim of fractal analyses is to offer a renewed approach of the classical problems of stability, variability, and flexibility in the behavior of complex systems. When such a system exhibits a cyclical activity, each cycle can be considered as a functional unit, and then cycle-to-cycle fluctuation represents a kind of "functional variability," directly related to the goal of the manifested activity. In contrast, continuous fluctuation corresponds to an "absolute variability," which could be meaningless at the macroscopic scale of successive cycles.

Special Case of Bounded Series

Another problem that can be highlighted raises from the bounded character of the dynamics of most behavioral variables. It is important to keep in mind that a pure fBm is typically unbounded: The fluctuations grow with the time interval length in a power-law way, and the expected displacement increases indefinitely with time. In other words, the diffusion with time of a pure fBm is unlimited. In contrast, behavioral time series are generally bounded within physiological limits. This is the case, for example, for the trajectory of the center-of-pressure during postural sway, which is obviously bounded within the area of support of the subject's feet. As a consequence, the diffusion process remains limited and the variance of such behavioral time series cannot exceed a ceiling value, and, at least beyond a critical time interval (necessary to reach this ceiling value), should become more or less independent to time. This should naturally yield to a crossover phenomenon in the relationship between variance (or displacement) and time interval, with persistence at short time intervals and anti-persistence at long time intervals. Several experiments showed evidence of such results in the fractal analysis of biological time series (Collins & De Luca, 1993; Treffner & Kelso, 1995, 1999) and one could hypothesize that this typical feature could be due to the bounded character of the series under study. This hypothesis was considered by Liebovitch and Yang (1997), who showed that a simulated bounded random walk yields similar results, with comparable crossover phenomena.

An elegant solution for this problem is to study the fractal properties of the integrated time series, rather than those of the original signals (Feder, 1988; Hurst, 1965; Peng, Havlin, Stanley, & Goldberger, 1995). If the original signal is constrained within physiological boundaries, the integrated series is not bounded and exhibits fractal properties that can be quantified on the basis of Equation 24.1. Such a procedure allows distinguishing between the uncorrelated noise (which gives a Brownian motion after integration), and a bounded fBm, which should exhibit after integration a higher diffusion than Brownian motion. In others words, the solution is to infer the fractal properties of the original signal from the diffusion properties of its integrated series. Delignières et al. (2003) showed that the application of such method on postural data avoided the appearance of a crossover phenomenon in the diffusion plot, and gave a more readable picture of the fractality of the system.

Series Length

Eke et al. (2000, 2002) showed that the accuracy of the estimation of fractal exponents is directly related to the length of the series. One of their main conclusions is that fractal methods cannot give reliable results with series shorter than 2^{12} data points, and in their papers, especially devoted to physiological research, they focused on results obtained with very long series (2^{17} data points). Such series cannot be collected in psychological research. The application of time series analyses supposes that the system under study remains unchanged during the whole window of observation, and in psychological experiments, the lengthening of the task raises evident problems of fatigue or lack of concentration (Madison, 2001). Generally, the use of series of 2^9 or 2^{10} data points was considered as an acceptable compromise between the requirements of time series analyses and the limitations of psychological experiments (see, e.g., Chen et al., 1997, 2001; Delignières et al., 2004b; Gilden, 1997, 2001; Musha et al., 1985; Yamada, 1996; Yamada & Yonera, 2001; Yamada, 1995).

In their evaluation of fractal methods, Delignières et al. (2004c) analyzed the performance of each method with particularly short series (i.e., 1024, 512, and 256 data points). They expected to find a dramatic increase of biases and variability with series shorter than 1024 data points. These results were generally present, but with rather moderate amplitudes. Only $^{low}PSD_{we}$ appeared severely affected by the shortening of series. This observation is very important, because of the difficulty to obtain long time series in psychological and behavioral experiments. These results suggest that a better estimate of H could be obtained, with a similar time on the experimental task, from the average of four exponents derived from distinct 256 data points series (with an appropriate period of rest between two successive sessions), than from a single session providing 1024 data points. This conclusion could open new perspectives of research in areas that were until now reticent for using this kind of analyses.

Research Goals

We would like to express some guidelines about the scientific goals that could be pursued through fractal analyses. Often, the aim of researchers is limited to evidencing the fractal character of fluctuation in the behavior of the system under study, and the papers are concluded by a general discussion about the potential interest of dynamical systems theory and/or self-organized criticality for a renewal of the models in the concerned domain.

A first problem that has to be explored is the reliability of the typical exponents provided by fractal methods. This is not only a psychometrical question (how reliable is the measure?), but a fundamental theoretical debate: can we consider the fractal properties of a time series an inherent characteristic of the system that produced the series, or are the obtained exponents only the reflection of a temporary emergent organization, which could be entirely different from one realization of the task to the other? This question was never clearly addressed in the literature, and generally a unique assessment of fractal exponent is performed for each participant. One of us recently tried to assess the reproducibility of the fractal exponents obtained in continuous tapping experiments (Lemoine, 2004). Participant had to perform continuous tapping according to four initially prescribed tempi, and the experiment was replicated after a delay of 3 months. Results showed a good reproducibility of fractal measures, especially for the highest tempi. On the other hand, fractal exponents appeared rather specific to each experimental condition (initial tempo), suggesting a close link between task requirements and fractal properties.

A second important research goal is to identify the experimental factors susceptible to alter the fractal properties of the collected series. We evoked in the introduction that a number of studies showed significant differences in fractal exponents when contrasting young and healthy participants with elderly or diseased patients (Gottschalk et al., 1995; Hausdorff et al., 1997; Peng et al., 1995; Yoshinaga et al., 2000). Generally, aging or disease led to specific alterations of $1/f$ noise, in the direction of white noise, or conversely in the direction of Brownian motion. Such alteration was interpreted as a disruption of the optimal compromise between order and chaos established by $1/f$ fractality, leading to a dramatic decrease of the adaptive capabilities of the system.

A more instructive line of research should aim at analyzing the effects of experimental factors on fractality in repeated measure designs. The goal of such a strategy is to highlight the possible temporary, or more definite alterations of fractality that could arise from the controlled manipulation of experimental factors. Such an experiment was recently conducted by Chen et al. (2001; see also Ding, Chen, & Kelso, 2002), by contrasting the exponents obtained by the same participants in two tapping conditions: a condition in which the taps were performed in synchrony with the beeps of the metronome, and a condition in which the tap were in syncopation with the beeps. Their results evidenced a significant difference between the two conditions, with a spectral exponent β of about 0.54 for synchronization, and 0.77 for syncopation. It is important to note that syncopation was previously proven to be intrinsically less stable than synchronization, and more difficult to perform. Moreover, the authors showed that it was possible to shift the value of the syncopation exponents toward the values observed for synchronization by inducing a conscious strategy of coordination. One of these strategies was to produce an oscillation of the finger twice faster than the metronome, with one "mimicked" tap on the beep, and one real tap off the beep. In this condition, the authors obtained exponent around 0.48, close to those observed in synchronization. In the same vein, Torre et al. (2004) showed that the series of relative phase collected in bimanual coordination were characterized by significantly different exponents, according to the required coordination: The estimation of H, performed with four methods ($^{low}PSD_{we}$, DFA, R/S analysis, and SWV), gave mean values ranging from 0.67 to 0.72 for in-phase series, and from 0.72 to 0.78 for anti-phase trials.

These two experiments suggest that fractal exponents could be sensitive to the characteristics of the task to perform (and especially its difficulty), and may also be altered by cognitive manipulations. This kind of result opens an interesting window, for a reappraisal of a number of classical concepts, including effort, activation, concentration, fatigue, and obviously learning.

Conclusion

Monofractal analyses were for a long time considered as a family of rather simple statistical tools, leading to intriguing results. Recent theoretical and methodological progresses showed that the application of these methods was not so easy, and necessitated methodological and statistical attention for producing effective results. The way seems now open to a more controlled use of such methods in psychological and behavioral research, in order to undertake a real experimental research program exploiting the fractal properties of most psychological and behavioral variables.

References

Bak, P., & Chen, K. (1991). Self-organized criticality. *Scientific American, 264,* 46–53.

Bassingthwaighte, J. B. (1988). Physiological heterogeneity: Fractals link determinism and randomness in structure and function. *News in Physiological Sciences, 3,* 5–10.

Box, G. E. P., & Jenkins, G. M. (1976) *Time series analysis: Forecasting and control.* Oakland: Holden-Day.

Caccia, D. C., Percival, D., Cannon, M. J., Raymond, G., & Bassingthwaighte, J. B. (1997). Analyzing exact fractal time series: Evaluating dispersional analysis and rescaled range methods. *Physica A, 246,* 609–632.

Cannon, M. J., Percival, D. B., Caccia, D. C., Raymond, G. M., & Bassingthwaighte, J. B. (1997). Evaluating scaled windowed variance methods for estimating the Hurst coefficient of time series. *Physica A, 241,* 606–626.

Chen, Y., Ding, M., & Kelso, J. A. S. (1997). Long memory processes ($1/f\alpha$ type) in human coordination. *Physical Review Letters, 79,* 4501–4504.

Chen, Y., Ding, M., & Kelso, J. A. S. (2001). Origins of timing errors in human sensorimotor coordination. *Journal of Motor Behavior, 33,* 3–8.

Collins, J. J., & De Luca, C. D. (1993). Open-loop and closed-loop control of posture: A random-walk analysis of center-of-pressure trajectories. *Experimental Brain Research, 95,* 308–318.

Davies, R. B., & Harte, D. S. (1987). Tests for Hurst effect. *Biometrika, 74,* 95–101.

Delignières, D., Deschamps, T., Legros, A., & Caillou, N. (2003). A methodological note on nonlinear time series analysis: Is Collins and De Luca (1993)'s open- and closed-loop model a statistical artifact? *Journal of Motor Behavior, 35,* 86–96.

Delignières, D., Fortes, M., & Ninot, G. (2004a). The fractal dynamics of self-esteem and physical self. *Nonlinear Dynamics in Psychology and Life Sciences, 8,* 479–510.

Delignières, D., Lemoine, L., & Torre, K. (2004b). Time intervals production in tapping and oscillatory motion. *Human Movement Science, 23,* 87–103.

Delignières, D., Damdani, S., Lemoine, L., Torre, K., Fortes, M., & Ninot, G. (2006). Fractal analysis for short time series: A reassessment of classical methods. *Journal of Mathematical Psychology, 50*, 525–544.

Diebolt, C., & Guirard, V. (2005). A note on long memory time series. *Quantity & Quality, 39*, 827–836.

Ding, M., Chen, Y., & Kelso, J. A. S. (2002). Statistical analysis of timing errors. *Brain and Cognition, 48*, 98–106.

Einstein, A. (1905). Über die von der molekularkinetischen Theorie der Wärme geforderte Bewegung von in ruhenden Flüssigkeiten suspendieren Teilchen. *Annalen der Physik, 322*, 549–560.

Eke, A., Herman, P., Bassingthwaighte, J. B., Raymond, G. M., Percival, D. B., Cannon, M., Balla, I., & Ikrényi, C. (2000). Physiological time series: distinguishing fractal noises from motions. *Pflügers Archives, 439*, 403–415.

Eke, A., Hermann, P., Kocsis, L., & Kozak, L. R. (2002). Fractal characterization of complexity in temporal physiological signals. *Physiological Measurement, 23*, R1–R38.

Farrell, S., Wagenmakers, E.-J., & Ratcliff, R. (2004). *ARFIMA time series modeling of serial correlations in human performance.* Submitted for publication.

Farrell, S., Wagenmakers, E.-J., & Ratcliff, R (2006). $1/f$ noise in human cognition: Is it ubiquitous, and what does it mean? *Psychonomic Bulletin & Review, 13*, 737–741.

Feder, J. (1988). *Fractals.* New York: Plenum Press.

Gilden, D. L. (1997). Fluctuations in the time required for elementary decisions. *Psychological Science, 8*, 296–301.

Gilden, D. L. (2001). Cognitive emissions of $1/f$ noise. *Psychological Review, 108*, 33–56.

Gilden, D. L., Thornton, T., & Mallon, M. W. (1995). $1/f$ noise in human cognition. *Science, 267*, 1837–1839.

Gottschalk, A., Bauer, M. S., & Whybrow, P. C. (1995). Evidence of chaotic mood variation in bipolar disorder. *Archives of General Psychiatry, 52*, 947–959.

Hausdorff, J. M., Mitchell, S: L., Firtion, R., Peng, C. K., Cudkowicz, M. E., Wei, J. Y., & Goldberger, A. L. (1997). Altered fractal dynamics of gait: Reduced stride-interval correlations with aging and Huntington's disease. *Journal of Applied Physiology, 82*, 262–269.

Hausdorff, J. M., Peng, C. K., Ladin, Z., Wei, J. Y., & Goldberger, A. R. (1995). Is walking a random walk? Evidence for long-range correlations in stride interval of human gait. *Journal of Applied Physiology, 78*, 349–358.

Hurst, H. E. (1965). *Long-term storage: An experimental study.* London: Constable.

Kelso, J. A. S. (1984). Phase transitions and critical behaviour in human bimanual coordination. *American Journal of Physiology: Regulatory, Integrative and Comparative Physiology, 15*, R1000–R1004.

Kelso, J. A. S., Holt, K. G., Rubin, P., & Kugler, P. N. (1981). Patterns of human interlimb coordination emerge from the properties of non-linear, limit cycle oscillatory processes: Theory and data. *Journal of Motor Behavior, 13*, 226–261.

Lemoine, L. (2004). *Reproductibilité des indices spectraux dans les tâches de tapping unimanuel* [Reproducibility of spectral indices in uni-manual tapping tasks]. Unpublished Masters Thesis, University Montpellier I, Montpellier, France.

Liebovitch, L. S., & Yang, W. (1997). Transition from persistent to antipersistent correlation in biological systems. *Physical Review E, 56*, 4557–4566.

Madison, G. (2001). Variability in isochronous tapping: Higher order dependencies as a function of intertap interval. *Journal of Experimental Psychology: Human Perception and performance, 27*, 411–422.

Mandelbrot, B. B., & van Ness, J. W. (1968). Fractional Brownian motions, fractional noises and applications. *SIAM Review, 10*, 422–437.

Marks-Tarlow, T. (1999). The self as a dynamical system. *Nonlinear Dynamics, Psychology, and Life Sciences, 3*, 311–345.

Musha, T., Katsurai, K., & Teramachi, Y. (1985). Fluctuations of human tapping intervals. *IEEE Transactions on Biomedical Engineering, BME-32*, 578–582.

Peng, C. K., Havlin, S., Stanley, H. E., & Goldberger, A. L. (1995). Quantification of scaling exponents and crossover phenomena in non-stationary heartbeat time series. *Chaos, 5*, 82–87.

Peng, C. K., Mietus, J., Hausdorff, J. M., Havlin, S., Stanley, H. E., & Goldberger, A. L. (1993). Long-range anti-correlations and non-Gaussian behavior of the heartbeat. *Physical Review Letter, 70*, 1343–1346.

Peters, B. T., Haddad, J. M., Heiderscheit, B. C., van Emmerik, R. E. A., & Hamill, J. (2003). Limitations in the use and interpretation of continuous relative phase. *Journal of Biomechanics, 36*, 271–274.

Rangarajan, G., & Ding, M. (2000). Integrated approach to the assessment of long range correlation in time series data. *Physical Review E, 61*, 4991–5001.

Schmidt, R. C., Beek, P. J., Treffner, P. J., & Turvey, M. T. (1991). Dynamical substructure of coordinated rhythmic movements. *Journal of Experimental Psychology: Human Perception and Performance, 17*, 635–651.

Slifkin, A. B., & Newell, K. M. (1998). Is variability in human performance a reflection of system noise? *Current Directions in Psychological Science, 7*, 170–177.

Thornton, T. L., & Gilden, D. L. (2005). Provenance of correlations in psychological data. *Psychonomic Bulletin & Review, 12*, 409–441.

Thornton, T., & Gilden, D. L. (2004). *Provenance of correlations in psychophysical data.* Submitted for publication.

Torre, K., Lemoine, L., & Delignières, D. (2004). *1/f fluctuation in bimanual coordination.* Submitted for publication.

Treffner, P. J., & Kelso, J. A. S. (1995). Functional stabilization on unstable fixed-points. In B. G. Bardy, R. J. Bootsma, & Y. Guiard (Eds.), *Studies in perception and action III* (pp. 83–86). Hillsdale, NJ: Lawrence Erlbaum Associates.

Treffner, P. J., & Kelso, J. A. S. (1999). Dynamic encounters: Long memory during functional stabilization. *Ecological Psychology, 11*, 103–137.

Van Orden, G. C., Holden, J. C., & Turvey, M. T. (2003). Self-organization of cognitive performance. *Journal of Experimental Psychology: General, 132*, 331–350.

Wagenmakers, E. J., Farrell, S., & Ratcliff, R. (2004). Estimation and interpretation of $1/f\alpha$ noise in human cognition. *Psychonomic Bulletin and Review, 11*, 579–615.

West, B. J., & Shlesinger, M. F. (1989). On the ubiquity of $1/f$ noise. *International Journal of Modern Physics B, 3*, 795–819.

Yamada, M. (1996). Temporal control mechanism in equalled interval tapping. *Applied Human Science, 15*, 105–110.

Yamada, M., & Yonera, S. (2001). Temporal control mechanism of repetitive tapping with simple rhythmic patterns. *Acoustical Science and Technology, 22*, 245–252.

Yamada, N. (1995). Nature of variability in rhythmical movement. *Human Movement Science, 14*, 371–384.

Yoshinaga, H., Miyazima, S., & Mitake, S. (2000). Fluctuation of biological rhythm in finger tapping. *Physica A, 280*, 582–586.

25

Frontiers of Nonlinear Methods

Robert A.M. Gregson

CONTENTS

As the field moves from a static representation of behavior to one that allows for and is interested in change, it becomes increasingly important to assess the kind of change we are trying to measure (Butner, Amazeen, & Mulvey, 2005, p. 159.).

This chapter is motivated by two issues. One is the question, "Where do we go from here?", which has been anticipated as a book title in the context of chaos theory (Hogan et al., 2003). The second is the shift to focusing on dynamical multistability in multiple attractor systems in various sciences including psychology (Feudel, 2008). Instabilities have become a central area of inquiry within nonlinear science (Schmidt, 2008).

Psychological research involves a three-way matching of data, experimental design, and statistical inference. For historical reasons that go back to the early nineteenth century, it has mostly been the practice to create situations in which there was stability in the observable relationships between stimuli and responses, and the assumptions in the statistical modeling implicit in the data analyses were also linear and stationary. In short, evidence of the dynamics, how things came to be what they were observed as, and what would happen if things were seriously destabilized, was filtered out and lost. What was recoverable and what was controllable became frozen in the experimenter's conceptual time.

Our understanding of nonlinear dynamics has increased, such that distinctions have emerged between different forms of chaos, for example, one may now read of robust chaos (Banerjee, Yorke, & Grebogi, 1998) or fuzzy chaos (Fridrich, 1994). Some of these constructs are as yet only mathematical and have no demonstrated mapping into biological or behavioral contexts, but real data that exhibit long memory or non-stationarity are ubiquitous, and, as we have seen from the previous chapters, are thus giving us strong signs that we may need to explore more than one sort of potential chaos from the varieties that have been defined.

As a consequence of this emerging depth of understanding, there are penalties to be paid if we oversimplify test procedures. One test that is relatively easy to compute is the largest Lyapunov exponent; if it is positive, then the system is expanding on one dimension and contracting in others, but this turns out not to be invariably a characteristic of chaos, and counter instances can be constructed. Similarly, the sign of the Schwarzian derivative is usually found to be negative in nonlinear dynamics, but exceptions can be found (Aguirregabiria, 2009). The problem is to find out if the circumstances that lead to apparent violations of convenient simple rules arise in the sorts of real data that we will meet in psychology.

So a contrast between an engineering interest in modeling control and predictability, and a biologically inspired interest in looking for stable relationships between physiological states and overt behavior, emerged over time. The mechanical sciences make relatively little use of statistical inference; outcomes when observed are reasonably well defined, and stochastic uncertainly takes a back seat. In contrast, much of statistical analysis in psychology is dominated by methods and assumptions derived from agricultural split-plot designs, originated in the 1920s, where processes are slow and simple, and outcomes are single-valued and socially useful. Important changes in the late twentieth century came about when computer simulation, on a large scale and with considerable complexity, came to be added to the empirical tools available, and dynamical theory in discrete time, including neural networks and cellular automata, became as important as differential equations. Problems that are ill-posed in traditional applied mathematics, with no closed solutions, became issues to be explored by the simultaneous simulation of many thousands of alternative scenarios. Some of the inbuilt algebra of simulation liberates the researcher from the assumption that models are linear (perhaps under transformations) with stochastic residuals in which the nonlinearities are masked, and allows the emergence of patterns of transients and dependence on critical conditions to be explored.

I want, in this final chapter, to address the question of what might happen next, but first to go back to the time of Newton. Newton's model of gravity was written for pairs of interacting bodies; its extension to three or more in a precise sense fails. This is called the three-body problem (Lodge, Walsh, & Kramer, 2003). We must be careful not to equate precisely the problem of multi-body interactions with multistability; the latter is one of a number of phenomena that arise more readily in complex networks, but is not to be equated with the network complexity as such.

> The equations representing the movements of the three gravitation-
> ally interacting bodies spawn no simple mathematical formula that can
> describe and predict the paths of all three bodies with unlimited accu-
> racy for all time.

(Peterson, 1993, p. 10)

It was Poincaré's graphical exploration of the three-body problem that laid foundations for what later came to be called chaotic dynamics. The three-body problem in physics involves one process that is initially defined on a pairwise relation between objects, and on all pairs within the set. The internal structure of each body is not specified, and can be replaced by a single point mass. Mass is in fact the only variable used to characterize the bodies, and the notion that a gas-giant might suddenly flip to being a rock-ice solid, or vice versa, is not in the Newtonian model of gravity. The chaos arises in the trajectories, or rather the limited predictability of the trajectories, in space and time, as the number of bodies involved increases, in theory without limit. The three-body problem is intractable without some simplifications, and it may be helpful to note that the use of symbolic dynamics has facilitated some progress in its study (Kaplan, 1999).

Levels of Multistability

Before considering multistability problems in psychological processes at various scales, an overview of multistability within mathematical models that have seen and are seeing application in various disciplines is pertinent. Feudel's extensive (2008) and timely review is drawn on here. In a formal sense, multistability involves, for a given mathematical model with fixed parameters, the coexistence of several attractors.

It is interesting that Attneave (1971) first commented on the notion of multistability when writing on the psychology of visual perception, an example of an idea that has spread from experimental psychology into mathematics, rather than as more usually the other way around (Zhou, Gao, White, Merk & Yao, 2004). We may draw historical significance from the fact that Attneave (1959) also was the pioneer of introducing information theory and, hence, entropy measures into psychology. Later work confirming Attneave has found multistability in human musical perception (Repp, 2007), and in vision in pigeons (Vetter, Haynes, & Pfaff, 2000). The perceptual multistability phenomenon was reexamined by Fürstenau (2004) explicitly as a chaotic attractor modeling problem.

Feudel classifies multistable systems, where the number of attractors can be arbitrarily large, into four types: (a) In *weakly dissipative systems*, the system is dominated by periodic attractors with mainly low periods; the bifurcation structure appears to be self-similar in parameter space, and suggests a search for fractals. (b) In *coupled systems*, simple examples are produced by coupled map lattices, and patterns emerge that are either coherent, ordered, partially ordered, or turbulent. (c) *Uncertain destination dynamics* are different from the first two in that infinitely many attractors can appear all of a sudden, and chaotic attractors dominate. (d) In delayed feedback systems, multistability is induced by a feedback loop, without which the system is monostable. For all these types, there is high sensitivity to perturbations, and if there are three or more attractors, then the basin boundaries become fractal.

Of particular interest in psychology is what happens when noise is introduced. Complex jumping between attractors can lead to two phases of motion (Engel, Fries, König, Bercht & Singer, 1999): regular motion, in the neighborhood of an attractor, and irregular motion, on the fractal boundaries of basins. There is hopping between attractors for low noise, but high noise masks the whole dynamics. Lyapunov exponents have a pronounced maximum for low noise but a Gaussian distribution with a higher maximum for large noise levels. Chaotic itinerancy has been observed where over time there is alternation between fully developed chaos and ordered behavior (Boker, 2002). Hopping behavior may be encoded coarsely by employing symbolic dynamics, and then computing entropies. Hopping goes with sudden increases in entropies, indicating higher complexity of the associated dynamics. Variations in higher-order entropies, mapped using bispectral kernel analyses, have been used in exploring instabilities in nonlinear psychometric time series (Gregson, 2006).

Brain Dynamics

In psychophysiology again we have a sort of three-body problem, but there are fundamental differences from the physical model, even though in their full analyses they also involve chaotic dynamics. What is exchanged between subprocesses is information, not gravity, and the structures involved are brain regions with fuzzy boundaries that may function variously as attractors, or repellers, or be dynamically transiently inert. The internal cellular structure of regions of the brain, treated as quasi-independent, itself can change with age; hence, comparisons between children and adults are important and of clinical and pedagogic interest (Szücs, Soltész, Jármi, & Csépe, 2007). A three-component system is, for example, the frontal–occipital–hippocampal network implicated in vision and memory (Sehatpour et al., 2008).

Hippocampal dynamics involve multiple maps and multistability emerging with age (Barnes, Suster, Shen, & McNaughton, 1997; Mizumori, Yeshenko, Gill, & Davis, 2004; Jackson & Redish, 2007). There are newer studies on variability between dynamic states in the electroencephalograph (EEG), arising both in sleep and in motor activity (Pfurtscheller & Solis-Escalante, 2009; Bruni et al., 2009), and neurophysiological transients that occur during the transitions between states.

Real behavior involves coupled attractors in the brain receiving input from an indeterminate number of external attractors generating stimuli in a changing environment. This environment can be thought of as physical or chemical, or social, and in feedback with the behavior of other individuals. It is these properties that make the modeling of hopping behavior an attractive idea when trying to capture the instability and variability of human behavior, and systems that can learn and forget, and make decisions, have to have multiple internal representations and an ability to quickly access different attractors in memory or perception.

Networks with Internal Delays

The relation between multistability and delay, as type (d) above, in processing or storing information is particularly important in representing neuropsychological data. The observed multistability at a behavioral or perceptual level is grounded in the neural network activity in the brain, when that network has a degree of connectivity, distributed delays between elements, and bounded activation functions. The models that can accommodate multistability are also informative in modeling memory, image processing, and pattern recognition, and any biological networks that allow elements to synchronize with each other. The starting point of analyses of this situation, which inevitably involves some heavy algebraic formulation, is often a Cohen–Grossberg activation equation, with delays introduced. In a simple form (e.g., see Ye, Michel, and Wang (1995), or Wang and Zou (2002)), we have

$$\frac{dx_i}{dt} = -a_i(x_i)\left[b_i(x_i) - \sum t_{ij}s_j(x_j(t - \tau_k)) \right] \tag{25.1}$$

where the summation is over $i = 1, 2, ..., n, j = 1, 2, ..., n$. x_i is the state variable associated with the ith neuron; a is an amplification function; b is arbitrary and bounded to keep the solutions of Equation 25.1 bounded; the matrix t_{ij} represents neuron interconnections; and τ_k is the delay at time k, which may be in discrete steps or continuous, but small enough for stability to be preserved.

What is called Markovian jumping is an additional structure, emerging where there are time-varying delays. Sheng and Yang (2009) show, by using a mix of both linear matrix inequality (LMI) and Lyapunov functions, that stability within each state can be achieved, and the transition matrix between states is computable.

Various investigators have explored Cohen–Grossberg networks (Sun & Wan, 2005) with delays, and used both LMIs and Lyapunov functions; the interplay of chaos, fuzziness, synchronization, distributed or multiple delays, coupling in discrete or distributed forms, synchronization at fast (exponential) rates, and stability in the presence of uncertainties with small-enough delays, has attracted a very extensive literature (Cao & Li, 2005; Li & Yang, 2009; Rong, 2005; Singh, 2007; Vandenberghe, Boyd, & Wu, 1998).

Magnitudes and the Number Line

To understand the distinctions between instability with contamination by high levels of white noise, and deterministic multistability, we need a topic within brain neuropsychology that has been studied in great depth and with precision, employing a mix of psychophysical, cognitive, functional magnetic resonance imaging (fMRI), and PET (positron emission tomography) methods. A candidate is the internal psychophysics of number and magnitude estimation, which has been explored extensively by Dehaene and his collaborators in a long series of studies (see Dehaene, 1997; Dehaene, Piazza, Pinel, & Cohen, 2003, and references therein).

The role of the parietal lobe in number processing had been confirmed by several investigators; Dehaene et al. (2003, p. 488) now postulate three regions with overlapping functions: "(1) a bilateral intraparietal system associated with a core quantity system, (2) a region of the left angular gyrus (AG) associated with the verbal processing of numbers, and (3) a posterior superior parietal system (PSPL) of spatial and nonspatial attention."

A current idea is that there is a "number line" in the horizontal segment of the intraparietal sulcus (HIPS) that holds a nonverbal representation of numerical quantity. Its activation is associated with the task of addition or subtraction, and is more active in performing approximate calculations than doing exact ones. The right hemisphere of HIPS is more active, but both hemispheres are always involved.

Which of the three regions dominates in its activity depends on the properties of the problem posed, and what is already available to be drawn on as a consequence of prior rote learning of rules or algorithms. Multiplication is processed differently from addition or subtraction. The number line concept implies that numerical quantity is on a quasispatial representation where numbers are organized by their proximity, and this representation also facilitates orientation in space and time (Dehaene et al., 2003, p. 498). There is some striking universality of evidence that the HIPS, AG, and PSPL are activated in different subjects from different cultures and different educational levels or mathematical competences. Clinical evidence, from premature babies with early left parietal injury, is that they exhibit developmental dyscalculia relatively frequently. There is evidence that some alexic patients can show an ability to read and process digits, and the converse, where number reading is impaired but word reading preserved, has also been reported. There is, however, a strong argument with detailed behavioral and neurophysiological evidence that numerical representation is not necessarily abstract but modality specific, and so numerosity has multiple representations (Kadosh, 2008). There is a recurrent complaint (Kadosh & Walsh, 2009) that the statistical methods used in many of the numerous studies on numerosity processing are insensitive, and indeed many of these studies use ANOVA (analysis of variance) only and look merely for evidence of nonadditivity to support or refute a case. There is no consideration of multistability inbuilt a priori into the statistical methodology (Hu & Johnson, 2009).

The whole picture is of a set of subsystems that fluctuate in their dominance with task demands, can in some cases be separately impaired, and can be augmented by learning; so their structure is not fixed in terms of their mutual connectivity or internal structure. In terms of multistability, the situation is perhaps nearest to type (b) listed above, but the mathematical physics is an oversimplification in a biological context.

Model Structure and Statistical Filtering Assumptions

There are problems of misidentification of system dynamics when irregular time series, apparently aperiodic and linking two or more attractors, are observed. Various statistical methods have come into use; dependencies between signals from different attractors are variously filtered in the time or spectral domains, by fuzzy networks, and by symbolic dynamics (Herrmann, 1997).

Butner et al. (2005) focus on the interaction of two simultaneously fluctuating variables, where each is assumed to have quasiperiodic characteristics. They introduce nonlinearity through damping, which is related to delayed feedback, that is, type (d) above, but is not extended by them to multistability. The terms coherence, covariation, and phase synchrony are all in use, and related, in considering methods to distinguish between real and spurious linkages of attractors.

It is clear that stability is a central problem in nonlinear dynamical analyses. Haken (1987) pioneered some of the work using instability as a central issue, and modeled pairwise relations between attractors where slaving of one by the parameters of another led to control. In neurobiology, there is another level of complication, where attractors in brain regions may cooperate and, at the same time, retain their functional autonomy (Tognoli & Kelso, 2009). The complication is that coupled dynamics, as revealed in synchronization of oscillatory outputs, can be ambiguous and arise either through real coupling or pseudocoupling, which is a product of both attractors sharing a common source of what is called volume conduction. The differences between spike synchronization and real covariations have lead to the creation of statistical methods to try and disentangle the sorts of coherence in the time series generated (Amjad, Halliday, Rosenberg, & Conway, 1997; Brody, 1999). It is not that only multistability can arise, but rather that various patterns, metastability, in-phase and anti-phase synchronization, and switching or multistability are all part of a complicated scenario where a multiplicity of attractors are involved and function together. Metastability (Tognoli & Kelso, 2009, p. 32) "refers to form of partial coordination that does not lock the dynamics of local areas into synchronized states. Rather, patterns of quasi-phase-locking (dwelling tendencies) are created that dynamically summon and release brain areas without requiring costly disengagement mechanisms." Kello and Van Orden (2009) show that metastability can lead to fractal and $1/f$ scaling; so if $1/f$ scaling is found, this may be used as a diagnostic to work back to identifying metastability. It appears that here we have processes at a substrate level to create sufficient time intervals for multistate phenomena at a behavioral level to show themselves. How such processes ever evolved in brain structures is itself a question that has been addressed (Kerszberg, Dehaene, & Changeux, 1992). Transient stabilization, and the opposite where destabilization arises from external disruptive stimuli (Gregson, 1992, p. 85), are both recurrent complicating features of psychophysiological processes that have to be modeled, and do not have equivalents in Newtonian physics where chaos was first found to be lurking.

Clusters and Levels in Social Structure

In studying the sociology of ideologies and their distribution in networks of partially interacting individuals, models that are what is called agent-based have come into use. Chaos, complexity, transients, and local multistability are also created in network models of social processes, though not with the same detailed consideration of biological substrates that is found in the psychophysiological work. A computer simulation study of networks of interacting and diverse beliefs, which could be about any sort of ideology apart from religion, has been developed in detail by Bainbridge (2006). This work explicitly uses nonlinear recursive functions in a fashion pioneered by Markovsky (1992) in modeling social network dynamics; so it is reasonable to expect multistabilities to emerge.

Where each individual can be represented by a vector of beliefs, encoded as binary (yes/no), social cooperation or rejection on a pairwise basis may be created based on one or more beliefs. This in turn creates larger patterns of local clique formation, and fluctuation in the extent of adherence to particular ideologies. In systems where there are only two major competing ideologies, each fuzzily defined, as in U.S. politics, third lesser groups do not come into extensive and persistent existence, but the larger ones tend to alternate in power. Interestingly, this approximates to the sort of perceptual multi-stability that Attneave originally considered. By contrast, in European politics (or Israel), there is permanently a multiplicity of parties (typically about five) and majority governments can only be formed as coalitions. Britain and Australia are political examples of an intermediate situation. The total scene, comprising bistable and multistable political democracies, resembles Feudel's type (b).

Consequent Resolution of Recording and Analysis Problems

If the alternative patterns of a fluctuating system are re-encoded as symbolic states, numerical information is dropped but the possibility of using state transition probability matrices is established. This leads to testing if the process is strictly Markov, or has Markovian dynamics embedded within it (Herrmann, 1997; Nuel, 2008). Strictly Markov processes have a limited memory of lag-one forwards or backwards, but examining departures from this transition to find long-memory processes can be done (Gregson, 1987). Multistability then becomes a recurrent cycle between two or more states, each with one unique symbolic representation.

If the process is ergodic, then long-term stability, as a probability vector of terminal state occupancies, may be used to distinguish random fluctuations from multistability over a finite subset of states. For the case of two states, the transition probability matrix, for multistability between them, is simple: If $p(i \mid j, \theta)$ denotes the probability of i being the current state, given j as the preceding state in the observed current time interval, given the parameter set θ, then there can be no absorbing states in the 2×2 i, j matrix, $p(i \mid i, \theta) < 1$, $p(j \mid j, \theta) < 1$, but it is not necessary that $p(i \mid j, \theta) = p(j \mid i, \theta)$, though in the reversible figures case used by Attneave this equality would be approximately true.

For three or more states, the transitions may involve cycling through a set of states, in various possible cyclic orders, but again there are no absorbing states for the parameters operating. In complicated dynamics, there can be local islands in the parameter space where multistability holds, but the probability of entering or staying in some of these islands can be vanishingly small (Garay & Chua, 2008). Multistability, either as persistent or as a transient phenomenon within a dynamical system, has been found to be ubiquitous at all levels of psychological processes, and hence it can be expedient to plan data collection in such a form that subsequent statistical analyses may identify multistability if it is occurring over time. The phenomena of catastrophes, bifurcations, phase shifts, and multistability can and do arise together within the same system as the controlling parameter values change, and graphical methods can usefully be employed to show the interplay of nonlinearities (Schmitz, Kolar-Anić, Anić, & Čupić, 2000). Bistable systems, with sigmoidal psychometric functions, which are typical in psychophysics, have also been explored graphically for evidence of multistability and bifurcations (Angeli, Ferrell, & Sontag, 2004; Angeli & Sontag, 2004). At the level of neurotransmitter regulation and genetics, iterative numerical integration has been used to identify multistability (Smolen, Baxter, & Byrne, 1998), and computer packages for this approach exist (Doedel et al., 2001). It is critical, as may be deduced from the fourfold typology summarized by Feudel, that we consider if the system under study is believed to consist of two, three, or more coupled maps, and if time delay is detected (Foss, Moss, & Milton, 1997; Astakhov, Seleznev, & Smirnov, 1997; Nikolaev, Shabunin, & Astakhov, 2005; Shabinin, Astakhov, Demidov, & Efimov, 2008).

Experimental Verification

We have seen that multistability can arise in a diversity of dynamic systems, and any statistical model used to support experimental design and analysis will be related to a system model, though not necessarily one of the four given here. To see how inappropriate experimental design can be misleading, and either suppress or falsely support the identification of multistability, consider a commonly used (almost invariably in some journals) paradigm in experimental psychology. In the simplest form, we have two main effects, *A* and *B*, and each operates at two levels, *A*, *a* and *B*, *b*. Data here are collected over samples of subjects, presumed independent, and on each subject there is a vector of scores *y* recorded at one time only.

The 2 × 2 experimental design, which will, if the effects are scaled in some metric, support an ANOVA data analysis under restricted conditions, has the four cells

$$AB \quad aB$$
$$Ab \quad ab$$

The linear model, *L*, implicit in ANOVA is written

$$\hat{y} = c + v_1 A + v_2 B + v_3 AB + e$$

where *e* is the residual error variance normally distributed, the *v* are scalar constants, with the first two moments of the predicted $\hat{y}$ sufficient and its variance independent of means and fairly constant, that is, the whole design is supposedly homoscedastic. Now let the response distributions in one or both of cells *AB* and *ab* be bimodal, the remaining cells being unimodal. Such bimodalities can arise more often than might be supposed (Liu & Hodges, 2003). *L* is violated; on performing a goodness-of-fit test on $y, \hat{y}$ should fail. The question we must now raise is, "Is this evidence of multistability?", and the answer is, "Not necessarily." Let us see why this is so.

Multistability is a within-subject process, over time, and the probabilities of hopping between states are asymptotic to unity over a long time and either symmetric in two states or are cyclic in three or more states. Observing any one subject at only one point in time will reveal him or her in only one state, and can reveal another subject, supposedly exchangeable in the experimental design, in the same or another state. But this can arise because the process has passed a bifurcation for some subjects under the stimulus loadings of *AB* or *ab* or both, or because there was unknown heterogeneity in the sample of subjects, due to exogenous or endogenous variables (Wichert, Fokianos & Strimmer, 2004).

If we admit the possibility of nonlinear dynamics, then the experimental design, with its inbuilt linear model, is ill-posed. A between-subject design cannot unambiguously identify a within-subject phenomenon. A different sort of experimental design, and data collection, with within-subject tests of sequential dynamics and non-stationarities is needed. To list potential situations, if bimodality arises in a cell of a factorial design, then

1. The subjects can be homogeneous and exhibiting multistability, but not in phase for shifts between states, and the phase shifts are not distributed continuously in time. This is improbable given the presumed independence of subjects.

2. There are two subject groups, and under some restricted range of system parameters, they respond bimodally consistently through time, but few individuals exhibit multistability.

3. If the stimulus conditions are interactive, so that some combinations of levels of *A* and *B* result in dominant $v_1 \gg v_2$ or $v_1 \ll v_2$, the model with fixed *v* being false, this can suppress bimodality by removing conditions for stimulus inputs for one variable in some *Ab* or *aB*. Variable *v* coefficients is a form of nonlinearity; it can arise if the system goes through a bifurcation that is reflected in the *v*.

A number of striking examples of between-subject variability in response times, which display both unimodality and bimodality in the same task, have been reported by Cousineau and Shiffrin (2004, p. 335); these arise from a combination of effects (2) and (3) above, and emphasize the risks that arise from invalidly pooling data over subjects so as to mask a heterogeneity of within-subject response states.

These examples, also reported by Reddi and Carpenter (2000), arise when the two factors *A* and *B* in a response time experiment are task complexity and pressure to perform; some subjects when faced with a difficult task and high demand shift to a response time frequency distribution that is different. The difference from multistability is that true multistability arises when the stimulus conditions are fixed, not like the two factors case, but within-subject processes have more than one stable mode.

Can multistability be present within some individual subjects when there is heteroscedascity across cells but no within-cell bimodality? Yes, sampling of multistable individuals increases within-cell variance for those parameter values associated with regions of multistability in the total nonlinear system. We then have to redesign the study to run over time in order to be able potentially to identify one of the four types of dynamics that can generate the multistability, but the interest then shifts to focusing on system identification among alternatives, and the multistability becomes just one symptom of the nonlinear dynamics present.

It would be possible, of course, to design an experiment that is a hybrid of the 2×2 factorial design and the individual time series sampling that is needed to identify when multistability occurs, and what stimulus conditions trigger it.

Using each subject in all four conditions, and balancing orders, requires a module of 24 subjects to balance out order effects between A, B conditions.

If this is done, then it becomes possible to distinguish between exogenous induction of shifts associated with the A, B parameters and, if observed, endogenous multistability, which is more formally defined as spontaneous involuntary switching of conscious awareness independent of changes in stimulus conditions. But the more interesting and important questions relate to what are measurable quantified properties of the multistable time series themselves, for which there are models and predictions. For example, it is known that the frequency distribution of the length of time intervals between switches follows a gamma or Poisson distribution (Levelt, 1967; Fürstenau, 2006), and that mean reversal times between two states are about 3–5 s. There are transitions between chaotic and limit cycle attractors representing the perceptual states. A number of competing models, of synergetic, stochastic, deterministic chaotic, and catastrophe types, have been advanced, and coupling between neurophysiological and behavioral levels is predicted in some detail (deGuzman & Kelso, 1991; Kelso et al., 1995; Zhou, Gao, White, Merk, & Yao, 2004). The model of Fürstenau (2006) involves delay and may thus be categorized in Feudel's type (d).

It is clear that multistability can be modeled by, and only by, nonlinear dynamics, and experimental designs must be sufficiently sensitive in scale and temporal resolution to reflect the subtleties of the behavioral dynamics. Given some real data that is noisy, the problem may be to get a best estimate of the location in time of the change-points between states. This problem can be approached by using the Modeling Markup Language (MML) framework mentioned in Chapter 1. Wallace and Dowe (2000) developed MML clustering of segments of time series in multistate systems, and Fitzgibbon, Allison, and Dowe (2000), and Fitzgibbon, Dowe, and Allison (2002) refined parameter estimation of change-points, and for this purpose redefined MML methods, showing a relation to the Kullback–Leibler loss function. Some confirmatory results, demonstrating that the best estimates of number and location of change-points are best performed by using Bayesian methods, have emerged (Kim, Yu, & Feuer, 2009).

A complication has to be noted, when a biological system is multistate at one level, say the perceptual, its change-points need not exactly coincide in time with those at a different level, such as the associated EEG patterns; one set of change-points can lead or lag another (Müller et al., 2005). The global spatiotemporal dynamics of the system have to be computed at three levels— field strength, rate of change, and covariance structure—to identify the dynamics. This can be done by using Omega statistics (Wackermann, 1996).

References

Aguirregabiria, J. M. (2009). Robust chaos with variable Lyapunov exponent in smooth one-dimensional maps. *Chaos, Solitons & Fractals, 42,* 2531–2539.

Amjad, A. M., Halliday, D. M., Rosenberg, J. R., & Conway, B. A. (1997). An extended difference of coherence test for comparing and combining several independent coherence estimates: Theory and application to the study of motor units and physiological tremor. *Journal of Neuroscience Methods, 73,* 69–79.

Angeli, D., Ferrell, Jr., J. E., & Sontag, E. D. (2004). Detection of multistability, bifurcations, and hysteresis in a large class of biological positive-feedback systems. *Proceedings of the National Academy of Sciences of the USA (PNAS), 101,* 1822–1827.

Angeli, D., & Sontag, E. D. (2004). Multi-stability in monotone input/output systems. *Systems Control Letters, 51*(3–4), 185–202.

Astakhov, S. A., Seleznev, Ye. P., & Smirnov, D. A. (1997, June 26–27). Multistability and transient processes in coupled period doubling systems. In *Proceedings: 5th International Specialist Workshop "Nonlinear Dynamics of Electronic Systems"* (pp. 437–442). Moscow.

Attneave, F. (1959). *Application of information theory to psychology.* New York: Holt, Rinehart and Winston.

Attneave, F. (1971). Multistability in perception. *Scientific American, 225,* 63–71.

Bainbridge, W. S. (2006). *God from the machine: Artificial intelligence models of religious cognition.* Lanham, NY: Altamira Press.

Banerjee, S., Yorke, J. A., & Grebogi, C. (1998). Robust chaos. *Physical Review Letters, 80,* 3049–3052.

Barnes, C. A., Suster, M. S., Shen, J., & McNaughton, B. L. (1997). Multistability of cognitive maps in the hippocampus of old rats. *Nature, 388*(6639), 272–275.

Boker, S. M. (2002). Consequences of continuity: The hunt for intrinsic properties within parameters of dynamics in psychological processes. *Multivariate Behavioral Research, 33,* 479–507.

Brody, C. D. (1999). Disambiguating different covariation types. *Neural Computation, 11,* 1527–1535.

Bruni, O., Novelli, L., Finotti, E., Luchetti, A., Uggeri, G., Aricò, D., & Ferri, R. (2009). All-night EEG power spectral analysis of the cyclic alternating pattern at different ages. *Clinical Neurophysiology, 120,* 248–256.

Butner, J., Amazeen, P. G., & Mulvey, G. M. (2005). Multilevel modeling of two cyclical processes: Extending differential structural equation modeling to nonlinear coupled systems. *Psychological Methods, 10*, 159–177.

Cao, J., & Li, X. (2005). Stability in delayed Cohen-Grossberg neural networks: LMI optimization. *Physica D, 212*, 54–65.

Cousineau, D., & Shiffrin, R. M. (2004). Termination of a visual search with large display size effects. *Spatial Vision, 17*, 327–352.

deGuzman, G. C., & Kelso, J. A. S. (1991). Multifrequency behavioural patterns and the phase attractor circle map. *Biological Cybernetics, 64*, 485–495.

Dehaene, S. (1997). *The number sense.* New York: Oxford University Press.

Dehaene, S., Piazza, M., Pinel, P., & Cohen, L. (2003). Three parietal circuits for number processing. *Cognitive Neuropsychology, 20*, 487–506.

Doedel, E. J., Paffenroth, R. C., Champneys, A. R., Fairgreve, T. F., Kuznetsov, Y. A., Oldeman, B., Sandstede, B., & Wang, X. J. (2001). AUTO 2000: Continuation and bifurcation software for ordinary differential equations. Pasadena, CA: Applied and Computational Mathematics, California Institute of Technology. Retrieved [2001] from https://sourceforge.net/projects/auto2000/.

Engel, A. K., Fries, P., König, P., Brecht, M., & Singer, W. (1999). Temporal binding, binocular rivalry, and consciousness. *Consciousness and Cognition, 8*, 128–151.

Feudel, U. (2008). Complex dynamics in multistable systems. *International Journal of Bifurcation and Chaos, 18*, 1607–1626.

Fitzgibbon, L. J., Allison, L., & Dowe, D. L. (2000). Minimum message length grouping of ordered data. In *Proceedings of the 11th International Conference on Algorithmic Learning Theory (ALT 2000)* (pp. 56–70). Sydney, Australia: Sydney University.

Fitzgibbon, L. J., Dowe, D. L., & Allison, L. (2002). Change-point estimation using new minimum message length approximations. In *Proceedings of the Seventh Pacific Rim International Conference on Artificial Intelligence (PRICAI-2002)*. Tokyo. Retrieved [2002] from www.csse.monash.edu.au.

Foss, J., Moss, F., & Milton, J. (1997). Noise, multistability, and delayed recurrent loops. *Physical Review E, 55*, 4536–4543.

Fridrich, J. (1994). On chaotic fuzzy systems: Fuzzified logistic mapping. *International Journal of General Systems, 22*, 369–380.

Fürstenau, N. (2004). A recursive attention-perception chaotic attractor model of cognitive multistability. In M. C. Lovett, C. D. Schunn, C. Lebiere, & P. Munro (Eds.), *Proceedings of the Sixth International Conference on Cognitive Modeling* (pp. 348–349). Mahwah, NJ: Lawrence Erlbaum.

Fürstenau, N. (2006). Modelling and simulation of spontaneous perception switching with ambiguous visual stimuli in augmented vision systems. *Lecture Notes in Computer Science/Artificial Intelligence LNAI, 4021*, 20–31.

Garay, B. M., & Chua, L. O. (2008). Isles of Eden and the Zuk theorem in Rd. *International Journal of Bifurcation and Chaos, 18*, 2951–2963.

Gregson, R. A. M. (1987). The time-series analysis of self-reported headache sequences. *Behavior Change, 4*, 6–13.

Gregson, R. A. M. (1992). *n-Dimensional nonlinear psychophysics.* Hillsdale, NJ: Lawrence Erlbaum Associates.

Gregson, R. A. M. (2006). *Informative psychometric filters.* Canberra: ANU e-Press.

Haken, H. (1987). *Advanced synergetics: Instability hierarchies of self-organizing systems and devices.* Berlin: Springer-Verlag.

Herrmann, C. S. (1997). Symbolic reasoning about numerical data: A hybrid approach. *Applied Intelligence, 7*, 339–354.

Hogan, S. J., Champneys, A. R., Krauskopf, B., Bernardo, M. D., Wilson, R. E., Osinga, H. M., & Homer, M. E. (Eds.). (2003). *Nonlinear dynamics and chaos: Where do we go from here?* Bristol: Institute of Physics.

Hu, J., & Johnson, V. E. (2009). Bayesian model selection using test statistics. *Journal of the Royal Statistical Society Series B, (Statistical Methodology), 71*, 143–158.

Jackson, J., & Redish, A. D. (2007). Network dynamics of hippocampal cell-assemblies resemble multiple spatial maps within single tasks. *Hippocampus, 17*, 1209–1229.

Kadosh, R. C. (2008). Numerical representation: Abstract or nonabstract? *The Quarterly Journal of Experimental Psychology, 61*, 1160–1168.

Kadosh, R. C., & Walsh, V. (2009). Numerical representation in the parietal lobe: Abstract or not abstract? *Behavioral and Brain Sciences, 32*, 313–328.

Kaplan, S. (1999). Symbolic dynamics of the collinear three-body problem. *Contemporary Mathematics, 246*, 143–162.

Kello, C. T., & Van Orden, G. C. (2009). Soft-assembly of sensorimotor function. *Nonlinear Dynamics, Psychology, and Life Sciences, 13*, 57–78.

Kelso, J. A. S., Case, P., Holroyd, T., Horvath, E., Raczaszek, J., Tuller, B., & Ding, M. (1995). Multistability and metastability in perceptual and brain dynamics. In P. Kruse & M. Stadler (Eds.), *Ambiguity in mind and nature* (pp. 255–273). Berlin: Springer-Verlag.

Kerszberg, M., Dehaene, S., & Changeux, J.-P. (1992). Stabilization of complex input-output functions in neural clusters formed by synapse selection. *Neural Networks, 5*, 403–413.

Kim, H.-J., Yu, B., & Feuer, E. J. (2009). Selecting the number of change-points in segmented line regression. *Statistica Sinica, 19*, 597–609.

Levelt, W. J. M. (1967). Note on the distribution of dominance times in binocular rivalry. *British Journal of Psychology, 58*, 143–145.

Li, C.-H., & Yang, S.-Y. (2009). Synchronisation in delayed Cohen-Grossberg neural networks with bounded external inputs. *IMA Journal of Applied Mathematics, 74*, 178–200.

Liu, J., & Hodges, J. S. (2003). Posterior bimodality in the balanced one-way random-effects model. *Journal of the Royal Statistical Society Series B, (Statistical Methodology), 65*, 247–255.

Lodge, G., Walsh, J. A., & Kramer, M. (2003). A trilinear three-body problem. *International Journal of Bifurcation and Chaos, 13*, 2141–2155.

Markovsky, B. (1992). Network exchange outcomes: Limits of predictability. *Social Networks, 14*, 267–286.

Mizumori, S. J., Yeshenko, O., Gill, K. M., & Davis, D. M. (2004). Parallel processing across neural systems: Implications for a multiple memory system hypothesis. *Neurobiology Learning and Memory, 82*, 278–298.

Müller, Th. J., Koening, Th., Wackermann, J., Kalus, P., Fallgatter, A., Strik, W., & Lehmann, D. (2005). Subsecond changes of global brain state in illusory multi-stable motion perception. *Journal of Neural Transmission, 112*, 565–576.

Nikolaev, S. M., Shabunin, A. V., & Astakhov, V. V. (2005). Multistability of partially synchronous regimes in a system of three coupled logistic maps. In *Proceedings of the International Conference PhysCon*. St. Petersburg. Retrieved [2005] from http://chaos.ssu.runnet.ru

Nuel, G. (2008). Pattern Markov chains: Optimal Markov chain embedding through deterministic finite automata. *Journal of Applied Probability, 45*, 226–243.

Peterson, I. (1993). *Newton's clock: Chaos in the solar system*. New York: W. H. Freeman & Co.

Pfurtscheller, G., & Solis-Escalante, T. (2009). Could the beta rebound in the EEG be suitable to realize a "brain switch"? *Clinical Neurophysiology, 120*, 24–29.

Reddi, B. A. J., & Carpenter, R. H. S. (2000). The influence of urgency on decision time. *Nature Neuroscience, 3*, 827–830.

Repp, B. H. (2007). Hearing a melody in different ways: Multistability of metrical interpretation, reflected in rate limits of sensorimotor synchronization. *Cognition, 102*, 434–454.

Rong, L. (2005). LMI-based criteria for robust stability of Cohen-Grossberg neural networks with delay. *Physics Letters A, 339*, 63–73.

Schmidt, J. C. (2008). From symmetry to complexity: On instabilities and the unity in diversity in nonlinear science. *International Journal of Bifurcation and Chaos, 18*, 897–910.

Schmitz, G., Kolar-Anić, L., Anić, S., & Čupić, Ž. (2000). The illustration of multistability. *Journal of Chemical Education, 77*, 1502–1505.

Sehatpour, P., Molholm, S., Schwartz, T. H., Mahoney, J. R., Mehta, A. D., Javitt, D. C., Stanton, P. K., & Foxe, J. J. (2008). A human intracranial study of long-range oscillatory coherence across a frontal-occipital-hippocampal brain network during visual object processing. *Proceedings of the National Academy of Sciences of the USA, 105*, 4399–4404.

Shabinin, A. V., Astakhov, V. V., Demidov, V. V., & Efimov, A. V. (2008). Multistability and synchronization of Chaos in maps with "internal" coupling. *Journal of Communications Technology and Electronics, 53*, 666–675.

Sheng, L., & Yang, H. (2009). Robust stability of uncertain Markovian jumping Cohen-Grossberg neural networks with mixed-time varying delays. *Chaos, Solitons and Fractals, 42*, 2120–2128.

Singh, V. (2007). Novel LMI condition of global robust stability of delayed neural networks. *Chaos, Solitons and Fractals, 34*, 503–508.

Smolen, P., Baxter, D. A., & Byrne, J. H. (1998). Frequency selectivity, multistability, and oscillations emerge from models of genetic regulatory systems. *American Journal of Physiology, 274*, C531–C542.

Sun, J., & Wan, L. (2005). Global exponential stability and periodic solutions of Cohen-Grossberg neural networks with continuously distributed delays. *Physica D, 208*, 1–20.

Szücs, D., Soltész, F., Jármi, E., & Csépe, V. (2007). The speed of magnitude processing and executive functions in controlled and automatic number comparison in children: An electroencephalography study. *Behavioral and Brain Functions, 3*, 3–23.

Tognoli, E., & Kelso, J. A. S. (2009). Brain coordination dynamics: True and false faces of phase synchrony and metastability. *Progress in Neurobiology, 87*, 31–40.

Vandenberghe, L., Boyd, S., & Wu, S.-P. (1998). Determinant maximization with linear matrix inequality constraints. *SIAM Journal on Matrix Analysis and Applications, 19*, 499–533.

Vetter, G., Haynes, J. D., & Pfaff, S. (2000). Evidence for multistability in the visual perception of pigeons. *Vision Research, 40*, 2177–2186.

Wackermann, J. (1996). Beyond mapping: Estimating complexity of mutichannel EEG recordings. *Acta Neurobiologiae Experimentalis, 56*, 197–208.

Wallace, C. S., & Dowe, D. L. (2000). MML clustering of multi-state, Poisson, von Mises circular and Gaussian distributions. *Journal of Statistics and Computing, 10*, 73–83.

Wang, L., & Zou, X. (2002). Exponential stability of Cohen-Grossberg neural networks. *Neural Networks, 15*, 415–422.

Wichert, S., Fokianos, K., & Strimmer, K. (2004). Identifying periodically expressed transcripts in microarray time series data. *Bioinformatics, 20*, 5–20.

Ye, H., Michel, A. N., & Wang, K. (1995). Qualitative analysis of Cohen-Grossberg neural networks with multiple delays. *Physical Review E, 51*, 2611–2618.

Zhao, Y., Sun, J., & Small, M. (2008) Evidence consistent with deterministic chaos in human cardiac data: Surrogate and nonlinear dynamical modelling. *International Journal of Bifurcation and Chaos, 18*, 141–160.

Zhou, Y. H., Gao, J. B., White, K. D., Merk, I., & Yao, K. (2004). Perceptual dominance time distributions in multistable visual perception. *Biological Cybernetics, 90*, 256–263.

Index

Printed and bound by CPI Group (UK) Ltd, Croydon, CR0 4YY

23/10/2024

01777697-0009